About the Author

David M. Potter (1910–1971) was born in Augusta, Georgia, and received his Ph.D. in history from Yale University. After beginning his teaching career at the University of Mississippi, Potter served as professor of history at Yale for nineteen years and at Stanford University, where he taught for the last ten years of his life. At the time of his death, Potter was the president of both the American Historical Association and the Organization of American Historians. He was posthumously awarded the 1977 Pulitzer Prize for history for *The Impending Crisis: 1848–1861.* Among his other books are *Lincoln and His Party in the Secession Crisis*, *People of Plenty*, and *The South and the Sectional Conflict*, which was a finalist for the 1969 National Book Award.

About the Editor

Don E. Fehrenbacher (1920–1997), David M. Potter's colleague at Stanford University, edited and completed *The Impending Crisis: 1848–1861.* He won the 1979 Pulitzer Prize for history for *The Dred Scott Case* and edited the Library of America's two-volume collection of Abraham Lincoln's speeches and writings.

THE IMPENDING CRISIS

America Before the Civil War: 1848–1861

By DAVID M. POTTER

Completed and Edited by DON E. FEHRENBACHER

ILLUSTRATED

HARPER PERENNIAL

NEW YORK • LONDON • TORONTO • SYDNEY • NEW DELHI • AUCKLAND

To my daughter,
Catherine Mary Potter

HARPER PERENNIAL

The map entitled "The Kansas-Nebraska Act, 1854" is in *The National Experience* by Blum, Catton *et al* © 1963 by Harcourt Brace Jovanovich, Inc. and is reproduced with their permission.

The map entitled "Election of 1860" is in *From These Beginnings . . . A Biographical Approach to American History* by Frederick Nash, copyright © 1973 by Roderick Nash. Reprinted by permission of Harper & Row, Publishers, Inc.

HarperCollins books may be purchased for educational, business, or sales promotional use. For information please write: Special Markets Department, HarperCollins Publishers, 10 East 53rd Street, New York, NY 10022.

Reprinted in Harper Perennial in 2011.

First Harper Colophon edition published 1976.

Library of Congress Cataloging-in-Publication Data is available upon request.

ISBN 978-0-06-131929-7

12 13 14 15 RRD 40

Contents

Abbreviations Used

AHA	American Historical Association
AHR	*American Historical Review*
AR	*Alabama Review*
CWH	*Civil War History*
IMH	*Indiana Magazine of History*
ISHS	Illinois State Historical Society
JAH	*Journal of American History*
JNH	*Journal of Negro History*
JSH	*Journal of Southern History*
KSHS	Kansas State Historical Society
LHQ	*Louisiana Historical Quarterly*
MHR	*Missouri Historical Review*
MVHA	Mississippi Valley Historical Association
MVHR	*Mississippi Valley Historical Review*
NHCR	*North Carolina Historical Review*
NEQ	*New England Quarterly*
NYH	*New York History*
PMHB	*Pennsylvania Magazine of History and Biography*
SAQ	*South Atlantic Quarterly*
SWHQ	*Southwestern Historical Quarterly*

CHAPTER 1

American Nationalism Achieves an Ominous Fulfillment

ON Saturday evening, February 19, 1848, a little after dusk, a special courier arrived in Washington at the end of a remarkably rapid journey from Mexico City. He had left the Mexican capital scarcely two weeks earlier, had hastened through the mountains and down to Vera Cruz, where he took ship to Mobile, and from there had pushed on in only four days to Washington. His first act upon arriving was to deliver to Mrs. Nicholas P. Trist two letters from her husband in Mexico, after which he went on to the house of Secretary of State James Buchanan. To Buchanan, he delivered a treaty which Trist had negotiated at Guadalupe Hidalgo on February 2 to terminate the war with Mexico.[1] By this treaty the United States was to acquire an area of more than 500,000 square miles, including what is now California, Nevada, and Utah, most of New Mexico and Arizona, and part of Wyoming and Colorado—next to the Louisiana Purchase, the largest single addition to the national domain.[2]

More than a century later, readers in Los Angeles, San Francisco, Salt Lake City, Phoenix, Tucson, Albuquerque, Santa Fe, and even

1. Norman A. Graebner, *Empire on the Pacific* (New York, 1955), p. 1; Milo Milton Quaife (ed.), *The Diary of James K. Polk* (4 vols.; Chicago, 1910), III, 345.

2. The area of the Mexican cession would measure 522,000 square miles if it did not include the portions of what is now New Mexico, Colorado, and Wyoming then claimed by Texas as having been part of the Republic of Texas, or 619,000 square miles with those portions included. The area of the continental United States is 3,022,000 square miles. U.S. Bureau of the Census, *Historical Statistics of the United States, 1789–1945* (Washington, 1949), p. 25; Thomas Donaldson, *The Public Domain* (Washington, 1884), pp. 12, 124, 134.

Las Vegas might suppose that such a vast acquisition would have been hailed with wild enthusiasm, but this was by no means true. On the contrary, James K. Polk, a purposeful man then in the third year of his presidency, found the treaty most unwelcome. True, its terms were closely in accord with what he had wanted when he sent Trist to Mexico in the preceding April. But since that time, a great deal had happened. In September, General in Chief Winfield Scott had marched victorious into Mexico City. The occupation of the capital had brought Mexico to a crisis during which Santa Anna resigned as president, leaving the government one step away from collapse and the country itself ripe for acquisition. These events had stimulated some of the eagle-screaming expansionists in the United States to enlarge their aspirations, and to join in a clamor, which had been growing ever since 1846, for the annexation of the entire Mexican republic. Even before these developments took place, Polk had prepared to raise the price of peace, and as he made plans for his annual message at the end of 1847, he drafted a statement threatening that "if Mexico protracted the war," more land cessions, in addition to California and the Southwest, "must be required as further indemnity." His political caution later led him to fall back upon more ambiguous language, but by 1848 his original goals in California and the Southwest, which had once seemed so bold and aspiring, now began to appear parochial and unimaginative.[3]

At the same time, too, when victory was swelling Polk's ambitions, his emissary of peace had fallen into deep disfavor. Nicholas Trist, whose only previous distinctions had been his marriage to a granddaughter of Thomas Jefferson and his office as chief clerk in the State Department, had been selected to go to Mexico because he seemed a loyal Democrat who would do as he was told and would leave any of the potential glory to be harvested by Secretary Buchanan or other luminaries. But he had greatly disappointed Polk. First, he had indicated to the Mexicans a willingness to consider yielding the area in south Texas between the Nueces and the Rio Grande, which his instructions had given him no discretion to do.

3. Quaife (ed.), *Polk Diary*, III, 161, 163, 216–217. On the movement to acquire all of Mexico, see Edward Gaylord Bourne, "The Proposed Absorption of Mexico in 1847–1848," in his *Essays in Historical Criticism* (New York, 1901), pp. 227–242; John D. P. Fuller, *The Movement for the Acquisition of All Mexico, 1846–1848* (Baltimore, 1936); Albert K. Weinberg, *Manifest Destiny* (Baltimore, 1935).

This alone determined Polk, in October, to hasten his recall, which had already been ordered simply because the president did not want to appear too anxious for peace.[4] Then, in December, the president learned that Trist, after initially quarreling bitterly with Winfield Scott, had become a warm friend of the Whig general in chief, and that the two of them had planned to use Scott's war fund to buy a treaty from the Mexican peace commissioners. It was the bane of Polk's presidency that his best generals were Whigs, whom he hated more than Mexicans, and he had no intention of countenancing a Democratic peace commissioner who would collaborate with them. Polk, now thoroughly aroused by the reports of the use of bribery, had begun to plan the recall of Scott and was restlessly awaiting the return of his dismissed emissary.[5]

Then the incredible happened. On January 15, a sixty-five-page letter arrived from Trist, who had not received the message of October 25 recalling him until he was already deep in negotiations for a treaty. He knew the administration wanted a treaty; he thought it was within his power to achieve peace and his moral duty not to waste this power. He believed that the letter recalling him was not binding because it was written without awareness of the circumstances in Mexico City. Thus the chief clerk, who had been appointed partly because of his expected pliancy, refused to be recalled and wrote on December 6 to inform the government that in his capacity as a private citizen he was continuing to negotiate a treaty of peace.

The administration could use this treaty or not, as it saw fit. For good measure, Trist lectured the president: he hinted that Polk planned a wrongful war of conquest; he implied that he and General Scott would save the administration in spite of itself; he denounced Polk's close friend Gideon Pillow as an "intriguer . . . of incompre-

4. For Trist's role as a kind of deputy for Buchanan with little discretionary power, see Quaife (ed.), *Polk Diary*, II, 465–468; for his recall, see *ibid.*, III, 185–199, and *Senate Executive Documents*, 30 Cong., 1 sess., No. 52 (Serial 509), pp. 91–95, 195.

5. Polk complained that he had been compelled from the beginning to conduct the war through the agency of two generals, Scott and Taylor, who were "hostile" to his administration. Quaife (ed.), *Polk Diary*, III, 58. For Trist's quarrel with Scott, see the correspondence in *Senate Executive Documents*, 30 Cong., 1 sess., No. 52 (Serial 509), pp. 120–127, 159–173. Eugene Irving McCormac, *James K. Polk* (Berkeley, 1922), pp. 509–512, quotes in full letters from the Trist papers on the reconciliation of Trist and Scott, and also letters of Trist to Scott, July 16, and Scott to Trist, July 17, 1847, showing clearly the intent to use bribes to secure a treaty.

hensible baseness of character." When Polk read this, his anger overflowed, and words of choking fury poured out on the pages of his diary: "His despatch is arrogant, impudent, and very insulting to his government, and even personally offensive to the President. ... It is manifest to me that he has become the tool of General Scott ... I have never in my life felt so indignant ... he is destitute of honour or principle ... a very base man."[6]

Polk wrote these words on January 15. Exactly five weeks later Mr. Trist's treaty arrived on his doorstep.

For two days, the president fought against the inevitable, but in fact he had no choice, and he knew it. For the Mexican War was highly unpopular throughout a large part of the country; it was regarded as a war of unjustified aggression on behalf of the evil institution of slavery; and Polk was denounced as a warmonger. The House of Representatives, under Whig control, had actually voted a resolution declaring its belief that the war had been "unnecessarily and unconstitutionally begun by the President of the United States";[7] the public was yearning for peace; and the treaty was, after all, an exact fulfillment of Polk's own terms as formulated ten months previously. He diagnosed his own predicament and stated it forcefully to his cabinet:

> If the treaty was now to be made, I should demand more territory, perhaps to make the Sierra Madre the line, yet it was doubtful whether this could be ever obtained by the consent of Mexico. I looked, too, to the consequences of its rejection. A majority of one branch of Congress is opposed to my administration; they have falsely charged that the war was brought on and is continued by me with a view to the conquest of Mexico; and if I were now to reject a Treaty made upon my own terms, as authorized in April last, with the unanimous approbation of the Cabinet, the probability is that Congress would not grant either men or money to prosecute the war. Should this be the result, the army now in Mexico would be constantly wasting and diminishing in numbers, and I might at last be compelled to withdraw them, and thus lose the two Provinces of New Mexico and Upper California, which were ceded to the U.S. by this Treaty. Should the opponents of my administration succeed in carrying the next Presidential election, the great probability is that the country would lose all the advantages secured by this Treaty.[8]

6. Trist's letter in *Senate Executive Documents,* 30 Cong., 1 sess., No. 52 (Serial 509), pp. 231–266; Polk's reaction in Quaife (ed.), *Polk Diary,* III, 300–301.

7. By a vote of 85 to 81, *Congressional Globe,* 30 Cong., 1 sess., p. 95.

8. Quaife (ed.), *Polk Diary,* III, 347–348.

There was nothing to do but to send Trist's document to the Senate.

The Senate received the treaty on February 23, but did not immediately begin deliberations upon it, for John Quincy Adams had been stricken on the floor of the House on the twenty-second, and congressional business was suspended until after his funeral.[9] But then, the Senate acted with remarkable promptness. Within less than two weeks, ratification was voted. But before the brief contest was over, events had demonstrated that complex and highly mixed attitudes lay behind the votes which saved the peace settlement. Two especially significant amendments came to a roll call, and these two revealed the crosscurrents in the Senate. On March 6, Jefferson Davis of Mississippi moved an amendment which would change the boundary in such a way as to include much of what is now northern Mexico. Since Mexico could hardly be expected to accept this change, a vote for the amendment was virtually a vote to continue the war, but the amendment nevertheless received the votes of eleven Democrats, including Stephen A. Douglas of Illinois, Daniel S. Dickinson of New York, Edward A. Hannegan of Indiana, William Allen of Ohio, and seven slave-state senators. On March 8, George E. Badger, a North Carolina Whig, offered an amendment which would have deleted all territorial acquisitions from the treaty. Since it was a foregone conclusion that the treaty could never command a two-thirds majority in this form, the introduction of this measure placed Whigs who opposed both annexation and war in the dilemma that to end the war, they would have to accept annexation, or to prevent annexation they would have to prolong the war. Nevertheless, fifteen Whigs voted for the Badger amendment. Eight of them, including Daniel Webster, came from New England, one from New Jersey, one from Ohio, three from the border states of Delaware, Maryland, and Kentucky, and one each from North Carolina and Georgia. From these two votes it was evident that enough senators were dissatisfied with the treaty to defeat it. But when the decisive vote came on March 10, the opposing groups could not combine. Expansionists who wanted to annex northern Mexico feared to reject a treaty which secured California and the Southwest, and on the question of ratification only five of the eleven who had voted for the Davis amendment now voted in the negative. If these five had been joined by the fifteen Whigs who wanted no

9. Samuel Flagg Bemis, *John Quincy Adams and the Union* (New York, 1956), pp. 534–538; *Senate Executive Documents*, 30 Cong., 1 sess., No. 52 (Serial 509), p. 4.

territorial acquisition, they would have formed a bloc larger than the one-third necessary to defeat the treaty, but the opponents of expansion feared to reject annexation when this meant also to reject peace, and only seven of the fifteen who had voted for the Badger amendment voted against ratification. On the question of ratification, two additional senators, Thomas Hart Benton of Missouri and Sidney Breese of Illinois, voted in opposition. Altogether twenty-six of the fifty-eight senators had at various times voted against basic features of the treaty, but it was nevertheless ratified by a vote of 38 to 14.[10] It was then hastily returned to Mexico and there approved by both houses of the Congress, so that ratifications could be exchanged on May 30.[11]

Thus, by the acts of a dismissed emissary, a disappointed president, and a divided Senate, the United States acquired California and the Southwest. This gigantic step in the growth of the American republic was not taken with enthusiasm by either president or Congress, but resulted from the fact that the elements in opposition could find no viable alternative and no basis on which they could combine. It was an ironic triumph for "Manifest Destiny," an ominous fulfillment for the impulses of American nationalism. It reflected a sinister dual quality in this nationalism, for at the same time when national forces, in the fullness of a very genuine vigor, were achieving an external triumph, the very triumph itself was subjecting their nationalism to internal stresses which, within thirteen years, would bring the nation to a supreme crisis.

Although serious potential divisions lay beneath the apparent unity of a triumphant nation in 1848, the fact remains that the appearances were indeed auspicious. Judging by material indications, no country on the planet had made such rapid strides in the fulfillment of national greatness and national unity as the United States at the midpoint of this century of nationalism in the Western world. Here was a country so young that many of the citizens were

10. The secrecy provisions of the executive sessions in which the treaty was approved were promptly lifted, and though the debates were not published, the journal of proceedings, showing roll-call divisions, was printed as *Senate Executive Documents*, 30 Cong., 1 sess., No. 52 (Serial 509). See pp. 18, 24, and 36 for the votes on the Davis and Badger amendments and on approval of the treaty. Historians (with the exception of George Lockhart Rives, *The United States and Mexico, 1821–1848* [2 vols.; New York, 1913], II, 630–637) have consistently neglected the story of senatorial approval while writing over and over again the story of Trist's mission.

11. *Senate Executive Documents*, 30 Cong., 1 sess., No. 60 (Serial 509).

older than the republic, yet in less than sixty years since the inauguration of George Washington, the population had almost doubled every twenty years, increasing from 4 million in 1790 to 23 million by 1850. The area of the country had increased from 890,000 to 2,997,000 square miles, and the march of empire, begun by thirteen states strung precariously along the Atlantic seaboard, had not slowed its pace until the United States became a transcontinental, two-ocean colossus, with the superb endowment of natural resources that would enable it, in the twentieth century, to assume a position of world leadership. Meanwhile, the original thirteen had increased to twenty-nine states, so that the majority of the states owed their existence to the creative act of the federal government. The strength of the infant Hercules seemed more impressive than ever before as Yankee volunteers patrolled the streets of Mexico City.

In terms of government, also, nationalism appeared to have made great strides. Andrew Jackson had shown that the president could be a national leader rather than a mere federal chairman of the board. A nationalistic Congress had adopted tariff laws to promote a nationally self-sufficient economy and internal improvement legislation to encourage a national system of transportation. In 1823, President Monroe had proclaimed for the United States a role in the Western Hemisphere which could be fulfilled only by a vigorous nation. Meanwhile, the federal courts were patiently laying the basis for a system of national law, a basis which John Marshall had proclaimed when he asserted, "The United States form, for many, and for most important purposes, a single nation. . . . In war, we are one people. In making peace, we are one people. In all commercial regulations we are one and the same people . . . America has chosen to be, in many respects, and to many purposes, a nation."[12]

By modern standards, the political structure of mid-nineteenth-century America was still quite inadequate for a vigorous nation. Andrew Jackson had avoided extensive use of federal power, sagely observing that the strength of the nation depended upon the devotion with which its citizens supported it and not upon the energy with which it performed governmental functions. He himself, by preventing recharter of the Bank of the United States, had effectively abandoned any effort to maintain a national monetary system.

12. Cohens *v.* Virginia, 6 Wheaton 413–414 (1821).

His party and the Whig party were both coalitions of local organizations rather than fully developed national political organizations.

But even if the political machinery did not bespeak a mature or complete nationality, there were nevertheless broad foundations of common experience and common culture on which American national unity was based. Students of the theory of nationalism generally agree that while nationalism itself is a subjective, psychological phenomenon—a matter of sentiment, will, feeling, loyalty—and not an objective phenomenon, capable of being measured by given ingredients, it is nevertheless true that a certain core of cultural conditions is conducive to the development of nationalism, and that among these conditions are "common descent, language, territory, political entity, customs and tradition, and religion."[13] Although no one of these components is indispensable, most of them are usually present in any fully developed nationality.

By all of these measures, the American people in the 1840s showed a considerable degree of homogeneity and cohesion. The great immigration from Ireland and Germany began during the decade, but most of the population, save for the Negroes in the South, was of British origin, seasoned by long residence in America. Ethnically, America has probably never shown a greater degree of sameness than at the time when the nation was dividing and moving toward civil war.[14]

American speech, already distinct from that of England, provided just such a medium of nationwide communication as Noah Webster had striven for when he made it his object, through speller and dictionary, to promote a "national language [as] . . . a

13. For discussion of conditions requisite to the growth of nationalism, see Hans Kohn, *The Idea of Nationalism* (New York, 1944), pp. 13–18 (quotation is from p. 14); Frederick Hertz, *Nationality in History and Politics* (London, 1944), pp. 7–8; Carlton J. H. Hayes, *Essays on Nationalism* (New York, 1926); Louis L. Snyder, *The Meaning of Nationalism* (New Brunswick, N.J., 1954), pp. 38, 67–69, 113; Karl W. Deutsch *et al.*, *Political Community and the North Atlantic Area* (Princeton, 1957), pp. 46–59. Most writers agree that the factors named in the text are somehow important to the development of nationalism, but they do not agree as to the nature of the process within which these factors operate.

14. In 1850, more than nine persons in ten were of native birth. The total population was 23,191,000, and the foreign-born numbered 2,244,000. More than 1,420,000 had come to America in the decade 1840–1850, which, even allowing for mortality and reverse migration, still indicates that there were scarcely 1 million foreign-born in the United States in 1840, when the population was about 17 million. *Historical Statistics* (1949), p. 32.

band of national union."[15] Yankee twang and southern drawl, to be sure, flavored the speech of diverse sections, but these were less serious as barriers to communication than the provincial dialects of Yorkshire and Somerset in England, or of Gascony and Alsace in France.

The problem of a common territory had been a source of conscious concern to American patriots, who had at one time feared that the mountain barriers between the Atlantic seaboard and the Ohio Valley would mold the people of these areas into separate groups, or that the vastness of the Louisiana region would spread the population too thin for any real cohesion. But the development, first of turnpikes and steamboats, later of canals and railroads, had furnished a means to conquer distance and thus to neutralize its dispersive effect. Many Americans were acutely conscious of this fact. Thus it was John C. Calhoun of South Carolina who, in 1817, on the floor of the House of Representatives, warned that "whatever impedes the intercourse of the extremes with . . . the centre of the republic, weakens the Union" and that "not even dissimilarity of language tends more [than distance] to estrange man from man"; therefore, he exhorted his colleagues, "let us, then, bind the republic together with a perfect system of roads and canals." By mid-century, the transportation system was still by no means perfect, but it had developed sufficiently to enable internal trade, which had been negligible at the time of the Revolution, to outstrip foreign trade by 1831, and to reach a volume three times as great as that of foreign trade by 1847. In fact, a regional division of labor had grown up, in which the South produced exports for the entire country, the Northwest supplied foodstuffs for the South and for growing urban and industrial centers in the East, and New England and the Middle States handled most of the commerce and manufacturing of the nation. These features of sectional differentiation involved friction at some levels. But they also led to economic

15. Noah Webster, *Dissertations on the English Language* (Boston, 1789), p. 397 and *passim*. On language as a factor, see Hertz, *Nationality*, pp. 78–89. The linguistic study of distinctive American speech has not lent itself to historical generalizations or to studies of historical development, but see H. L. Mencken, *The American Language* (4th ed.; New York, 1936), pp. 104–163, and *The American Language, Supplement* (New York, 1945), pp. 151–226. On the related subject of literary nationalism, see Hans Kohn, *American Nationalism* (New York, 1957), pp. 41–89, with valuable citations.

interdependence, and they contributed to making the area of the republic a common territory in a functional sense.[16]

In religion, all sections of the United States responded to the fervor of evangelical Protestant Christianity and to the ethics of a gospel which, by promising damnation for sin and salvation for repentance and virtue, emphasized the responsibility of the individual. Hard work and self-denial were virtues; idleness and self-indulgence were vices, and this was no less true in the backwoods of Mississippi than in the most rockbound strongholds of Yankee Puritanism—though lapses from grace might take more extravagant form and call for more emotional repentance in the backwoods. Enclaves of aristocratic Anglicanism and of intellectualized Unitarianism existed, but were minor, at least in numbers, while Catholi-

16. For Calhoun's speech, see Richard K. Crallé (ed.), *Works of John C. Calhoun* (6 vols.; New York, 1854–57), II, 188–192. For other contemporary statements showing an awareness of the importance of communications for national unity see Merle Curti, *The Roots of American Loyalty* (New York, 1946), pp. 113–118. For expressions of concern lest too extensive a territory should prevent the development of national unity see *ibid.*, p. 32; Fisher Ames to Christopher Gore, Oct. 3, 1803, on the acquisition of Louisiana ("We rush like a comet into infinite space"), in Seth Ames (ed.), *Works of Fisher Ames* (2 vols.; Boston, 1854), I, 323–324; warnings of James Jackson on opening Louisiana to settlement, Feb. 1804, as reported in Memorandum by William Plumer, in Everett Somerville Brown, *The Constitutional History of the Louisiana Purchase* (Berkeley, 1920), pp. 226, 228, 230; speech of Josiah Quincy, Jan. 14, 1811, on statehood for Louisiana in *Annals of Congress,* 11 Cong., 3 sess., cols. 534, 537 ("The Constitution . . . never can be strained to lap over all the wilderness of the West"); letter of Jefferson to John Breckinridge, Aug. 12, 1803 ("Federalists see in this acquisition [of Louisiana] the formation of a new Confederacy . . . and a separation . . . from us"), in Andrew A. Lipscomb (ed.), *The Writings of Thomas Jefferson* (20 vols.; Washington, 1903–04), X, 409; report of Major Stephen Long on his explorations and his discovery of a great American Desert, unfit for settlement but "of infinite, importance . . . inasmuch as it is calculated to serve as a barrier to prevent too great an extension of our population westward," in Reuben Gold Thwaites (ed.), *Early Western Travels* (32 vols.; Cleveland, 1904–07), XVII, 148.

It is remarkable that Calhoun and others, with their idea of strengthening the Union by improving communications, had arrived in an operative way at almost exactly the same functional approach to nationalism advanced at the conceptual level by Karl W. Deutsch. Deutsch argues that nationalism is best measured not in terms of "common attributes," which present certain fallacies and circularities of argument, but in terms of actual volume and intensity of communications. A high incidence of communication bespeaks a "complementarity" and a tendency toward national unity among the people who are involved in it, and a lower incidence between these people and others indicates the limits of the national unit. Viewed in this functional sense, common language and common religion are significant because they enhance communication within the in-group and diminish the communication between the in-group and other groups. Deutsch, *Nationalism and Social Communication: An Inquiry into the Foundations of Nationality* (New York, 1953).

cism still seemed exotic and suspect to most Americans.[17] When, in the demoralizing hour of Lincoln's death, James A. Garfield affirmed that "God reigns and the Government at Washington lives," it was generally understood that the deity in question was a Protestant God quite as certainly as the government was a democratic republic.

The fact that it was a democratic republic was a further bond uniting the American people in a political communion. Travelers from abroad were forcibly reminded of the strength of the political ties which bound the citizens of the United States together, for the Americans boasted of them incessantly. After being asked how he liked "our institutions," the traveler seldom had time to reply before his questioner would launch into a vainglorious harangue on the decadence of monarchies, the merits of a system in which the people were sovereign, and the superiority of republicanism, American style. So strong was the belief in American political values that it was hardly deemed excessive for Andrew Jackson in his farewell address to say that Americans were "the guardians of freedom to preserve it for . . . the human race," or for James K. Polk to call the Federal Union "this most admirable and wisest system of well-regulated government among men ever devised by human minds."[18]

If common political ideals and loyalties bound the American people together, a common culture and a common tradition reinforced the political ties. Here was a body of more than 20 million people who had no privileged aristocracy and, except for the Negroes, no

17. "Protestantism was, in truth, a patriotic touchstone . . . the Bible figured as a sacred patriotic symbol." Curti, *Roots*, pp. 77–79. On religion as a factor, see Hertz, *Nationality*, pp. 98–145. In 1850 the census reported all church buildings and the numbers of people whom they would accommodate. The enumeration showed 37 Jewish churches, 1,227 Roman Catholic ones, and 36,534 Protestant ones. The Jewish churches were estimated to accommodate 19,000 people; the Catholic, 676,000; the Protestant, 14,000,000. It was recognized that the Catholics had greater numbers in proportion to their church accommodations, but even when heavily discounted, the figures are revealing, and are more reliable than many church statistics. Compiled by the author from J. D. B. De Bow, *Statistical View of the United States: Compendium of the Seventh Census* (Washington, 1854). Independent estimates in 1844 by Robert Baird, *Religion in America* (New York, 1844), pp. 264, 271, 283, placed the numbers thus: Protestants, 15,364,000; Catholics, 1,300,000; Jews, 50,000.

18. James D. Richardson (ed.), *A Compilation of the Messages and Papers of the Presidents* (11 vols.; New York, 1907), II, 1527; III, 2225. For notable specimens of extravagant glorification of America, see Curti, *Roots*, pp. 30–64; Weinberg, *Manifest Destiny*, pp. 107–111, 117–119, 127, 171, 194, 202–207.

proletariat and peasantry. True, the South had a tradition of planter leadership, and in New England deference was still due to ministers, magistrates, sea captains, and East India merchants. But in both areas, elite leadership had to be democratically exercised, as the Federalists discovered when Jefferson carried every state in New England except Connecticut in 1804, and as the Whig gentry of the plantation South learned when the hell-roaring Jacksonians swept them out of office and kept them out until they learned to match hard cider and log cabins against buckskin and plain hickory.[19] Although urban working men were beginning to be a significant factor in the population, the vast majority of Americans still lived by the cultivation of the soil, and their lives were patterned by the rhythms and rigors of nature.[20] Pitting their muscle against the elements, these men were independent, aggressively individualistic, and fiercely hostile to external controls. Prizing the opportunity to become unequal in personal achievement and hating the inequality of pretension to status, they cherished an unsleeping distrust of public authority and glorified the virtues of simplicity, frugality, liberty, and self-reliance. Despite the nuances of regional difference, Americans conformed to this basic pattern from one end of the Union to the other. The fact that Negroes were largely excluded from this pattern constituted a great exception but did not seriously weaken the prevalence of these attitudes otherwise.

With a body of common values to unite them, they shared pride in the memories of the War for Independence. As the Revolutionary generation passed from the scene, Americans became increasingly

19. Fletcher M. Green, "Democracy in the Old South," *JSH*, XII (1946), 3–23; Green, *Constitutional Development in the South Atlantic States, 1776–1860* (Chapel Hill, 1930).

20. In 1850 the population of the United States was 15 percent urban and 85 percent rural. The population of the North (the free states plus Missouri) was 20 percent urban and 80 percent rural; that of the South (the slave states except Missouri) was 8 percent urban and 92 percent rural. (California is not included here.) "Urban" means living in towns of more than 2,500 population. U.S. Bureau of the Census, *Sixteenth Census of the United States: Population* (4 vols.; Washington, 1942), I, Tables 7 and 8. Also in 1850, the number of free males gainfully employed was reported at 5,371,000, with 2,400,000 in agriculture and 944,000 in manufacturing establishments which produced more than $500 per year output. The slave states' total of 1,569,000 included 957,000 in agriculture and 160,000 in such manufacturing; the free states' total of 3,802,000 included 1,572,000 in agriculture and 784,000 in such manufacturing labor. In only four states (Massachusetts, Rhode Island, Connecticut, and New Jersey) did the numbers in manufacturing exceed those in agriculture. Compiled by the author from De Bow, *Compendium of the Seventh Census.*

conscious of their heritage from the men of an age which had come to be regarded as heroic. It was this consciousness which made Lafayette's visit a national festival in 1824–1825; which prompted the completion of the Bunker Hill monument in 1843 and the beginning of the Washington Monument in 1848; which inspired a South Carolinian in 1854 to found the Mount Vernon Ladies Association to preserve Washington's home as a national shrine; and which caused men to uncover their heads in the presence of a bronze bell that had pealed for independence in 1776. Deep patriotic sentiment had inspired Everett and Webster to famous orations which echoed in countless schoolhouses; it had enshrined the Constitution as a "palladium of all our liberties," to be venerated and not merely to be admired; and it had apotheosized George Washington, who was certainly no democrat, but who neatly avoided classification as an aristocrat by being transferred to the category of a god. It had made the twenty-second of February and the fourth of July national holidays at a time when Thanksgiving still remained a regional festival and Christmas still seemed too popish to be countenanced by Yankees of the true persuasion. On these days of gargantuan eating and drinking, Americans poured forth torrents of overblown rhetoric to voice the boundless innocence and pride with which they loved their country.[21]

The exuberant nationalism of the forties has long been recognized by historians, but what has often been overlooked is that this sentiment seemed to prevail in the South as vigorously as elsewhere. Although southerners consistently subscribed to the constitutional doctrine that the United States was a federation and not a nation, they sometimes forgot their political metaphysics in moments of enthusiasm, and allowed unguarded expressions to escape. Thomas Jefferson himself had done so in 1785, when he said, "The interests of the states . . . ought to be made joint in every possible instance, in order to cultivate the idea of our being one Nation."[22] In the early

21. On American nationalism considered with reference to theoretical concepts of nationalism, see Kohn, *The Idea of Nationalism,* pp. 263–325; Kohn, *American Nationalism;* Curti, *Roots;* Wesley Frank Craven, *The Legend of the Founding Fathers* (New York, 1956); David M. Potter, "The Historian's Use of Nationalism and Vice Versa," in Potter, *The South and the Sectional Conflict* (Baton Rouge, 1968), pp. 34–83; Paul C. Nagel, *One Nation Indivisible: The Union in American Thought, 1776–1861* (New York, 1964)—the last especially rich in illustrative evidence.

22. Jefferson to James Monroe, June 17, 1785, in Lipscomb (ed.), *Writings of Jefferson,* V, 14. For another use of the term "nation," XV, 46.

period of the Republic exuberant nationalism had been as prevalent in the South as elsewhere, and even after sectional dissensions became acute, nationalistic attitudes continued to find expression. Thus it was at Charleston, South Carolina, in 1845, that young Edwin De Leon, later to be a stalwart of secession, launched the ultranationalistic "Young America" movement by proclaiming that the United States was in the full flush of "exulting manhood," and that if there were a young Italy, a young Ireland, et cetera, there should be a "young America" also.[23] At Raleigh, North Carolina, in 1849, a local editor could boast on the Fourth of July, "There are but few places in the Union, where, in proportion to means and population, the day is celebrated with more lively enthusiasm."[24] In 1854 the *Southern Quarterly Review* rejoiced in "our position as the leading power of the Western world," and in January 1861 *De Bow's Review*, of New Orleans, proclaimed that European immigration to America might result in "a race of men nobler than any which has hitherto worked to adorn God's beautiful earth." Even a southern fire-eater like Pierre Soulé of Louisiana was capable of invoking in 1852 "reverence for the institutions of our country, that devout faith in their efficacy which looks to their promulgation throughout the world." Stephen R. Mallory of Florida, in 1859, used a nonfederative figure of speech when he exulted that it was "no more possible for this country to pause in its career than for the free and untrammeled eagle to cease to soar." Two years later he was secretary of the navy in a government at war with the United States.[25]

Against this background of basic homogeneity, common ideals, integrating policies, increasing cohesiveness, rapid growth of the republic, and ardent national loyalties, the Treaty of Guadalupe Hidalgo seemed a crowning fulfillment of American nationalism.

It was a timely moment for such a triumph, for nationalism, in the early months of 1848, appeared to be coming into its own throughout the Western world. In Europe, where nationalism was distinguished by a markedly revolutionary flavor, a new surge of national-

23. Edwin De Leon, *The Position and Duties of "Young America"* (Charleston, 1845); Merle E. Curti, " 'Young America,' " *AHR* XXXII (1926), 34.

24. Quoted from Raleigh *Register*, July 7, 1849, by Fletcher M. Green, "Listen to the Eagle Scream: One Hundred Years of the Fourth of July in North Carolina (1776–1876)," *NCHR*, XXXI (1954), 318.

25. Quotations from *Southern Quarterly Review* and from Mallory in Weinberg, *Manifest Destiny*, pp. 199–207; from *De Bow's Review* in Curti, *Roots*, p. 72; from Soulé in Curti, " 'Young America,' " p. 39.

ism really began on February 24. On that day, while Polk, in Washington, was waiting for the Senate to take up Trist's treaty, a mob in Paris, milling about the Tuileries, frightened Louis Philippe into abdicating the throne of France and making way not, as he supposed, for his grandson, but for a republic. On that very same day, copies of a thirty-page octavo pamphlet written in German, by Karl Marx, then in Brussels, lay fresh from the press in a London warehouse. This *Communist Manifesto,* as it was called, was published four days later, but the vast explosion which it ultimately set off was long delayed, and thus the most momentous event of 1848 exercised no perceptible effect during that year. Instead, it was the tumult in Paris that detonated a string of revolutions which, in rapid succession, drove Metternich from Vienna, the Habsburg emperor from his throne, and the pope from the Vatican. In the brief interval while America was waiting for Mexico to ratify the treaty of peace, nationalism scored repeated triumphs in various parts of Europe. In Italy in March, patriots from all parts of the peninsula united under the King of Piedmont and drove the Austrians into the mountain defenses of the Quadrilateral. Cavour, Mazzini, and Garibaldi were all at work. In Hungary in April, the Magyars under Louis Kossuth demanded and were promised a separate Hungarian ministry for their country. In Frankfurt in May, liberal Germans who had deposed the king of Bavaria and overawed the king of Prussia met in a parliament to frame a constitution which would bring liberal nationalism to all of Germany. Meanwhile, Denmark had already moved peaceably from absolutism to constitutional government. In Prague, also in May, the St. Wenceslas Committee affirmed the historic rights of Bohemia and called for a pan-Slavic congress to unite the Slavic people. In Poland, in Croatia, in Serbia, nationalism was stirring into life.

But this tide fell as swiftly as it had risen. The last American troops left Mexico in August. By the time they did so, a French army had smashed an insurrection of the workers during the terrible Days of June, at the barricades in Paris, and France had ceased to be a generative force for liberal nationalism in Europe; by the end of the year, Napoleon the Little would be at the head of the government. In Italy, the Piedmontese troops had sustained a crushing defeat at Custozza, and Milan had fallen again under Austrian control; within another year, the king would abdicate, the last desperate struggles of the Italian people would be suppressed at Rome and Venice, and

Garibaldi would be on his way to exile as a candlemaker on Staten Island. In Germany, the Frankfurt Parliament had begun to let its nationalistic energies trickle away in a footless war with Denmark and in futile academic debate; in another year, its members would learn that no one wanted an imperial crown conferred by them, and the remnant of their body, transferred to Stuttgart, would be ignominiously locked out of their hall. As an aftermath of '48 in Germany, Carl Schurz came to America, Karl Marx went to the British Museum to become a preceptor rather than a practitioner of revolution; and Otto von Bismarck began to plan for a national unification, which would be based on blood and iron and not on liberal reform. In Bohemia, Marshal Windischgrätz had made quick work of the Pan-Slavic Congress, and at Budapest, the Magyars were beset by the very force which they had themselves invoked, as Croatian and Serbian nationalists revolted against Hungarian control; Kossuth was soon to become a hero, sublime in his defeat, a lionized exile on a triumphal American tour which took him in 1852 to dinner at the White House with President Fillmore, while his high-spirited entourage was breaking up the furniture at Brown's Hotel.[26]

After a transient hour of glory, seasoned with incredibly romantic episodes of heroism and drama, liberal nationalism in Europe had met with disasters from which it never recovered. The fact that it had done so made the success of the national experiment in America all the more crucial to the fate of democratic nationalism in the modern world. This was the essential truth which Abraham Lincoln later affirmed at Gettysburg when, without once mentioning the word "America," he defined the Civil War as a testing to determine whether "this nation or any nation so conceived and so dedicated can long endure."

But although liberal nationalism seemed to enjoy an ascendancy in America at the end of 1848 which contrasted dramatically with its debacle in Europe at that time, it in fact faced challenges in the New World almost as portentous as those which had overwhelmed it in the Old. The American victory over Mexico and the acquisition of the Southwest had sealed the triumph of national expansion, but

26. Priscilla S. Robertson, *Revolutions of 1848: A Social History* (New York, 1960); Arnold Whitridge, *Men in Crisis: The Revolutions of 1848* (New York, 1949); Carl Wittke, "The German Forty-Eighters in America," *AHR*, LIII (1948), 711–725.

it had also triggered the release of forces of sectional dissension. Much of the national harmony had rested upon the existence of a kind of balance between the northern and southern parts of the United States. The decision to fight the war had disturbed this balance, and the acquisition of a new empire which each section desired to dominate endangered the balance further. Thus, the events which marked the culmination of six decades of exhilarating national growth at the same time marked the beginning of sectional strife which for a quarter of a century would subject American nationalism to its severest testing. Perhaps it may even be said that the developments which gave American nationalism the strength to survive also generated a supreme threat to its survival.

CHAPTER 2

Portents of a Sectional Rift

IF a climax in the early growth of American nationalism was symbolized by the treaty which made the United States a transcontinental republic, the emergence of the sectionalism which almost destroyed the nation was symbolized by an amendment to an appropriation bill which was never enacted. Both symbols appeared unexpectedly as the work of obscure men—a repudiated emissary who had previously been a clerk in the State Department, and a freshman representative from Pennsylvania named Wilmot. The curious overlap and interplay of the national and sectional forces is suggested by the fact that Wilmot's amendment, which raised the curtain on the sectional drama, came on August 8, 1846, almost two years before Trist's treaty signalized an apogee of nationalism that sectional forces had already begun to threaten.

The eighth of August was a Saturday. The first session of James K. Polk's first Congress had voted to adjourn on the following Monday, and both houses were in the usual end-of-session turmoil. At this eleventh hour, Polk made a belated decision to swallow an unpalatable necessity. For many weeks, even before the Mexican War had begun in the preceding May, he had been maneuvering to obtain funds to be used in negotiating a treaty by which the United States would acquire territory from Mexico. Not wishing to reveal his objectives prematurely, he had first sought to arrange for an appropriation to be voted in secret executive session by the Senate, after which it could be sent to the House and adopted without debate. But the Whigs had at last made it clear that publicity would be the price of their support. Thereupon, Polk decided to disclose his intentions, and about noon on August 8 he sent to the House a public message, expressing the hope that "a cession of territory . . . may be made" by Mexico, for which "we ought to pay them a fair equivalent," and requesting an appropriation of $2 million with which to negotiate.[1]

Thus the president avowed a purpose which everyone had privately understood but no one had publicly known.[2] By delaying the avowal until the eve of adjournment, he left only a few hours in which protest might be expressed, but also a minimum of time for getting his measure enacted. The time factor, however, did not daunt the Democratic floor leaders. They set the machinery of party control into motion at once, and the House voted to take up the proposed appropriation that very evening, under a rule which would limit debate to two hours altogether with no member allowed more than ten minutes.[3]

When the House reassembled after dinner, the members—some partially intoxicated—straggled in reluctantly, only half reconciled to the idea of a hard session on one of Washington's sultriest August nights. Ice water and fans were in heavy requisition, and there were no bevies of ladies to grace the House as they did on days when major oratorical displays were anticipated. But an air of expectancy began to develop as the session got under way.[4]

Seasoned political practitioners perhaps sensed that some sort of upset might occur. For seven months, the administration had been driving its measures through Congress with an extremely firm rein, and with little regard for the feelings of rank-and-file members. The declaration of war with Mexico, the measures for supporting the war, the Oregon treaty, the tariff reduction drafted by Secretary of the Treasury Robert J. Walker, the president's veto of a river and harbors bill which would have provided pork dear to many congressmen—all had been accompanied by the cracking of the whip of party discipline, all had

1. Milo Milton Quaife (ed.), *The Diary of James K. Polk* (4 vols.; Chicago, 1910), II, 70–73; James D. Richardson (ed.), *A Compilation of the Messages and Papers of the Presidents* (11 vols.; New York, 1907), IV, 456.

2. There had, of course, been constant speculation about Polk's annexationist plans; e.g., see Baltimore *American*, July 9, 1846.

3. *Congressional Globe*, 29 Cong., 1 sess., pp. 1211–1213.

4. New York *Herald*, Aug. 11, 1846.

aroused resentment in various quarters, and all had been assailed in speeches on the floor. On some divisions, northern Democrats had broken party ranks to vote against the administration. No major opposition had yet materialized within the party, but the mood of many Democrats was angry, and the question of territorial acquisitions remained touchy.

As the session opened, Hugh White, a Whig from New York, launched the debates by assailing the expansionist plans of the administration, suggesting that the ulterior purpose was to extend the area of slavery, and challenging the northern Democrats to amend the bill so as to exclude slavery from any newly acquired area. Next came Robert C. Winthrop of Massachusetts, one of the big guns of the Whig battery, who predictably spoke in opposition. Two other speakers defended Polk, and then David Wilmot, still in his first term from Pennsylvania's Bradford District, joined the clamor of those seeking the floor.[5]

With debate so stringently limited, the chairman of the Committee of the Whole must have wondered whether to recognize Wilmot or some other claimant. If he did, he may have recalled that the Pennsylvanian had been an exceptionally faithful administration man. Wilmot had voted for measures to carry through the annexation of Texas, already decided by the previous Congress; he had supported the Oregon compromise with its embarrassing retreat from demands for the boundary at 54° 40′; and most important, he had gone down the line for the administration's tariff reduction when every other Democrat from Pennsylvania crossed party lines to vote against it.[6] The chair recognized Mr. Wilmot.

Within the allotted ten minutes, Wilmot made a place for himself in history. His very first sentence unexpectedly condemned Polk for not acting more openly. As for expansion, Wilmot approved of it, and where it involved a region like Texas in which slavery already existed, he had not protested against it. But if free territory were acquired, "God forbid that we should be the means of planting this institution upon it."

Thus far, Wilmot had merely complained vocally against the administration, and a little such protest could always be condoned by

5. *Congressional Globe,* 29 Cong., 1 sess., pp. 1213–1214.

6. Charles Buxton Going, *David Wilmot, Free Soiler* (New York, 1924), pp. 61–93; Richard R. Stenberg, "The Motivation of the Wilmot Proviso," *MVHR,* XVIII (1932), 535–541.

party regulars if a congressman needed to strengthen his position in his home district in an election year. But Wilmot now turned from discussion to action. Invoking the language of the Northwest Ordinance, he offered an amendment to the appropriation: "that, as an express and fundamental condition to the acquisition of any territory from the Republic of Mexico . . . neither slavery nor involuntary servitude shall ever exist in any part of said territory, except for crime, whereof the party shall first be duly convicted." This was the Wilmot Proviso.[7]

The word spread quickly that Wilmot had raised the standard of

7. *Congressional Globe,* 29 Cong., 1 sess., p. 1217; Chaplain W. Morrison, *Democratic Politics and Sectionalism: The Wilmot Proviso Controversy* (Chapel Hill, 1967), pp. 16–18. The authorship of the Proviso was later questioned. Jacob Brinkerhoff, an Ohio Democrat, in a letter of Sept. 16, 1846, in the Columbus *Statesman,* Oct. 2, 1846, claimed that he had written the draft of the Proviso (cited in Stenberg, "Motivation of Wilmot Proviso") and later renewed this claim even more emphatically in a letter to Henry Wilson, April 4, 1868 (*New York Times,* April 23, 1868), less than three weeks after Wilmot's death. It was also asserted that a copy of the Proviso in Brinkerhoff's handwriting was deposited in the Library of Congress after his death in 1880 but disappeared about 1890. William Henry Smith, *A Political History of Slavery* (New York, 1903), I, 83, 84. According to this story, a group of northern Democrats including Preston King, Hannibal Hamlin, Gideon Welles, Brinkerhoff, and Wilmot consulted on free-soil strategy and chose Wilmot to be their spokesman because he was more likely to gain recognition from the chair, since he had regularly supported administration measures.

For a long period, historians accepted the Brinkerhoff claim (von Holst, McMaster, A. B. Hart, G. P. Garrison, Channing), but Milo Milton Quaife, *The Doctrine of Non-Intervention with Slavery in the Territories* (Chicago, 1910), pp. 13–16, and Going, *David Wilmot,* pp. 117–141, have shown that Wilmot made detailed statements concerning his authorship in speeches at Tioga, Pennsylvania, and Albany, New York, Oct. 21 and 29, 1847, which were never challenged although there were certain men who were in position to have recognized false statements on this subject; and that the original manuscript of the resolution introduced in Congress is in the files of the 29th Congress and is in Wilmot's own hand, with corrections also in his own hand. It appears that neither Brinkerhoff nor anyone else had a better claim to the authorship than Wilmot. Yet Eric Foner, "The Wilmot Proviso Revisited," *JAH,* LVI (1969), 264, points to evidence that each of the antislavery congressmen in the planning group wrote out his own copy of the Proviso, and that each tried to get the Speaker's recognition. The language of the Proviso, after all, was that of the Northwest Ordinance, and thus in a sense Wilmot, as he himself later declared, "was but the copyist of Jefferson." *Congressional Globe,* 30 Cong., 1 sess., appendix, p. 1076. However it was that Wilmot came to be the front man, the strategy was the work of a group, and according to Foner, p. 265, "It is hard to resist the conclusion that the New York Van Burenites, and especially [Preston] King were the leading spirits of the group." It was King and not Wilmot, Foner adds, who reintroduced the Proviso in the next session of Congress. For the political background of the Proviso, with emphasis on Democratic factionalism, see also Charles Sellers, *James K. Polk: Continentalist, 1843–1846* (Princeton, 1966), pp. 476–484.

revolt, and members of the administration hastened to the lobby of the House. Soon, no fewer than three cabinet members were in attendance. But despite the growing excitement, debate was still rigorously limited, and within less than two hours, the House voted. As it did so, William W. Wick, an Indiana Democrat, tried to substitute a different amendment which would have applied the Missouri Compromise line of 36° 30′ to any new territory, but this was defeated 89 to 54. Wilmot's amendment then passed by a vote of 80 to 64, with every negative vote except three coming from the slave states. Now it developed that the southern members, who had been the bill's warmest supporters, would rather have it killed altogether than to accept it with its exclusion of slavery, and they moved to table. On this motion, an ominous development occurred. The roll call produced a division not between Whigs and Democrats, but between northerners and southerners. Seventy-four southerners and four northerners voted to table; ninety-one northerners and three southerners voted against tabling. The bill itself, as amended, was then carried by a vote of 85 to 80, with the two sides again divided almost wholly on sectional lines, and was sent to the Senate.[8]

Since the next day was Sunday, the Senate did not take up the measure until August 10, the last day of the session, and in fact did not get to it until an hour before the time set for adjournment. Time now became the critical factor in the strategy of administration leaders. They planned to strike Wilmot's amendment out of the House bill and rush the measure back to the House, where the shortage of time would force the representatives to take it without the amendment. But timing is a game at which two can play, and Senator John Davis of Massachusetts, a Whig and a friend of the amendment, apparently conceived of the idea of talking until it would be too late to return the bill, whereupon the senators would be forced to take it as the House offered it—with the amendment. But if this was indeed his purpose, he miscalculated, and taking the floor on the motion to strike out the House's amendment, he kept on talking until the clock showed eight minutes left in the session. At that point he was interrupted by the information that the House, whose clock was faster, had already adjourned, and the session had expired.[9]

8. *Congressional Globe,* 29 Cong., 1 sess., pp. 1217–1218; New York *Herald,* Aug. 11, 1846; Quaife, *Doctrine of Non-Intervention,* p. 16.

9. *Congressional Globe,* 29 Cong., 1 sess., pp. 1220–1221. The question of why Davis, who favored the amended bill, should have talked it to death has been a subject of

The Two-Million Bill had expired too, and Mr. Wilmot's amendment had apparently expired with it, but in fact the brief resolution had already begun to realign the structure of American politics. The Boston *Whig* correctly observed, "As if by magic, it brought to a head the great question which is about to divide the American people."[10]

The episode had occurred so suddenly and ended so abortively that its full significance was not perceived until much later. But in that age of fairly strict party discipline, it must have been shocking to see northern Democrats deserting the administration, not in detachments, but as a solid body.

It could not have happened, of course, without a background of antagonism within the Democratic party. In retrospect, it was evident that incipient divisions had existed for a long time. Ever since the days when Martin Van Buren had jockeyed to displace John C. Calhoun in the favor of Andrew Jackson, there had been northern and southern wings within the party, but Jackson himself had warned sternly against such divisions, and Calhoun's defection to the Whigs had largely destroyed his sectional influence among southern Democrats. The party had remained well united under the adroit leadership of Van Buren, and its defeat in 1840 had only sharpened the determination to reestablish Democratic control in 1844. But then the question of Texas annexation came to the fore, and as it did, three fateful things happened to the Democratic party. First, the southern Democrats sabotaged the renomination of Van

---

minor disagreement among historians. James G. Blaine, *Twenty Years of Congress* (2 vols.; Norwich, Conn., 1884–86), I, 68, suggested that Davis intended to defeat the whole measure, but the fact is that he was in favor of the bill as amended. H. E. von Holst, *Constitutional and Political History of the United States* (8 vols.; Chicago, 1877–92), III, 287–288, thought that Davis was merely foolish and that his "unseasonable loquacity" defeated his own purposes. But Going, *David Wilmot*, p. 103, shows that Davis himself explained his strategy, as it is described above, in a speech on Feb. 25, 1847. See *Congressional Globe*, 29 Cong., 2 sess., pp. 508–509. A different result would have changed the course of American history, and there has been much speculation as to whether the measure could have passed, with the Proviso in it, if it had come to a vote. Wilmot himself (*ibid.*, appendix, p. 315) said it would have. Henry Wilson, *History of the Rise and Fall of the Slave Power in America* (3 vols.; Boston, 1872–77), II, 17, discusses the belief of Salmon P. Chase and of Jacob Brinkerhoff that it would have passed, but Wilson then proceeds to an analysis indicating that Brinkerhoff's calculations were wrong. Polk said that it would have been defeated (Quaife [ed.], *Polk Diary*, II, 75–76), and his overall policy indicates that he would have vetoed it if Congress had passed it.

10. Boston *Whig*, Aug. 15, 1846, quoted in Frank Otto Gatell, *John Gorham Palfrey and the New England Conscience* (Cambridge, Mass., 1963), pp. 130–131.

Buren and did this in circumstances that left great bitterness. Before the Texas question became ascendant, many southern delegates to the Democratic convention had already been chosen and had been pledged to Van Buren in a way that left them no option but to vote for him. Then, when Van Buren came out against annexation, they sought a means to block his nomination and found it in the application of a rule, used in the convention of 1832, requiring the vote of two-thirds of the delegates for a nomination. This strategy not only blocked the will of the majority, but what was worse, it meant that a number of men pledged to Van Buren were voting for a rule designed to prevent his nomination. In their eyes this seemed a legitimate means of recovering the freedom of action they had lost by a premature pledge. But to the friends of Van Buren it appeared that the leader of the party was being treacherously slain in the house of his friends, and that the sinister force at work was something coming to be called the Slave Power. Many northern Democrats never forgot or forgave, as the party regulars would learn when Van Buren ran as a third-party candidate four years later.[11]

The only thing holding the party together in this crisis was the fact that the nomination itself appeared to be less a victory for Van Buren's enemies than a compromise between his enemies and his friends. The prize did not go to Lewis Cass, Van Buren's chief rival in the balloting; nor did the platform endorse the annexation of slaveholding Texas alone. Instead the convention nominated a dark horse, James K. Polk of Tennessee, and this southern candidate's southern managers had wit enough to arrange that his name should first be brought forward by delegations from Massachusetts and Pennsylvania. Also, the convention balanced the objective of expansion in Texas with that of expansion in Oregon. Thus, new free territory would offset new slave territory, and the nationwide impulses of expansion could be fulfilled without arousing sectional

11. For the disruptive effects of the revolt against Van Buren and of the question of the annexation of Texas, see especially James C. N. Paul, *Rift in the Democracy* (Philadelphia, 1951). Also, Charles M. Wiltse, *John C. Calhoun, Sectionalist* (Indianapolis, 1951), pp. 60–186; Justin H. Smith, *The Annexation of Texas* (New York, 1911), pp. 234–257; James P. Shenton, *Robert John Walker: A Politician from Jackson to Lincoln* (New York, 1961), pp. 22–50; Foner, "Wilmot Proviso Revisited," pp. 267–273. In the long run, the two-thirds rule was Calhoun's doctrine of the concurrent majority applied within the Democratic party, and it accounted for the peculiar relationship between the South and the Democratic party not only in the antebellum era but until the administration of Franklin Roosevelt, when southern leaders, temporarily forgetful of their historic status as a minority section, permitted it to be repealed. See David M. Potter, *The South and the Concurrent Majority* (Baton Rouge, 1972).

jealousies. With a disingenuous claim that both areas properly belonged to the United States already, the platform called for "the Re-occupation of Oregon and the Re-annexation of Texas—the whole of the territory of Oregon." Running on this platform, Polk gained close but decisive majorities in every state of the lower South, carried every state of the Northwest except Ohio, and also won Maine, New Hampshire, New York, and Pennsylvania, thus capturing the presidency with well-distributed bisectional support.

The Oregon question did not receive any great emphasis in this campaign, and the use of the aggressive slogan "Fifty-four forty or fight," which many historians have attributed to the campaign, actually came later.[12] But northern Democrats had every reason to expect that Polk would push for free territory in Oregon as vigorously as he would for slave territory in Texas. As matters developed, however, they saw the claim to all of Oregon sidetracked while Texas annexation was voted in February 1845, troops were sent to the farthest edge of the disputed zone between the Nueces and the Rio Grande in January 1846, and war with Mexico was declared in May. They supported these steps with great solidarity, but in June Polk submitted to the Senate a treaty dividing the Oregon country about equally between Britain and the United States, along the 49th parallel. At this point the pent-up resentments of the northern Democrats burst out in a flood of bitter incrimination. "Texas and Oregon were born the same instant, nursed and cradled in the same cradle—the Baltimore Convention," and no one hesitated about Oregon until Texas was admitted, exclaimed Senator Hannegan of Indiana; but then "the peculiar friends of Texas turned, and were doing all they could to strangle Oregon!" Representative John Wentworth of Illinois took note of predictions, then circulating, that the South, "having used the West to get Texas, would now abandon it, and go against Oregon." On the question of ratification, the northern Democratic senators, then, for the first time, openly rebelled against the administration. Twelve of them voted against the treaty and only three voted in favor. Their vote was barely offset by that of slave-state Democrats, sixteen of whom voted in favor and two of whom opposed. The Democratic president secured the desired ratification only because every Whig in the Senate supported him, and the final vote was 41 to 14. But it was a dearly

12. Edwin A. Miles, " 'Fifty-four Forty or Fight'—An American Political Legend," *MVHR*, XLIV (1957), 291–309.

bought victory, as Polk soon learned. The Oregon Compromise left many northern Democrats with a sense of betrayal; it signaled the first open breach in Congress between southern and northern wings of the Democratic party; and it destroyed the bisectional basis for expansion. Thus it was the second fateful cause of rift in the Democracy.[13]

The third apple of discord was the tariff. Here again, Polk's excessively adroit campaign methods made trouble for his administration. During the campaign he had written an ambiguous letter to John K. Kane of Philadelphia in which he did not quite say that he favored a protective tariff, but did express approval of "protection to all the great interests of the whole Union . . . including manufactures." With this document in hand, Pennsylvania Democratic leaders had been able to convince the voters, and perhaps even themselves, that Polk would not reduce duties, and they had carried the state for him against Clay. But when he appointed Robert J. Walker, a man of free-trade convictions, as his secretary of the treasury, and when Walker produced an administration-sponsored measure that was one of the few real tariff reductions in American history, northern Democrats again felt betrayed. In July 1846, Walker's bill passed the House by a vote of 114 to 95 with seventeen northern Democrats joining the Whigs who voted solidly against it. In the Senate,

13. *Congressional Globe,* 29 Cong., 1 sess., pp. 110, 460 (Hannegan); 205–206 (Wentworth), cited in Avery O. Craven, *The Growth of Southern Nationalism, 1848–1861* (Baton Rouge, 1953), pp. 30–32. See also Morrison, *Democratic Politics,* pp. 11–13. On the Oregon settlement generally, see Frederick Merk, *The Oregon Question* (Cambridge, Mass., 1967); and the preceding volume in the New American Nation series, Glyndon G. Van Deusen, *The Jacksonian Era, 1828–1848* (New York, 1959), pp. 209–213. Clark E. Persinger, "The 'Bargain of 1844' as the Origin of the Wilmot Proviso," AHA *Annual Report, 1911,* I, 189–195, stated the thesis that Texas and Oregon had been quid pro quos in a "bargain" in 1844, and that the revolt against the Oregon Treaty and the move to ban slavery in the Southwest were reactions by northerners to the violation of the agreement. This argument reflected an important underlying truth, but was stated too restrictively in two ways: First, evidence is scant of an explicit agreement by either section to support annexation in one area in return for annexation in another; for instance, Ohio could not have been party to such a deal, for she did not vote for Polk either in the nomination or in the election; for Ohio, there was no quo for which to expect a quid. Probably Texas and Oregon were linked simply in the sense of giving expansion a balanced bisectional character, instead of skewing it to the advantage of one section. Second, Persinger suggests that the Wilmot Proviso was not motivated by antislavery ideals, but simply by a desire to retaliate on the southerners for compromising in Oregon. Yet the evidence is clear that most antislavery men were antislavery before Polk ever agreed on the 49th parallel. Persinger oversimplifies a very complex motivation.

it passed by a single vote, 28 to 27, with three northern Democrats in opposition and one Whig, under the duress of instructions from his state legislature, in support. Northern opponents were quick to note that the measure could not have passed without the votes of the two new senators from Texas.[14]

Twenty-six months after the defeat of Van Buren in the Baltimore convention, seven weeks after the adoption of the Oregon compromise, scarcely more than a week after the enactment of the Walker Tariff, David Wilmot offered his proviso. The reaction of the northern Democrats showed that many of them had scores to settle. In this sense, the Wilmot Proviso may be explained in terms of party politics, as the climax to a series of intraparty rivalries which took a sectional form within the Democratic organization.

In the perspective of more than a century, however, these party squabbles seem less important in themselves than as indications of a deep rift among the American people. If politicians chose to revolt on the slavery issue rather than the tariff issue or the Oregon boundary issue, this in itself reflected their awareness that there was a public opinion on the slavery question which would make it a strategic focus for their action. A sharp division was developing along sectional lines, and this division was finding expression in the alignments of national politics. This politicizing of sectionalism may seem too obvious to be worth formal analysis, but it is important to recognize that at an earlier time, important sectional dissimilarities had existed without taking a chronic political form, and sectional division always could and sometimes did take other forms, such as the economic rivalry between New Orleans and Buffalo for the trade of the upper Mississippi Valley, or the later cultural separatism by which southerners sought to develop a literature, a publishing industry, and an educational system independent of those of the North. But instead of developing primarily in an economic or cultural context, the sectionalism of the mid-century expressed itself primarily in political strife. The sectional leaders were party chief-

14. Stenberg, "Motivation of Wilmot Proviso," makes a good statement of the importance of the tariff issue. See also Edward Stanwood, *American Tariff Controversies in the Nineteenth Century* (2 vols.; Boston, 1903), II, 75–77; Shenton, *Walker*, pp. 52–53; and especially Sellers, *Polk*, pp. 116–123, 451–468. Polk's letter to Kane was published in *Niles' Register*, LXVI (June 22, 1844), 259. For a case history of another northern Democrat who broke with the administration see Don E. Fehrenbacher, *Chicago Giant: A Biography of "Long John" Wentworth* (Madison, 1957), chap. IV: "The Making of an Insurgent."

tains; the sectional battles took place in Congress and in conventions and in legislatures; the power they fought for was political control; and their objectives were measures political, such as acts of Congress, the organization of territories, the admission of states. The fact that the sectional impulse operated within a political medium is significant, for it meant that the conditions and circumstances of the medium had an important effect upon the way in which the sectional force worked itself out. For instance, the frequency of American elections meant the constant exploitation of sectional tensions for the purpose of arousing the voters; a system with fewer appeals to the ballot box might have seen a less chronic practice of sectional agitation. Another political feature which conditioned the operation of the sectional force in a significant way was the dissimilarity of the bases of representation in the Senate and in the House. This system tended to make southern influence dominant in one branch and northern in the other, which in turn meant that deadlocks tended to develop in Congress, thus prolonging situations of sectional strife. Also, the interplay between sectionalism and the party system was of vital importance. It is commonly supposed that the existence of two national parties, each with both a northern and a southern wing, exercised a unifying effect which offset the disruptive tendencies of sectionalism. In a sense, this may be true: certainly it is true that each sectional wing tried to cooperate with the other wing of its own party. The extremism of both northern and southern Democrats, for instance, was tempered by their association with one another. But on the other hand, the intrasectional rivalry of parties caused each sectional wing to compete against the other party's corresponding sectional wing in expressions of sectional zeal: southern Democrats and southern Whigs tried to exceed one another in their proofs of devotion to slavery; northern Democrats and northern Whigs in their commitment to free soil. And each tried to discredit its rival within the section by suggesting that this rival had sold out to its counterpart in the other section. Southern Whigs insisted that the southern Democrats were allied with free-soilers; when Taylor ran for the presidency in 1848, northern Democrats capitalized on the fact that northern Whigs had accepted a Louisiana slaveholder as their leader.

Another crucial feature of the political system which also shaped the operation of sectionalism was the prevailing acceptance of the concepts of the negative state and of strong constitutional limitations on the power of the central government. These limitations meant, in effect, that Congress could do little about slavery except to talk about it. While serving as a sounding board for ceaseless sectional recrimination, Congress lacked power to act as an effective arbiter of sectional disputes, and in fact could not even address itself directly to the question of slavery.

Since the sectional impulse took a political form and the circumstances of politics conditioned the operation of sectionalism, this book, a study of sectional conflict, will deal primarily with political events. But as a preliminary, it is well to recognize that sectionalism was not initially or intrinsically a political phenomenon, and it is important to consider sectionalism in its prepolitical form. What originally differentiated North from South? How did dissimilarities become sources of tension? What part was played by cultural disparities, by economic rivalry, by ideological disagreement? And above all, what was the role of slavery in producing sectional conflict?

Viewing sectionalism in its most general terms, one may observe that in a country with the extent and the physical diversity of the United States, regional differentials necessarily exist, and they may lead to dissimilarities that clearly distinguish one region from another, or to conflicts of interest that bring regional groups into rivalry with one another. Such a process is always at work, more or less, and is usually balanced by other, unifying forces, so that the sectional tendencies do not become disruptive. But sectionalism has been chronic in American history. At times, the divisions between East and West have seemed even deeper and more serious than those between North and South. In this sense, it can be argued that the North-South division which ended in the Civil War was nothing unique, but was only the most acute manifestation of a phenomenon which has appeared again and again.[15]

There remains, however, the problem of why the sectionalism of the 1850s was so much more disruptive than any other sectional

15. On the general concept of sectionalism, see Frederick Jackson Turner, *The Significance of Sections in American History* (New York, 1932); Merrill Jensen (ed.), *Regionalism in America* (Madison, 1951); David M. Potter and Thomas G. Manning (eds.), *Nationalism and Sectionalism in America, 1775–1877* (New York, 1949); Potter, "The Historian's Use of Nationalism and Vice Versa," in Potter, *The South and the Sectional Conflict* (Baton Rouge, 1968).

strife in American history. Here was the one instance where the unifying forces failed to counterbalance the divisive tendencies, where the intensity of sectional feeling was scarcely mitigated in any way. What accounts for this unique failure?

Explanation of the uncontrolled growth of sectionalism during the 1850s has been one of the major problems of American historical scholarship. The refinements of interpretation have been endless, but broadly speaking, there has been one school of thought which regards the presence of Negro slavery in the South and its absence in the North as the essence of the sectional controversy, with the result that the term "sectional conflict" becomes little more than a euphemism for a fight about slavery. Opposing this view, other historians have argued that the commitment of the North to Negro equality was minimal, that the prolonged struggle over slavery in the territories scarcely touched the vital question of the servitude of more than 3 million human chattels, and therefore that there was not enough antislavery in the "antislavery" movement to justify an explanation of the sectional conflict primarily in terms of the slavery issue. Such writers have offered two alternative explanations —one which sees the struggle as a clash of profoundly dissimilar cultures, whose disparities transcended the difference over slavery; the other which sees it as a clash between economic interests of an emerging industrialism on the one hand and of plantation agriculture on the other.

Proponents of the cultural explanation of sectionalism argue essentially that the people of the North and the people of the South were at odds not merely because they disagreed about the servitude of the Negro, but because they lived in different cultural worlds. As they see it, the cotton and tobacco plantations, the isolated backwoods settlements, and the subsistence farms of the South were all part of a rural and agricultural way of life, static in its rate of change, decentralized and more or less primitive in its social and economic organization, and personal in its relationships. Southerners placed a premium on the values of loyalty, courtesy, and physical courage— these being the accustomed virtues of simple, agricultural societies with primitive technology, in which intelligence and skills are not important to the economy. By contrast, the North and West, although still agricultural and rural by statistical measurement, had begun to respond to the dynamic forces of industrialization, mass transportation, and modern technology; and to anticipate the mobile, fluid, equalitarian, highly organized, and impersonal culture of cities and machines. Their values of enterprise, adaptability, and capacity to excel in competition were not the values of the South. In the eyes of some scholars, the sum of these differences was so great that North and South had become, in fact, separate cultures, or, as it is said, distinct civilizations. Any union of the two, lacking a basis of homogeneity, must be artificial and, as it were, fictitious. If North and South clashed politically, it was because of this general incompatibility and not because of disagreement over slavery or any other single, specific issue. The two cultures still would have clashed even if all the Negroes had been free. As for slavery, of course the southern system of chattel labor was static and archaic, while the northern system of wage labor was fluid and competitive. But each, in its own way, could be brutally exploitative, and the dissimilarities between them did not, in themselves, separate the two societies but were merely reflections or aspects of a broader and deeper duality. Further, the cultural explanation asserts, slavery was not itself the determinative fact in the life of the Negro. The controlling feature—the thing that made him what he was—was not his legal status as a chattel but his economic status as a hoer and picker of cotton. He was an unskilled worker in the production of a raw material for the world market, and all such workers, whether slave or free, led lives of deprivation. Exponents of this view point out that even after emancipation, the daily life of the Negro did not change appreciably for nearly seventy years, and indeed it never did become very different until he ceased to work in the cotton fields.[16]

16. For sharp, unqualified statements of the cultural antithesis of North and South, see Edward Channing, *A History of the United States* (6 vols.; New York, 1905–25), VI, 3–4; James Truslow Adams, *The Epic of America* (Boston, 1931), pp. 250–255. To offset the image of the South as a purely aristocratic society, see Thomas J. Wertenbaker, *Patrician and Plebeian in Virginia* (Charlottesville, Va., 1910), which refuted the idea that the Virginia planters were of noble English families and that the cavalier origins of Virginia presented a contrast with Puritan New England; Fletcher M. Green, "Democracy in the Old South," *JSH*, XII (1946), 3–23, Frank Lawrence Owsley, *Plain Folk of the Old South* (Baton Rouge, 1949). The latter two show how effectively the planters were challenged within the South politically and how limited their social control was.

There are many treatments which describe southern antebellum society, some of them specifically contrasting it with northern society, but without specifically evaluating the effect of the distinctive features in causing sectional strife: see William E. Dodd, *The Cotton Kingdom* (New Haven, 1919); Ulrich Bonnell Phillips, *Life and Labor in the Old South* (Boston, 1929); Arthur Charles Cole, *The Irrepressible Conflict, 1850–1865* (New York, 1934); W. J. Cash, *The Mind of the South* (New York, 1941); Avery

The outstanding weakness of this cultural interpretation is that it exaggerates the points of diversity between North and South, minimizes the similarities, and leaves out of account all the commonalities and shared values of the two sections which have been discussed in the preceding chapter. These features had proved their reality and their importance by nourishing the strong nationalism which was in full vigor by the 1840s. Further, any explanation which emphasizes the traditionalism of the South is likely to lose sight of the intensely commercial and acquisitive features of the cotton economy.

The economic explanation of sectionalism avoids this difficulty, for it does not emphasize dissimilarities, and instead of attributing conflict to unlikeness, explains it as a result of the collision of interests. Deriving as it does from a vein of economic determinism, it argues that two regions with dissimilar economies will develop diverse economic objectives, which will lead in turn to a conflict over policies. When such conflict occurs in a repetitive pattern, along geographical lines, the phenomenon is sectionalism.

Concretely, the southern economy, which was based on cotton and tobacco, shipped its produce by river and ocean to be sold in a world market, and it needed generous credit terms to operate. The northern and western economy of manufacture, diversified agriculture, and grain production, shipped by turnpike or canal to domestic markets, and its mercantile interests had accumulated enough capital to be wary of inflationary, cheap credit. As a result of these differences, the South, with no domestic sales to protect, opposed protective tariffs, while the North and West supported them. The South opposed public appropriations to improve the means of transport, while the landlocked Northwest consistently supported them. The South opposed controls on banking by a central authority, while the centers of capital favored such controls. These points of rivalry and others like them made for chronic fric-

---

Craven, *The Coming of the Civil War* (New York, 1942), chaps. 1–5; Allan Nevins, *Ordeal of the Union* (2 vols., New York, 1947), I, 412–544; John Hope Franklin, *The Militant South, 1800–1861* (Cambridge, Mass., 1956); J. G. Randall and David Donald, *The Civil War and Reconstruction* (3rd ed.; Boston, 1969), chaps. 1–3; Clement Eaton, *The Growth of Southern Civilization, 1790–1860* (New York, 1961); William R. Taylor, *Cavalier and Yankee: The Old South and American National Character* (New York, 1961). For the view that economic circumstances were more important than legal status (slavery) in shaping the conditions of life of the Negro, see Craven, *Coming of Civil War*, pp. 74–93.

tion which divided the opposing forces along lines that recurred with enough regularity to harden into barriers of sectional division.[17]

So long as the opposing sections were evenly balanced, and their growth rate was stabilized, they might have gone on peaceably, it is argued, in the equilibrium of a union where neither need fear domination by the other. Indeed, North and South had been fairly evenly balanced when the states of both regions had ratified the "more perfect union" of 1787. But scarcely more than a generation later, the economic transformations of the industrial age set the North upon a more rapid rate of growth than the South, with the result that the North drew steadily further ahead of the South in population, wealth, and productivity. This was reflected by an increasing northern preponderance in Congress. Soon the South began to show, psychologically, the signs of fear that it would be overpowered. This awareness of minority status stimulated the southern sense of solidarity, apartness, and defensiveness, and caused the elaboration of the perennial southern political doctrines of states' rights.[18] At the same time, the unforeseen thrust of American expansion westward, first to the Rockies and then to the Pacific, opened the prospect of a race between the sections to dominate the new regions and to create states which would either perpetuate or upset the balance that still endured in the Senate between the two sections. When this occurred, the South began to resent northern success in the race for physical growth, and the North to resent the determination of the South to preserve its political parity although it had lost the numerical basis for a claim to equality. According to this analysis, the sectional conflict was really a struggle for power.

The flaw in the economic explanation, when it is rigidly applied, is that history can show many instances in which economic diversities and conflicts existed without producing the separatist tendencies of acute sectionalism. Economic dissimilarities may, in just the

17. A classic formulation of this interpretation is in Charles A. and Mary R. Beard, *The Rise of American Civilization* (2 vols.; New York, 1927), II, 3–7, 36–38, 39–41, 105–106. See also Robert R. Russel, *Economic Aspects of Southern Sectionalism, 1840–1861* (Urbana, Ill., 1924); Frederick Jackson Turner, *The United States, 1830–1850* (New York, 1935).

18. On the sectionalizing effect of differential growth rates, see Jesse T. Carpenter, *The South as a Conscious Minority, 1789–1861* (New York, 1930), pp. 7–33. Before 1850, the northern preponderance in Congress was limited to the House of Representatives.

opposite way, promote harmony between two regions, if each supplements the other, and if their combined resources can give them self-sufficiency.[19] For example, in the United States, the Middle West and the East have had very dissimilar economies, and their interests have often clashed violently, but since the diverse economies could be made to supplement one another in important ways, a separatist sectionalism never developed in the Middle West. Could not the economy of the South have been drawn into some similar interdependence? In the United States in the forties, the South's cotton exports paid for the imports of the entire country, and it is an arbitrary theory which would deny that North and South might have found roles, to some degree complementary, in an economy of national self-sufficiency.

It is possible to join the cultural and the economic explanations in one overall analysis that begins by demonstrating the existence of social dissimilarities which, in themselves, do not necessarily cause friction, and then goes on to show how these dissimilarities are translated into specific conflicts of interest. But though the two may be treated as complementary in this way, they differ basically in emphasis. At bottom, the cultural explanation assumes that people quarrel when they are unlike one another; the economic explanation assumes that no matter how much alike they may be, they will quarrel if the advantage of one is the disadvantage of the other. One argues that important cultural dissimilarities cause strife; the other that strife causes the opposing groups to rationalize their hostility to one another by exaggerating unimportant dissimilarities. One explains sectionalism as a conflict of values; the other, as a conflict of interests. One sees it as a struggle for identity; the other as a struggle for power.

Both explanations agree in minimizing slavery as a cause of sectional division, but again they differ in their reasons for doing so. The cultural explanation denies that the difference between chattel and wage labor systems was enough to produce the immense disparity that developed between North and South, and it argues instead that the broad cultural difference between two societies—one stressing status and fixity, the other equality and fluidity—was reflected in the divergence of their labor systems. In short, the pro-

19. J. G. Randall develops this point in "The Civil War Restudied," *JSH*, VI (1940), 441–449.

found cultural division between two fundamentally dissimilar systems transcended slavery. The economic approach, on the other hand, questions the primacy of the slavery factor on quite a different basis. It approaches the problem with deterministic assumptions that men are motivated by interests rather than ideals, that they contend for power rather than principles, and that moral arguments are usually mere rationalizations or secondary "projections," used by contending interest groups to convince themselves or the public that they have right on their side. With such assumptions, spokesmen of the economic explanation have measured very skeptically the exact differences between northern and southern attitudes toward slavery and the Negro, and they have questioned the intensity of sectional disagreement on these subjects. Such terms as "free" and "slave," "antislavery" and "proslavery," suggest a complete antithesis, but at the level of concrete policy and conduct, the people of the North did not propose to emancipate the slaves, and they did not themselves accord equality to the Negro.

The "free" Negro of the northern states of course escaped chattel servitude, but he did not escape segregation, or discrimination, and he enjoyed few civil rights. North of Maryland, free Negroes were disfranchised in all of the free states except the four of upper New England; in no state before 1860 were they permitted to serve on juries; everywhere they were either segregated in separate public schools or excluded from public schools altogether, except in parts of Massachusetts after 1845; they were segregated in residence and in employment and occupied the bottom levels of income; and at least four states—Ohio, Indiana, Illinois, and Oregon—adopted laws to prohibit or discourage Negroes from coming within their borders.[20]

Ironically, even the antislavery movement was not in any clear-cut sense a pro-Negro movement but actually had an anti-Negro aspect

20. Leon F. Litwack, *North of Slavery: The Negro in the Free States, 1790–1860* (Chicago, 1961). There was a certain amount of proslavery sentiment in the North —see Howard C. Perkins, "The Defense of Slavery in the Northern Press on the Eve of the Civil War," *JSH*, IX (1943), 501–503. But the real point is that even antislavery men showed some anti-Negro sentiment. Louis Filler, *The Crusade Against Slavery, 1830–1860* (New York, 1960), pp. 224–225; Eugene H. Berwanger, *The Frontier Against Slavery: Western Anti-Negro Prejudice and the Slavery Extension Controversy* (Urbana, Ill., 1967); James A. Rawley, *Race and Politics: "Bleeding Kansas" and the Coming of the Civil War* (Philadelphia, 1969), pp. 11–15, 258–274. Rawley offers the thesis that racism rather than slavery was the fundamental cause of the Civil War.

and was designed in part to get rid of the Negro. For several decades, the chief agency which advocated emancipation also advocated "colonization," or as it might now be called, deportation. When the militant abolitionists came on the scene in the 1830s, they launched a bitter fight against the colonizationists, but to the general public this seemed merely an intramural, doctrinal dispute. Most antislavery men were colonizationists, as were Thomas Jefferson and Abraham Lincoln. Lincoln advocated colonization throughout his career and actually put it into operation on an experimental basis by sending a shipload of Negroes to an island off the coast of Haiti in 1863. In 1862, Lincoln had told a delegation of Negroes that "it is better for us to be separated," and that they ought to emigrate.[21] The prevalence of attitudes like these even among antislavery men seems to justify the conclusion that while slavery was sectional, Negrophobia was national.

Historians who question the real primacy of the slavery issue in the sectional conflict have found their clinching argument in the peculiar focus and objectives of the free-soil movement, which came to overshadow the abolition movement politically in the North. Instead of dealing with slaves where they were in bondage—in the southern states—the free-soil movement dealt with them where they did not exist—in the territories; instead of proposing to free them, it proposed to keep them (and free Negroes as well) out of the new areas where they might compete with white settlers. Only a handful of militant abolitionists proposed to free any of the several million Negroes who were held in slavery, and these few were persecuted and reviled for their uncompromising zeal or extremism; they failed to build a popular movement such as a large political party, and they remained, to the end, a tiny minority. The vast majority of "antislavery" Whigs or Democrats or, later, Republicans, even including men like Lincoln, concentrated all their effort on keeping slavery out of the new territories, while proclaiming that they never would interfere with slavery in the states. Their attitude

21. On colonization, P. J. Staudenraus, *The African Colonization Movement, 1816–1865* (New York, 1961); Frederic Bancroft, "The Colonization of American Negroes, 1801–1865," in Jacob E. Cooke, *Frederic Bancroft, Historian* (Norman, Okla., 1957), pp. 145–258; Brainerd Dyer, "The Persistence of the Idea of Negro Colonization," *Pacific Historical Review*, XII (1943), 53–65. On Lincoln's views on colonization, J. G. Randall, *Lincoln the President* (4 vols.; New York, 1945–55), II, 137–148; Benjamin Quarles, *Lincoln and the Negro* (New York, 1962), pp. 108–123; Roy P. Basler (ed.), *The Collected Works of Abraham Lincoln* (8 vols.; New Brunswick, N.J., 1953), V, 370–375.

has lent itself to the contention that the northern motive was more one of hostility to slaveowners than of humanitarian concern for the slaves, and that slavery was objectionable—to paraphrase Macaulay—not because it gave pain to the slaves but because it gave pleasure to the slaveowners. It enabled the planters to keep up an aristocratic tone which was invidious and offensive to plain American democrats. Through the three-fifths clause in the Constitution, it gave the planters extra representation and therefore extra strength in Congress.[22] When the time came for opening new territories, northern whites did not want to share these either with slaveholders or with slaves—did not want to compete with slave labor or to permit any further extension of the political power of the planters. If this meant keeping slaveholders out and also keeping Negroes out, it would be hard to say which exclusion the free-soilers would welcome more. David Wilmot himself made it brutally clear in 1847 that, in waging his campaign for free territories, his concern was entirely for the free white laborers of the North and not at all for the fettered Negro slaves of the South.[23]

These anomalies in the antislavery movement and these profound differences between the moral position of free-soilers and that of abolitionists deserve emphasis if a complex position is to be realistically understood.[24] But while a recognition of the paradoxical ele-

22. Albert F. Simpson, "The Political Significance of Slave Representation, 1787–1821," *JSH,* VII (1941), 315–342; Glover Moore, *The Missouri Controversy, 1819–1821* (Lexington, Ky., 1953), p. 11. Representative George Rathbun of New York complained that the representation of slaves gave undue political power to the South and asserted that if the South would give up this advantage, he would be willing to give up his free-soilism. This led David Kaufman of Texas to say that the objection to slavery was "not because it was a sin; not at all; but simply because it was to the South an element of political power." *Congressional Globe,* 29 Cong., 2 sess., pp. 364–365; appendix, p. 152, cited in Craven, *Growth of Southern Nationalism,* pp. 39–40.

23. *Congressional Globe,* 29 Cong., 2 sess., appendix, pp. 315–317. See also Berwanger, *The Frontier Against Slavery;* Going, *David Wilmot,* p. 174 n.; Eric Foner, *Free Soil, Free Labor, Free Men: The Ideology of the Republican Party before the Civil War* (New York, 1970), pp. 261–300.

24. Jefferson Davis said in the Senate in 1860, "What do you propose, gentlemen of the Free-Soil party? Do you propose to better the condition of the slave? Not at all. What then do you propose? You say that you are opposed to the expansion of slavery. . . . Is the slave to be benefited by it? Not at all. It is not humanity that influences you . . . it is that you may have an opportunity of cheating us that you want to limit slave territory. . . . It is that you may have a majority in the Congress of the United States and convert the Government into an engine of Northern aggrandizement . . . you want by an unjust system of legislation to promote the industry of the New England states at the expense of the people of the South and their industry." Quoted in Beard and Beard, *Rise of American Civilization,* II, 5–6.

ments is necessary, there remains much tangible evidence that the people of the North did differ profoundly from those of the South in their attitudes toward slavery, if not toward the Negro. This difference had been increasing ever since the beginning of the nineteenth century, and it had grown to major proportions.

During the Colonial period, there had scarcely been any difference in sectional opinions concerning the morality of slavery, though there had been a vast difference in the degree to which the northern and the southern colonies depended upon slave labor. Eighteenth-century morality had hardly regarded slavery as presenting an ethical problem,[25] and the institution had existed with legal sanction in all the colonies. Later, when the War of Independence came, and with it the revolutionary ideals of liberty, equality, and the rights of man, both North and South had moved in unison to condemn slavery as an evil. The upper South had witnessed a formidable movement for the voluntary manumission of slaves by their masters, and societies for the emancipation and colonization of slaves had flourished in the South for more than a generation after the Revolution. Southern and northern congressmen alike had joined in voting to abolish the importation of slaves after the year 1808. Slavery was barred from the Old Northwest by the Ordinance of 1787; it was confined, even within the South, mostly to the limited areas of tobacco culture and rice culture, both of which were static. At this point, it seemed to many men in both sections only a question of time until the institution would wither and die.[26]

25. Lawrence W. Towner, "The Sewall-Saffin Dialogue on Slavery," *William and Mary Quarterly*, 3rd series, XXI (1964), 40–52, concludes that slavery as such "experienced little opposition until the decades of the Revolution." For the intellectual origins of antislavery, David Brion Davis, *The Problem of Slavery in Western Culture* (Ithaca, N.Y., 1966), chaps. 10–14; for the Revolution as a turning point in attitudes toward slavery, Winthrop D. Jordan, *White Over Black: American Attitudes toward the Negro* (Chapel Hill, 1968), chap. 7.

26. On the development and character of the early antislavery movement and its strength in the South see Stephen B. Weeks, "Anti-Slavery Sentiment in the South," Southern History Association *Publications*, II (1898), 87–130; Mary Stoughton Locke, *Anti-Slavery in America, 1619–1808* (Boston, 1901); Alice Dana Adams, *The Neglected Period of Anti-Slavery in America, 1808–1831* (Boston, 1908); Robert McColley, *Slavery and Jeffersonian Virginia* (Urbana, Ill., 1964); Donald L. Robinson, *Slavery in the Structure of American Politics, 1765–1820* (New York, 1971); Arthur Zilversmit, *The First Emancipation: The Abolition of Slavery in the North* (Chicago, 1967); Clement Eaton, *Freedom of Thought in the Old South* (Durham, N.C., 1940), pp. 1–26; Gordon E. Finnie, "The Antislavery Movement in the Upper South before 1840," *JSH*, XXXV (1969), 319–342.

But despite the presence of a certain amount of antislavery sentiment in the post-Revolutionary South, there is considerable reason to doubt that the antislavery philosophy of the Age of Reason ever extended very far beyond the intellectuals in the South or very deep into the lower South even among the intellectuals. In any case, as the cotton economy, with its demand for slave labor, fastened itself upon the region, and as the center of southern population and leadership shifted southward from Virginia to South Carolina, a reaction set in. By 1832, the southern antislavery movement had vanished and the South had begun to formulate a doctrine that slavery was permanent, morally right, and socially desirable. As the abolitionists grew abusive, the South became increasingly defensive. When David Walker in 1829 published a pamphlet advocating insurrection, and when the bloody uprising of Nat Turner followed in 1831, many southerners interpreted it as proof that such advocacy was taking effect. The South reacted by adopting the proslavery doctrine as a matter of creed, not subject to doubt. Open discussion of slavery fell under a taboo, and the South established what has been called an "intellectual blockade."[27]

Meanwhile, the states north of Maryland and Delaware had abolished slavery, either by immediate or by gradual steps. These states showed a consistent aversion to slavery long before the militant abolition movement began. But in the 1830s there arose a group of reformers—the abolitionists—who made an issue of slavery and aroused a widespread public sentiment against it. Where previous critics of slavery had been content with gradualism, with voluntary manumission by slaveholders, and with persuasion as a method, the

27. Eaton, *Freedom of Thought*, pp. 27–161; Ulrich Bonnell Phillips, *American Negro Slavery* (New York, 1918), pp. 132–149, on early defense of slavery in the South; Theodore M. Whitfield, *Slavery Agitation in Virginia, 1829–1832* (Baltimore, 1930); Joseph Clarke Robert, *The Road from Monticello: A Study of the Virginia Slavery Debate of 1832* (Durham, N.C., 1941); Kenneth M. Stampp, "The Fate of the Southern Antislavery Movement," *JNH*, XXVIII (1943), 10–22; Russel B. Nye, *Fettered Freedom: Civil Liberties and the Slavery Controversy, 1830–1860* (East Lansing, Mich., 1949); Joseph Cephas Carroll, *Slave Insurrections in the United States, 1800–1865* (Boston, 1938); Herbert Aptheker, *American Negro Slave Revolts* (New York, 1943); William Sumner Jenkins, *Pro-Slavery Thought in the Old South* (Chapel Hill, 1935); Richard N. Current, "John C. Calhoun, Philosopher of Reaction," *Antioch Review*, III (1943), 223–234; William W. Freehling, *Prelude to Civil War: The Nullification Controversy in South Carolina, 1816–1836* (New York, 1965), which develops the thesis that the Nullification movement, "although ostensibly aimed at lowering the tariff, was also an attempt to check the abolitionists" (p. xii).

abolitionists demanded immediate action by coercive means, and they resorted to unbridled denunciation of slaveowners. Abolitionism was nourished by a pervasive humanitarianism which made this whole era a period of reform; it was stimulated by the fervor of a great evangelical revival; and it was encouraged by the British abolition of West Indian slavery in 1837. The abolitionists preached their cause from hundreds of pulpits, flooded the mail with pamphlets, sent numerous lecturers into the field, and organized scores of local antislavery societies, as well as two national associations. The militant William Lloyd Garrison is best remembered of the abolitionists, together with his supporters, Wendell Phillips, John Greenleaf Whittier, and Theodore Parker, but the more moderate Tappan brothers in New York, the talented ex-slave Frederick Douglass, and the dedicated and eloquent preacher Theodore Dwight Weld in the Ohio region, supported by James G. Birney and the Grimké sisters, all helped to galvanize public opposition to slavery on moral grounds. At times, the abolitionists were denounced and persecuted, but by the 1840s they had found a few voices in Congress, including no less a person than former President John Quincy Adams, and by 1845, they had been able to force repeal of the "gag rule" which prevented the discussion of antislavery petitions on the floor of Congress. Thus the antislavery movement by the mid-forties had proven itself a powerful force in American life.[28] This

28. On the antislavery movement in general, and for an excellent bibliography of the copious literature, see Filler, *Crusade Against Slavery.* Also the following not cited by Filler or published subsequently: Benjamin P. Thomas, *Theodore Weld, Crusader for Freedom* (New Brunswick, N.J., 1950); Ralph Korngold, *Two Friends of Man: The Story of William Lloyd Garrison and Wendell Phillips* (Boston, 1950); Russel B. Nye, *William Lloyd Garrison and the Humanitarian Reformers* (Boston, 1955); John L. Thomas, *The Liberator: William Lloyd Garrison* (Boston, 1963); Walter M. Merrill, *Against Wind and Tide: A Biography of William Lloyd Garrison* (Cambridge, Mass., 1963); Aileen S. Kraditor, *Means and Ends in American Abolitionism: Garrison and His Critics on Strategy and Tactics, 1834–1850* (New York, 1969); Irving H. Bartlett, *Wendell Phillips, Brahmin Radical* (Boston, 1961); Tilden G. Edelstein, *Strange Enthusiasm: A Life of Thomas Wentworth Higginson* (New Haven, 1968); Bertram Wyatt-Brown, *Lewis Tappan and the Evangelical War Against Slavery* (Cleveland, 1969); Merton Dillon, *Benjamin Lundy* (Urbana, Ill., 1966); Gerda Lerner, *The Grimké Sisters from South Carolina* (Boston, 1967); Martin Duberman, *James Russell Lowell* (Boston, 1966); Milton Meltzer, *Tongue of Flame: The Life of Lydia Maria Child* (New York, 1965); James Brewer Stewart, *Joshua R. Giddings and the Tactics of Radical Politics* (Cleveland, 1970); Gatell, *John Gorham Palfrey;* Edward Magdol, *Owen Lovejoy, Abolitionist in Congress* (New Brunswick, N.J., 1967); David Donald, *Charles Sumner and the Coming of the Civil War* (New York, 1961); Richard H. Sewell, *John P. Hale and the Politics of Abolition* (Cambridge, Mass., 1965); Dwight Lowell Dumond, *Antislavery: The Crusade for Freedom in America* (Ann Arbor,

had happened partly because it was increasingly clear that slavery was not in the process of extinction and the issue would not take care of itself. More fundamentally, it had happened because so many people sensed that slavery presented a giant contradiction to the two most basic of American values—equality and freedom—and to the Christian concept of the brotherhood of man. The reaction against slavery in terms of these values cannot be dismissed as a mere rationalized defense of northern industrial interests, for some of the harshest critics of slavery also opposed the exploitative elements in the northern system of factory labor, while some of the northern industrial magnates, such as the "cotton Whigs" of the Massachusetts textile industry, were conciliatory toward the South in their attitudes concerning slavery.[29]

Thus, from this point of view, a conflict of values, rather than a conflict of interests or a conflict of cultures, lay at the root of the sectional schism.

These three explanations—cultural, economic, and ideological—have long been the standard formulas for explaining the sectional conflict. Each has been defended as though it were necessarily incompatible with the other two. But culture, economic interest, and values may all reflect the same fundamental forces at work in a society, in which case each will appear as an aspect of the other. Diversity of culture may naturally produce both diversity of interests and diversity of values. Further, the differences between a slaveholding and a nonslaveholding society would be reflected in all three aspects. Slavery presented an inescapable ethical question which precipitated a sharp conflict of values. It

---

Mich., 1961), an important work of vast erudition, but of deficient perspective; Lawrence Lader, *The Bold Brahmins: New England's War Against Slavery, 1831–1863* (New York, 1961); Martin Duberman (ed.), *The Antislavery Vanguard: New Essays on the Abolitionists* (Princeton, 1965); Clifford S. Griffin, *Their Brothers' Keepers: Moral Stewardship in the United States, 1800–1865* (New Brunswick, N.J., 1960); Benjamin Quarles, *Black Abolitionists* (New York, 1969); Hans L. Trefousse, *The Radical Republicans: Lincoln's Vanguard for Racial Justice* (New York, 1969); Gerald Sorin, *Abolitionism: A New Perspective* (New York, 1972). On the gag rule, Robert P. Ludlum, "The Anti-Slavery 'Gag Rule,' History and Argument," *JNH*, XXVI (1941), 203–243; Nye, *Fettered Freedom*, pp. 32–54; Samuel Flagg Bemis, *John Quincy Adams and the Union* (New York, 1956), pp. 326–383, 416–448; James M. McPherson, "The Fight Against the Gag Rule: Joshua Leavitt and Antislavery Insurgency in the Whig Party, 1839–1842," *JNH*, XLVIII (1963), 177–195.

29. Philip S. Foner, *Business and Slavery: The New York Merchants and the Irrepressible Conflict* (Chapel Hill, 1941), pp. 1–168; Thomas H. O'Connor, *Lords of the Loom: The Cotton Whigs and the Coming of the Civil War* (New York, 1968).

constituted a vast economic interest, and indeed the Emancipation Proclamation was the largest confiscation of property in American history. The stakes were large in the rivalry of slavery and freedom for ascendancy in the territories. Also, slavery was basic to the cultural divergence of North and South, because it was inextricably fused into the key elements of southern life—the staple crop and plantation system, the social and political ascendancy of the planter class, the authoritarian system of social control. Similarly, slavery shaped southern economic features in such a way as to accentuate their clash with those of the North. The southern commitment to the use of slave labor inhibited economic diversification and industrialization and strengthened the tyranny of King Cotton. Had it not done so, the economic differentials of the two sections would have been less clear-cut, and would not have met in such head-on collision.

The importance of slavery in all three of these aspects is evident further in its polarizing effect upon the sections. No other sectional factor could have brought about this effect in the same way. Culturally, the dualism of a democratic North and an aristocratic South was not complete, for the North had its quota of blue-bloods and grandees who felt an affinity with those of the South, and the South had its backwoods democrats, who resented the lordly airs of the planters. Similarly, the glib antithesis of a dynamic "commercial" North and a static "feudal" South cannot conceal the profoundly commercial and capitalistic impulses of the plantation system. But slavery really had a polarizing effect, for the North had no slaveholders—at least, not of resident slaves—and the South had virtually no abolitionists. Economically, also, the dualism was not complete, for the North had shipping interests which opposed protection, prairie farmers who wanted cheap credit, and Boston merchants who did not want to pay for canals and roads for the benefit of their rivals in New York. Northern politicians, while supporting the primary interests of their section, had also to heed these secondary interests, and to avoid antagonizing them unduly. But nowhere north of the Mason-Dixon line and the Ohio River were there any slaveholding interests, at least in a direct sense, and northern politicians found more to gain by denouncing slaveholders than by conciliating them. Conversely, the South had Charleston and New Orleans bankers who wanted conservative credit policies, landlocked Appalachian communities that yearned for subsidized roads, and aspiring local manufacturers who believed that the South had an industrial future which the tariff would help to realize. Southern politicians had to accommodate themselves to these secondary interests. But the South after 1830 had few white inhabitants who did not shudder with alarm at the thought of the servile insurrection which antislavery might produce, and southern politicians found that they gained many votes and lost few by stigmatizing as an abolitionist anyone who entertained any misgivings about slavery.

Thus in cultural and economic matters, as well as in terms of values, slavery had an effect which no other sectional factor exercised in isolating North and South from each other. As they became isolated, instead of reacting to each other as they were in actuality, each reacted to a distorted mental image of the other—the North to an image of a southern world of lascivious and sadistic slavedrivers; the South to the image of a northern world of cunning Yankee traders and of rabid abolitionists plotting slave insurrections. This process of substituting stereotypes for realities could be very damaging indeed to the spirit of union, for it caused both northerners and southerners to lose sight of how much alike they were and how many values they shared. It also had an effect of changing men's attitudes toward the disagreements which are always certain to arise in politics: ordinary, resolvable disputes were converted into questions of principle, involving rigid, unnegotiable dogma. Abstractions, such as the question of the legal status of slavery in areas in which there were no slaves and to which no one intended to take any, became points of honor and focuses of contention which rocked the government to its foundation. Thus the slavery issue gave a false clarity and simplicity to sectional diversities which were otherwise qualified and diffuse. One might say that the issue structured and polarized many random, unoriented points of conflict on which sectional interest diverged. It transformed political action from a process of accommodation to a mode of combat. Once this divisive tendency set in, sectional rivalry increased the tensions of the slavery issue and the slavery issue embittered sectional rivalries, in a reciprocating process which the majority of Americans found themselves unable to check even though they deplored it.[30]

30. This polarizing effect of the slavery issue was clearly recognized and very often mentioned by contemporaries. E.g., James K. Polk wrote on Jan. 22, 1848, "It [the slavery question] is brought forward at the North by a few ultra Northern members to advance the prospects of their favourite [candidate for the Presidency]. No sooner

From this viewpoint, the centrality of the slavery issue appears clear. Slavery, in one aspect or another, pervaded all of the aspects of sectionalism. But the recognition of this fact has often been obscured by fallacies in the prevailing analysis of northern attitudes toward slavery. Noting the conspicuous hostility of the northern public toward the abolitionists, the northern acceptance of slavery in the southern states, and the northern emphasis on keeping slaves out of the territories, historians have tried to understand northern attitudes by asking a simple question: Did the people of the North *really* oppose slavery? rather than a complex one: What was the rank of antislavery in the hierarchy of northern values?

If the question is posed in the simple form, as it usually is, the difficulty of an affirmative answer is obvious. There were too many situations in which the northern public would not support antislavery activism. This inescapable fact has been emphasized both by prosouthern historians, eager to demonstrate the lack of northern idealism, and by liberal historians, disillusioned that nineteenth-century antislavery falls far short of twentieth-century expectations. But if the question is posed in the complex form—that is, as an inquiry into the relationship between antislavery and other values—it will give room for recognition of the often-neglected truth that politics is usually less concerned with the attainment of one value than with the reconciliation of a number of them. The problem for Americans who, in the age of Lincoln, wanted slaves to be free was not simply that southerners wanted the opposite, but that they themselves cherished a conflicting

is it introduced than a few ultra Southern members are manifestly well satisfied that it has been brought forward, because by seizing upon it, they hope to array a Southern party in favour of their favourite." Quaife (ed.), *Polk Diary*, II, 348, also II, 457–459; IV, 33–34. Stephen A. Douglas, addressing Henry S. Foote of Mississippi in Congress on April 20, 1848, said that Senator Hale, a free-soiler, "is to be upheld at the North, because he is the champion of abolition; and you are to be upheld at the South, because you are the champions who meet him; so that it comes to this, that between those two ultra parties, we of the North who belong to neither are thrust aside." *Congressional Globe*, 30 Cong., 1 sess., appendix, pp. 506–507. Thomas Hart Benton of Missouri almost made a career of accusing Calhoun of using the slavery question to array section against section. Noting the antithesis of the abolitionist opposition to slavery in all territories and Calhoun's claim that slavery could go into any territories, Benton said, "So true it is that extremes meet, and that all fanaticism, for or against any dogma, terminates at the same point of intolerance and defiance." Speech at St. Louis, 1847, *Niles' Register*, LXXII (June 5, 1847), 222–223. See Frank L. Owsley, "The Fundamental Cause of the Civil War: Egocentric Sectionalism," *JSH*, VII (1941), 3–18.

value: they wanted the Constitution, which protected slavery, to be honored, and the Union, which was a fellowship with slaveholders, to be preserved. Thus they were committed to values that could not logically be reconciled.

The question for them was not a choice of alternatives—antislavery or proslavery—but a ranking of values: How far ought the harmony of the Union to be sacrificed to the principle of freedom, how far ought their feeling against slavery to be restrained by their veneration for the Union? How much should morality yield to patriotism, or vice versa? The difference between "antislavery men" and "conciliationists" in the North was not a question of what they thought about slavery alone, but of how they ranked these priorities.[31] A few took the position that Union was not worth saving unless it embodied the principle of freedom, and thus they gave the slavery issue a clear priority. They agreed with John P. Hale of New Hampshire when he declared, "If this Union, with all its advantages, has no other cement than the blood of human slavery, let it perish." A few others took the clear-cut view that the Union was infinitely more important than the slavery issue and must not be jeopardized by it. Like John Chipman of Michigan, they would have said, "When gentlemen pretending to love their country would place the consideration of the nominal liberation of a handful of degraded Africans in the one scale, and this Union in the other, and make the latter kick the beam, he would not give a fig for their patriotism."[32] But

31. The confusion in thinking about both northern and southern attitudes has been reflected in a lack of precision in the terminology which is applied to political groups. Those who gave a priority to Union are often designated as "moderates," with connotations of approval; those who gave a priority to the slavery issue, either as antislavery men or as vigorous defenders of the southern system, are designated as "extremists," with connotations of disapproval. In a strictly logical sense, this approaches absurdity, for those who were "moderate" about slavery were "extreme" about the Union, quite as much as those who were "moderate" about the Union were "extreme" about slavery. When there are two reference points—the Union and slavery—it is purely arbitrary to make one, rather than the other, the measure of extremism.

On the other hand, there is a good logical case for calling men who try to reconcile opposing values "moderates" (e.g., Lincoln in the North and the proslavery Unionists in the South), while defining "extremist" to mean a person who pursues one value to the exclusion of all others (e.g., Garrison in the North or "fire-eating" secessionists in the South). In this book, the two words are used sparingly, and always with reference to a plurality or a singularity of values, and not with reference to whether the Union value or the slavery value received the priority.

32. *Congressional Globe,* 30 Cong., 1 sess., p. 805 (Hale, May 31, 1848); 29 Cong., 2 sess., appendix, p. 322 (Chipman, Feb. 8, 1847).

most people were profoundly unwilling to sacrifice one value to the other.

Functionally, there is a standard way for preserving two or more values which cannot coexist logically in the same context: they must be kept in separate contexts. And this is what the northern public had learned to do, thus finding a way both to oppose slavery and to cherish a Constitution and a Union which protected it.[33] They placed their antislavery feelings in a context of state action, accepting personal responsibility for slavery within their own particular states. By abolishing slavery in each northern state, they had been true to antislavery principles in the state context. Meanwhile they placed their patriotism in a context of inherited obligation to carry out solemn promises given in the Constitution as an inducement to the South to adhere to the Union. By emphasizing the sanctity of a fixed obligation, they eliminated the element of volition or of personal responsibility for slavery at the federal level, and thus were true to the value of Union in this context.

In both contexts, the circumstances made their treatment seem realistic. Their concept of the Union as a rather loose association of states, each with a high degree of autonomy, was historically accurate, and made it easier for them to disclaim personal responsibility for slavery in distant states which had adopted it before they were born. This attitude was so deeply rooted that, on the eve of the Civil War, Abraham Lincoln, who thoroughly disapproved of slavery, was willing to amend the Constitution to guarantee its protection in states which chose to retain it. While thus convincing themselves that they were not responsible for slavery in the South, antislavery people also persuaded themselves—again plausibly—that in countenancing slavery in the South they had not betrayed the long-range goal of freedom, but that if they merely kept slavery from spreading into new areas it would eventually die out—as Lincoln hopefully expressed it, it would "be put in the course of ultimate extinction." At the federal level, their concept of the Constitution as an exchange of promises, by which each party made great

33. Dumond, *Antislavery*, pp. 174, 294–295, 367–370, argues that the Constitution did not protect slavery. The abolitionists were divided on this question, but Garrison and Phillips thought that it did (Filler, *Crusade Against Slavery*, pp. 205–207). Regardless of what anyone may now conclude, the point here is that the northern public believed (correctly in my opinion) that the Constitution protected slavery, and it was the belief that was operative.

concessions in return for great advantages, was also historically realistic. But the constitutional obligations not only inhibited them from attacking slavery in the southern states; it also gave a perfect excuse to those who did not really want to attack it because they knew that such an assault would endanger the Union. This constitutional obligation proved a psychological lifesaver to a man like John Quincy Adams, who was genuinely both a great antislavery leader and a great champion of the Union. Adams was too much a Protestant to realize that he was taking absolution from the Founding Fathers for his sin of temporizing with slavery, when he declared that protection of the institution was "written in the bond," and that while lamenting the fact he must nevertheless "faithfully perform its obligations."[34]

By these means, the people of the North who disliked slavery but felt patriotic devotion to the Union under the Constitution found a way to be antislavery men and Unionists at the same time. One had only to keep the two contexts apart. If leaders in the North did not overtly recognize this fact, many of them sensed it, and it is significant that the man who ultimately became the greatest figure in the antislavery movement was not one who was most ardent, but one who most successfully kept the two contexts apart. Abraham Lincoln could say that "if slavery is not wrong then nothing is wrong," but he could also pledge himself to enforce the fugitive-slave clause of the Constitution and to defer the goal of emancipation into the remote future.

Anything that tended to expose the incompatibility of these values by bringing them to the same level and forcing them to confront one another in the same context was, of course, extremely threatening to the tranquillity of the northern mind. This was why the abolitionists incurred so much hostility. It is often supposed that their unpopularity stemmed from their opposition to slavery, but they were disliked in fact because they insisted upon the necessity to choose between the principle of antislavery and the principle of Union. Garrison, perhaps the most hated of the abolitionists, was also the one who asserted this necessity most explicitly. He admitted that the Constitution protected slavery, but instead of going on to the usual conclusion that this fact justified inaction, he contended

34. [Thomas Hart Benton], *Abridgment of the Debates of Congress . . .* (New York, 1860), XIII, 33.

that it damned the Constitution. To ensure that no one would misunderstand what he meant, he burned a copy of the Constitution in public, denouncing it as "a covenant with death and an agreement with Hell."[35] Garrison frankly, almost gladly, proclaimed the inescapability of choosing either slavery or disunion. The northern public hated him as much for his insistence that these were necessary alternatives as they did for the alternative which he chose.[36]

When Garrison exposed the people's dilemma at the ideological level, they could still evade it by ostracizing him, placing a taboo on his ideas, and clinging to the devices by which they had kept the principle of antislavery and the principle of Union from colliding either in the realm of public affairs or in their own minds. But it was a precarious intellectual arrangement, and when Polk's Two-Million Bill exposed the dilemma at the operative level, the fragile adjustment broke down. Once the question of acquiring land from Mexico was raised, the threat of the antislavery principle to a Union which joined nonslaveholders with slaveholders, and the threat of such a Union to the ideal of antislavery, could no longer be evaded.

The slavery problem, which had been so carefully diffused and localized, could not now be kept from coming to a sharp focus as a national issue when it was presented in terms of a question whether the American flag would carry slavery to a land which had been free under the flag of "benighted" Mexico. It could no longer be sequestered behind constitutional sanctions and inhibitions when it arose in an area where most northerners believed that Congress had the power and an obligation to act. Men who had comforted themselves with the thought that they were not implicated by slavery in the southern states could not escape a sense of personal responsibility for slavery in the common territories.

Thus slavery suddenly emerged as a transcendent sectional issue in its own right, and as a catalyst of all sectional antagonisms, politi-

35. Filler, *Crusade Against Slavery*, pp. 178, 205–206, 216, 258–259, cites substantial evidence for his conclusion (p. 303) that "disunion sentiments were not a Garrisonian vagary but a popular Northern view" and that this fact "has been obscured for decades."

36. Perhaps one of the most serious deficiencies in the historical literature of this period is the lack of an analysis of the growth of a popular dislike of slavery, as distinguished from the growth of an abolitionist willingness to take steps against it. Nearly all histories of "antislavery" are in fact histories of the abolitionist movement, which never enjoyed the support of a public that nevertheless heartily disliked slavery.

cal, economic, and cultural. By removing the frail devices which had kept this issue from coming to a head, Polk's bill and Wilmot's amendment opened the floodgates of sectionalism, for now all the pent-up moral indignation which had been walled in by the constitutional inhibition could be vented into the territorial question.[37] As this happened, the slavery question would grow to dominate national politics, and Congress would become for fifteen years the arena of a continuous battle watched by millions of aroused sectional partisans. No other issue in American history has so monopolized the political scene. As early as 1848, the ubiquity of the slavery question reminded Thomas Hart Benton of the plague of frogs described in the Bible. "You could not look upon the table but there were frogs, you could not sit down at the banquet but there were frogs, you could not go to the bridal couch and lift the sheets but there were frogs!" So, too, was it with "this black question, forever on the table, on the nuptial couch, everywhere!"[38] Benton survived for ten years after this statement, but he did not live to see the end of the plague.

Thus, in circumstances which have puzzled so many Americans of the twentieth century, the slavery question became the sectional question, the sectional question became the slavery question, and both became the territorial question. By this transposition, they entered the arena of politics and there became subject to all the escalation and intensification which the political medium could give to them. By this transposition, also, the slavery question became cryptic. Instead of being fought out on the direct and intelligible alternatives of emancipation versus continued servitude, it became a contest over the technicalities of legal doctrine concerning the relation of Congress and the states to territories, organized or unorganized. Instead of being challenged where it prevailed, slavery was challenged where it did not exist. Instead of proclaiming the goal of emancipation, the opponents of slavery began the long battle in a way which prevented them from admitting the goal even to them-

37. On the function of the territorial question in providing an outlet for antislavery impulses which were otherwise inhibited by constitutional sanctions, see Arthur M. Schlesinger, Jr., "The Causes of the Civil War: A Note on Historical Sentimentalism," *Partisan Review*, XVI (1949), 969–981, reprinted in Schlesinger, *The Politics of Hope* (Boston, 1963), pp. 34–47; Potter and Manning (eds.), *Nationalism and Sectionalism*, pp. 215–216.

38. *Congressional Globe*, 30 Cong., 1 sess., appendix, p. 686.

selves. Certainly it was not dreamed of in the philosophy of David Wilmot. But from the sultry August night in 1846 when Wilmot caught the chairman's eye, the slavery question steadily widened the sectional rift until an April dawn in 1861 when the batteries along the Charleston waterfront opened fire on Fort Sumter and brought the vigorous force of American nationalism to its supreme crisis.

CHAPTER 3

Forging the Territorial Shears

IF American sectionalism entered a new phase in 1846, it was neither because North and South clashed for the first time nor because the issue of slavery for the first time assumed importance. As early as the Confederation, North and South had been at odds over the taxation of imports and exports, over the degree of risk to be run in seeking navigation rights at the mouth of the Mississippi, and over the taxation of slave property. Once the government under the Constitution went into effect, bitter sectional conflicts raged over the assumption of state debts, the chartering of a central bank, and other matters. This sectional rivalry tended to become institutionalized in the opposing Federalist and Jeffersonian Republican organizations, and it became so serious that Washington issued a solemn warning against sectionalism in his Farewell Address. Later, as the Jeffersonians enjoyed a quarter-century of domination in national politics, they became more nationalistic in their outlook, while Federalist nationalism withered. But no matter which region embraced nationalism and which particularism, sectional conflict remained a recurrent phenomenon.[1]

1. For sectionalism before 1820, see John Richard Alden, *The First South* (Baton Rouge, 1961); Edmund Cody Burnett, *The Continental Congress* (New York, 1941), pp. 28, 78, 237–240, 248–258, 433–438, 595–706; Glover Moore, *The Missouri Controversy, 1819–1821* (Lexington, Ky., 1953), pp. 1–32; Staughton Lynd, "The Abolitionist Critique of the United States Constitution," in Martin Duberman (ed.), *The Antislavery Vanguard* (Princeton, 1965), pp. 209–239; Donald L. Robinson, *Slavery in the Structure of American Politics, 1765–1820* (New York, 1971).

From the outset, slavery had been the most serious cause of sectional conflict. In the constitutional convention, questions of taxing slave property and of counting it in the basis of representation had engendered intense friction. These quarrels were adjusted, if not resolved, by the three-fifths compromise and other provisions of the Constitution. But more often than not, sectional disagreements were adjourned rather than reconciled. If friction did decrease, it was less because of sectional agreement on the moral question of slavery than because of the general understanding that slavery was primarily a state problem rather than a federal one. Minor contests, sometimes very stubbornly fought, took place over slavery in the District of Columbia, suppression of the international slave trade, and rendition of fugitive slaves.[2] Later, similar battles were fought over the disposition of antislavery petitions in Congress and the annexation of Texas as a slave state.[3]

But these were marginal affairs. On the central issue of slavery itself, the locus of decision was the states, which had abolished slavery throughout New England and the Middle Atlantic region while perpetuating it from Delaware south. In the late twentieth century, when federal authority seems to reach everywhere and to be invoked for every purpose, it is difficult to realize that during much of the nineteenth century, state government rather than federal government symbolized public authority for most citizens. Thus, for several decades after the founding of the Republic, the

2. Russel B. Nye, *Fettered Freedom: Civil Liberties and the Slavery Controversy, 1830–1860* (East Lansing, Mich., 1949); William R. Leslie, "The Fugitive Slave Clause, 1787–1842" (Ph.D. dissertation, University of Michigan, 1945); W. E. Burghardt DuBois, *The Suppression of the African Slave Trade to the United States of America, 1638–1870* (New York, 1896); Hugh G. Soulsby, *The Right of Search and the Slave Trade in Anglo-American Relations, 1814–1862* (Baltimore, 1933); Richard W. Van Alstyne, "The British Right of Search and the African Slave Trade," *Journal of Modern History*, II (1930), 37–47; Harral E. Landry, "Slavery and the Slave Trade in Atlantic Diplomacy, 1850–1861," *JSH*, XXVII (1961), 184–207; Warren S. Howard, *American Slavers and the Federal Law, 1837–1862* (Berkeley, 1963); Peter Duignan and Clarence Clendenen, *The United States and the African Slave Trade, 1619–1862* (Stanford, 1963); Stanley W. Campbell, *The Slave Catchers: Enforcement of the Fugitive Slave Law, 1850–1860* (New York, 1970); Alfred G. Harris, "Lincoln and the Question of Slavery in the District of Columbia," *Lincoln Herald*, LI (1949), 17–21; LII (1950), 2–16; LIII (1952), 11–18; LIV (1953), 12–21.

3. For the gag rule, see chap. 2, note 28. There is apparently no full account of the development of the concept of the "slave power" or "slavocracy," but see Nye, *Fettered Freedom*, pp. 217–249; Chauncey S. Boucher, "*In Re* That Aggressive Slavocracy," *MVHR*, VIII (1921), 13–79. For the Texas question as a slavery question, see titles cited in chap. 2, note 11.

question of slavery did not naturally come within the federal orbit, and it was only by some special contrivance that even an aspect of it could be brought into the congressional arena. It was this fact and not any agreement on the substantive question that drew the fuse of the explosive issue.

There was one contingency, however, which did transfer the slavery question at once and inescapably to the federal level. This was when the federal government held jurisdiction over western lands, not yet organized or admitted as states, in which the status of slavery was indeterminate. There had been such lands in 1787, but Congress had decided, with only a minimum of sectional disagreement, to exclude slavery from the Northwest Territory by the Ordinance of 1787. South of the Ohio River, Kentucky entered the Union as a slave state without ever being a federal territory, and the western lands that constituted most of the Southwest Territory, or later the Alabama and Mississippi territories, were ceded by North Carolina and Georgia with stipulations that Congress must not disturb the existing status of slavery in those areas. Thus Congress was deprived of what might have been a discord-breeding authority, and the status of slavery was settled throughout the then existing area of the United States.

The disruptive potentialities of territory in an indeterminate status did not become fully apparent until 1820. Missouri had applied for admission as a slave state, thus raising the question of slavery for the whole area of the Louisiana Purchase and presenting the imminent possibility that slave states would outnumber free states in the Union. A violent political convulsion followed, ending with a compromise that settled the territorial issue for another quarter of a century.[4] During that interval, the bitterness over the gag rule against antislavery petitions and the decade-long struggle over the annexation of Texas (which, like Kentucky, skipped the territorial phase) showed what disruptive forces were ready to burst forth. But the potentiality did not again become an actuality until the prospect of acquiring land from Mexico revived the issue of slavery in the territories, thus returning the problem of slavery to the federal level and making Congress the area of combat for the whole complex of sectional antagonisms. As this situation developed, everyone in politics needed a defined position on the ter-

4. Moore, *The Missouri Controversy.*

ritorial status of slavery, even more than he needed a position on slavery itself. Militants on both sides wanted arguments to justify complete restriction or complete nonrestriction, as the case might be, while those politicians seeking to preserve a measure of national harmony needed formulas to prevent complete victory for either side.

For fifteen years between 1846 and 1861, countless speeches, resolutions, editorials, and party platforms set forth a wide variety of proposals for resolving the territorial issue. But essentially there were four basic positions. Significantly, all four were put forward within sixteen months after the territorial question reemerged to prominence in 1846. For more than a decade thereafter they remained the fixed rallying points of a shifting political warfare. Sometimes opportunists followed a weaving path among the available choices, and in the election of 1848 both major parties contrived to evade them. But sooner or later, almost everyone in public life committed himself to one of the four basic formulas.

The first of these was David Wilmot's—that Congress possessed power to regulate slavery in the territories and should use it for the total exclusion of the institution. This free-soil formula was, in a sense, older than the Constitution, having received its first sanction in the Jefferson-inspired Ordinance of 1787, which declared: "There shall be neither slavery nor involuntary servitude in the said territory, otherwise than in the punishment of crimes whereof the party shall have been duly convicted." This was the language which Wilmot adopted, and so it has sometimes been said that Thomas Jefferson was the real author of the Wilmot Proviso.[5]

Once the Northwest Ordinance was adopted under the Confederation, it remained the basic policy for the Old Northwest under the Constitution. Congress reaffirmed it on August 2, 1789, and again as the successive territories of the region were erected—Indiana (1800), Michigan (1805), Illinois (1809), and Wisconsin (1836).[6] Thus Presidents Washington, John Adams, Jefferson, Monroe, and Jackson all assented to the principle that Congress possessed a constitutional power to prohibit slavery in the territories.

But not everyone who believed that the power existed believed

5. Text of slavery clause in Clarence Edwin Carter (ed.), *The Territorial Papers of the United States* (Washington, 1934–), II, 49; also see chap. 2, note 7.

6. *Ibid.*, II, 203 (Act of 1789); III, 86–88 (Indiana); X, 5–7 (Michigan); *U.S. Statutes at Large,* II, 514–516 (Illinois); V, 10–16 (Wisconsin).

also that it ought to be exercised. Some political leaders embraced the view that the power of Congress should be used in a way that would recognize the claims of both sections. Accordingly, Congress accepted cessions of western land from North Carolina in 1790 and from Georgia in 1802, with the condition that "no regulation made or to be made, by Congress, shall tend to emancipate slaves." It organized the Southwest Territory (out of the North Carolina cession) in 1791 and the Mississippi Territory (originally the northern zone of West Florida, to which the Georgia cession was later added) in 1798, both without restrictions upon slavery. Meanwhile, it admitted Kentucky (separated from Virginia) as a slave state in 1792.[7] In sum, the federal government did not maintain a uniform policy concerning slavery in the territories, but instead practiced a kind of partition by which the Ohio River became a boundary between free territory to the north and slave territory to the south.

At first this practice was more an expedient or a reflex than a deliberate policy, but it assumed a formal character at the time of the Missouri crisis, which resembled the crisis of the late 1840s in more ways than one. In each case, a free-state congressman offered a motion in the House to exclude slavery from some part of the trans-Mississippi West. Both motions caused divisions along strictly sectional lines; both passed the House and failed to pass the Senate. Each precipitated a crisis that was not settled until a later session of Congress. Each inspired the formulation of an alternative plan making some kind of territorial adjustment between proslavery and antislavery interests. In 1820, Congress adopted the compromise proposed by Senator Jesse Thomas of Illinois, admitting Missouri as a slave state and dividing the rest of the Louisiana Purchase (except the state of Louisiana, already admitted) along latitude 36° 30′, with slavery prohibited north of that line. By 1846, this compromise formula had become both familiar and traditional, and within a few minutes after Wilmot introduced his proviso, Representative William W. Wick of Indiana offered a resolution to extend the 36°

7. The various acts of cession by the states, acceptance by Congress, and organization by Congress, all stipulating the protection of slavery or omitting any regulation of it, are in Carter (ed.), *Territorial Papers,* IV, 7, 13 (North Carolina); V, 145 (Georgia); IV, 18 (Southwest Territory); V, 20 (Mississippi Territory); William Waller Hening, *Statutes at Large . . . Virginia* (13 vols.; New York, 1823), XII, 37–40, 240–243, 788–791 (separation of Kentucky from Virginia); *U.S. Statutes at Large,* I, 189 (admission of Kentucky to statehood).

30′ line into the prospective Mexican cession.

This principle of territorial division had thus become the second basic formula, and the sanction of solemn agreement between opposing parties was later claimed for it. Actually, it was adopted only because Henry Clay and other compromisers skillfully used two separate majorities to get it passed—one a solid bloc of southerners, with a sprinkling of northern support, to defeat restrictions on slavery in Missouri; the other a solid bloc of northerners, together with slightly more than half of the southern members, to exclude slavery north of 36° 30′ in the remainder of the Louisiana Purchase. But despite the lack of a clear mandate which would have been necessary to a real covenant, and despite the limitation of this settlement to the Louisiana Purchase, the Compromise had brought peace, and consequently the line 36° 30′ later took on a certain aura of sanctity. It was probably for this reason that Wick put it forward so promptly on the night of the Wilmot Proviso.[8]

In the four years between 1846 and 1850, the proposal to extend the Missouri Compromise received a large measure of influential support. The administration rallied to it: Polk, as party leader, urged Democrats in Congress to support it; and the secretary of state, James Buchanan, made it his primary issue in a bid for the Democratic nomination in 1848. In Congress, southern Democrats, although questioning its constitutionality, voted repeatedly to apply it as a basis of settlement, and Stephen A. Douglas, later a champion of popular sovereignty, became its sponsor in the Senate. In July 1848, the 36° 30′ line almost became the basis of a compromise proposed by John M. Clayton of Delaware, which had the backing of all the forces of conciliation. To many people in both parties and both sections, the Missouri Compromise seemed to offer the best hope of peaceable adjustment.[9]

To a surprising degree, historians have overlooked the strength of the movement to extend the Missouri Compromise line, and it has become, in a sense, the forgotten alternative of the sectional controversy. History has made heroes of the free-soilers like Lincoln. Douglas has had a body of admirers who argue that popular sovereignty offered the most realistic way of restricting slavery with-

8. For the Wick Resolution, see above, p. 22. On the crisis and settlement of 1820, Moore, *The Missouri Controversy*, supersedes all previous treatments.

9. For efforts between 1846 and 1861 to extend the Missouri Compromise line, see below, pp. 65–66, 69–76, 531–534, 547–551.

out precipitating civil war. And Calhoun has been accorded much respect for the intellectual acumen with which he saw through the superficialities of compromise. But the champions of 36° 30′ are forgotten, and even James Buchanan's biographers scarcely recognize his role as an advocate of the Missouri Compromise principle.[10]

No doubt this neglect arises primarily from the fact that the proposal to divide the new territory, like the old, along a geographical line was the first of the four alternatives to be discarded in the late 1840s. Perhaps, too, historians have felt that as a simple, unadulterated bargain by which both parties would have given up part of what they believed in, it lacked the ideological rationale to make it interesting. But in a situation in which there were apparently no rational solutions acceptable to both sides, it had already proved to be a remarkably effective irrational solution. If in the end it failed to provide either a nonviolent answer to the slavery question or an enduring peace, no other alternative succeeded better. With it, for more than thirty years, the country had avoided the twin dangers of disruption and war.

Whatever its philosophical defects, the Missouri formula had one ostensible merit that proved more disadvantageous than all its faults. It was free of ambiguity; it spelled out clearly what each side would gain and lose. Thus it did not offer either side the hope of gaining ground by favorable construction of ambiguous language.

While President Polk was supporting the Missouri Compromise plan, the chief aspirant to the presidential succession came forward with a proposal possessing all the charms of ambiguity. The aspirant was Lewis Cass of Michigan, and his "Nicholson letter" of December 1847 formulated the doctrine of what was later called popular sovereignty as a third major position on the territorial question. Without taking a decisive stand on the question of whether Congress possessed power to regulate slavery in the territories, Cass held that if such power existed, it ought not to be exercised, but that slavery should be left to the control—at a stage not clearly specified —of the territorial government. His doctrine was based upon the plausible and thoroughly democratic premise that citizens of the territories had just as much capacity for self-government as citizens

10. George Ticknor Curtis, *Life of James Buchanan* (2 vols.; New York, 1883), does not mention it at all; Philip Shriver Klein, *President James Buchanan* (University Park, Pa., 1962), pp. 200–201, deals with it summarily and without emphasis.

of the states. If it was consistent with democracy to permit the citizens of each state to settle the slavery question for themselves, it would be equally consistent with democracy to permit the citizens of a territory also "to regulate their own internal concerns in their own way." For good measure, this was not only a matter of sound policy, but also of constitutional obligation: Cass did "not see in the Constitution any grant of the requisite power [to regulate slavery] to Congress," and he believed that such regulation would be "despotic" and of "doubtful and invidious authority."

On its face, this position seemed simple and enticing: by invoking the principle of local self-government, against which no one would argue, it promised to remove a very troublesome question from Congress and to make possible a consensus within the badly divided Democratic party. It seemed impartial, in that it challenged both northern and southern partisans to accept the verdict of the local majority.

But either by contrivance or by chance, the popular sovereignty formula held a deeply hidden and fundamental ambiguity: it did not specify at what stage of their political evolution the people of a territory were entitled to regulate slavery. If they could regulate while still in the territorial stage, then there could be "free" territories, just as there were "free" states; but if they could regulate only when framing a constitution to apply for statehood, then slavery would be legal throughout the territorial period, and the effect would be the same as legally opening the territory to slavery. Cass's letter lent itself to the inference that territorial legislatures might exclude slavery during the territorial stage. But his statement that he favored leaving to the people of a territory "the right to regulate it [slavery] for themselves, under the general principles of the Constitution," said far less than it appeared to say, for all that it amounted to ultimately was a proposal to give the territorial governments as much power as the Constitution would allow, without specifying what the extent of this power might be. Cass did state that he saw nothing in the Constitution which gave Congress power to exclude slavery, and this statement implicitly raised a question whether Congress could confer upon territorial legislatures powers that it did not itself possess. But Cass refrained from exploring this implication also. The doctrine of congressional non-intervention, as he first formulated it, was more a device to get the territorial question out of Congress than a solution to place it definitely in the hands of the territorial legislatures.[11]

The doctrine of popular sovereignty need not have been so ambiguous. To give it a clearer meaning, Cass needed only to do at the outset what both he and Douglas did later—that is, to assert his belief in the constitutionality as well as the desirability of a system by which the territorial legislatures, rather than Congress, would regulate slavery in the territories. But for nearly two years, Cass avoided this clarification and preserved the ambiguity. This equivocation made the doctrine especially enticing to politicians, for it allowed northern Democrats to promise their constituents that popular sovereignty would enable the pioneer legislatures to keep the territories free, while southern Democrats could assure proslavery audiences that popular sovereignty would kill the Wilmot Proviso and would give slavery a chance to win a foothold before the question of slavery exclusion could arise at the end of the territorial period. Each wing of the party, of course, understood what the other was up to, condoned it as a political expedient for getting Democrats elected, and hoped to impose its own interpretation after the elections were won. But two years beyond each election there was always another election, and a clear confrontation of the meaning of popular sovereignty was repeatedly avoided. The territorial issue, difficult at best and badly needing to be faced with candor and understanding on both sides, thus remained for more than a decade an object of sophistry, evasion, and constitutional hair-splitting, as well as of disagreement.

While middle-ground alternatives to the Wilmot Proviso were being developed by the administration and by Cass, leaders within the slave states had already formulated a fourth major position which was the logical antithesis of the free-soil position. This was the contention that Congress did not possess constitutional power to regulate slavery in the territories and, therefore, that slavery could not be excluded from a territory prior to admission to statehood. Like all major southern doctrines for more than a generation,

11. Cass to Nicholson, Dec. 24, 1847, in Washington *Union*, Dec. 30, 1847. For details concerning the enunciation of the doctrine in 1847–48, see below, pp. 71–72; for the background of popular sovereignty, see Allen Johnson, "Genesis of Popular Sovereignty," *Iowa Journal of History and Politics*, III (1905), 3–19; for the specific evolution of the doctrine before Cass took it up and for the built-in ambiguity see Milo Milton Quaife, *The Doctrine of Non-Intervention with Slavery in the Territories* (Chicago, 1910), pp. 45–55, 59–77.

this one was more effectively stated by John C. Calhoun than by anyone else. Thus, the accepted formulation appeared in a set of resolutions which Calhoun introduced in the Senate on February 19, 1847. Essentially, these resolutions argued that the territories of the United States were the common property of the several states, which held them as co-owners; that citizens of any given state had the same rights under the Constitution as the citizens of other states to take their property—meaning slaves—into the common territories, and that discrimination between the rights of the citizens of various states in this respect would violate the Constitution; therefore, any law by Congress (or by a local legislature acting under authority from Congress) which impaired the rights of citizens to hold their property (slaves) in the territories would be unconstitutional and void.[12]

According to this reasoning, the Wilmot Proviso would be unconstitutional, and so, for that matter, would the exercise of popular sovereignty by a territorial legislature. These implications, Calhoun intended. But further, his argument plainly meant that the Missouri Compromise was unconstitutional also, since it embodied a congressional act depriving citizens of the right to carry slaves into the territories north of 36° 30′. This challenge to the constitutionality of the compromise of 1820 was not new. In fact a substantial number of southerners—especially strict constructionists from Virginia—had voted against the act for constitutional reasons when it was originally adopted. But despite his theory, Calhoun was only half-hearted in challenging the 36° 30′ line. Embarrassing evidence was brought to light that he had himself supported it in 1820, as a member of Monroe's cabinet, and in any case he regarded it as a fair operating arrangement. In fact, the twenty-ninth Congress witnessed the odd spectacle of Calhoun's loyal follower, Armistead Burt, proposing the extension of the Missouri line in the House, at almost the same time when Calhoun himself was enunciating a doctrine which implicitly challenged the line's constitutionality in the Senate. At this point, he would have been willing to abandon consistency and accept the Missouri line, if the North had been prepared to extend it. But as he saw the Burt proposal voted down by a solid northern majority, his position hardened, and he later became adamant in his insistence that the South must accept noth-

12. *Congressional Globe,* 29 Cong., 2 sess., p. 455.

ing less than the full recognition of its literal rights in all the territories.[13] By 1848, many southerners were asserting that they would never lend their support to a presidential candidate or to a party which advocated any federal law affecting "mediately or immediately" the institution of slavery.[14]

Calhoun never pressed his resolutions to a vote, and indeed he had no reason to, for they were certain to be defeated. He had no way of knowing that ten years later, long after any hope of their adoption in Congress had been abandoned, they would be adopted, in somewhat modified form, by the Supreme Court in the Dred Scott decision.[15] What he sought primarily was to state a southern position which would serve as a counterpoise, to unify the South, as the free-soil position was already unifying the North. Historians have not taken sufficient note of the fact that in this effort, Calhoun gained one of the few clear-cut successes of his career. Most of his life was spent in attempts to create political solidarity among southerners, and most of these attempts failed. But the doctrine that Congress could neither exclude slavery from a territory itself nor grant power to a territorial government to do so became one of the cardinal tenets of southern orthodoxy and operated as one of the key elements of southern unity in the crises that were to follow.

The four doctrines championed by Wilmot, Buchanan, Cass, and Calhoun soon became so many converters to be used by men who needed to discuss the slavery question in terms of something other than slavery. In their legal subtleties and constitutional refinements, these doctrines appear today as political circumlocutions, exercises in a kind of constitutional scholasticism designed to concentrate attention upon slavery where it did not exist and to avoid contact with the real issue of slavery in the states. But Thomas Hart Benton

13. See below, pp. 65–66; on the position of Calhoun and other southerners in 1820 concerning the power of Congress to regulate slavery in the territories, see Moore, *The Missouri Controversy*, pp. 46, 63, 122, which finds a majority of southern congressmen conceding the congressional power, but "almost half of them were unwilling to concede even this as a matter of principle, although somewhat more than half would vote for it in the form of a 'hoss trade' compromise," and also Charles M. Wiltse, *John C. Calhoun, Nationalist, 1782–1828* (Indianapolis, 1944), p. 196; Wiltse, *John C. Calhoun, Sectionalist, 1840–1850* (Indianapolis, 1951), pp. 352–353.

14. Resolutions of a meeting at Lowndes, South Carolina, April 14, 1847, quoted in Quaife, *Doctrine of Non-Intervention*, p. 34.

15. See below, pp. 276–277.

characterized two of the doctrines in a figure of speech which illuminated their functional reality and their historical importance: Wilmot's doctrine and Calhoun's doctrine, he said, were like the two blades of a pair of shears: neither blade, by itself, would cut very effectively; but the two together could sever the bonds of Union.[16]

Buchanan's proposal and Cass's concept were intended to prevent the cutting action of the two blades, and for some years they did so. But in all the prolonged and involved legislative battles that embittered the years between 1846 and 1861, the devices to inhibit sectionalism never succeeded for very long. Time and again, the forces which were trying to resist sectional polarization temporarily rallied their followers under the banner of the Missouri Compromise or of popular sovereignty. But invariably the divisions returned, after a while, to the polarities of free soil and of Calhoun's doctrine. The shears continued to cut, deeper and deeper. Thus, although the dispute was to be waged with many variations and many diversionary thrusts, the contest always came back to one of these four doctrinal bases. The dialectic of the crisis of 1860 had been articulated by December of 1847.

16. Thomas Hart Benton, *Thirty Years' View . . . 1820 to 1850* (2 vols.; New York, 1854–56), II, 695–696.

CHAPTER 4

The Deadlock of 1846–1850

DURING the latter half of 1846, American interests in the Far West advanced rapidly. The ratification of the Oregon Treaty in June opened the way for exclusive settlement by Americans south of 49°, and the pioneers promptly organized a provisional government for a future territory. In Alta California, where the Bear Flag Revolt signalized the end of Mexican rule, John C. Frémont, Stephen W. Kearny, and Commodores John D. Sloat and Robert F. Stockton brought the entire region under American control.

In these circumstances, President Polk wanted more than ever to complete the formal processes of acquisition and organization. When Congress met in December he recommended that Oregon be organized as a territory and he again asked for an appropriation ($3 million instead of the $2 million that he had failed to obtain in August) with which to acquire title to land from Mexico. Within another year, he would be able to ask for bills to organize territorial governments in California and New Mexico.[1]

Despite Whig grumbling, Polk had reason to feel optimistic. The Oregon difficulty had been settled; the war was going well; and the grave nature of the flare-up at the end of the preceding session was not yet recognized. Polk, who temperamentally dis-

1. James D. Richardson (ed.), *A Compilation of the Messages and Papers of the Presidents* (11 vols.; New York, 1907), IV, 457–458, 587–593, 638–639; Eugene Irving McCormac, *James K. Polk* (Berkeley, 1922), p. 61.

trusted the motives of anyone who opposed him, regarded David Wilmot and his supporters simply as disgruntled patronage seekers trying to exploit their nuisance value. At this time, the territorial question did not seem ominous. Geographical alignments in Congress had not crystallized; sectional dissension had not become chronic; sectional dogma had not exercised its rigidifying effect. Moreover, Polk regarded sectional rivalries as irrelevant to expansion in the Far West. No one doubted that the Oregon region would be free. As for California, the public hailed the prospective acquisition and, after the discovery of gold in January 1848, thought of it in terms of sourdoughs rather than slaves, prospectors rather than planters.

Polk could scarcely have foreseen that he stood at the beginning of a sectional impasse on the territorial question. But in fact, Oregon was not organized and the clamor of settlers for a government was not satisfied until the end of a bitter fight extending through two sessions of Congress. As for New Mexico, with its resident population of 60,000, and California, with the riptide of migration it began to receive from the Gold Rush, the urgent political needs of these two went unheeded until late in 1850. Not until Polk and his successor were both in their graves did Congress break the prolonged four-year deadlock which had blocked all action to organize California and New Mexico.

In December 1847, a few days after the new session began, Polk had a little talk with David Wilmot, and the Pennsylvanian promised his party chief not to introduce the Proviso again, though he said he would have to vote for it if someone else introduced it.[2] All prospects of a truce, however, were soon blasted. On January 4, Preston King of New York introduced in the House a bill which revived the Proviso and sought to attach it to the administration's Three-Million Bill. The fact that King was one of the principal lieutenants of Martin Van Buren meant that the northern Democrats had now embarked on a deliberate and continuing revolt. The problem involved far more than getting one recalcitrant freshman

2. Polk recorded his opinion of this "mischievous and foolish amendment" and his two interviews with Wilmot in his diary for Aug. 10, Dec. 23, 31, 1846. Milo Milton Quaife (ed.), *The Diary of James K. Polk* (4 vols.; Chicago, 1910), II, 75, 288–290, 299. Wilmot gave his version of the interview on Feb. 17, 1849, in *Congressional Globe,* 30 Cong., 2 sess., appendix, p. 139.

congressman back into line.[3] Ten days later, the House Committee on Territories reported an Oregon Bill which excluded slavery from the proposed territory by applying the Ordinance of 1787. This measure aroused the opposition of southern congressmen, not because they had any thought of slaves being taken into Oregon,[4] but because it embodied the principle of congressional exclusion, which, if extended, would keep slaves out of other areas as well. Also, southern leaders did not want to concede a free status for Oregon unless they could obtain in return a slave status for the Southwest. As a consequence, the question of slavery in Oregon, which had no practical significance, nevertheless assumed considerable strategic importance as a bargaining point in the legislative struggle between the sections. In opposition to the committee's proposal, Armistead Burt of South Carolina, acting at the suggestion of Calhoun, proposed to amend the Oregon Bill by stipulating that the region should be free, "inasmuch as the whole of the said territory lies north of 36° 30′ . . . the line of the Missouri Compromise."[5] This seemed to imply that acquisitions from Mexico lying south of 36° 30′ would, on the same basis, be open to slavery. In other words, it would have changed the rationale of Oregon's freedom from a free-soil principle to a principle of compromise. As a follower of Calhoun, Burt took care to explain that he did not concede the constitutionality of the Missouri Compromise, but that he considered it a workable agreement which both North and South had accepted, and which would be an effective means of laying the controversy to rest (Carolinians, who also regarded the tariff as unconstitutional, were quite accustomed to peaceful coexistence

3. *Congressional Globe*, 29 Cong., 2 sess., p. 105; Chaplain W. Morrison, *Democratic Politics and Sectionalism: The Wilmot Proviso Controversy* (Chapel Hill, 1967), pp. 31–32, 187. Morrison offers evidence that some of the older Barnburners tried to stop King's motion. Despite Wilmot's promise to Polk not to renew the Proviso, he did renew it in the House, Feb. 1, 1847; see *Congressional Globe*, 29 Cong., 2 sess., p. 303; Charles Buxton Going, *David Wilmot, Free-Soiler* (New York, 1924), pp. 159–181.

4. Robert Barnwell Rhett said of Oregon: "It is not probable that a single planter would ever desire to set his foot within its limits" (*Congressional Globe*, 29 Cong., 2 sess., appendix, p. 246). John J. Crittenden to John M. Clayton, Dec. 19, 1848: "No sensible man would carry his slaves there [to California] if he could," quoted in George Rawlings Poage, *Henry Clay and the Whig Party* (Chapel Hill, 1936), p. 193.

5. On the Oregon Bill and the Burt amendment, *Congressional Globe*, 29 Cong., 2 sess., pp. 178–180 and appendix, pp. 116–119. For Calhoun's sponsorship of this amendment, see his remarks in the Senate, Feb. 19, 1847, in *ibid.*, 29 Cong., 2 sess., p. 454. For southern reasons for making an issue of free soil in Oregon, see James M. Mason, in Senate, *ibid.*, 30 Cong., 1 sess., p. 903.

with what they regarded as unconstitutional legislation). The House promptly rejected Burt's proposal by a strictly sectional vote of 82 to 113.[6] This action served almost as a signal to Calhoun's followers to move to a more advanced position, and on January 15, Robert Barnwell Rhett made a strong speech denying that Congress had any power to regulate slavery in the territories.[7] Calhoun himself continued to hold his fire, but on February 15 the House voted again to apply the Wilmot Proviso to the Three-Million Bill, and four days later Calhoun countered by presenting the resolutions in which he gave definitive formulation to his doctrine of noninterference with slavery in the territories—the second blade of the shears.[8]

The possibility of securing any kind of agreement had, by now, been thoroughly demolished. Before the Congress automatically expired on March 4, the Senate, where the South had greater strength, refused to insert the Wilmot Proviso into the Three-Million Bill, and a Senate committee took the slavery restriction out of the Oregon Bill. On the floor of the Congress, northern senators rolled up a majority to table the Oregon Bill rather than accept it in such a form. At the last moment, and after great turmoil, a handful of northerners did vote with a solid array of southern representatives to adopt the Three-Million Bill without the Wilmot Proviso, and thus President Polk had at least one piece of legislation to show for the session.[9] But in fact, nothing had been settled. The adoption of the Three-Million Bill meant only that some of the opponents of slavery had agreed to wait until territory was acquired before making an issue of the exclusion of slavery

6. *Congressional Globe,* 29 Cong., 2 sess., pp. 187–188. In affirmative, 6 from free states, 76 from slave states; in negative, 113 from free states. Milo Milton Quaife, *The Doctrine of Non-Intervention with Slavery in the Territories* (Chicago, 1910), p. 24. On Feb. 15, Stephen A. Douglas, in the House, moved to apply the Missouri Compromise line to the territory to be acquired from Mexico. This was defeated 82 to 109. *Congressional Globe,* 29 Cong., 2 sess., p. 424.

7. *Ibid.*, appendix, pp. 244–247.

8. House adoption of Proviso, by vote of 115 (114 free state, 1 slave state) to 106 (18 free state, 88 slave state), *ibid.*, 29 Cong., 2 sess., p. 425; Calhoun Resolutions, *ibid.*, p. 455.

9. Senate defeat of Proviso in Three-Million Bill, March 1, 1847, *ibid.*, p. 555. The vote was 21 (1 slave state, 20 free state) to 31 (26 slave state, 5 free state). Senate tabling of Oregon Bill, 26 to 18, *ibid.*, p. 571. On March 3, the House, in committee of the whole, inserted the Proviso into the Three-Million Bill by a vote of 90 to 80, but took it out again when the bill was reported to the House, by a vote of 102 (79 slave state, 23 free state) to 96 (1 slave state, 95 free state). The House then adopted the bill, 115 to 81. *Ibid.*, p. 573.

from it;[10] but only a few would concede even this much; and the contest was certain to be resumed as soon as the acquisitions were in hand. Meanwhile, it was ominous that even when men agreed that Oregon should be free, they could not disentangle this agreement from their disagreement about the territories in general, and therefore could not put it into effect. Congress was beginning to lose its character as a meeting place for working out problems and to become a cockpit in which rival groups could match their best fighters against one another. This tendency showed up in the hardening of the antithetical positions that slavery should be sanctioned in all of the territories or that it should be prohibited in all of them. It showed also in the fact that eleven northern state legislatures sent to Congress resolutions demanding the Wilmot Proviso, and six southern legislatures or party conventions adopted resolutions demanding an open field for slavery; these actions placed the representatives of the states in the position of troops sent to defeat the enemy rather than negotiators sent to adjust differences.[11] But the most serious symptom of all was that solutions could not be found, and legislation could not be enacted. Oregon still remained unorganized.

This protracted struggle inevitably provoked bitter wrangling over the general question of slavery, which crowded out all others, said Polk, "day after day and week after week." It "meets you," declared Senator Thomas Corwin of Ohio, "in every step you take, it threatens you which way soever you go." Corwin trembled, he said, at the possible consequences of the acquisition of the Southwest, for he anticipated that an issue had arisen on which neither side would yield, and this clearly portended danger to the Union.[12]

The debates of the winter of '46–'47 did much to justify such apprehensions. On the northern side, Columbus Delano of Ohio breathed fire at the southerners, warning them, "We will establish

10. Some northern members decided to let the bill pass without the amendment because it did not provide for the acquisition of territory, but only looked to future acquisition. On a measure involving actual acquisition, they would have insisted on the Proviso. Daniel Dickinson, in Senate, *ibid.*, pp. 553–554.

11. For summary of resolutions of the state legislatures and conventions, with citations, see Herman V. Ames (ed.), *State Documents on Federal Relations*, No. VI, *Slavery and the Union, 1845–1861* (Philadelphia, 1906), p. 3.

12. Quaife (ed.), *Polk Diary*, II, 334 (Jan. 16, 1847); *Congressional Globe*, 29 Cong., 2 sess., appendix, p. 218 (Feb. 11, 1847).

a cordon of free states that shall surround you; and then we will light up the fires of liberty on every side until they melt your present chains and render all your people free." Southerners, in turn, responded by stating openly that they preferred disunion to a union which stigmatized them and threatened the institution of slavery. Henry Hilliard of Alabama, James A. Seddon of Virginia, Henry Bedinger of Virginia, Robert Roberts of Mississippi, and Barnwell Rhett and Andrew P. Butler of South Carolina all asserted, with varying degrees of passion and solemnity, that they would break up the Union if the protection of the Union were denied them.[13]

This constant dissension troubled many thoughtful men, and not least of these was the president. Both alarmed and disgusted by the turn of events, Polk regarded the question of slavery in the territories as an abstraction. Slaveowners might proclaim their right to take slaves to the territories, but no one, in fact, would be foolish enough to waste the value of slave labor by taking it where factors of climate and terrain were unfavorable to its use. "In these provinces slavery could probably never exist."[14] Since he could not

13. *Congressional Globe,* 29 Cong., 2 sess., pp. 119–120 (Hilliard), appendix, pp. 76–80 (Seddon), 86 (Bedinger), 134–136 (Roberts), 246 (Rhett, Butler), 281 (Delano); Avery O. Craven, *The Growth of Southern Nationalism, 1848–1861* (Baton Rouge, 1953), has a good discussion, including several of these citations.

14. Polk was apparently the first significant figure to develop the idea, later made famous in Webster's Seventh of March speech (see below, p. 101), that the legal exclusion of slavery in the territories was redundant because physical conditions would exclude it. Quaife (ed.), *Polk Diary,* II, 289 (Dec. 23, 1846), 308 (Jan. 5, 1847); IV, 345 (Feb. 20, 1849). Message to Congress, Dec. 5, 1848, in Richardson (ed.), *Messages and Papers,* IV, 640.

The validity of this view has continued a subject of dispute among historians. Charles W. Ramsdell, "The Natural Limits of Slavery Expansion," *MVHR,* XVI (1929), 151–171, argued that slavery could flourish only in a plantation economy; that the plantation crops could not spread into the West because of their requirements of soil and moisture; and that slavery had very nearly reached the boundaries of its potential expansion, within the limits of the United States. By implication, this meant that legal restrictions on its expansion were irrelevant, which is what Polk said in 1846. More recently, a number of writers, including Kenneth Stampp, have cogently pointed out that neither Negro labor nor slave labor was necessarily confined to the cultivation of cotton or other staple crops, and that unfree Negro labor might have been used, as free Negro labor now is, for varied economic activities in varied climatic environments throughout the United States. For a good review of the question and statement of this argument, see Harry V. Jaffa, *Crisis of the House Divided: An Interpretation of the Issues in the Lincoln-Douglas Debates* (New York, 1959), pp. 387–404. While Jaffa would seem to be correct *economically* about what unfree Negro labor can be used for, one wonders whether Polk and Webster were not correct *culturally* about what white slaveowners would and would not try to use their slaves for. The planter class had equated slavery with cotton, and could no more conceive of slavery apart from the staple crops than they could conceive of the staple crops apart from slavery.

regard it as a real question, Polk felt that the men who pressed it were unpatriotic and irresponsible. Because of their "mischievous and wicked" behavior, the slavery question was "assuming a fearful and most important aspect" that would have "terrible consequences to the country." It could "not fail to destroy the Democratic Party" and it might "ultimately threaten the Union itself."[15]

Because of his fear for the Union, Polk began to seek a formula that would restore sectional harmony, and he found what seemed a possible solution in the revival of the Missouri Compromise. On the day after Preston King reintroduced the Proviso, Polk consulted his cabinet. When he did so, Secretary of State Buchanan "expressed his willingness to extend the Missouri Compromise West to the Pacific. All the members of the cabinet agreed with him." At first, Polk was reluctant to commit himself to this policy, but within less than a fortnight, he submitted the question once more for a full and deliberate discussion, after which he took the opinion of each member of the cabinet separately. "The Cabinet were unanimous . . . in opinion that . . . the line of the Missouri Compromise should extend West to the Pacific and apply to . . . territory [acquired from Mexico]."[16]

For a long interval after this, however, Polk held back from committing himself to the Missouri formula. In February, when Representatives Burt and Douglas offered resolutions applying the Missouri line as a basis of settlement, he did not intervene to support them, and they were voted down.[17] In March Congress adjourned and the question was left in abeyance.

In August, it began to appear that the administration was going to throw its full weight behind the extension of the Missouri Compromise, for James Buchanan declared his support of it as his principal issue in a bid for the presidency. Polk had announced earlier that he would not run for reelection, and Buchanan, as secretary of state, was widely regarded as the administration's candidate. It was the custom in that era to enter the race by writing a public letter in reply to questions put by some "correspondent," who was in fact a sup-

15. Quaife (ed.), *Polk Diary*, II, 305 (Jan. 4, 1847); III, 501–503 (June 24, 1848); IV, 33–34 (July 28, 1848); 67 (Aug. 12, 1848); 250–251 (Dec. 22, 1848). Polk to Cass, Aug. 24, 1848, in Polk papers, as quoted by Charles A. McCoy, *Polk and the Presidency* (Austin, Tex., 1960), p. 159; Message to Congress, Dec. 5, 1848, in Richardson (ed.), *Messages and Papers*, IV, 563–564, 639–642.

16. Quaife (ed.), *Polk Diary*, II, 309 (Jan. 5, 1847), 335 (Jan. 16).

17. See above, p. 65–66.

porter of the candidate. Thus Buchanan's was the first of a series of letters written by himself, Lewis Cass, Martin Van Buren, Henry Clay, and Zachary Taylor, in 1847–1848. It took the form of a "reply" to the Democrats of Berks County, who had invited him to attend their "Harvest Home" festival, and it recognized the primacy of the territorial question by dealing exclusively with this issue. After alluding to the great tradition of compromise in and under the Constitution, Buchanan went on to advocate a continuation of the spirit of mutual concession, and finally he asserted that the extension of the Missouri Compromise over any territory acquired from Mexico would be the best possible application of this spirit. The harmony of the states and even the security of the Union required it.[18]

By taking the lead, Buchanan gained for himself a kind of prior claim to the Missouri Compromise as his own issue. At the same time, the administration's newspaper, the Washington *Union*, launched a vigorous campaign to win public support for Buchanan's plan. Soon the press throughout the country was bursting with comment upon the "Buchanan movement."[19]

Still, the president himself would not take a clear-cut position on the territorial question. When the new Congress met in December, he warned solemnly of the danger which continued strife would bring to the Union,[20] but he did not say what he thought the basis of adjustment ought to be. Polk's hesitation was probably accentuated by the fact that a new candidate, with another plan for compromise, was entering the contest for the Democratic nomination. This was Lewis Cass, a Democratic stalwart who had fought with valor in the War of 1812, served in Jackson's cabinet, been minister to France, and had then been sent by Michigan to the Senate. In his early life, a man of some dash, learning, and literary talent, Cass had become increasingly a candidate for the presidency, and as such an astute and calculating political opportunist. The free-soilers already hated him as an accomplice of the men who had defeated Van Buren in 1844, and he needed to retrieve his position among the northern Democrats. On first reaction, he had favored the Wilmot Proviso,

18. Buchanan to Charles Kessler *et al.*, Aug. 25, 1847, in John Bassett Moore (ed.), *The Works of James Buchanan* (12 vols.; Philadelphia, 1908–11), VII, 385, 387; Quaife (ed.), *Polk Diary*, III, 142 (Aug. 25, 1847). On Buchanan's candidacy, see Philip Shriver Klein, *President James Buchanan* (University Park, Pa., 1962), pp. 194–205.
19. Washington *Union*, Aug. 31, 1847.
20. Richardson (ed.), *Messages and Papers*, IV, 563–564 (Dec. 7, 1847).

but later he perceived its explosive nature, and in December 1847, four months after Buchanan's Harvest Home letter, he entered the campaign with his Nicholson letter, in which he put forward the doctrine of popular sovereignty. Even before this letter was published, but while it was circulating privately among Democratic leaders, one of his supporters, Daniel S. Dickinson of New York, had introduced in the Senate a set of resolutions to give congressional approval to popular sovereignty.[21]

At the end of December 1847, the Washington *Union* published the Nicholson letter and pronounced it to be sound Democratic doctrine. Since the *Union* was the administration's organ, this declaration seemed to mark both an abandonment of the clear-cut support previously given to the Missouri Compromise and a certain measure of vacillation on the part of the president. It was one of the few subjects on which Polk ever seemed indecisive. Thus, although the forces of compromise appeared stronger than in the previous Congress, they were now divided into two camps, and the only significant compromise plans were championed by rivals for the Democratic nomination, so that the president could not support either plan without seeming to take sides against one of the candidates. Consequently, Polk held aloof or steered a veering course for several months.

The Dickinson resolutions had set the stage for another full-scale debate, and though they were presented as a compromise measure, John C. Calhoun quickly reached the conclusion that they were no compromise at all. If a territorial government might outlaw slavery at any time, he thought, no slaveholder would dare to carry slaves into such a territory, and in the absence of migration by slaveholders, all territories would become free.[22] Thus, in operative terms the Dickinson proposals would produce a free-soil result, quite as much as the Wilmot Proviso. The southerners, therefore, began to press

21. On Cass's original support for the Wilmot Proviso, see Quaife (ed.), *Polk Diary*, II, 291–292 (Dec. 23, 1846); speech of Cass in Senate, March 1, 1847, *Congressional Globe*, 29 Cong., 2 sess., pp. 548–551; Frank B. Woodford, *Lewis Cass* (New Brunswick, N.J., 1950), pp. 245–352; on Cass and popular sovereignty, see pp. 57–59 above; Dickinson was a supporter of Cass, and probably introduced his resolutions in cooperation with Cass. As is shown in text, these resolutions were less ambiguous than the Nicholson letter in asserting that the territorial legislature could act on slavery *during* the territorial stage, and that may be why Cass did not introduce them himself.

22. Charles M. Wiltse, *John C. Calhoun, Sectionalist* (Indianapolis, 1951), p. 326, citing Calhoun to Connor, Dec. 16, 1847, Connor Papers, and Calhoun to Elmore, Dec. 22, 1847, Elmore Papers.

Dickinson hard, especially since he had asserted in January that "the people of a territory have, in all that appertains to their internal condition, the same sovereign rights as the people of a state." In February he yielded to this pressure and accepted an amendment which left doubt as to whether the territories could exclude slavery prior to their admission to statehood.[23] Thus he restored the ambiguity which Cass had maintained from the beginning. Despite this concession, he was assailed from both flanks, with a free-soil substitute offered by John P. Hale of New Hampshire and a nonintervention substitute presented by Calhoun's supporter David Yulee of Florida.[24] On February 17 the resolutions and the substitutes were all tabled without a roll call.[25] What may have gone on behind the scenes is not clear, but the readiness with which the advocates of popular sovereignty gave up their fight suggests that they knew they lacked strength to adopt Dickinson's resolutions.

Three months later, the Democrats nominated Cass for the presidency and thus, in a sense, made the equivocal policy of popular sovereignty a party doctrine. At this point, when Polk might naturally have been expected to move decisively to the popular sovereignty position, he at last turned somewhat inexplicably to a full support of the Missouri Compromise. He was deeply impressed with the need for finding a basis of settlement, because in June New York's Free Soil Democrats had bolted the party and nominated Martin Van Buren for the presidency on a Wilmot Proviso platform. Also, an Indian war had broken out in Oregon, thus underscoring the acute need for organization in that region. Meanwhile Congress seemed hopelessly deadlocked. At this point, Polk and his cabinet, on June 24, 1848, agreed unanimously that "the adoption of the Missouri Compromise was the only means of allaying the excitement and settling the question." Polk requested Senator Hannegan of Indiana to "bring forward and press the adoption of the Missouri Compromise line," as a solution for Oregon; soon he had urged half a dozen senators to support this plan, and on June 27 he actually dictated to Senator Bright of Indiana a proposal which Bright introduced that same day, amending the Oregon Bill by specifying the

23. January statement in *Congressional Globe*, 30 Cong., 1 sess., p. 159; amendment in *ibid.*, p. 773, and appendix, p. 306.
24. *Ibid.*, 30 Cong., 1 sess., p. 160.
25. *Ibid.*, p. 374.

application of the 36° 30′ line as the basis for the exclusion of slavery. The administration at last stood firmly behind extension of the Missouri Compromise.[26]

Near the end of the session, the Senate finally adopted the Bright measure, and the House defeated it. But in the long interval before this happened, tensions in Congress steadily grew worse. Debate became increasingly angry, and the prospects of agreement approached the vanishing point. A sense of grave crisis hung over Washington, and at this point the sages of the Senate put their heads together and decided to see what a special committee could accomplish.

Accordingly, on July 12, John Middleton Clayton moved to refer the territorial question to a select committee of eight, divided equally between Whigs and Democrats, with each of these groups of four in turn divided equally between North and South. Clayton represented Delaware, the least southern of the slave states, and he was in himself a personification of the middle-ground position. His motion carried by a vote of 31 to 14, with all of the opposition votes cast by northern senators and nine by New Englanders.[27]

By now, the impasse between the sections had reached a dramatic point. It appeared to be the worst crisis since South Carolina had adopted her Ordinance of Nullification, and public attention was focused intently on events in Congress. The Committee set to work in an atmosphere of emergency, laboring hard and earnestly. Early in its deliberations, David Atchison of Missouri moved that a settlement should be sought in the spirit of the Missouri Compromise, and this was carried by a vote of 5 to 3, with John C. Calhoun voting with the majority. But when it came to applying the spirit of the Missouri Compromise in specific form, the proponents could attain nothing better than a tie vote of 4 to 4. After two unsuccessful attempts to implement the initial decision, the committee abandoned the Missouri Compromise and turned instead to a plan which would leave the status of slavery in the territories to be determined

26. Quaife (ed.), *Polk Diary*, III, 501–505 (June 24–27, 1848); IV, 12–13 (July 10), 21 (July 16), 65–67 (Aug. 10–11). Later, when defeated on this issue in Congress, Polk ceased to advocate the Missouri Compromise so emphatically, *ibid.*, IV, 207 (Nov. 23), but he never did abandon it entirely (below, p. 77). For the introduction of his proposal in Congress, *Congressional Globe*, 30 Cong., 1 sess., pp. 875–876.

27. *Congressional Globe*, 30 Cong., 1 sess., pp. 927–928.

by the courts.[28] Under the bill that Clayton reported on July 18, Congress would establish no restrictions on slavery; Oregon, when organized, would retain the laws against slavery which the provisional government had adopted; and the territorial legislatures of California and New Mexico would be denied any authority to make laws concerning slavery. The crucial feature of the bill, however, was the provision that any slave coming into these territories might sue in the federal courts to determine the legal status of slavery in the area to which he had been brought.[29]

It was an ominous indication of the depth of the disagreement which had developed by this time that Congress could no longer reach an accord on any measure whose terms were understood in the same way by both parties. In short, it could not reach a genuine meeting of minds. In the absence of such an understanding, it could only vote for measures ambiguous enough in their meaning or uncertain enough in their operation to gain support from men who hoped for opposite results. This kind of dualism was what gave tactical attractiveness to popular sovereignty, and it appeared also in the Clayton Compromise, for the Clayton Committee's proposal was designed to attract support both from southerners who believed that the courts would uphold the Calhoun doctrine and from northerners who believed that Mexican law forbidding slavery would apply. But any formula ambiguous enough to excite hope on both sides was also capable of arousing fear on both. Thus, Alexander H. Stephens of Georgia pessimistically believed that a judicial solution would favor the North, while northerners like John P. Hale and Thomas Corwin feared southern domination of the judiciary. Congressmen have seldom been more confused as to what they were doing, and it is perhaps an indication of the desperate mood prevailing that senators, in spite of their bewilderment, pushed hard for enactment. They debated strenuously for a little more than a week, and then, on July 27, at the end of a twenty-one-hour session, adopted the Clayton Compromise by a vote of 33 to 22. Slave-state

28. Report of Clayton, for the Committee, *ibid.*, p. 950; Quaife (ed.), *Polk Diary*, IV, 17–22 (July 14–17, 1848); Calhoun to Thomas G. Clemson, July 23, 1848, in J. Franklin Jameson, ed., "Correspondence of John C. Calhoun," AHA *Annual Report*, 1899, II, 760. For a good secondary account see Wiltse, *Calhoun, Sectionalist*, pp. 349–353.

29. Text of the Clayton bill fills ten columns in *Congressional Globe*, 30 Cong., 1 sess., pp. 1002–1005. Explanation by Clayton, *ibid.*, p. 950.

senators cast 23 votes in support and 3 in opposition, while free-state senators cast 10 votes in support and 19 in opposition.[30]

After this travail, the Senate was stunned by an extraordinary anticlimax in the House. On the day after the Senate drama was completed, before a word of debate had been uttered in the House, Alexander H. Stephens moved to table the Clayton bill. He and seven southern Whigs voted with 112 free-state congressmen to table, while seventy-six southerners and twenty-one northerners voted at least to discuss the measure. Thus, the Clayton proposal was killed, leaving the forces of compromise shattered and exhausted.[31]

On August 11, when the session was nearing its last gasp, one final effort to revive the Missouri Compromise took place in the Senate. During debate on the Oregon Bill, Stephen A. Douglas proposed an amendment similar to the one that Polk had dictated to Jesse Bright, but more explicit in admitting slavery south of 36° 30′ as well as in prohibiting it north of that line. The Senate adopted the Douglas amendment, and again there seemed a substantial prospect of an important compromise.[32]

But once more, the House abruptly destroyed the prospect. It defeated the Douglas amendment the day after the Senate had adopted it, by a vote of 82 to 121, with the division along a strictly sectional line.[33] Two days earlier a great convention at Buffalo had launched a new party, the Free Soil party, with Martin Van Buren as its presidential candidate, and many northern congressmen were now afraid to have a record of supporting the Missouri Compromise when they stood for reelection.

On August 12, with adjournment only two days away, the Senate took up the Oregon Bill again. Benton of Missouri moved to recede

30. Belief that the result would favor the South was expressed by Hale of New Hampshire, Upham of Vermont, Phelps of Vermont, Niles of Connecticut, Hamlin of Maine, Corwin of Ohio, and Miller of New Jersey. *Ibid.*, pp. 988, 989, 992, 994, appendix, pp. 1161, 1188. On the other side, Foote of Mississippi, Badger of North Carolina, and Stephens of Georgia (with a cogently reasoned argument) believed that the result would favor the North, *ibid.*, 30 Cong., 1 sess., pp. 998–1001; appendix, pp. 1103–1107. Letter of Stephens, Aug. 30, 1848, to the editor of the Milledgeville *Federal Union*, in Ulrich Bonnell Phillips (ed.), *The Correspondence of Robert Toombs, Alexander H. Stephens and Howell Cobb*, AHA, *Annual Report*, 1911, II, 117–124.

31. *Congressional Globe*, 30 Cong., 1 sess., pp. 1006–1007. For a statement by Stephens of his reason for this action, *ibid.*, appendix, pp. 1103–1107.

32. *Ibid.*, pp. 1061–1063.

33. *Ibid.*, pp. 1062–1063.

from the amendment, and a long, angry session followed. During a night of violent speeches, punctuated by one challenge to a duel, the closely matched forces fought for advantage, but after a twenty-four-hour session, the motion to recede was carried by a vote of 29 to 25. Three slave-state senators—Benton, Sam Houston of Texas, and Presley Spruance of Delaware—had voted with the majority. The bill itself, containing an exclusion of slavery, was then passed. In response to this action, Polk sent a message stating that he accepted the free status of Oregon only because it was north of the 36° 30′ line and would have been free even if the line had been applied. But he signed the measure, and Oregon became a territory as Congress adjourned.[34]

After nearly nine months of fierce contention, the net result seemed to be that popular sovereignty had failed, the Missouri Compromise had failed, and the Clayton Compromise had failed. California and New Mexico were still unorganized, and the Oregon Act seemed the only constructive accomplishment of the session. Yet the fact was that until the organization of Oregon, there had always remained a substantial likelihood that sooner or later the Missouri Compromise would be revived and applied to all of the Far West. If it were applied to both Oregon and the Mexican Cession, the area guaranteed to freedom would far exceed the area opened to slavery, and as long as this was true, there was always a substantial possibility that the 36° 30′ line would attract enough northern support in the House to gain for it the majority which it had already commanded in the Senate. But once the free-soil status of the Oregon territory was fixed, and only the Mexican Cession remained in dispute, a victory for 36° 30′ offered far more advantage to the South, opening the area of what is now New Mexico, Arizona, and southern California to slavery. Utah, Nevada, and northern California would have been free, but they were not a focus of attention as was Oregon. Polk might continue to support the Missouri formula, commending it to Congress in his last annual message,[35] but the Oregon Act had killed it as a political possibility.

The session ended in futility partly because it had been held in the shadow of a presidential election. In July and August 1848, with the voting only three months away, the Whigs had no desire to help

34. *Ibid.*, pp. 1074–1078; Polk, Message, Aug. 14, 1848, in Richardson (ed.), *Messages and Papers*, IV, 606–610; Quaife (ed.), *Polk Diary*, IV, 67–78.

35. Polk, Message, Dec. 5, 1848 in Richardson (ed.), *Messages and Papers*, IV, 641.

the Democrats achieve a triumphant solution of the sectional impasse. Also, northern congressmen did not want to face reelection under the accusation that they had betrayed the principle of free soil. President Polk was not alone in believing, therefore, that "if no Presidential election had been pending . . . the [Clayton] compromise bill would have passed in the House."[36]

If the election itself had served in any way to clarify the issues or to secure a mandate from the voters, its demoralizing effect upon Congress might have been offset by the establishment of some firmer basis for a solution. But both of the major parties, in their quest for votes, resorted to evasion and double-dealing rather than to conciliation or to the advocacy of any clear policy. In the end, the only intelligible positions taken during the campaign were those of political irregulars who revolted against the two major parties.

The campaign entered its all-out phase in May, with the meeting of the Democratic convention at Baltimore. In a revealing indication of priorities, the convention decided to nominate candidates first and draft a platform afterward. In the balloting, Lewis Cass, who enjoyed substantial support from both the North and the South, assumed an early lead over James Buchanan and minor candidates, and on the fourth ballot, he was nominated. Thus popular sovereignty triumphed over the Missouri Compromise, and though Buchanan had intended to borrow the strength of the 36° 30′ line for his candidacy, the actual result was to transmit some of the weakness of his candidacy to the principle of compromise that he advocated. The convention adopted a platform which evaded the territorial question by denouncing abolitionists and alluding piously to the "principles and compromises of the Constitution." In the campaign that followed, the Democrats used one campaign biography of Cass for circulation in the North and another for circulation in the South.[37]

Not all Democrats were satisfied with evasion. In New York, the

36. Quaife (ed.), *Polk Diary*, IV, 34–35 (July 28, 1848), 60 (Aug. 7).

37. On the election of 1848 generally, Joseph G. Rayback, *Free Soil: The Election of 1848* (Lexington, Ky., 1970); Holman Hamilton, "Election of 1848," in Arthur M. Schlesinger, Jr., *et al.* (eds.), *History of American Presidential Elections, 1789–1968* (4 vols.; New York, 1971), II, 865–896. On the Democratic convention, *The Proceedings of the Democratic National Convention, held at Baltimore, May 22, 1848* (n.p. [1848]); Woodford, *Cass*, pp. 248–258; Klein, *Buchanan*, pp. 204–205; Quaife (ed.), *Polk Diary*, III, 463–470; Allan Nevins, *Ordeal of the Union* (2 vols.; New York, 1947), I, 192–194; Lee F. Crippen, *Simon Cameron, Ante Bellum Years* (Oxford, Ohio, 1942), pp. 91–109; also works cited in notes 38 and 39. On the two biographies of Cass, see speech of Willie P. Mangum in Senate, July 3, 1848, *Congressional Globe*, 30 Cong., 1 sess., pp. 892–893.

free-soilish followers of Martin Van Buren had long been engaged in a bitter contest with a faction more tolerant toward slavery. The two groups had nicknamed each other "Barnburners" (presumably because, like a legendary farmer who foolishly burned down his barn to rid it of rats, they were willing to destroy the party in an effort to get rid of slavery) and "Hunkers" (apparently because of their alleged hankering or hunger for public office and spoils).

At the national convention, the Barnburner delegation found New York's seats contested by the Hunkers, and when the convention offered to seat both factions and divide the state's vote between them, both refused, and the Barnburners walked out in protest.

Later, the Barnburners were further antagonized by the action of the convention in adopting a platform that was evasive on the slavery question, and in nominating Cass, who had helped defeat Van Buren in 1844 and had deserted the Wilmot Proviso in 1847. After their return to New York, they called a convention to meet at Utica in June and there raised the standard of revolt, nominating Van Buren for the presidency. Their movement began strictly as a Democratic party affair, confined to the state of New York.[38] But other antislavery groups were eagerly watching the turn of events. Wilmot Proviso Democrats throughout the North proved responsive.[39] Also, antislavery Whigs could already anticipate that their party would probably nominate a Louisiana slaveholder for the presidency, and they were ripe for revolt. In Massachusetts, the small but able group of "conscience Whigs" who had followed John Quincy Adams until his death in the preceding February, and who now began to look to his son, Charles Francis, saw the possibilities of a coalition.[40] In Ohio, Indiana, and Illinois, the ardent antislavery

38. On the party divisions in New York: Rayback, *Free Soil,* pp. 60–77; William Trimble, "Diverging Tendencies in the New York Democracy in the Period of the Loco Focos," *AHR,* XXIV (1919), 396–421; Herbert D. A. Donovan, *The Barnburners* (New York, 1925); Stewart Mitchell, *Horatio Seymour of New York* (Cambridge, Mass., 1938); John Arthur Garraty, *Silas Wright* (New York, 1949); Ivor Debenham Spencer, *The Victor and the Spoils: A Life of William L. Marcy* (Providence, R.I., 1959); Jabez D. Hammond, *The History of Political Parties in the State of New York* (2 vols.; Buffalo, 1850).

39. William O. Lynch, "Antislavery Tendencies of the Democratic Party in the Northwest, 1848–1850," *MVHR,* XI (1924), 319–331; William Ernest Smith, *The Francis Preston Blair Family in Politics* (2 vols.; New York, 1933), I, 216–243.

40. On the free-soil movement in Massachusetts, see David Donald, *Charles Sumner and the Coming of the Civil War* (New York, 1960), pp. 130–204; Martin B. Duberman, *Charles Francis Adams, 1807–1886* (Boston, 1961), pp. 160–174; Frank Otto Gatell, *John Gorham Palfrey and the New England Conscience* (Cambridge, Mass., 1963),

Whigs who had sent Joshua R. Giddings and George W. Julian to Congress also wanted an alternative to the candidacy of Zachary Taylor.[41] Along with these, there were also the Liberty party men —almost pure abolitionists—who had run James G. Birney for the presidency in 1840 and 1844. They had already nominated Senator John P. Hale of New Hampshire for 1848, but many of them responded quickly to the possibility of a broadly based antislavery party which conceivably had a chance to win the election.[42]

It required great circumspection and finesse to bring these three groups together, for none wanted to appear to have abandoned its own standard to rally to another, and each group remained deeply and justifiably suspicious of the other. For a generation, the Democrats had hated John Quincy Adams as an acid, self-righteous, relentless adversary of their Old Hero and of everything Jacksonian. The Whigs had portrayed Van Buren as a sinister "little magician" —the Machiavelli of American politics—until they believed it themselves. The Barnburners cared more about squaring old political accounts with Lewis Cass and Hunkers like William L. Marcy than they did about the evils of slavery, and they shunned abolitionists as if they were diseased. The Liberty men, on the other hand, were haunted by the fear that the lofty principle of antislavery would be prostituted to sordid political ends, and they remembered uneasily Van Buren's long years of victorious alliance with the slaveholders. James Russell Lowell's Birdofredum Sawin expressed their view exactly when he said:

> I used to vote for Martin, but, I swan, I'm clean disgusted,—
> He aint the man thet I can say is fittin' to be trusted;
> He aint half antislav'ry 'nough, nor I aint sure, ez some be,
> He'd go in fer abolishin' the Deestrick o' Columby;

---

pp. 121–176; Gatell, "Conscience and Judgment: The Bolt of the Massachusetts Conscience Whigs," *The Historian,* XXI (1958), 18–45; Kinley J. Brauer, *Cotton versus Conscience: Massachusetts Whig Politics and Southwestern Expansion, 1843–1848* (Berkeley, 1963).

41. The most recent biographies are Patrick W. Riddleberger, *George Washington Julian, Radical Republican* (Indianapolis, 1966); James Brewer Stewart, *Joshua R. Giddings and the Tactics of Radical Politics* (Cleveland, 1970). On the Northwest generally, there is still much of value in Theodore Clarke Smith, *The Liberty and Free Soil Parties in the Northwest* (New York, 1897).

42. Richard H. Sewell, *John P. Hale and the Politics of Abolition* (Cambridge, Mass., 1965), pp. 89–104. On Birney, Betty Fladeland, *James Gillespie Birney: Slaveholder to Abolitionist* (Ithaca, N.Y., 1955).

An', now I come to recollect, it kin' o' makes me sick 'z
A horse, to think o' wut he wuz in eighteen thirty-six.[43]

Despite all these obstacles, however, the coalition was achieved. An intricate sequence of prior meetings prepared the way for the three groups to come together on an equal footing, and when Taylor was nominated in June, the coalition set in motion the plans that had already been made for a Free Soil convention at Buffalo in August.

Just before the Buffalo convention, the Liberty party voluntarily dissolved, thus liquidating Hale's candidacy. The way was clear now for each main element in the coalition to receive some vital concession. The Barnburners gained their main objective when Van Buren was nominated on a ballot in which Hale ran a strong second; the Whigs were recognized by the selection of Charles Francis Adams as the vice-presidential candidate; and the Liberty men were compensated for the defeat of Hale by the insertion into the platform of a plank declaring that the national government ought to abolish slavery whenever such action became constitutional. The convention adopted a resounding pledge to "fight on and fight ever" for "free soil, free speech, free labor and free men."[44]

While the Barnburners were assailing the national Democratic party on its northern flank, militant southerners had made a less successful effort to organize a revolt on the other wing. Before the convention met, John C. Calhoun had argued persistently that the South ought to hold aloof from the nominations—a strategy that lost some of its appeal because of the widespread suspicion that Calhoun nursed presidential ambitions of his own. The southern wing of the party had not followed Calhoun in this matter, but Alabama Democrats had adopted an "Alabama Platform" declaring that they would not support any candidate unless he repudiated the idea that either Congress or a territorial legislature could exclude slavery from a territory. Accordingly, William Lowndes Yancey of Alabama proposed in the convention a plank incorporating this position, and when it was voted down, 36 to 216, he attempted to

43. James Russell Lowell, *The Biglow Papers* (first series; Cambridge, Mass., 1848), No. 9; Arthur Voss, "Backgrounds of Lowell's Satire in 'The Biglow Papers,' " *NEQ*, XXIII (1950), 47–64.

44. Oliver C. Gardiner, *The Great Issue or the Three Presidential Candidates* (New York, 1848); *Oliver Dyer's Phonographic Report of the Proceedings of the National Free Soil Convention at Buffalo, N.Y., August 9th and 10th, 1848* (Buffalo, 1848); Rayback, *Free Soil*, pp. 201–230; Morrison, *Democratic Politics*, pp. 148–157.

lead a walkout of the southern delegates. This maneuver failed and Yancey himself left with one follower—a Quixote with his Sancho Panza.[45] It was not the last time that Yancey would walk out of a Democratic convention, but it was the last time he would do so virtually alone. In 1848, however, he was an isolated figure, and both Alabama and South Carolina cast their electoral votes for Lewis Cass.

The ambiguity of Cass's position might well have won him the election in a more normal year, but the Whigs showed a talent for evasion that made the Democrats seem decisive by comparison. In June, the Whigs passed over their party leader, Henry Clay, and nominated Zachary Taylor, a war hero who had never voted and had not been a Whig, a Louisiana planter who owned more than a hundred slaves but whose nomination had been engineered in part by two prominent antislavery Whigs from New York—Thurlow Weed and William H. Seward. While the Democrats had adopted a platform whose meaning no one could be sure about, the Whigs found a way to be evasive without equivocation: they adopted no platform at all. Taylor's strength in the convention came overwhelmingly from the South, but as events were later to disclose, Seward and Weed understood perfectly well what they were doing. The vice-presidential candidacy was awarded, with customary casualness, to win the support of a disaffected faction; Seward's opponents within the Whig party in New York obtained the nomination for one of their leaders, Millard Fillmore of Buffalo.[46]

After a campaign in which most participants furiously avoided the issues, Taylor carried electoral majorities in both the slave states and the free states, and won the election. He polled 1,360,000

45. Wiltse, *Calhoun, Sectionalist,* pp. 308–311, 359–373; John Witherspoon Du Bose, *The Life and Times of William Lowndes Yancey* (2 vols.; Birmingham, Ala., 1892), I, 212–221. For interpretations which stress Calhoun's own ambitions, see Gerald M. Capers, *John C. Calhoun, Opportunist* (Gainesville, Fla., 1960), pp. 226–234; Joseph G. Rayback, "The Presidential Ambitions of John C. Calhoun, 1844–1848," *JSH,* XIV (1948), 331–356.

46. On the Whig candidacy, Holman Hamilton, *Zachary Taylor, Soldier in the White House* (Indianapolis, 1951), pp. 38–133, is an especially full and able account. But valuable also are Brainerd Dyer, *Zachary Taylor* (Baton Rouge, 1946), pp. 265–301; Rayback, *Free Soil,* pp. 145–170, 194–200. Poage, *Clay and Whig Party,* pp. 152–182; Glyndon G. Van Deusen, *Life of Henry Clay* (Boston, 1937), pp. 383–393; Van Deusen, *Thurlow Weed, Wizard of the Lobby* (Boston, 1947), pp. 154–170; Robert J. Rayback, *Millard Fillmore* (Buffalo, 1959), pp. 182–191; Albert D. Kirwan, *John J. Crittenden: The Struggle for the Union* (Lexington, Ky., 1962), pp. 200–234.

popular votes, while Cass polled 1,220,000 and Van Buren 291,000. The results of Van Buren's candidacy were especially confusing, for he carried enough normally Democratic votes in New York to throw the state to Taylor, but enough normally Whig districts in Ohio to throw the state to Cass. He did not carry any state, but he ran ahead of Cass in New York, Massachusetts, and Vermont. His vote was large enough to make all northern Democrats extremely respectful of the Free Soilers, but small enough to discourage his followers from continuing their third-party organization, so that in 1852 most of them returned to the Democratic ranks.[47]

In terms of issues, the results were even more confusing. In the Democratic convention, Cass had defeated Buchanan's Missouri Compromise position and the South's doctrine of noninterference; in the election, Taylor had defeated Cass's popular sovereignty and Van Buren's free-soilism. Only the future could tell the significance of the triumph of the Louisiana slaveholder, supported as he was by antislavery men like Seward, Abraham Lincoln, and Benjamin F. Wade.

If the first session of the thirtieth Congress had done nothing because it was awaiting the result of the election, the second session, meeting in December, did nothing because it was waiting to learn the meaning of the result. By this time, everyone had given up hope of passing any kind of measure that would give territorial status to California or New Mexico, and Stephen A. Douglas, an unusually resourceful legislator, tried to by-pass the territorial question altogether by a far-fetched proposal to admit the entire Mexican Cession as one vast state,[48] which would, like other states, decide the slavery question for itself. Douglas's effort failed, and late in the session, Senator Isaac P. Walker of Wisconsin attempted another by-passing maneuver which would have dealt with the area acquired from Mexico by abrogating the Mexican laws (including those

47. The most detailed election returns are in W. Dean Burnham, *Presidential Ballots, 1836–1892* (Baltimore, 1955). Joseph G. Rayback, "The American Workingman and the Antislavery Crusade," *Journal of Economic History*, III (1943), 152–163, analyzes attitudes of laboring-class voters in the election of 1848. Norman A. Graebner, "1848: Southern Politics at the Crossroads," *The Historian*, XXV (1962), 14–35, gives a mature evaluation of the significance of this election, showing especially the disruptive interplay of sectional rivalries and party rivalries, as each party, in each section, tried to discredit the opposite party in the same section by proving that it had betrayed the interests of the section to its party affiliates in the opposite section.

48. *Congressional Globe*, 30 Cong., 2 sess., pp. 21, 190–196, 262, 381, 685; Quaife (ed.), *Polk Diary*, IV, 236–237 (Dec. 14, 1848), 286–289 (Jan. 16, 1849); good summary in Allen Johnson, *Stephen A. Douglas* (New York, 1908), pp. 134–142.

against slavery) and authorizing the president to make necessary regulations. In an effort to force its passage, this measure was attached, as an amendment, to the appropriation bill for the civil operations of the government. At first the stratagem appeared to succeed, for the Senate adopted the amendment, but in the end, instead of securing the passage of the rider, this ruse had the effect of endangering the bill to which the rider was attached. The House rejected the amendment, and for a time it appeared that there would be no funds to conduct the government. But for the second time the Senate receded, and for the second time Congress adjourned without making any provision for California and the Southwest.[49]

The principal activity of this Congress took place outside the formal sessions, for it was at this time that John C. Calhoun made his supreme effort to bring southern Whigs and southern Democrats into a united southern front in Congress. For twenty years, Calhoun had believed in southern unity, and when he witnessed repeated votes in which northern Democrats and northern Whigs combined to form majorities for the Wilmot Proviso, he began again to look for means of uniting the South. For instance, he worked earnestly to establish a prosouthern newspaper in Washington; he sought to persuade the South to hold aloof from the major candidates in the election of 1848; and as early as March 8, 1847, in a speech at Charleston, he made a cogent public appeal for united southern action. In the North, he said, only 5 percent of the voters were abolitionists, but these few, constituting a balance of power between Whigs and Democrats, could dictate policy to both. So long as the South remained divided, the parties would ignore the South and heed the free-soilers. But a united South could force both parties to respect southern rights.[50]

As the crisis worsened, Calhoun had become increasingly eager to find an occasion for applying his ideas. Early in the session, fourteen slave-state senators, including Calhoun, set up a committee to examine the possibilities of issuing a united address to the southern people or calling a meeting of southern congressmen.[51]

49. *Congressional Globe,* 30 Cong., 2 sess., pp. 595, 664, 691. The Senate adoption was by a vote of 29 to 27; the House rejection, by 100 to 114; the Senate recession, 38 to 7.

50. Wiltse, *Calhoun, Sectionalist,* pp. 303–311, 313–314, 341–344, 369–373.

51. *Ibid.*, p. 378. The resolution of the fourteen senators is in Charles Henry Ambler (ed.), "Correspondence of Robert M. T. Hunter, 1826–1876," AHA *Annual Report,* 1916, II, p. 104, but is there misdated. Wiltse has established the correct historical context.

These plans were still undeveloped when events in the House provided the southerners with a basis for action. On December 13, John G. Palfrey of Massachusetts asked in the House for permission to introduce a bill for the abolition of slavery in the District of Columbia; this was defeated 82 to 69, but 69 northern members voted for it with only 21 against. On the same day, the House adopted by a vote of 106 to 80 a motion charging the committee on territories to apply the Wilmot Proviso to New Mexico and California. On December 18, Joshua R. Giddings offered a bill providing for a plebiscite on the continuation of slavery in the District of Columbia. Although 79 northern congressmen voted to sustain Giddings, the bill was tabled by a vote of 106 to 79. But on December 21 the House adopted by a vote of 98 to 88 a resolution of Daniel Gott, a Whig of New York, calling for prohibition of the slave trade in the District of Columbia.[52] It began to seem to southern congressmen that Calhoun's warnings were coming true, and on the next evening, eighteen southern senators and fifty-one representatives held a meeting in the Senate chamber.

To Calhoun's hopeful eyes, this may have looked like the birth of a southern party at last, and to suspicious free-soilers, it may have seemed proof of the solidarity of the slaveocracy. But in fact, the movement began to lose momentum before the chairman called the meeting to order. Thomas H. Bayly of Virginia had come armed with resolutions reciting the grievances of the South, but no action was taken and the whole matter was referred to a Committee of

52. On the Palfrey resolution, *Congressional Globe,* 30 Cong., 2 sess., p. 38; on the Wilmot Proviso resolution, *ibid.*, p. 39; on the Giddings resolution, *ibid.*, pp. 55–56; on the Gott resolution, *ibid.*, pp. 83–84. Abraham Lincoln, congressman from Illinois, on Jan. 10, 1849 (*ibid.*, p. 212), announced that he would offer an amendment to the bill abolishing slavery in the District. This amendment would have made the bill less drastic, by specifying that abolition (1) would go into effect only if approved by vote of the inhabitants of the District, (2) would be gradual, applying only to children of slaves, born after the passage of the act, and (3) would not apply to the slaves of officials in Washington on public business. Lincoln's plan encouraged voluntary emancipation by providing for the government to pay for any slave whom an owner in the District was willing to set free. It also called for the rendition of fugitive slaves escaping into the District. It was this final clause which led Wendell Phillips later to speak of Lincoln as "the slave hound of Illinois." The extent to which Lincoln was acting with the antislavery bloc in Congress, and to which his proposal contributed to the southern reaction of Calhoun's Address, is the subject of some disagreement. See Wiltse, *Calhoun, Sectionalist,* p. 381; Albert J. Beveridge, *Abraham Lincoln, 1809–1858* (4 vols.; Boston, 1928); II, 184–188; Donald W. Riddle, *Congressman Abraham Lincoln* (Urbana, Ill., 1957), pp. 170–175.

Fifteen. The committee appointed Calhoun and four others to prepare an Address to the People of the Southern States. Calhoun, accordingly, wrote an able and restrained statement, setting forth southern views as to the northern transgressions against the slave system and the rights of the South. "If you become united," southerners were told, "the North will be brought to a pause." But when the committee reported this address, some of the southern congressmen criticized it as being too drastic. Though efforts to vote it down were defeated, it was sent back to committee for modifications. These modifications were made, and at a third general meeting the address was adopted, but this did not occur until many Whigs had first tried to adopt a substitute address and, failing in that, had withdrawn from the meeting.[53]

Calhoun got forty-eight signatures to the address, but this was more a defeat than a victory, for there were 121 southern congressmen. Seventy-three had not signed, and the movement for a united South had conspicuously failed. It failed in part because northern congressmen, in a well-timed strategic retreat, had reconsidered the Gott resolution for abolition of the slave trade in the District.[54] It failed, also, because too many southerners distrusted Calhoun. An ancient enemy of the Jackson Democrats, a one-time Whig, he had reopened old hostilities by opposing the Mexican War, and many southerners, including President Polk, regarded him as a disunionist. Howell Cobb called him an "old reprobate" who wanted to organize "a Southern party of which he shall be head and soul." But it failed most of all because the southern Whigs had no incentive to support it. Two months earlier they had elected Taylor to the presidency; in their hour of victory they saw no reason to join their vanquished opponents. Robert Toombs of Georgia denounced Calhoun's movement as a bold attempt to disorganize the Southern Whigs and warned that the southern Democrats were backing Calhoun "not on the conviction that Genl. [Taylor] can *not* settle our

53. Washington *Union*, Jan. 28, 1849; Niles' *Register*, LXXV, 45–46, 84, 100–101; Wiltse, *Calhoun, Sectionalist*, pp. 378–388; Quaife (ed.), *Polk Diary*, IV, 249–251 (Dec. 22, 1848), 284–288 (Jan. 16, 1849), 306 (Jan. 23, 1849). The Address of the Southern Congressmen, in Washington *Union*, Feb. 4, 1849, or Richard K. Crallé (ed.), *Works of John C. Calhoun* (6 vols.; New York, 1874–88), VI, 290–313.

54. *Congressional Globe*, 30 Cong., 2 sess., pp. 210–216. Reconsideration was voted 119 to 81. On Jan. 31, the Committee on the District reported a bill abolishing the slave trade in the District. A motion to table this was defeated 72 to 117 (*ibid.*, pp. 415–416), but the bill itself never came to a vote.

sectional difficulties, but that he *can* do it. They do not wish it settled." While forty-six out of seventy-three southern Democrats signed the address, only two out of forty-eight southern Whigs signed it. The Whigs had gone into the meeting only "to control and crush it," and Robert Toombs could justifiably boast that Calhoun had been completely frustrated in his "miserable" effort to form a Southern party.[55]

Fundamentally, however, neither the tactical retreat of the free-soilers nor the southern distrust of Calhoun would have neutralized the southern movement if most southerners had not believed that the incoming administration of a Louisiana slaveholder would solve their problems. They regarded Taylor as their man. They had nominated him over northern opposition, and many northern Whigs had bolted the party after he was chosen. Trusting him as a southerner, they had not even asked him to state his position on the territories.[56] Only gradually did they learn that they had played an incredible trick on themselves.

Taylor's position was simple and not unreasonable, but one which southern political leaders simply could not imagine that a Louisiana slaveholder and a father-in-law of Jefferson Davis would take. He believed that slavery in the South ought to be defended, but did not think it was necessary to contest the territorial question in order to do so.[57] Moreover, he was a political innocent, and he soon fell under the tutelage of William H. Seward, one of the most

55. Quaife (ed.), *Polk Diary,* IV, 251, 288; Cobb to Mrs. Cobb, Feb. 8, 9, 1849 in Robert P. Brooks (ed.), "Howell Cobb Papers," *Georgia Historical Quarterly,* V (1921), 38 ("If it would please our Heavenly Father to take Calhoun & Benton *home,* I should look upon it as a national blessing"); Toombs to Crittenden, Jan. 3, Jan. 22, 1849 in Phillips (ed.), *Toombs, Stephens, Cobb Correspondence,* pp. 139–142; Wiltse, *Calhoun, Sectionalist,* p. 388; Henry T. Shanks, *The Secession Movement in Virginia, 1847–1861* (Richmond, 1934), p. 24. For a compendium of citations showing southern distrust of the radicalism of South Carolina, see Harold S. Schultz, *Nationalism and Sectionalism in South Carolina, 1852–1860* (Durham, N.C., 1950), p. 19, n. 34.

56. "General Taylor and the Wilmot Proviso," a campaign pamphlet, quotes numerous southern sources expressing confidence that Taylor would be loyal to the interests of slavery. One of these said: "To expect that at home we were so distrustful of each other as to ask pledges on the subject, would be to admit that the institution of slavery does not, of itself, create the bond that unites all who live under its influence." Also see Cleo Hearon, "Mississippi and the Compromise of 1850," Mississippi Historical Society *Publications,* XIV (1914), pp. 30–31.

57. See discussion in Hamilton, *Taylor, Soldier in the White House,* pp. 43–47. Poage, *Clay and Whig Party,* p. 195, suggests that Taylor saw an opportunity to bring all the free-soilers and antislavery Democrats into the northern wing of the Whig party, thus greatly enhancing its political power.

dextrous political operators then in practice. Seward was not in the cabinet, but soon after Taylor's inauguration, he began attending cabinet meetings. This was a man whom southern Whigs regarded as one of the most extreme antislavery figures in their party.[58]

As early as December 1848, some southern Whigs began to realize that Taylor would probably not oppose the Wilmot Proviso. At first, they denied these reports, at least publicly, and tried to conceal their own uneasiness. But soon they lapsed into bitter silence.[59] Taylor's inaugural address was as cryptic as his campaign had been, but in April 1849 he showed his hand when he sent Thomas Butler King, a Georgian, to California, with instructions to encourage the organization of a government which would apply directly for statehood. If this took place, it would enable California to by-pass the territorial stage, during which southerners claimed a protected status for slavery, and to reach at once the stage at which slavery could be prohibited by state action. In August, in a speech at Mercer, Pennsylvania, the president stated flatly, "The people of the North need have no apprehension of the further extension of slavery." In December, Robert Toombs of Georgia sought from Taylor private assurances that he would veto the Wilmot Proviso if it passed Congress. Toombs had worked for Taylor's nomination against Clay's because he believed that Clay had "sold himself body and soul to the Northern anti-slavery Whigs." When the man he had helped to elect now told him, "If Congress sees fit to pass it [the Proviso], I will not veto it," his disillusionment was complete.[60]

Taylor's totally unforeseen affinity with the antislavery Whigs precipitated a very severe, even violent, reaction in the South. Momentarily, it seemed to unite the South more than Calhoun had ever been able to do. But at the same time when it brought south-

58. Hamilton, *Taylor, Soldier in the White House,* pp. 168–170; Glyndon G. Van Deusen, *William Henry Seward* (New York, 1967), p. 114; Frederick W. Seward, *Seward at Washington as Senator and Secretary of State* (2 vols.; New York, 1891), I, 100–103, 107–108.

59. Arthur Charles Cole, *The Whig Party in the South* (Washington, 1913), p. 138, quoting A. H. Stephens, Dec. 6, 1848, and other correspondents of John J. Crittenden, Jan. 1849; on southern denials of Taylor's heresy, see Craven, *Growth,* p. 60.

60. Hamilton, *Taylor, Soldier in the White House,* p. 225; on the King mission, *ibid.,* pp. 176–180, and also *House Executive Documents,* 31 Cong., 1 sess., No. 17 (Serial 573); *Senate Executive Documents,* 31 Cong., 1 sess., No. 52 (Serial 561). Toombs to James Thomas, April 16, 1848, in Phillips (ed.), *Toombs, Stephens, Cobb Correspondence,* pp. 103–104; Toombs to Crittenden, April 25, 1850, in Mrs. Chapman Coleman, *The Life of John J. Crittenden* (2 vols.; Philadelphia, 1871), I, 364–366.

erners of all parties together in agreeing to resist the forces of antislavery, it divided them on the means of resistance, since some were prepared to threaten dissolution of the Union and others favored more conventional modes of opposition. Ultimately the southern militants learned that their talk of disunion did more to divide the South than to unite it, but as Taylor's first session of Congress drew near, the temper of the South seemed more angry than ever before.

In fact, the people of the South had responded to Calhoun's Address of the previous winter far more heartily than had the southerners in Congress, whose party ties, both Whig and Democratic, made them reluctant to act in a context which separated them from their northern affiliates. As early as January 1849, the Virginia legislature adopted resolutions urging that if the Wilmot Proviso or some other antislavery measure were passed, the governor should convene a special session to consider "the mode and measure of redress." During the same month, the Florida legislature pledged its state to join other southern states in their common defense, "through a Southern convention or otherwise." Missouri likewise was ready to cooperate "with the slave-holding states in such measures as may be deemed necessary for our mutual protection." In May a statewide meeting of South Carolinians at Columbia announced that the Palmetto State would willingly "enter into council and take . . . firm, united, and concerted action with other Southern and South Western states." These Carolinians would willingly have led the movement themselves, but knowing the reputation as hotspurs which the nullification movement had given them in the South, they purposely refrained from taking the initiative; yet even as they did so, their governor was putting the state on a military footing. In October a large, bipartisan, unofficial southern state convention, meeting at Jackson, Mississippi, called on the southern states to send delegates to a formally constituted convention to meet at Nashville on June 3, 1850. In Alabama and Georgia, the Democratic state conventions went on record as wanting special meetings of their state legislatures if antislavery legislation should be adopted in Congress. By the time Congress met, it was clear that a number of the southern states would send official delegates to the proposed convention at Nashville. It was also clear that the southern opponents of Calhoun's plan had been thrown on the defensive, for a number of them were defeated in off-year elections during the summer and fall of 1849. At about this time, Howell Cobb reported that in Georgia, Zachary Taylor's name had become a byword and a reproach.[61]

The strength of this reaction led the more militant southerners to discuss the possibility of disunion quite freely. "My soul sickens at the threats to dissolve the Union," lamented John M. Clayton in January. In December, William A. Richardson of Illinois said, "There is a bad state of things here, and, as little as it is thought about, I fear this Union is in danger. . . . It is appalling to hear gentlemen, Members of Congress sworn to support the Constitution, talk and talk earnestly for a dissolution of the Union."[62]

This militancy was later to cause a revulsion among many southerners and to stimulate a resurgence of Unionism in both sections. But no one could foresee this in the midst of the crisis. What men knew at the time was that the incessant and inconclusive political strife seemed to grow worse rather than better. The Democrats had failed to find a solution, and the dissension already evident among the Whigs presaged their failure also. Southerners and free-soilers resorted increasingly to vituperation and the urging of drastic measures. Southern disunionism had a rendezvous at Nashville. Anarchy threatened in California, partly because all efforts to extend the rule of law over the new El Dorado had failed in a paralyzed Congress. Thus, in December 1849, a new president and a new Congress, no better than their predecessors, faced a steadily worsening crisis.

61. For an overview of attitudes in the southern states, see Craven, *Growth,* pp. 59–65; Nevins, *Ordeal,* I, 240–252; Wiltse, *Calhoun, Sectionalist,* pp. 394–410. For state resolutions, *Senate Miscellaneous Documents,* 30 Cong., 2 sess., Nos. 41, 51, 58 (Serial 533); *House Miscellaneous Documents,* 30 Cong., 2 sess., No. 54 (Serial 544); *Senate Miscellaneous Documents,* 31 Cong., 1 sess., No. 24 (Serial 563). For the action of the separate southern states in 1849, see Shanks, *Secession Movement in Virginia,* pp. 18–28; Dorothy Dodd, "The Secession Movement in Florida, 1850–1861," *Florida Historical Quarterly,* XII (1933), 3–24; Herbert J. Doherty, Jr., "Florida and the Crisis of 1850," *JSH,* XIX (1953), 32–47; Philip May Hamer, *The Secession and Co-operation Movements in South Carolina, 1848 to 1852,* in Washington University *Studies,* Vol. V (1918); Hearon, "Mississippi and the Compromise of 1850," an able account of the movement in general, as well as in Mississippi; James Kimmins Greer, "Louisiana Politics, 1845–1861," *LHQ,* XII (1929), 381–425, 555–610, and XIII (1930), 67–116, 257–303, 444–483, 617–654; Richard Harrison Shryock, *Georgia and the Union in 1850* (Durham, N.C., 1926), pp. 178–263; Clarence Phillips Denman, *The Secession Movement in Alabama* (Montgomery, Ala., 1933), pp. 22–30. On the unpopularity of Taylor in the South, Cobb to Buchanan, June 17, 1849, quoted in Nevins, *Ordeal,* I, 241.

62. Clayton to John J. Crittenden, Jan. 23, 1849, in Cole, *Whig Party in South,* p. 150; Richardson to D. T. Berry, Dec. 16, 1849, and to unidentified correspondent, Feb. 19, 1850, in George Fort Milton, *The Eve of Conflict: Stephen A. Douglas and the Needless War* (Boston, 1934), pp. 50, 57–58.

CHAPTER 5

The Armistice of 1850

ZACHARY Taylor's first Congress met on December 3, 1849. From the moment it convened, matters went badly. To begin with, the House fell into a deadlock on the election of a Speaker, partly because neither Whigs nor Democrats held a clear majority, since there were ten Free Soilers in a closely divided House. Also, sectional divisions within each party caused a dispersal of votes. Some northern Democrats would not support the Democratic candidate because he was a southerner, Howell Cobb of Georgia, and some southern Whigs would not support the Whig candidate because he was a northerner, Robert C. Winthrop of Massachusetts, and because the Whig caucus would not go on record against the Wilmot Proviso. For three weeks, amid scenes of rancor and incipient violence, the House continued in fruitless balloting while the Senate met and adjourned from day to day and the president's message remained unread. Finally, after fifty-nine ballots, the House squeezed through a resolution to elect on a plurality, and Cobb, still without a majority, was elected. But though a Speaker might be chosen by a plurality, legislation could not be passed on that basis, and the prolonged wrangle had demonstrated the paralysis that could result when party loyalties were strong enough to neutralize a sectional majority and at the same time sectional loyalties were strong enough to neutralize a party majority. The contest had also placed the House where the Senate was already—in the hands of the opposition party.[1]

1. *Congressional Globe*, 31 Cong., 1 sess., pp. 1–66; Robert P. Brooks, "Howell Cobb and the Crisis of 1850," in *MVHR*, IV (1917), 279–284; Myrta Lockett Avary (ed.),

At this juncture, Taylor sent in his message to Congress. Written in ponderous rhetoric conspicuously unlike Taylor's personal style, this document contained at least one howler inviting Democratic ridicule: "We are at peace with all the nations of the world, and seek to maintain our cherished relations with the rest of mankind." But it was straightforward and vigorous in meaning, and it set forth, without evasion, a policy for the lands in the Mexican Cession. The people of California, Taylor said, had met to form a state government for themselves after Congress failed to provide for them; they would soon apply for statehood, and if their proposed constitution should conform to the Constitution of the United States, "I recommend their application to the favorable consideration of Congress." As for New Mexico, its people also would "at no very distant period" apply for statehood, and when they did, Taylor implied, they should receive the same treatment. As for the sectional quarrel, he deplored it, reminded Congress of Washington's warnings against "characterizing parties by geographical discriminations," and stated flatly his purpose to put down any attempt at disunion: "Whatever dangers may threaten it [the Union] I shall stand by it and maintain it in its integrity."[2]

Regarded purely as a solution for the troubled questions of California and the Southwest, Taylor's policy showed a certain skill in its design. Since California and New Mexico seemed certain to be free states, the North had every reason to respond affirmatively; but the Wilmot Proviso, which had become anathema to the South, would at the same time be avoided. Southerners had insisted repeatedly that they did not expect slavery to go into the Southwest but that they objected to having Congress make an invidious distinction between their institutions and those of the North. If they meant what they said, Taylor's formula might satisfy them. Moreover, his plan could claim some impressive precedents. Stephen A. Douglas had proposed in the previous Congress to avoid the Wilmot question by bringing the Mexican Cession directly to statehood, without its going through the territorial phase. William B. Preston of Vir-

---

Recollections of Alexander H. Stephens (New York, 1910), pp. 21–27; Ulrich Bonnell Phillips, *The Life of Robert Toombs* (New York, 1913), pp. 64–72; Holman Hamilton, *Zachary Taylor, Soldier in the White House* (Indianapolis, 1951), pp. 243–253; William Y. Thompson, *Robert Toombs of Georgia* (Baton Rouge, 1966), pp. 57–59.

2. James D. Richardson, ed., *A Compilation of the Messages and Papers of the Presidents* (11 vols.; New York, 1907), V, 9–24. For the plural authorship of the message, the "rest of mankind" (deleted in the published form), and public reactions, see Hamilton, *Taylor, Soldier in the White House*, pp. 254–259.

ginia, who was Taylor's secretary of the navy, had sponsored such a proposal in the previous Congress, and at that time leading southern Whigs had shown themselves ready to support it.[3]

A defender of the president's message might point out that, by leaving the question of slavery to local decision, Taylor was really adopting a kind of popular sovereignty. It could be argued, also, that by breaking away from the compulsive dispute over the Wilmot Proviso, Taylor would give the southerners a chance to turn away from an issue offering only an abstract gain to the pursuit of more tangible goals—for instance, John Bell of Tennessee thought they should turn their attention to dividing Texas to make more than one slave state.[4]

But while Taylor's proposal may have offered a possible solution

3. For the Douglas proposal, see above, p 82; and for Preston's bill, *Congressional Globe,* 30 Cong., 2 sess., pp. 477–480; Hamilton, *Taylor, Soldier in the White House,* pp. 165, 409; Charles M. Wiltse, *John C. Calhoun, Sectionalist* (Indianapolis, 1951), pp. 390–391. The extent of earlier southern support for the kind of plan which Taylor brought forward in December is shown by a letter of Robert Toombs to John J. Crittenden, Jan. 22, 1849: "This morning, Preston will move to make the territorial bills the special order. . . . We shall then attempt to erect all of California and that portion of New Mexico lying west of the Sierra into a state as soon as she forms a constitution and asks it, which we think the present state of anxiety there will soon drive her to do. . . . The principle I act upon is this: It cannot be a slave country! we have only the point of honor to serve, and this will serve it and rescue the country from all danger of agitation. The Southern Whigs are now nearly unanimous in favor of it." Mrs. Chapman Coleman, *The Life of John J. Crittenden* (2 vols.; Philadelphia, 1871), I, 335–336. The extent to which Taylor's plan conformed to this earlier southern Whig position has not been recognized by historians other than Holman Hamilton, and even he, after making the point briefly but clearly in *Taylor, Soldier in the White House,* pp. 257–258, disregards it in his *Prologue to Conflict: The Crisis and Compromise of 1850* (Lexington, Ky., 1964). Why the southern Whigs changed their ground has also not been adequately considered. One contributory factor was the action of the northerners in amending Preston's bill to provide that the new state could not be admitted with slavery. This amendment was not part of Taylor's formula, but it had antagonized the southerners and diminished their belief in the general desirability or feasibility of making concessions to adversaries who would concede nothing (Phillips, *Life of Toombs,* pp. 63–64). Also, Taylor was not merely advocating that one new state should be left free to act on the slavery question; he was actively promoting a free state program for two new states, and his plan aimed at slavery exclusion almost as much as Wilmot's had done. See also Arthur Charles Cole, *The Whig Party in the South* (Washington, 1913), pp. 144–145.

4. John M. Clayton, John J. Crittenden, and Alexander H. Stephens had foreseen that the South must lose California and that to be beaten "in the least offensive and injurious form" was the most they could expect. See citations in Cole, *Whig Party in the South,* pp. 155–162. For the views of John Bell, see *Congressional Globe,* 31 Cong., 1 sess., pp. 436–439; Joseph H. Parks, "John Bell and the Compromise of 1850," *JSH,* IX (1943), 328–356.

for the purely territorial aspect of the sectional dispute, the dispute itself had been worsening rapidly and taking on a new and more serious dimension. During the prolonged and angry territorial deadlock, southerners had grown increasingly to believe that the issue raised by the Wilmot Proviso was merely a symptom of a far more serious danger to them. The long-standing sectional equilibrium within the Union was disappearing and the South was declining into a minority status, outnumbered in population, long since outnumbered and outvoted in the House, and protected only by the balance in the Senate. But there was not one slave territory waiting to be converted into another slave state, while all of the upper part of the Louisiana Purchase, all of the Oregon country, and now all of the Mexican Cession stood ready to spawn free states in profusion.[5] The president, pretending to wish only that California should decide the question of slavery for herself, had in fact helped to make California a free state. His chief adviser was a man who had declared bluntly that "slavery . . . can be and must be abolished, and you and I can and must do it."[6] When such men had made enough of these potential free states into actual ones, they would move, by constitutional steps, to carry out their threat. Southerners believed, with fearful conviction, that abolition would literally destroy southern society. It would subject "the two races to the greatest calamity, and the section to poverty, desolation, and wretchedness"; in attacking slavery, the North had decided "to make war on a domestic institution, upon which are staked our property, our social organization and our peace and safety."[7] When a northern congressman spoke openly of a servile war,[8] as preferable to the extension of slavery, he touched southerners on the quick. Regardless of whether their fears were realistic or fantastic, the dominating fact is that they believed that abolition would produce a "holocaust of blood," and

5. Stephen A. Douglas, in Senate, March 13, 1850, predicted seventeen new free states from the area then under the flag. *Congressional Globe*, 31 Cong., 1 sess., appendix, p. 371. The states were not parceled out as Douglas predicted, but the total number finally did amount to exactly seventeen.

6. Speech at Cleveland, Oct. 26, 1848, in George E. Baker (ed.), *The Works of William H. Seward* (5 vols., Boston, 1887–90) III, 301.

7. Calhoun, in Senate, March 4, 1850, *Congressional Globe*, 31 Cong., 1 sess., pp. 451–455; Address by Southern Congressmen to the People of the Southern States, May 6, 1850, text in M. W. Cluskey, *The Political Text-Book* (Philadelphia, 1860), pp. 606–609.

8. Horace Mann, in *Congressional Globe*, 31 Cong., 1 sess., appendix, p. 224.

they resisted anything which might lead to abolition as if they were resisting the holocaust itself.

With apprehensions such as these, many southerners had come to believe that they faced a crucial choice: they must somehow stabilize their position in the Union, with safeguards to preserve the security of the slave system, or they must secede before their minority position made them impotent. Though the form of Taylor's plan did not offend them as much as Wilmot's, they saw that it would exclude them from the Southwest just as decisively. They now wanted, therefore, not just a settlement of the territorial question, but a broad sectional adjustment. What Taylor faced was not the territorial deadlock which had frustrated Polk, but a crisis of Union.

Signs of southern alienation from the Union seemed abundant and, to many observers, alarming. Calhoun himself wrote happily that he had never known the South so "united . . . bold, and decided." Many southern members of Congress, he said, "avow themselves disunionists." During the speakership contest, Robert Toombs had held the floor against the uproar of an excited House to trumpet his defiance:

"I do not hesitate to avow before this House and the Country, and in the presence of the living God, that if, by your legislation, you seek to drive us from the territories of California and New Mexico, purchased by the common blood and treasure of the whole people, and to abolish slavery in this District, thereby attempting to fix a national degradation upon half the states of this Confederacy, *I am for disunion.*"[9]

While southern congressmen were voicing their militancy, the southern states prepared to take action. In October 1849 a large bipartisan convention in Mississippi had called for a meeting at Nashville the following June of representatives of all the slaveholding states. In December the South Carolina legislature took up this proposal, appointing delegates to the Nashville meeting. In February and March, Georgia, Texas, Virginia, and Mississippi also voted to participate, and some other states adopted resolutions express-

9. Calhoun to Thomas G. Clemson, Dec. 8, 1849, and to various correspondents, in J. Franklin Jameson (ed.), *Correspondence of John C. Calhoun*, in AHA *Annual Report*, 1899, II, 776 and *passim;* Toombs in House, Dec. 13, 1849, in *Congressional Globe*, 31 Cong., 1 sess., pp. 27–28; Hermann V. Ames, "John C. Calhoun and the Secession Movement of 1850," in American Antiquarian Society *Proceedings*, New Series, XXVIII (1918), 19–50.

ing approval of the convention and determination not to submit to the Wilmot Proviso, but abstaining from sending delegates. The Nashville project enjoyed enough public support in the South, especially among Democrats, to make it formidable, and to indicate that if Congress adopted a free-soil policy, a severe crisis would follow.[10]

For this crisis Zachary Taylor had a straightforward Jacksonian response. Taylor intended to make no concessions to those who talked of a disruption of the Union, but to uphold it against all adversaries, convinced that they would yield if confronted by strong policy. Since his theory was never tested, it can never be either proved or disproved, but one thing is clear, though frequently overlooked: Taylor had a definite and positive position. If he was correct in believing that the South would have yielded to a firm uncompromising attitude at this stage, before its separatist impulses had been hardened by a decade of contention, then the refusal of Congress to follow his policy cost the republic ten years of avoidable strife ending in a titanic civil war. If he was wrong, his policy would have forced the North to face the supreme test of war for the Union before it had attained the preponderance of strength, or the technological sinews, or the conviction of national unity which enabled it to win the war that finally came in 1861.

Even if Taylor's position had been effectively defended in debate, it would have faced grim opposition, for many of the members of Congress believed that the danger of disunion was acute and that the need for concessions was urgent. But Taylor's views never received an adequate presentation. Taylor himself could not articulate them well, being politically naïve and unskillful with words. As a party leader, he was negligible, for he had not been a Whig before his election, and the great Whigs, Webster and Clay, still did not regard him as one. Worst of all, he had no effective floor leaders to represent the administration in Congress. Of the two of his follow-

10. On the southern rights movement between Oct. 1849 and the summer of 1850, see titles in chap. 4, note 61, and chap. 5, note 25. Also see Joseph Carlyle Sitterson, *The Secession Movement in North Carolina* (Chapel Hill, 1939), pp. 49–71; Laura A. White, *Robert Barnwell Rhett, Father of Secession* (New York, 1931), pp. 103–134; Lewy Dorman, *Party Politics in Alabama from 1850 through 1860* (Wetumpka, Ala., 1935), pp. 34–44; Melvin J. White, *The Secession Movement in the United States 1847–1852* (New Orleans, 1916); Elsie M. Lewis, "From Nationalism to Disunion: A Study in the Secession Movement in Arkansas, 1850–1861" (Ph.D. dissertation, University of Chicago, 1946); Edwin L. Williams, "Florida in the Union, 1845–1861" (Ph.D. dissertation, University of North Carolina, 1951).

ers who could best have upheld his policy in Congress, one, John M. Clayton, had moved from the Senate into the cabinet, while the other, John J. Crittenden, had left Washington to become governor of Kentucky.[11] Of course there was William H. Seward, but at the crucial moment Seward chose to bespeak his personal views rather than the administration's, and in any case, Seward had incurred the distrust of southerners, who saw him as a backstair manipulator of Old Rough and Ready, and as the Iago of the Whig party. Even to northerners, his advanced views on slavery made him suspect. Few presidents ever needed an effective spokesman in Congress as acutely as did Taylor; none ever lacked such a spokesman more conspicuously.

If Taylor's policy suffered from a lack of adequate exposition, it suffered even more from the mounting evidence of the disaffection of the South. It appeared that every medium of southern expression was sending the same message. From the pulpit, from the editorial sanctum, from state legislatures, from party conventions, from mass meetings, from southern congressmen there poured out a steady stream of sermons, editorials, resolutions, speeches, and joint statements, all warning of the immediate possibility of disunion. Most historians have come to the conclusion that the danger was too great to be averted by anything short of a sweeping compromise,[12] and most of the public men at the time were deeply impressed by the gravity of the crisis. One of these was Henry Clay of Kentucky. Keenly aware of his reputation as a pacificator gained in the crises of 1820 and 1833, Clay had begun to lay plans for a major compromise even before Taylor's message went to Congress. In doing this,

11. Crittenden refused a cabinet post primarily because he feared that his position would be misconstrued if he accepted. After supporting Clay for many years, he had thrown his support to Taylor in 1848, and he did not want to be open to the charge that he had done this to advance his own interests. Albert D. Kirwan, *John J. Crittenden* (Lexington, Ky., 1962), pp. 235–241. Balie Peyton of Tennessee wrote to Crittenden, Aug. 29, 1848, concerning Clay's return to the Senate after being passed over in the election, that "impelled by a morbid state of feeling, he will play hell and break things . . . and unless the old general [Taylor] obeyed his orders in all things, he would make war upon him too." Quoted in George Rawlings Poage, *Henry Clay and the Whig Party* (Chapel Hill, 1936), p. 190.

12. For a striking statement of this view, see Herbert Darling Foster's excellent "Webster's Seventh of March Speech and the Secession Movement, 1850," *AHR*, XXVII (1922), 245–270; also, Albert J. Beveridge, *Abraham Lincoln, 1809–1858* (4 vols.; Boston, 1928), III, 71–78; for a contrary view, see Hamilton, *Taylor, Soldier in the White House*, pp. 330–344, 405–410.

Clay was not merely moving into a leaderless situation or a political vacuum, as so many historians have suggested. Rather, he was challenging Taylor's leadership of the Whig party and was preparing an alternative to Taylor's policy.[13]

On a stormy night in January, the old Kentuckian called on Daniel Webster, and Webster agreed to support him in working for a compromise.[14] Eight days after this interview, Clay rose in the Senate, and with a brief speech which held his famed oratorical powers carefully in reserve, he introduced a series of eight resolutions designed to provide a comprehensive settlement of all the various points of political contention involving slavery.[15] As events soon showed, Clay had successfully seized the initiative. The limited attention previously given to Taylor's plans now shifted, and Clay stood clearly in the spotlight.

For the next six months Congress deliberated over Clay's proposals, in one form or another, and ended by enacting most of them in an important legislative settlement which history has dubbed the Compromise of 1850. The story of these deliberations, and of the great debate which ran through them, has become one of the classic and inevitable set pieces in American historical writing. The gravity of the crisis, the uncertainty as to the outcome, and the brilliant effects of oratory in the grand manner all combined to create scenes of stunning dramatic effect. The stage was the Old Senate Chamber so rich in historic associations. (It is always understood but seldom mentioned that the House also adopted the Compromise; no oil paintings depict that part of the story.) The theme was a heroic one —the preservation of the Union. The suspense was overpowering and long-sustained as protagonist and antagonist battled in evenly

13. E.g., see the treatment in Beveridge, *Lincoln*, III, 76: "Nobody paid the slightest attention to what he [Taylor] had said;" Allan Nevins, *Ordeal of the Union* (2 vols., New York, 1947), I, 257–260: "The President's plan was for several reasons quite unrealistic," in a chapter entitled "Clay to the Rescue." Hamilton, *Taylor, Soldier in the White House*, broke sharply with prevailing views when he argued that Taylor had a definite policy of greater merit than Clay's. For Clay's role, see Poage, *Clay and the Whig Party*, pp. 191–192, 204–205; Glyndon G. Van Deusen, *Life of Henry Clay* (Boston, 1937), pp. 394–413.

14. For Clay's call on Webster, Jan. 21, 1850, see George Ticknor Curtis, *Life of Daniel Webster* (2 vols., New York, 1870), II, 396–398.

15. Clay's resolutions, *Congressional Globe*, 31 Cong., 1 sess., pp. 244–245. The general history of the Compromise has been admirably developed by Hamilton in his writings cited in notes 1, 3, 32, 36, 37. Two other excellent analytical narrations are by Nevins and Poage, cited in notes 13 and 11, respectively.

matched combat to settle the destiny of the republic. And then there were the dramatis personae. Here, for the last time together, appeared a triumvirate of old men, relics of a golden age, who still towered like giants above the creatures of a later time: Webster, the kind of senator that Richard Wagner might have created at the height of his powers; Calhoun, the most majestic champion of error since Milton's Satan in *Paradise Lost;* and Clay, the old Conciliator, who had already saved the Union twice and now came out of retirement to save it with his silver voice and his master touch once again before he died. Besides these, there was an able supporting cast—Seward, Bell, Douglas, Benton, Cass, Davis, Chase—who would have been stars on any other stage. And not only the men, but the stage effects. Philip Guedalla once said of the elder Pitt, "He was lit, he was draped, he was almost set to music." But Pitt's dramatic touches seemed contrived and sometimes forced. Not so the heightening effects of 1850. Calhoun stood visibly in the shadow of death and spoke audibly in a voice from beyond the grave; they would bury him before they voted. The Jove-like Webster never seemed greater than when he launched into his classic speech of the seventh of March: "Mr. President, I wish to speak today not as a Massachusetts man, not as a Northern man, but as an American. . . . I speak today for the preservation of the Union. 'Hear me for my cause.' " Clay, still an embodiment of grace, wit, and eloquence at the age of seventy-two, knew how to invoke, in his swan song, the same magic with which he had charmed even his enemies for nearly forty years.

If it is not taken too literally, there is a great deal of truth in this legend of 1850. Clay, Webster, Calhoun, and the others held to a superb standard of debate, and if they did not say very much that had not been said before, they expressed it somewhat better than it had ever been expressed before. Clay and Webster served in a crucial way as spokesmen for the Union, but in an even more significant way as symbols for the cause which they bespoke. They appealed to the best sentiments of their fellow-countrymen, and the Union was saved. If it came to a more impassable crisis later, that was another story. Most of all, by the very act of dramatizing the issue, they called into play the emotions which prepared the American people for conciliation, and in this respect the drama was also the reality. In a larger sense the warnings of Calhoun, the concessions of Webster, and the appeals of Clay for harmony were the stuff of which the adjustment was made.

But in another sense, it is important to recognize, along with the oratory, some prosaic and often neglected features of the settlement—its concrete terms, the meaning of its various items, the complex process of enactment, and the significance of parliamentary tactics leading to adoption. For these features will show the measure of failure as well as the measure of success in the great compromise effort.

Henry Clay designed his eight resolutions to cast a wide net over all the points at which sectional dissension touched the orbit of federal authority. First, he confronted the territorial question by proposing the admission of California to statehood on her own terms as to slavery—which meant as a free state—and the establishment of territorial governments in the rest of the Mexican Cession, "without the adoption of any restriction or condition on the subject of slavery"—which might mean either popular sovereignty, exercised by the territorial legislatures, or the Calhoun doctrine of obligatory constitutional extension, but certainly meant no congressional exclusion—no Wilmot Proviso. Next the resolutions took up a rapidly developing and acrimonious dispute over the boundary of the state of Texas. The Lone Star State, in her grandiose days as a republic, had claimed the upper Rio Grande as a western boundary, which would have made more than half of the present state of New Mexico part of the slave state of Texas. Clay proposed to solve this problem by fixing the boundary at approximately the present limits of Texas, thus keeping New Mexico intact and solacing the Texans by taking over the public debt of Texas—a measure that would have the important collateral effect of rallying the by no means negligible influence of the Texas bondholders to the support of the compromise. A further point of friction arose in connection with slavery in the District of Columbia. On this point, Clay proposed to abolish the slave trade, but to reaffirm the continuation of slavery itself, as long as it continued in Maryland, unless both the state of Maryland and the people of the District should agree to terminate it. Finally, the resolutions affirmed the immunity of the interstate slave trade from congressional interference, and proposed a fugitive slave law to enforce more effectively the constitutional provision that a "Person held to Service or Labour in one state . . . escaping into another . . . shall be delivered up on Claim of the party to whom such Service or Labour may be due."

As a compromise, Clay's proposals gave most of the material

concessions to the North: California would become a free state by law; the rest of the Mexican cession was supposedly unsuited to slavery, and so its organization on a neutral basis would presumably lead to freedom; most of the disputed area east of the Rio Grande would go with New Mexico rather than Texas; the slave trade in the District of Columbia would be abolished. The South, while gaining few tangible advantages, would secure at least a formal recognition of the "rights" of slavery—that is, a reaffirmation of the existence of slavery in the District of Columbia; a stronger implementation of the constitutional right to recover fugitive slaves; and a territorial settlement rejecting the Wilmot Proviso. Furthermore, the whole territorial issue would be defused, since the remaining unorganized area was already covered by the Missouri Compromise.[16] These provisions contributed nothing to the strength of the "slave power," but symbolically they were important to the South; implicitly they promised what no act of legislation really could promise—namely that the crusade against slavery would die down for lack of issues on which to feed.

On February 5, Clay launched the famous full-dress debate in the Senate, with a detailed exposition of his resolutions—an exposition impressive chiefly for his moving portrayal of the danger to the Union, his earnest prediction that disunion must lead to war, and his poignant appeal for a spirit of conciliation. Later in the month, Sam Houston of Texas made an important address, supporting Clay in a general way; Jefferson Davis voiced a militant southern belief that there were no physical reasons why slavery should not flourish in California, and that the proposals were not fair to the South; and Jacob Miller of New Jersey spoke for the administration in demanding that California be admitted forthwith, on its merits, and without regard to contingent questions.[17]

Then on March 4, Calhoun came from his sickbed to present a speech which was read for him by Senator James M. Mason of Virginia. This address dealt with the problem of Union at a high intellectual level, analyzing the social and cultural factors which had fostered the growth of American nationalism. "It is a great mistake

16. Stephen A. Douglas later claimed that the legislation of 1850 had implicitly repealed the act of 1820, but the evidence seems to me to prove that this possibility was not recognized in 1850. See below, pp. 156–158.

17. *Congressional Globe,* 31 Cong., 1 sess., appendix, pp. 115–127 (Clay), 97–102 (Houston), 149–157 (Davis), 310–318 (Miller).

to suppose that disunion can be effected at a single blow. The cords which bind these states together in one common Union are far too numerous and powerful for that. . . . The cords . . . are not only many but various in character. Some are spiritual or ecclesiastical; some political, others social. Some appertain to the benefit conferred by the Union, and others to the feeling of duty and obligation. . . . Already the agitation of the slavery question has snapped some of the most important, and has greatly weakened all the others, as I shall proceed to show." Once the cords were snapped, Calhoun argued, only force would remain to hold the Union together and then disunion must ensue. He also spoke of the significance of equilibrium as an essential factor in the union of North and South, and of the means by which equilibrium could be maintained. Although he did not develop the idea in his speech, he was perhaps thinking of a constitutional amendment by which the presidency would become a dual office, and North and South would each have one executive, with full powers of veto. In direct terms, therefore, Calhoun's speech seemingly had no bearing upon the pending resolutions, for he ignored Clay's plan, warned of a deeper problem, advocated solutions that could not possibly gain adoption, and virtually predicted disunion. Yet, in a sense, his speech contributed powerfully to the achievement of compromise, for the three decades of sectional controversy never witnessed a clearer or more solemn warning of the deep discontent of the South and the basic dangers that confronted the Union.[18]

Three days later, Calhoun came to the Senate for almost the last time,[19] to hear Daniel Webster. Webster's address gained in importance from the fact that no one knew what his position would be, and he had been regarded as a free-soiler. To advocate concessions to slavery would bring down upon him the fury of the abolitionists; nevertheless, Webster bared his head to the storm. Although he argued with all his force that there could be no disunion without

18. *Ibid.*, pp. 451–455. Valuable discussion of Calhoun's position in Avery O. Craven, *The Coming of the Civil War* (New York, 1942), pp. 252–258; Craven, *The Growth of Southern Nationalism, 1848–1861* (Baton Rouge, 1953), pp. 74–76; Wiltse, *Calhoun, Sectionalist,* pp. 458–465 (with a good critical review of the question of whether Calhoun intended to propose a dual executive); Hamilton, *Prologue to Conflict,* pp. 71–74 (reactions to the speech).

19. Calhoun heard Seward speak on March 11, and he engaged in a sharp exchange with Lewis Cass and Henry Foote on March 13, his last visit to the Senate. New York *Tribune,* March 16, 1850; *Congressional Globe,* 31 Cong., 1 sess., pp. 519–520.

civil war, he conceded that the South had legitimate grievances for which redress ought to be given. Restating with superb effectiveness an idea that had been advanced by Polk, Clay, and many others, he argued that it was supererogatory to insult the South by discriminating against the South's institution in an area where physical conditions would exclude it in any case: "I would not take pains to reaffirm an ordinance of nature nor to re-enact the will of God. And I would put in no Wilmot Proviso for the purpose of a taunt or a reproach. I would put into it no evidence of the votes of superior power to wound the pride, even whether a just pride, a rational pride, or an irrational pride—to wound the pride of the gentlemen who belong to the Southern States." The debate would still run on into April, but when Webster sat down, his auditors knew that he had made the supreme peace offering and the climactic appeal for conciliation.[20]

William H. Seward replied on March 11. As the ablest and closest of Taylor's supporters, Seward ought to have devoted his effort to an exposition and defense of the presidential program, which had not had an adequate spokesman in Congress, but instead he used the occasion to state essentially his personal view, that legislative compromise was "radically wrong and essentially vicious." In a context more sober than is usually recognized, he also made the arresting remark that "there is a higher law than the Constitution," thus giving the impression of a disregard for constitutional obligations, and leaving some doubt as to whether he was floor leader for Zachary Taylor or for God. Historians ever since have recognized the importance of the "Higher Law" speech, but they have overlooked the fact that Seward threw away the opportunity to make a vigorous defense of Taylor's program. The president himself undoubtedly recognized this fact with keen regret.[21]

20. *Congressional Globe,* 31 Cong., 1 sess., appendix, pp. 269–276; Foster, "Webster's Seventh of March Speech"; Claude M. Fuess, *Daniel Webster* (2 vols.; Boston, 1930), II, 198–227; Craven, *Growth,* p. 77; Hamilton, *Prologue to Conflict,* pp. 76–81; on northern reactions, which condemned Webster severely, Godfrey Tryggve Anderson, "The Slavery Issue as a Factor in Massachusetts Politics, from the Compromise of 1850 to the Outbreak of the Civil War" (Ph.D. dissertation, University of Chicago, 1944).

21. *Congressional Globe,* 31 Cong., 1 sess., appendix, pp. 260–269; Glyndon G. Van Deusen, *William Henry Seward* (New York, 1967), pp. 122–128 (shows how "higher law" passage was misunderstood). Seward's enemies exaggerated the extent of Taylor's rift with Seward; the administration organ, the Washington *Republic,* went further than Taylor wanted to go in repudiating Seward, but the evidence indicates

At the end of March, Calhoun died. In April, Clay secured the appointment of a Select Committee of Thirteen, of which he was chairman, to consider his own and other compromise proposals. In the committee, he took up a plan originated by Henry S. Foote, which he had at first opposed, to incorporate most of the proposals that were recommended into one comprehensive, overall bill. By vigorous effort he secured the adoption of this plan. The comprehensive measure which resulted soon became known, somewhat derisively, as the "Omnibus" bill because it was a vehicle on which any specific provision could ride. Clearly this embodied a definite strategy: in plain terms, Clay was betting that the supporters of compromise would form a majority and that if the entire compromise plan were put into a single legislative package, this majority would vote for it, whereas if it were brought up piecemeal, single measures would be voted upon ad hoc, on their own merits and not necessarily as part of the compromise, in which case some of them might be defeated. Or to state the strategy in other terms, Clay was counting on the Omnibus as a device to induce congressmen to vote for items they did not favor by linking such items with others they did favor. The admission of California was said to be the towrope by which the entire compromise would be pulled through. The Omnibus also offered the tactical advantage of enabling all parties to be sure that they would receive the concessions promised to them at the same time that they yielded the concessions requested of them; it avoided the awkwardness of asking one side to make concessions before the other did so, and to trust the good faith of the other to reciprocate later.

The committee accepted Clay's strategy, and in May he reported his measures to the Senate.[22] Until this time he had kept up a flimsy pretense that his proposals were in accord with the spirit of Taylor's

---

that Taylor was disappointed by the "Higher Law" speech. Hamilton, *Taylor, Soldier in the White House,* pp. 321–322; Hamilton, *Prologue to Conflict,* pp. 84–86; Poage, *Clay and Whig Party,* p. 215, minimizes the extent of this break. On the higher law, see John P. Lynch, "The Higher Law Argument in American History, 1850–1860" (M.A. thesis, Columbia University, 1947).

22. *Congressional Globe,* 31 Cong., 1 sess., pp. 769–774; for excellent discussion of the formation of the committee, Poage, *Clay and Whig Party,* pp. 211–226; also Cleo Hearon, "Mississippi and the Compromise of 1850," Mississippi Historical Society *Publications,* XIV (1914), p. 114, n. 34. The Omnibus included the California, New Mexico, Utah, and Texas Boundary measures, but not the District of Columbia and Fugitive Slave proposals.

program, but on May 21 he openly challenged the administration. The Washington *Republic*, replying for the president, declared it was "a national misfortune" that Clay had disturbed the unity of support for Taylor's plan, and it accused the Kentucky senator of the "ambition of appropriating to himself the glory of a third compromise."[23] At this point, the Whig party faced a bitter internal struggle.

On June 3 the delegates of nine southern states met in convention at Nashville. Here was the end product of years of effort on the part of militant southerners to secure a united South. During the preceding winter, when southerners felt the duress of the Wilmot Proviso about to be imposed, they had looked to this meeting of the southern states as the beginning of a new era for the South. No longer would defenders of southern rights have cause to regret the impotence of a single state in acting by itself. No longer would partisan divisions between Whig and Democrat neutralize the mighty power of the region acting as a unit. For the first time, the southern states, by standing together, would compel a recognition of their rights within the Union or would move by concerted action to go out of it.

To the more optimistic champions of southern rights, it was a glorious day. Five states—Virginia, South Carolina, Georgia, Mississippi, and Texas—had sent official delegations chosen in formal elections held in accordance with acts of the state legislatures. Four others—Florida, Alabama, Arkansas, and Tennessee—were unofficially represented by delegates named in party conventions, legislative caucuses, or otherwise. But southerners of a realistic turn of mind must have observed the portentous absence of six of the slave states. Not only Delaware, Maryland, Kentucky, and Missouri but also North Carolina and Louisiana were unrepresented. The effort to create a united South had failed again, for the same reasons which had caused it to fail previously and would cause it to fail thereafter. Southerners were almost wholly united in their purpose to maintain southern rights, but they were deeply divided as to how these rights ought to be protected. A minority beginning to be called fire-eaters believed that the South, with its slave system, could never be safe in a union with the North, which was growing steadily more op-

23. *Congressional Globe,* 31 Cong., 1 sess., appendix, pp. 614–615, 1091–1093; Washington *Republic,* May 27, 1850, quoted in Hamilton, *Taylor, Soldier in the White House,* p. 337.

posed to slavery, and they wanted to secede from the Union. Robert Barnwell Rhett, editor of the Charleston *Mercury,* William L. Yancey of Alabama, and Edmund Ruffin of Virginia were among the most prominent of this group.[24] But a majority of southerners, continuing to hope for safety for slavery within the Union, regarded the idea of disunion as disloyal, if not actually treasonable, and they deprecated the tactics of the fire-eaters. They usually took care to disclaim disunionist intentions, as they did at the opening of the Nashville Convention. But always some hothead would do what Rhett did immediately after this convention—that is, issue a secessionist pronunciamento which implicated them all. The distrust that southern rights unionists and secessionists felt toward one another continued to prevent the creation of a united South, even in 1861. Further, the majority of southerners were so reluctant to take a strong position that even temporary unity was never attained except in circumstances of acute crisis, and it withered away at the slightest indication that the danger to the South might be averted. Thus Calhoun's effort to secure a southern address in the previous Congress had derived its vitality from the vote in the House in favor of abolition of slavery in the District of Columbia, and had lost vigor as soon as that vote was reconsidered. Similarly, the movement to call a southern convention at Nashville had stemmed from the prospect that an overbearing majority would force the Wilmot Proviso upon the outnumbered South. But by the time the delegates met, the Committee of Thirteen had reported Clay's compromise to the Senate, and all that the Nashville Convention could do was to await the outcome. It sat for nine days, adopted resolutions proclaiming the rights of the South and endorsing the Missouri Compromise line, and adjourned with an agreement to reassemble if its demands were not met.[25]

24. White, *Robert Barnwell Rhett;* John Witherspoon Du Bose, *The Life and Times of William Lowndes Yancey* (2 vols.; Birmingham, Ala., 1892); Avery Craven, *Edmund Ruffin, Southerner: A Study in Secession* (New York, 1932); citations in notes 10, 25; Ulrich Bonnell Phillips, *The Course of the South to Secession* (New York, 1939), pp. 128–149.

25. On the Nashville Convention, *Resolution, Address, and Journal of Proceedings of the Southern Convention held at Nashville . . . June 3rd to 12th . . . 1850* (Nashville, 1850); Adelaide R. Hasse, "The Southern Convention of 1850," New York Public Library, *Bulletin,* XIV (1910), 239 (for bibliography); St. George L. Sioussat, "Tennessee, the Compromise of 1850, and the Nashville Convention," *MVHR,* II (1915), 316–326, perhaps the best overall account; Dallas Tabor Herndon, "The Nashville Convention of 1850," Alabama Historical Society *Transactions,* V (Montgomery, 1906), 216–237; Craven, *Growth,* pp. 92–98, for southern newspaper comment on the convention;

Meanwhile, the focus of crisis had suddenly shifted from Nashville to Austin, where a collision between the state of Texas and the government of the United States suddenly loomed up as a possibility. The question of Texas's title to the eastern side of the upper Rio Grande valley (in what is now New Mexico) was a complex one. It needed careful negotiation between the United States and Texas, and there should have been no hurry about it. But as part of his plan for settling the territorial controversy, Taylor was anxious to make New Mexico a state. In May, therefore, he sent agents to Santa Fe to sponsor a constitutional convention, and these agents treated the disputed area as part of New Mexico in their plans of organization. Although Taylor disclaimed any intention to resort to unilateral action, it appeared that he intended to create a situation where the disputed area would be functioning as part of New Mexico. When this fact became apparent, Texas almost exploded. Angry protests denounced the "partition" of Texas. The governor, encouraged by southern rights men throughout the South, breathed defiance at Taylor, took steps to organize the disputed region into Texas counties, and laid plans to send Texas troops to drive the federal garrison out of New Mexico. Sam Houston declared in the Senate that Texas would never condone disunion, but if Texas soldiers had to fight the United States army to defend territory belonging to Texas, of course they would do so.

By the end of June, southerners learned that the convention in New Mexico had adopted a constitution which was on its way to Washington. In a last-ditch effort to arrest what they regarded as Taylor's mad course, the southern Whigs sent a committee to remonstrate with him. But he remained adamant and made it clear that he intended to go ahead with his plans for admitting New Mexico as a state with the disputed region at least provisionally included. He would use force, he indicated, to suppress any resistance that his action might provoke.[26] It was in this matter, rather

---

titles cited, chap. 4, note 61, and chap. 5, note 10, above, for attitudes of various states.

26. The standard work on the Texas–New Mexico dispute is in William Campbell Binkley, *The Expansionist Movement in Texas, 1836–1850* (Berkeley, 1925), pp. 152–218. Also, see Kenneth F. Neighbours, "The Taylor-Neighbors Struggle over the Upper Rio Grande Region of Texas in 1850," in *SWHQ*, LXI (1958), 431–463; Loomis Morton Ganaway, *New Mexico and the Sectional Controversy, 1846–1861* (Albuquerque, 1944), pp. 26–34, 46–58; William A. Keleher, *Turmoil in New Mexico, 1846–1868* (Santa Fe, 1952). The essential point in this dispute was that Texas had

than in his attitude toward the Compromise as a whole, that Taylor most clearly demonstrated his indifference to the dangers of the situation. The overall territorial problem was not of his making, and his proposed solution won approval from some able contemporaries during the crisis and from some able historians long after. But he himself precipitated the New Mexico crisis by his hasty attempt to award a disputed area to a prospective new state before the long-standing and strongly backed claims of an adjoining state had been settled. Events were soon to show that the risk Taylor took was unnecessary, and that the kind of boundary settlement he desired could readily be attained by the use of tact, money, and forbearance. But Taylor, refusing to see this possibility, persisted in a policy which, if it had continued to the end, might well have started a war.

Thus while Clay was trying to heal one crisis, it appeared that another, even more explosive one had sprung up elsewhere. But on the night of July 4, Taylor fell ill, and five days later he died. Death had come to Calhoun as a kind of culmination and almost by appointment, but it came to Taylor abruptly and irrelevantly as one of those extraneous events which suddenly and in an irrational way alter the course of history. The two deaths were alike, however, in that they probably contributed to the ultimate success of Clay's proposals.

On July 31, almost before the new administration of Millard Fillmore had taken hold, Clay's Omnibus came up for action in the Senate. The extremely delicate state of the Texas–New Mexico situation led Senator James A. Pearce, who was managing the bill on the floor, to undertake a difficult parliamentary maneuver. In previous sessions, the New Mexico clauses in the Omnibus Bill had been amended in a way that favored the claims of Texas in eastern New Mexico. Pearce wanted to get rid of this amendment, and he naïvely acceded to a suggestion to do this in two steps—first by removing the New Mexico section from the bill, and then by reinserting it without the unwelcome amendment. The first step, to

---

a serious claim to the area east of the Rio Grande. But instead of negotiating a settlement of this claim, Taylor was trying to force the issue by sponsoring admission of New Mexico to statehood with boundaries to include the disputed area. For Taylor's encouragement of the statehood movement in New Mexico, see Ganaway, pp. 46–50; for the constitutional convention in New Mexico, *Senate Executive Documents*, 31 Cong., 1 sess., Nos. 74, 76 (Serial 562). For the impact of this boundary crisis at large, Hamilton, *Prologue to Conflict, passim.*

delete, succeeded, but when Pearce moved to reinsert his substitute provisions he discovered that he had set a trap for himself. First, he lost on reinserting the Texas boundary settlement, 28 to 29; then he lost on reinserting the provisions for territorial government in New Mexico, 25 to 28. At this point the adversaries of compromise gleefully seized the initiative and moved to strike out the admission of California. Some southerners who had expected to vote for California as part of the Omnibus were afraid to do so before the other items had been voted, and California also was deleted. Utah now remained as the only passenger in the Omnibus, and this pitiful remnant was permitted to pass 32 to 18.[27]

At the end of six months of sustained effort, Clay's compromise had been defeated. After the vote, jubilant opponents of conciliation were described as being in transports of delight—Jefferson Davis grinning, Seward dancing about, William L. Dayton laughing, and Thomas Hart Benton triumphant that at last he had routed Clay. But Cass was unhappy, and Robert C. Winthrop, who had succeeded Webster when the latter went into Fillmore's cabinet, was a picture of dejection. Clay himself sat "melancholy as Caius Marius over the ruins of Carthage."[28] In fact, Clay was quite spent; he had worked constantly, denying himself the social pleasures that meant so much to him, and addressing the Senate seventy times altogether in defense of his plan. Two days later, feeling all of his seventy-three years, he left for Newport to recuperate.[29]

At this point, Stephen A. Douglas stepped from the wings, where he had been waiting purposefully for many weeks, and took over the management of the Compromise measures. Douglas had refused to serve on the Committee of Thirteen, because he had never believed in the Omnibus and wanted to keep himself uncommitted, and Clay had agreed to this measure of insurance. Despite the defeat of July 31, Douglas felt a justified optimism. He knew that the shadow of Taylor's veto had always hung over Clay's Omnibus, but now Presi-

27. Bernard C. Steiner, "James Alfred Pearce," *Maryland Historical Magazine,* XVIII (1923), p. 349; Poage, *Clay and Whig Party,* pp. 254–257, citing New York *Express,* Aug. 1, 2, and 5, 1850; for a thorough analysis of the voting in this complex and crucial episode, Hamilton, *Prologue to Conflict,* pp. 109–117; *Congressional Globe,* 31 Cong., 1 sess., pp. 1490–1491, appendix, pp. 1470–1488.

28. New York *Express,* Aug. 2, 1850, cited by Poage, *Clay and Whig Party,* p. 258.

29. Calvin Colton, *The Last Seven Years of the Life of Henry Clay* (New York, 1856), pp. 200–201; Allan Nevins (ed.), *The Diary of Philip Hone 1828–1851* (2 vols.; New York, 1927), II, 900.

dent Fillmore, the antislavery Whig who had once seemed to southerners the only blemish on the Taylor ticket, was proving friendly to compromise. The reverse side of a coin that was ironic on both sides began to show as the southern Whigs found in this New York vice-president a savior from the fate which their own Louisiana planter, under the influence of a New York senator, had designed for them. Further, Douglas knew that Clay had failed to capture control of the Whig party and was in no position to lead; the crucial votes would be Democratic votes, and Douglas was the man to marshal them. Most of all, however, Douglas had a strategy entirely different from Clay's. Where Clay depended upon the existence of a majority in favor of compromise, and hence lumped the several measures together to buttress one another and to make the issue one of compromise in general, Douglas was astute enough to recognize that there was no workable majority in favor of compromise. But there were strong sectional blocs, in some cases northern, in others southern, in favor of each of the measures separately, and there was a bloc in favor of compromise. This compromise bloc, voting first with one sectional bloc and then with the other, could form majorities for each of the measures, and all of them could thus be enacted.[30] Thus Douglas remembered something that Clay had forgotten, for by such strategy Clay had brought about the adoption of the Missouri settlement in 1820.[31] Douglas also knew that the real obstacle was neither in the

30. The question of tactics for getting the Compromise enacted is an intricate one, and the principle of the Omnibus, which Clay had accepted only reluctantly, had much to be said for it. One great problem at the beginning of the session lay in the fact that many congressmen were willing to vote for Clay's settlement but afraid to make concessions to the opposite section before they received concessions for their own. Southerners, especially, feared to admit California as a free state before action was taken on Utah and New Mexico, since Taylor appeared likely to veto the latter bills. Taylor's death thus made the Omnibus less necessary. Even so, the final enactment in the House employed the principle of the Omnibus in a "little omnibus." See discussion in Poage, *Clay and Whig Party*, pp. 262–264.

On Aug. 3, 1850, Douglas wrote to Charles H. Lanphier and George Walker, "When they [the separate bills] are all passed, you see, they will be collectively Mr. Clay's Compromise," in George Fort Milton, *The Eve of Conflict: Stephen A. Douglas and the Needless War* (Boston, 1934), p. 74. Benton, who opposed the Omnibus, though supporting the individual measures, compared the separate items to "cats and dogs that had been tied together by their tails four months, scratching and biting, [but], being loose again, every one of them ran off to his own hole and was quiet." *Congressional Globe*, 31 Cong., 1 sess., p. 1829.

31. Glover Moore, *The Missouri Controversy, 1819–1821* (Lexington, Ky., 1953), pp. 94–112. Moore makes a comment on the act of 1820 which is exactly applicable to the measures of 1850: "One of the most noticeable things about the passage of

White House nor in the Senate, but in the House of Representatives, and as early as the preceding February, he had begun to concert strategic plans with House leaders of both parties.[32]

Before the Illinois senator swung into action, Millard Fillmore had already moved decisively to the support of the Compromise. Immediately upon taking office, Fillmore accepted the resignation of his predecessor's entire cabinet—he was the only vice-presidential successor who has ever done this. By bringing Webster in as secretary of state, he threw his support behind the Compromise, and the weight of the administration soon made itself felt among the Whigs. On August 6 he delivered a long message concerning the Texas–New Mexico boundary, which demonstrated how wholly needless the boundary crisis had been. Fillmore made it as clear as Taylor ever could have that the United States would use force if necessary to prevent any unilateral action by Texas against New Mexico, but he also implicitly promised that he would refrain from any unilateral action himself and that he would insist upon "some act of Congress to which the consent of the state of Texas may be necessary or . . . some appropriate mode of legal adjudication." Going beyond this pledge not to force the issue of the boundary, Fillmore also eloquently omitted all mention of statehood for New Mexico, and when the proposed state constitution reached Washington in official form, he quietly killed it. Thus, Fillmore settled a very inflamed crisis—in some ways more explosive than the one on which Clay had been working—and settled it with such adroitness and seeming ease that history has scarcely recognized the magnitude of his achievement.[33]

---

the Missouri Compromise was that the House never voted 'yea' or 'nay' on the compromise as a whole. Such a vote could have produced only a negative decision, since all but eighteen of the Northerners, as well as a substantial minority of the Southerners . . . probably would have voted 'nay.' "

32. Not only was Douglas floor manager for the compromise in the Senate, carrying it to enactment after the defeat of the Omnibus; he was also author of the California, New Mexico, and Utah bills, which Clay had adopted with his consent; he was instrumental in establishing liaison between Senate and House; and he had planned from an early stage to be in position to take over the leadership if the Omnibus should be defeated; on his role, see Frank H. Hodder, "The Authorship of the Compromise of 1850," *MVHR*, XXII (1936), 525–536; George D. Harmon, "Douglas and the Compromise of 1850," ISHS *Journal*, XXI (1929), 453–499; Hamilton, *Prologue to Conflict*, pp. 183–184; on his influence in the House, Hamilton, "The 'Cave of the Winds' and the Compromise of 1850," *JSH*, XXIII (1957), 341.

33. Fillmore showed a decent hesitancy in accepting the cabinet resignations, but the fact remains that he accepted them. For his support of the Compromise, see Robert J. Rayback, *Millard Fillmore* (Buffalo, 1959), pp. 224–247. On his treatment of the Texas–New Mexico crisis, see messages to Congress, Aug. 6, Sept. 9, 1850,

In August, the rare parliamentary virtuosity of Douglas began to bear fruit in a Senate which was much changed since the stately speeches of the previous winter. Calhoun was dead; Webster was in the cabinet, and Clay was at Newport licking his wounds. On August 9 the Senate passed a new Texas boundary bill which gave that state 33,333 square miles more than the Omnibus had allowed and which also made the arrangement contingent upon the consent of Texas, but which did not yield the disputed area east of the Rio Grande.[34] Within two weeks after this first action, bills for the admission of California, for the establishment of territorial government in New Mexico, and for the enforcement of the fugitive slave provision of the Constitution were also passed.[35] The Senate then put aside the

in Richardson (ed.), *Messages and Papers,* V, 67–73, 75.

The later stages in the history of the New Mexico statehood project have been badly neglected. New Mexico's convention drafted a constitution for statehood on May 15–25. If this document was forwarded at once to Washington (on a journey which normally took about six weeks) without awaiting ratification by the voters, which took place on June 20, it may have reached Washington before Taylor's death, and Taylor, while ill, may have instructed the cabinet to meet on it and prepare a message recommending the Constitution to Congress, as was rumored in Washington, and reported in the Washington *Union,* July 23, 1850. Nevins, *Ordeal,* I, 332, which has the only adequate statement on this intended message, shows that Thomas Hart Benton later sought to elicit information from Secretary of State Clayton concerning it, but the result is not known. Abundant evidence exists that southern congressmen were much alarmed lest Taylor use military force to support his New Mexico program. They protested strongly, and there has been elaborate controversy as to the exact form which their protests took; critical summary in Hamilton, *Taylor, Soldier in the White House,* pp. 380–381.

The text of the New Mexico constitution was in Washington by July 25. On that day, Seward used part of it when he introduced a motion in the Senate for the admission of New Mexico as a state. The motion was defeated 42 to 1 (*Congressional Globe,* 31 Cong. 1 sess., appendix, pp. 1442–1447). The officially ratified document was sent from New Mexico on July 15, and could hardly have reached Washington before Sept. 1. The New Mexico territorial measure passed the Senate on Aug. 15 and the House on Sept. 6. On Sept. 9, Fillmore transmitted the New Mexico constitution for statehood to Congress in a laconic, seven-line message in which he said, "Congress having just passed a bill providing a Territorial government for New Mexico, I do not deem it advisable to submit any recommendation on the subject of a State government." Two days later, an emissary from New Mexico sent to Congress an appeal for admission to statehood, but Congress adjourned on Sept. 30 without considering the matter. *Senate Executive Documents,* 31 Cong., 1 sess., Nos. 74, 76 (Serial 562).

34. *Congressional Globe,* 31 Cong., 1 sess., pp. 1540–1556; appendix, pp. 1517, 1561–1581; on the acceptance of the boundary settlement by Texas, Llerena Friend, *Sam Houston, the Great Designer* (Austin, 1954), pp. 209–213; Binkley, *Expansionist Movement in Texas,* pp. 215–218.

35. *Congressional Globe,* 31 Cong., 1 sess., pp. 1573, 1589, 1647, 1660; Hamilton, *Prologue to Conflict,* p. 141. For the legislative provisions of the Fugitive Slave Act, see below, p. 131.

District of Columbia bill until the House could act. But it had not long to wait, for the House moved even more swiftly than the Senate. On September 6 it took up a "little Omnibus" which combined settlement of the Texas boundary with territorial government for New Mexico and adopted it 108 to 97.

The Wilmot Proviso had at last been abandoned by the House. Significantly, this was done in a bill vigorously supported by an influential lobby because it allotted $5 million for payment at par of some heavily depreciated Texas securities.[36] Statehood for California, territorial status for Utah, and passage of the fugitive slave bill followed in order within nine days. On September 16 and 17, Senate and House passed the bill to abolish the slave trade in the District of Columbia. Meanwhile, President Fillmore was signing the measures as fast as they arrived at his desk, and thus by September 17 the long struggle had come to an end.[37] Douglas's strategy had achieved a complete success. His skill becomes especially evident upon analysis of the roll calls on the successive bills, which reveal that the voting was largely along sectional lines. Southern majorities opposed two of the measures —the admission of California and the abolition of the slave trade in the District of Columbia, while northern majorities opposed the Fugitive Slave Act and the organization of the New Mexico and Utah territories without the Wilmot Proviso. It was a highly significant and curiously overlooked fact that on all of the crucial roll calls by which the six measures of compromise passed in both the Senate and the House, only once in one house did a northern majority and a southern majority join in support of a bill. On the New Mexico bill in the Senate, northern senators voted 11 to 10 in favor of what southern senators also favored by a vote of 16 to 0. But otherwise, North and South voted always at odds. The House did not vote on New Mexico as a separate bill, but on the vote to attach New Mexico to the Texas

36. The definitive analysis of the highly significant role of the Texas bonds is in Holman Hamilton's "Texas Bonds and Northern Profits," *MVHR*, XLIII (1957), 579–594, and Hamilton, *Prologue to Conflict*, pp. 118–132. But see also a somewhat erratic account, Elgin Williams, *The Animating Pursuits of Speculation: Land Traffic in the Annexation of Texas* (New York, 1949).

37. *Congressional Globe*, 31 Cong., 1 sess., pp. 1502–1837, *passim*. For enactment in the Senate, see Holman Hamilton, "Democratic Senate Leadership and the Compromise of 1850," *MVHR*, XLI (1954), 403–418; in the House, Hamilton, "The 'Cave of the Winds' "; overall, Hamilton, *Prologue to Conflict*, pp. 133–165.

boundary bill the North rolled up a majority of 23 votes against; this was offset by a majority of 31 southerners in favor. On the combined measure, a majority of 9 northern votes against was offset by a southern majority of 22 in favor. Meanwhile, in the Senate, the Texas boundary bill by itself gained the support of northern senators by a vote of 18 to 8, while southerners were divided equally at 12–12. On other measures, the contrasts were even more pronounced. The Utah bill passed despite northern votes of 11 to 16 against it in the Senate and 41 to 70 against it in the House. The Fugitive Slave Law passed primarily because northern abstainers skulked in the corridors while every southern congressman who voted cast his vote in favor, thus overriding adverse northern tallies of 3 to 12 in the Senate and 31 to 76 in the House. On the other hand, unanimous northern majorities carried the California bill through, although southerners opposed by votes of 6 to 18 in the Senate and 27 to 56 in the House. Similarly, the District of Columbia bill commanded unanimous support from northerners and thus overrode southern opposition, which voted 6 to 19 and 4 to 49 against the measure.

Consistently, the preponderant strength of one section opposed the preponderant strength of the other; yet in each case the measure passed. This was because, as Douglas had perceived, there were small blocs of advocates of compromise ready to exert a balance of power. In the Senate, four senators voted for the compromise measure every time, and eight others did so four times while abstaining on a fifth measure; in the House 28 members gave support five times and 35 did so four times out of five.[38]

These facts raise a question of whether the so-called Compromise of 1850 was really a compromise at all. If a compromise is an agreement between adversaries, by which each consents to certain terms desired by the other, and if the majority vote of a section is necessary to register the consent of that section, then it must be said that North and South did not consent to each other's terms, and that there was really no compromise—a truce perhaps, an armistice, certainly a settlement, but not a true compromise. Still, after four

38. Extremely elaborate analyses of votes have been made by Poage, *Clay and Whig Party*, *passim*, who says that he analyzed 110 roll calls (p. 213, n. 27), and by Hamilton, *Prologue to Conflict*.

years of deadlock, any positive action seemed a great accomplishment. California had at last been admitted and the Southwest need no longer remain unorganized. For the first time since 1846 Congress could meet without confronting questions which automatically precipitated sectional clashes.[39]

After the crucial votes in the House, congressmen began to relax, and the final days of the session witnessed scenes of great conviviality and jubilation. Crowds thronged the streets of Washington and serenaded the Compromise leaders. On one glorious night, the word went abroad that it was the duty of every patriot to get drunk. Before the next morning many a citizen had proved his patriotism, and Senators Foote, Douglas, and others were reported stricken with a variety of implausible maladies—headaches, heat prostration, or overindulgence in fruit.[40]

If one should ask, more than a century later, exactly what they thought they were celebrating, it is impossible to find a categorical answer. Partly, no doubt, they rejoiced to see the end of the longest and toughest session through which any American Congress had ever sat. Partly, they were relieved that a disaster which they dreaded had not materialized, for Daniel Webster was not alone in believing that "if General Taylor had lived, we should have had civil war."[41] Partly, they were glad to believe that the eternal territorial question, the everlasting Wilmot Proviso, the omnipresent slavery issue, would not now brood over all their transactions, and they felt as did Stephen A. Douglas, who declared that he had "resolved never to make another speech upon the slavery question in the Houses of Congress," or as did Lewis Cass, who said, "I do not believe any party could now be built up in relation to this question of slavery. I think the question is settled in the public mind. I do not think it worthwhile to make speeches upon it."[42]

But though they could, with some assurance, celebrate a settle-

39. Despite the superb research and interpretation which Poage, Nevins, and most of all Hamilton have devoted to the crisis of 1850, the fact that the settlement was not in the true sense a compromise has not, in my opinion, been adequately developed by any of them.

40. New York *Herald*, Sept. 8, 10, 1850; New York *Tribune*, Sept. 10, 1850; Nevins, *Ordeal*, I, 343.

41. Henry W. Hilliard, *Politics and Pen Pictures* (New York, 1892), p. 231. For other contemporary statements of the same tenor, see Nevins, *Ordeal*, I, 345.

42. *Congressional Globe*, 31 Cong., 1 sess., p. 1859 (Cass); 32 Cong., 1 sess., appendix, p. 65 (Douglas); Nevins, *Ordeal*, I, 345, 349.

ment, it was not entirely clear what the settlement was. Most of the measures, to be sure, seemed explicit, and the admission of California, the boundary of Texas, the fugitive slave provisions, and the provisions concerning slavery and the slave trade in the District of Columbia were plain enough. But with the exception of California, these matters had not presented major problems. The great problem, the central issue, had been the question of slavery in the territories. What had the settlement done about it?

The answer, of course, was that New Mexico and Utah had been organized as territories without any restriction as to slavery. Clearly there was no Wilmot Proviso; equally clearly there was no geographical line. But did this mean an acceptance of the southern doctrine of the mandatory constitutional extension of slavery, or did it imply popular sovereignty in the sense which would leave the status of slavery to the territorial legislature? When Clay's Omnibus came from committee, it had held a seeming answer to this question, for it specifically prohibited the territorial legislatures from passing any law "in respect to African slavery." Some northerners hoped that this meant that the law of Mexico, which had prohibited slavery, would remain in effect, but it seems reasonably clear that the South stood to benefit most, for this clause left a situation in which Congress would not itself exclude slavery from a territory and would not permit the territorial legislature to do so. But this provision, in an amended form, was struck out before the defeat of the Omnibus, with both Clay and Douglas working to eliminate it.[43] Even before this happened, when Douglas was reporting the territorial bills from committee he had made the highly suggestive remark that there was disagreement in the committee on some points in regard to which each member reserved the right of stating his own opinion and of acting in accordance therewith. Apparently this meant that the ambiguity of Cass's original popular sovereignty was still to be kept alive, although Douglas did not resort to it personally and Cass had ceased to do so.[44] To them, and to other

43. *Congressional Globe,* 31 Cong., 1 sess., pp. 1463–1473. The original phrase, "in respect to African slavery," had been amended to "establishing or prohibiting African slavery," so that if the courts declared slavery legal, the territorial legislature could then pass legislation regulating or supporting it. But this amendment was struck out, July 30, on motion of Moses Cotton of New Hampshire, by a vote of 32–20.

44. Douglas statement, March 25, 1850, in *ibid.*, p. 592; for the position of Cass and Douglas on the power of the territorial legislature to regulate, *ibid.*, pp. 398–399, 1114.

northern congressmen, the "nonintervention" of Congress meant that the territorial legislature might exclude slavery from the territory, but to southern congressmen it meant that slavery could not be excluded, at least until statehood. Only the support of both of these groups provided the narrow margins by which the territorial bills were adopted, and if the meaning had been explicit the measures would have failed. Douglas, clearly understanding the situation, regarded the ambiguity as benign, and left it undisturbed. But the question would have to be decided somehow, and a general awareness of this fact probably inspired the incorporation of amendments extending the Constitution over all territories and providing for appeals on the slavery question to the Supreme Court. These amendments would have the double effect of invalidating local Mexican law, which prohibited slavery, if such law was inconsistent with the Constitution, and also of giving the federal courts effective jurisdiction on the question of whether a territorial legislature could constitutionally restrict slavery. Insofar as the territorial question was evaded by leaving it to the courts, the settlement of 1850, for all its apparent concreteness, closely resembled the Clayton Compromise of two years before, which, Thomas Corwin had said, proposed to enact a lawsuit instead of a law. The true meaning of the acts of 1850 would have become evident if the territorial legislature of either New Mexico or Utah had adopted a statute excluding slavery, whereupon a court action no doubt would have challenged its constitutionality. But since neither territory took such action, many historians have lost sight of this aspect of the Compromise.

After the measures of 1850 were adopted it was quite possible for Douglas to go back to Chicago and declare that the settlement recognized the "right" of the people to regulate "their own internal concerns and domestic institutions in their own way," while Robert Toombs, who had cooperated closely with Douglas's lieutenants in the House, could return to Georgia and tell his constituents that they had regained the principle so unwisely bargained away in 1820, the right of the people of any state to hold slaves in the common territories. Before the Congress adjourned, Salmon P. Chase was already saying, acidly, "The question of slavery in the territories has been avoided. It has not been settled."[45]

45. For the amendments extending the Constitution and providing for judicial appeal, *ibid.*, pp. 1144–1146, 1212, 1379–1380, 1585, appendix, pp. 897–902; the

If a man like Chase, viewing the settlement at short range, observed that the territorial problem had not been solved, a twentieth-century reader, viewing it at long range, may observe that the two great problems of slavery and the Union had not been solved either. Because of these omissions, the verdict on the measures of 1850 has been a subject of continuing controversy among historians, involving in part a disagreement concerning values, and in part a disagreement concerning the possible alternatives in 1850. In terms of values, writers who attached high importance to the preservation of the Union or to the maintenance of peace have tended to regard the Compromise as constructive because it helped to conserve these two values, while writers who attached high importance to the eradication of slavery have usually condemned the Compromise as tending to perpetuate slavery. Since the historian has no special competence to appraise the relative priority of these values, which is more a question of ethics than of history, he cannot contribute much to the resolution of a disagreement concerning them, except to note that the most successful statesmanship has usually sought pragmatically to reconcile values rather than to follow rigid logic in sacrific-

---

comparison of the interpretations by Douglas and by Toombs is from Allen Johnson, *Stephen A. Douglas* (New York, 1908), pp. 189–190; statement by Chase in *Congressional Globe,* 31 Cong., 1 sess., p. 1859; John Bell, senator from Tennessee, said, "The crisis is not past; nor can perfect harmony be restored to the country until the North shall cease to vex the South upon the subject of slavery," Memphis *Daily Eagle,* Sept. 27, 1850, quoted in Joseph Howard Parks, *John Bell of Tennessee* (Baton Rouge, 1950), p. 262.

In stating that the Compromise left a deliberate ambiguity on the territorial question I reluctantly take issue with an able analysis by Robert R. Russel, "What Was the Compromise of 1850?" *JSH,* XXII (1956), 292–309. Russel argues vigorously that when Congress repealed the clause which forbade territorial legislatures to act on the question of slavery, "it was understood by all concerned that the legislatures were left entirely free to legislate on slavery." But it is questionable either that they were left free or that the arrangement was so understood. Although a denial of a given power had been removed, this removal did not mean that the power had been conferred; it merely left a question as to the constitutionality of the exercise of such power in the absence of any congressional action one way or the other. As to what was understood, Clay himself recognized that there was a deliberate ambiguity when he stated, "The bill is silent; it is non-active upon the subject of slavery. The bill admits that if slavery is there [under the Constitution], there it remains. The bill admits that if slavery is not there, there it is not" (May 21, *Congressional Globe,* appendix, p. 614); "We cannot settle the question [of the status of slavery in New Mexico] because of the great diversity of opinion which exists" (June 7, *ibid.,* p. 1155). Douglas recognized it when he made the statement quoted above, about the differences in the Committee on Territories. Pierre Soulé recognized it when he said, "We all know that we do not understand this 11th section alike. We know that its import in different minds amounts to absolute antagonism. If we are not deceiving one another, we are deceiving our constituents" (*ibid.,* appendix, p. 631).

ing one value for the sake of another. But as a sifter of evidence, the historian should be able to contribute something to the resolution of disagreement concerning the nature of the alternatives in 1850. Both North and South unwillingly made concessions because the compromisers were convinced that the immediate alternatives to compromise were disunion or war, or perhaps both. This reading of the alternatives involved beliefs as well as facts, and historians, of course, do not agree on them as if they were factual. Some historians argue that Taylor's policy of firmness would have dissolved the crisis and averted the dangers of secessionism while it was still incipient, and before its partial victory in 1850 and the ensuing decade of dissension made it unmanageable. Others contend that the disruptive forces in 1850 were extremely powerful and that the compromise gave the Union another indispensable decade to grow in strength and cohesiveness before it faced a test which, even in 1860, was almost too much for it.

No historian can declare with certitude that either of these appraisals of the situation is correct. What then can he say? He can say that in 1832 and again in 1861, people also faced crises in which some thought the danger of disunion was exaggerated, that it would die down if firmly handled and not encouraged by "appeasement." In 1832 this proved at least partially right, though concessions were certainly made; in 1861 it turned out to be wrong. Were the dangers of 1850 more like those of 1832 or of 1861? In my opinion, the evidence, on balance, indicates that by 1850 southern resistance to the free-soil position was so strong and widespread that if the Union were to be preserved, the South had either to be conciliated or to be coerced. It is true that the disunionists in the South began to lose ground to the southern moderates long before the Compromise was enacted, but I believe this was because compromise was confidently expected and the South distinctly preferred compromise to disunion.

If agreement on this point were possible, which it is not, what conclusion would follow as to the merit of the policy of conciliation in 1850, in terms of the values of peace and of union and even of antislavery? As for peace, the pacification of 1850 lasted fewer than ten years, and it can easily be dismissed as a mere stopgap or deferral of war. But no peace has been eternal, and no peacemaker, including Henry Clay, is responsible for subsequent acts, such as the Kansas-Nebraska Act and the Dred Scott decision, by which a well-designed but fragile peace may later be destroyed. As for Union, the supreme challenge to the Union was not ultimately avoided; it was only delayed. But the decade of delay was also a decade of growth in physical strength, cohesiveness, and technological resources which enabled the Union to face the supreme challenge far more effectively. (This is to say nothing of the relative advantage, which no one could have foreseen, of having Abraham Lincoln rather than Millard Fillmore in the White House at the moment when greatness in leadership was vitally needed.) Even as for antislavery, it is difficult to see that the Compromise ultimately served the purpose of the antislavery idealists less well than it served those who cared primarily for peace and union, though it is easy to see why antislavery men found the medicine more distasteful. If, as Lincoln believed, the cause of freedom was linked with the cause of Union, a policy which dealt recklessly with the destiny of the Union could hardly have promoted the cause of freedom.

These conclusions seem all the more tenable when one considers the evidence as to what the concessions to the South actually cost. The number of fugitive slaves returned to their masters was relatively small, and virtually no slaves were carried into Utah or New Mexico.[46] The lively historical debate as to whether slaves *might have been* carried there should not obscure the fact that virtually none *were.* Thus the North paid little in a tangible way for a ten-year deferral which ultimately proved favorable to the causes of both antislavery and Union. But the settlement was nevertheless criticized, at the time and by historians afterward, chiefly from the antislavery standpoint, because, of course, it was not intended to be a deferral—it was intended to be a permanent settlement to save a union which would remain half slave and half free. From the antislavery point of view, the settlement might later be vindicated in terms of results, but it could never be vindicated in terms of intentions.

On the other side of the coin, it is ironical that historians sympathizing with the Confederacy have seldom deplored the settlement, though it caused a fatal ten-year delay in the assertion of southern independence. By 1850, some southerners like Calhoun perceived

46. Stanley W. Campbell, *The Slave Catchers: Enforcement of the Fugitive Slave Law, 1850–1860* (Chapel Hill, 1968), although demonstrating that the law was well enforced when invoked, is able to document returns of only approximately 300 fugitives over the ten-year period (nearly half of them without judicial process)—Table 12, p. 207. This amounts to two slaves per year per slave state.

that time was running against them and that they would lose by temporizing. The events of the next two decades showed how realistic they had been. Men like Robert Toombs, who talked so fiercely about secession and then accepted the Compromise, were setting the stage for Appomattox. The ultimate irony is that prosouthern historians have not been logical enough to condemn Toombs and the other southern unionists for a compromise that apparently proved ruinous to the South, while antislavery historians have castigated Webster for a compromise that eventually worked to the advantage of the antislavery cause.

Hindsight has long since shown that the Compromise of 1850 did not bring either the security for the Union which many hoped for or the security for slavery which others feared. But at the time, this was not yet evident. Realistic men like Douglas and Chase knew that North and South had not really acted in accord and that the arrangements for Utah and New Mexico did not really answer the territorial question. But if the measures were not themselves a compromise, might they yet become a compromise? Daniel S. Dickinson hoped so, and he remarked that "neither the Committee of Thirteen, nor any other committee, nor Congress have settled these questions. They were settled by the healthy influence of public opinion."[47] At the very least, this Congress, through the leadership of Henry Clay, Daniel Webster, Millard Fillmore, and Stephen A. Douglas, had averted a crisis, and it had reached a settlement of issues which four preceding sessions of Congress had been unable to handle.[48] It remained to be seen whether the American people, North and South, would, by their sanction, convert this settlement into a compromise.

47. *Congressional Globe,* 31 Cong., 1 sess., p. 1829.

48. For a provocative comparison of Webster's views with those of the free-soilers, see Major L. Wilson, "Of Time and the Union: Webster and His Critics in the Crisis of 1850," *CWH,* XIV (1968), 293–306.

CHAPTER 6

Fire-Eaters, Fugitives, and Finality

MILLARD Fillmore's message to Congress in December 1850 announced a doctrine soon adopted as an article of faith among defenders of the Compromise. The measures enacted at the previous session, said the president, had settled some extremely dangerous and exciting subjects. The legislation was "in its character final and irrevocable," and he recommended that Congress adhere to it as "a final settlement."[1]

"Finality" quickly became a watchword, echoing through the halls of Congress for the next two sessions. Stephen A. Douglas sounded it also when he termed the Compromise a "final settlement" and appealed to his colleagues to regard the slavery question as laid to rest. "Let us cease agitating," he said, "stop the debate, and drop the subject." Before long, the advocates of finality went beyond a mere promise to uphold the settlement, and urged the proscription of anyone who would not accept it. In Congress, ten free-state members and thirty-four from the slave states put their signatures to a pledge that they would never give political support to anyone not positively committed to stand by the Compromise. Meanwhile, Senator Henry S. Foote introduced resolutions affirming congressional endorsement of finality. He could not push these through at the short session, but similar resolutions passed the House in the

1. Message to Congress, Dec. 2, 1850, in James D. Richardson (ed.), *A Compilation of the Messages and Papers of the Presidents, 1789–1902* (11 vols.; New York, 1907), V, 93.

session following by a vote of 103 to 74, with 54 northerners and 20 southerners recorded in the negative.[2] At the same session, commitment to finality also arose in both party caucuses, but the Democrats, for tactical reasons, avoided taking the pledge, and the Whigs adopted a resolution in a caucus so thinly attended that the adoption meant very little.[3]

But while the unionists were making "finality" a part of their creed, there was serious question whether the Compromise measures would in fact be accepted by either the North or the South. Northern reactions were uncertain because antislavery men, who had made a flaming issue of the legality of slavery in remote territories, now faced a Fugitive Slave Act which was far more offensive to them than any territorial situation. In the South, on the other hand, radical leaders for many months had been urging the people to prepare for disunion, and there was some doubt whether the momentum of the secession movement could be arrested by the rather limited concessions which the Compromise offered.

The actual danger of disunion in 1850 remains a matter of controversy among historians. Some hold that several states stood on the verge of secession; others, that there was more noise than substance in the secession excitement.[4] Dispute on this important point has to some extent diverted attention from the equally significant fact that the idea of secession as a possible recourse first won widespread acceptance in the South during the prolonged deadlock of 1846–1850. At the time of Wilmot's Proviso, the doctrine that each state

2. *Congressional Globe,* 32 Cong., 1 sess., pp. 976–983, and appendix, pp. 65–68; *National Intelligencer,* Jan. 29, 1851. Allan Nevins, *Ordeal of the Union* (2 vols.; New York, 1947), I, 346–352, 396–404, has a good account of the general acceptance of the Compromise.

3. Washington *Republic,* Dec. 3, 1851; *National Intelligencer,* Dec. 2, 1851; *Congressional Globe,* 32 Cong., 1 sess., pp. 6–11. In April 1852 the Whig caucus evaded a reaffirmation of its earlier vote. *National Intelligencer,* May 8, 1852; Arthur Charles Cole, *The Whig Party in the South* (Washington, 1913), pp. 234–237.

4. James Ford Rhodes reviewed this question in his *History of the United States* (7 vols.; New York, 1892–1906), I, 130–138, and stated, "I think that little danger of an overt act of secession existed," though much of his evidence seemed contrary to his conclusion. Several writers exploring the situation in the South adduced evidence which ran counter to Rhodes, but probably the most important item in reversing Rhodes's verdict was an essay, with very extensive evidence, by Herbert Darling Foster in 1922 (above, chap. 5, note 12). After Foster, the gravity of the crisis was scarcely questioned until Holman Hamilton (above, *ibid.*) again put forward the argument that the danger to the Union had been exaggerated. For literature on the situation in 1850, see chap. 4, note 61; for the situation in 1851–52, Nevins, *Ordeal,* I, 354–379; Avery O. Craven, *The Growth of Southern Nationalism, 1848–1861* (Baton Rouge, 1953), pp. 83–115; Cole, *Whig Party in the South,* pp. 174–244.

retained its full sovereignty was already prevalent, stemming from the Virginia and Kentucky resolutions of 1798–1799 and the much more elaborate theories of Calhoun. Also, a few lonely spirits like Robert J. Turnbull and Thomas Cooper in South Carolina, and later Beverly Tucker and Edmund Ruffin in Virginia, Robert Barnwell Rhett in South Carolina, and William L. Yancey in Alabama, had actually advocated secession as a line of action.[5] But even the most ardent defenders of states' rights had usually recognized that the concept of disunionism bore a stigma, and they had avoided it. Calhoun, for example, had defended nullification with the claim that it would prevent disunion. It was a bitter memory for South Carolinians that their state had stood alone during the Nullification crisis, and they knew that the rest of the South continued to distrust their efforts at sectional leadership.[6] As late as 1846, southerners generally associated disunion with treason. Whatever attraction it may have had as an abstraction, they shrank from secession with patriotic revulsion. Only under the stress of strong emotion did they mention it as a contingency—usually with a disclaimer of some kind by which they crossed themselves, so to speak, to atone for possible sin. Often they used a euphemism such as "resistance to the last extremity," or they accompanied the word "disunion" with a saving phrase, such as, "May God forever avert the necessity," or with the suggestion that they admitted this awful possibility to consideration only because the alternative was something even worse, like "degradation" or "dishonor." The fact that this kind of language became conventionalized is itself a mark of the fact that the proslavery South was still a pro-Union South as late as 1846.[7]

5. On the intellectual origins of the doctrine of secession and its gradual acceptance as a theory, see Ulrich Bonnell Phillips, "The Literary Movement for Secession," in *Studies in Southern History and Politics Inscribed to William Archibald Dunning* (New York, 1914), pp. 33–60; Dumas Malone, *The Public Life of Thomas Cooper, 1783–1839* (New Haven, 1926), pp. 281–336; Arthur C. Cole, "The South and the Right of Secession in the Early Fifties," *MVHR*, I (1914–15), 376–399; Jesse T. Carpenter, *The South as a Conscious Minority, 1789–1861* (New York, 1930), pp. 171–220.

6. At the time of the Bluffton Movement in South Carolina (1844), Calhoun had vigorously opposed Rhett, Hammond, and other disunionists. Charles M. Wiltse, *John C. Calhoun, Sectionalist* (Indianapolis, 1951), pp. 187–198. The way in which men could advocate secession if their demands were not met and at the same time insist that they were unionists is neatly illustrated by a speech of Albert G. Brown of Mississippi, Nov. 2, 1850, quoted in Cleo Hearon, "Mississippi and the Compromise of 1850," Mississippi Historical Society *Publications*, XIV (1914), 168.

7. For quotations showing the use of these locutions and others like them, see *ibid.*, pp. 89, 123, 155; Henry T. Shanks, *The Secession Movement in Virginia, 1847–1861* (Richmond, 1934), pp. 23–24; Joseph Carlyle Sitterson, *The Secession Movement in*

But the four years leading to 1850 had witnessed an immense change. By December 1849, Alexander H. Stephens was telling his brother, "I find the feeling among the Southern members for a dissolution of the Union—if the antislavery [measures] should be pressed to extremity—is becoming much more general than at first. Men are now beginning to talk of it seriously, who, twelve months ago, hardly permitted themselves to think of it." At almost the same time, James J. Pettigrew of North Carolina agreed: "I am amazed when I see the rapid strides that the spirit of disunion has made in all quarters in the course of a year. No one considers it at all startling to discuss the matter in a calm tone, whereas a few years ago it was necessary to be worked up into a furious passion before the word could be uttered."[8] These observations, indeed, seemed justified by language such as that of Senator Jeremiah Clemens of Alabama when he denounced "the wretched silk-worms who, in peaceful times, earn a cheap reputation for patriotism by professing unbounded love for the Union," or such as Robert Toombs employed when he asserted that the political equality of the South was "worth a thousand such unions as we have," and spoke of swearing his children "to eternal hostility to your foul domination."[9]

As matters turned out, both Clemens and Toombs decided to support the Compromise, but many other southerners agreed with Calhoun that Clay's measures were superficial. They believed that the South must ultimately choose between the Federal Union and the institution of slavery.[10] Some were therefore ready to reject the

---

North Carolina (Chapel Hill, 1939), pp. 42, 48, 93; Avery O. Craven, *The Coming of the Civil War* (New York, 1942), p. 262; Craven, *Growth*, p. 47.

8. Stephens to Linton Stephens, Dec. 5, 1849, in Richard Malcolm Johnston and William Hand Browne, *Life of Alexander H. Stephens* (Philadelphia, 1883), p. 239; Pettigrew, Jan. 8, 1850, in Pettigrew Papers, quoted in Sitterson, *Secession Movement in North Carolina*, p. 55. In May 1851, James L. Orr told a South Carolina audience, "Five years ago disunion would not have been tolerated in South Carolina, but now there is not one union man in this vast assembly," Nevins, *Ordeal*, I, 372, quoting Columbia *Transcript*, May 31, 1851.

9. *Congressional Globe*, 31 Cong., 1 sess., p. 1216 (Toombs); appendix, pp. 52–54 (Clemens).

10. James M. Mason said, "this pseudo compromise . . . will . . . be found fatal either to the Union . . . or to the institution of slavery." Virginia Mason, *The Public Life and Diplomatic Correspondence of James M. Mason* (New York, 1906), pp. 84–85. Alexander H. Stephens wrote to his brother, Jan. 21, 1850, "The present crisis may pass, the present adjustment may be made, but the great question of the permanence of slavery in the Southern states will be far from being settled thereby. And in my opinion the crisis of that question is not far ahead." Johnston and Browne, *Life of Stephens*, p. 244. The Columbus, Georgia, *Sentinel* said in an editorial, "A momentary quiet has hushed the voice of agitation, but there is no peace. There can be none

Compromise and press for secession. In the states of Georgia, Mississippi, and South Carolina, these secessionist groups were formidable, and the first aftermath of the Compromise was a fierce struggle between them and the unionists.

The secessionists prepared to strike without delay. On September 23, 1850, Governor George W. Towns of Georgia, acting under instructions adopted earlier by the state legislature, called for the election in November of a special state convention to meet in December. Late in September, Governor John A. Quitman of Mississippi called his state legislature into a special session in November. Meanwhile, Governor Whitemarsh B. Seabrook of South Carolina refrained from calling a special legislative session only because Towns had cautioned him that undue haste on his part might arouse Georgia's traditional fear of South Carolina's extremism, but the legislature had a regular session in November in any case, and few people doubted that South Carolina was ready to go as far as the farthest. In fact, all three governors made it clear that they expected their respective states to secede.[11]

If Mississippi and South Carolina had moved first, it is likely that

---

as long as slaveholders and abolitionists live under a common government." Quoted in Ulrich Bonnell Phillips, *The Life of Robert Toombs* (New York, 1913), p. 102.

11. On the struggle between the Southern Rights party and the Unionists in Georgia, see Richard Harrison Shryock, *Georgia and the Union in 1850* (Durham, N.C., 1926), pp. 295–363; Phillips, *Life of Toombs*, pp. 89–115; Percy Scott Flippin, *Herschel V. Johnson of Georgia, State Rights Unionist* (Richmond, 1931), pp. 33–53; Horace Montgomery, "The Crisis of 1850 and Its Effect on Political Parties in Georgia," *Georgia Historical Quarterly*, XXIV (1940), 293–322; Montgomery, *Cracker Parties* (Baton Rouge, 1950), pp. 19–71; James Z. Rabun, "Alexander H. Stephens, 1812–1861" (Ph.D. dissertation, University of Chicago, 1948); Helen I. Greene, "Politics in Georgia, 1830–1854" (Ph.D. dissertation, University of Chicago, 1946); John T. Hubbell, "Three Georgia Unionists and the Compromise of 1850," *Georgia Historical Quarterly*, LI (1967), 307–323; William Y. Thompson, *Robert Toombs of Georgia* (Baton Rouge, 1966), pp. 71–76. For Mississippi: Hearon, "Mississippi and the Compromise of 1850," pp. 148–227; J. F. H. Claiborne, *Life and Correspondence of John A. Quitman* (2 vols.; New York, 1860), II, 114–185; James Byrne Ranck, *Albert Gallatin Brown, Radical Southern Nationalist* (New York, 1937), pp. 74–100. For South Carolina: Chauncey S. Boucher, "The Secession and Cooperation Movement in South Carolina, 1848 to 1852," Washington University *Studies*, V (1918), 92–138; Philip May Hamer, *The Secession Movement in South Carolina, 1847–1852* (Allentown, Pa., 1918), pp. 62–143; N[athaniel] W. Stephenson, "Southern Nationalism in South Carolina in 1851," *AHR*, XXXVI (1931), 314–335; Laura A. White, *Robert Barnwell Rhett, Father of Secession* (New York, 1931), pp. 103–134; Lillian Adele Kibler, *Benjamin F. Perry, South Carolina Unionist* (Durham, N.C., 1946), pp. 239–277; Harold S. Schultz, *Nationalism and Sectionalism in South Carolina, 1852–1860* (Durham, N.C., 1950), pp. 19–51. See also the more general treatments in Craven, *Growth*, pp. 83–141; Nevins, *Ordeal*, I, 354–379; Cole, *Whig Party in the South*, pp. 174–244; Melvin J. White, *The Secession Movement in the United States, 1847–1852* (New Orleans, 1916).

secession would have been adopted in both states,[12] but the sequence of events completely frustrated the secessionists. First of all, the second session of the Nashville Convention, which had agreed in June to meet again after the adjournment of Congress, assembled on November 11. At once it became evident that the unionists were boycotting the convention, for only fifty-nine irregularly chosen delegates from seven states appeared, whereas more than a hundred delegates from nine states had attended originally. This left the secessionists in control, but their sense of futility was such that, although they passed resolutions denouncing the Compromise and asserting the right of secession, they did not recommend any action except southern abstention from the national party conventions and the scheduling of another southern convention.[13]

Less than a week after they adjourned, Georgia elected delegates to the state convention. Toombs and Stephens, two key leaders in this strategic state, had flirted with disunionism while Taylor was in the White House but then had turned back to support of the Union after enactment of the Compromise. Together with Howell Cobb, they now hastily organized a Whig-Democratic coalition of "Constitutional Unionists" to oppose the Southern Rights Democrats, led by Governor Towns and Herschel V. Johnson. The powerful Toombs-Stephens-Cobb combination enjoyed the advantage of high cotton prices and widespread prosperity. There was too much contentment for a secession movement to take root. In the election, they defeated the secessionists by a smashing majority of 46,000 to 24,000, which meant Unionist domination of the convention.

The election in Georgia fell on the same day that the South Carolina legislature convened and precisely one week after the opening session of the Mississippi legislature. Thus, in both these states, the secessionists found that they had not acted quickly enough. Events at Nashville and in Georgia had shown that there was not a united South ready to secede, and already the issue before them had undergone a subtle but decisive transformation. Two months earlier, their choice had appeared to be one between North

12. Seabrook to Quitman, Sept. 20, Oct. 23, 1850, quoted in Nevins, *Ordeal*, I, 363.

13. On the second meeting of the Nashville Convention, see citations in chap. 5, note 25, above. Also, Hearon, "Mississippi and the Compromise of 1850," p. 175; *National Intelligencer*, Nov. 28, 1850; *Journal of the Proceedings of the Southern Convention, at Its Adjourned Session, Held at Nashville, Tenn., Nov. 11, 1850 and Subsequent Days* (Nashville, 1850).

and South, between the alternatives of submission and resistance. But now, the secessionists had to choose either deferring their action in the hope of later "cooperation" with other southern states or acting immediately and alone in the one or two states which they could control. Both courses presented hazards: "cooperation" might immobilize them and rob them of their initiative; separate state action might divide the South and alienate them from one another. Because of this dilemma, they found their unity gone and their ranks divided into two hostile factions—one of "separate state actionists," who were denounced by their opponents as revolutionists ready to destroy the unity of the South; the other of "cooperationists," who were in turn branded "submissionists," afraid to resist northern subjugation of the South. Instead of secessionists opposing unionists, immediate secessionists were now arrayed against cooperative secessionists, while the latter received support from unionist allies.[14]

As a result, the secession program faltered. In December, the state actionists in the South Carolina legislature attempted to authorize a state convention, but the cooperationists defeated this proposal. The deadlock that followed was broken only by a compromise calling for both a southern "congress" *and* a state convention. The convention delegates were to be elected in February 1851, but the governor could not convene them until the meeting of the southern congress was assured, and delegates to the congress would not be elected until the following October! In the Mississippi legislature, the state actionists were stronger, and they forced through an act providing for a state convention in November. The opposition in Mississippi consisted more of unionists and less of cooperative secessionists than in South Carolina. This group, though overridden at almost every point, did manage to defeat a bill giving the governor discretion to call the convention into session at an earlier date.

In February, South Carolina elected delegates to a state convention, and the separate state actionists won by a wide margin, but with the balloting so light that it indicated a pronounced apathy among the voters. As late as May, the governor of South Carolina

14. In addition to the works cited in note 11, see *Debates and Proceedings of the Georgia Convention, 1850* (Milledgeville, 1850); *Journals of the Conventions of the People of South Carolina held in 1832, 1833, and 1852* (Columbia, 1860); *Journal of the Convention of the State of Mississippi and the Act Calling the Same, 1851* (Jackson, 1851).

said, "There is now not the slightest doubt but that . . . the state will secede." But in fact, the secessionists had surrendered the initiative and lost their opportunity. Setting the date for action nearly twelve months away allowed time for tempers to cool and for opponents to organize.

Even earlier, the reaction had already set in. On December 14 the Georgia convention approved resolutions which, as the "Georgia Platform," became the cornerstone of southern policy for several years. These resolutions began by declaring that Georgia did "not wholly approve" of the Compromise, but that she would "abide by it as a permanent adjustment of this sectional controversy." They then proceeded to state categorically the basis on which Georgia would remain in the Union:

> The state of Georgia will and ought to resist even (as a last resort) to the disruption of every tie that binds her to the Union, any action of Congress upon the subject of slavery in the District of Columbia, or in places subject to the jurisdiction of Congress incompatible with the safety and domestic tranquility, the rights and honor of the slave-holding states, or any refusal to admit as a state any territory hereafter applying, because of the existence of slavery therein, or any act, prohibiting the introduction of slaves into the territories of Utah and New Mexico, or any act repealing or materially modifying the laws now in force for the recovery of fugitive slaves.
>
> It is the deliberate opinion of this Convention that upon a faithful execution of the *Fugitive Slave Law* . . . depends the preservation of our much beloved Union.[15]

The Georgia Platform epitomized the attitude of the great majority of southerners in 1850. They still cherished their "beloved Union" and would not part from it lightly. They did not quite like the Compromise, especially the admission of California, but they appreciated the fact that the Compromise had buried the Wilmot Proviso and therefore they would abide by it. Their acquiescence, however, was emphatically conditional and not absolute; they would resist any future step which endangered what they regarded as the safety, the rights, or the honor of the slave states. And, far from renouncing the right of secession, they stated plainly that if the conditions which they had laid down should be violated, they would resist "even to the disruption of every tie binding Georgia to the Union."

By its skillful fusion of the two principles of Unionism and of

15. *Debates and Proceedings of the Georgia Convention, 1850*, pp. 7–9.

Southern Rights, both still dear to the citizens of the South, the Georgia Platform placed the immediate secessionists in an almost untenable position. The events of 1851 showed just how untenable it was, for in the four key states of Mississippi, Alabama, Georgia, and even South Carolina, the secessionists suffered staggering reverses. In the fall of 1851, Georgians ratified the Georgia Platform by giving the Unionist candidate for the governorship an 18,000 majority over the Southern Rights candidate. Alabama elected Unionist congressmen after a spirited campaign in which William L. Yancey threw all his oratorical talents into a campaign for "Southern Rights." Mississippi elected as governor the Unionist Henry S. Foote over the Southern Rights candidate, Jefferson Davis. The Mississippi convention that met in January 1852 voted to accept the Compromise. It listed certain actions to be resisted as constituting "intolerable oppression" but also declared that secession was "utterly unsanctioned by the Federal Constitution." In South Carolina at the same time, the cooperationists defeated the separate state actionists, 25,000 to 17,000. This was the election of delegates to the southern congress, which, it was now clear, would never meet, but both factions in South Carolina had agreed to treat the election as a plebiscite and to abide by the results. The state actionists carried out this agreement in good faith when the state convention, in which they held a majority, finally met in April 1852. They simply declared that federal violations of the rights of South Carolina justified secession and that the state refrained from acting accordingly "from considerations of expediency only." Thereupon, Robert Barnwell Rhett, who was as fanatical in his personal integrity as he was in his devotion to states' rights, resigned the seat in the Senate to which he had been elected after the death of Calhoun. Nothing could have more dramatically symbolized the fact that the first concerted attempt to take the South out of the Union had failed. This did not mean that the South had accepted either the finality of the Compromise or the permanence of the Union. Rather, it had accepted the Union if the Compromise was in fact final.[16]

But to many people at a distance, the only clearly evident fact was

16. See works cited in note 11, and also, for Alabama, Lewy Dorman, *Party Politics in Alabama from 1850 through 1860* (Wetumpka, Ala., 1935), pp. 65–76; John Witherspoon Du Bose, *The Life and Times of William Lowndes Yancey* (Birmingham, Ala., 1892); G. F. Mellen, "Henry W. Hilliard and William L. Yancey," *Sewanee Review*, XVII (1909), 32–50; Clarence P. Denman, *The Secession Movement in Alabama* (Montgomery, Ala., 1933), pp. 45–64.

that all the noise of the secession movement throughout the South had resulted in even less action than South Carolina alone had provided in 1832. Accordingly, many northerners formed a stereotyped conclusion: the talk of secession was "gasconade"; no one actually intended to secede; the only real purpose was to frighten northern "Union-lovers" into making concessions. This belief became a fixed idea, especially among Republicans, and was to play an important part in nourishing northern illusions that the situation was not serious, when one by one the southern states began to secede a decade later.

Perhaps nothing will reveal the futility of the Compromise of 1850 so much as a simple recognition of the strange way in which it was expected to achieve its end. The purpose was to put a stop to the agitation of the slavery question. But to accomplish this, the compromisers adopted a law to activate the recapture of fugitive slaves. Here was a firebrand vastly more inflammatory than the Wilmot Proviso. The Proviso had dealt with a hypothetical slave who might never materialize; the Fugitive Slave Act, on the contrary, dealt with hundreds of flesh-and-blood people who had risked their lives to gain their liberty, and who might now be tracked down by slave-catchers. The Proviso had related to a remote, unpeopled region beyond the wide Missouri; the Fugitive Slave Act was concerned with men and women in the back streets of New York, Philadelphia, Boston, and many a town and hamlet. The Proviso turned upon an abstruse constitutional question, but the Fugitive Slave Act involved an issue with immense emotional impact. No dramatic image ever brought the Proviso to life as Eliza's flight across the ice of the Ohio River brought to life the plight of human beings who had escaped from slavery. Yet in a supreme effort to avert the dangers of the Proviso and restore sectional harmony, the wise men of 1850 enacted a law for the rendition of fugitive slaves.[17]

Any measure which required the sending of men from freedom into slavery would have caused a strong revulsion at best, but the Fugitive Slave Law, as enacted, contained a number of gratuitously

17. Although the Fugitive Slave Act passed after a surprisingly small amount of debate, the pattern of voting gave some indication of the storm to follow. Among northern members of Congress, only about one in five voted for the measure (34 out of 154). Thirty-two absented themselves, including those northern champions of the Compromise, Stephen A. Douglas and Lewis Cass.

obnoxious provisions. First, it denied the alleged fugitive any right to jury trial, not even guaranteeing it in the jurisdiction from which he had escaped. Second, it permitted his case to be removed from the ordinary judicial tribunals and tried before a commissioner appointed by the courts. Third, it allowed the commissioner a $10 fee in cases in which the alleged fugitive was delivered to the claimant, but only a $5 fee in cases when he was set free. Finally, it empowered federal marshals to summon all citizens to aid in enforcement of the Act.[18] In the eyes of many northerners this meant that the federal government had not only gone into the business of man-hunting itself but also required every freeborn American to become a man-hunter on occasion.

To appreciate the full impact of this measure, one must recognize that it was far more than a law to overtake slaves in the act of running away. It was also a device to recover slaves who had run away in the past. Thus, under the Act, many cases arose involving Negroes who had lived as peaceful residents of free-state communities for many years. For example, in February 1851, in Madison, Indiana, a Negro named Mitchum was torn from his wife and children and delivered to a person from whom it was claimed that he had run away nineteen years before.[19] Moreover, the law left all free Negroes with inadequate safeguards against claims that they were fugitives, and it exposed them to the danger of kidnapping. For years, this danger of being dragged away into slavery had made the life of the free Negro precarious, and it was undoubtedly accentuated by the new law. Many Negroes were seized and carried off forcibly, without any judicial process, and in one case, after the

18. *U.S. Statutes at Large,* IX, 462–465. The stated reason for the difference in fees was that remanding an alleged fugitive required more paper work than freeing him. The major published study of the Fugitive Slave Law is Stanley W. Campbell, *The Slave Catchers: Enforcement of the Fugitive Slave Law, 1850–1860* (Chapel Hill, 1968), which includes chapters on the legislative history and the question of constitutionality. For a defense of the Act's constitutionality (criticized by Campbell, p. 41), see Allen Johnson, "The Constitutionality of the Fugitive Slave Acts," *Yale Law Journal,* XXXI (1921), 161–182.

19. Samuel J. May, *The Fugitive Slave Law and Its Victims* (rev. ed.; New York, 1861), p. 15. At about the same time in Philadelphia, one Euphemia Williams was claimed as a slave who had run away twenty-two years earlier. Her six children, all born in freedom, were also claimed, but the tribunal declared her not to be the fugitive described. *Ibid.*, p. 14, but Campbell, *Slave Catchers,* p. 199, says that Mitchum was remanded at Vernon, Indiana, on March 7, 1851, and that the woman's name was Tamar Williams.

authorities had delivered up a free Negro as a fugitive, he was saved only by the fact that the claimant, upon receiving this man, truthfully admitted that he was not the slave who had run away.[20]

Cases of mistaken identity and other injustices, added to the basic reality that even an undoubted slave in the overt act of running away was a pitiable figure, inspired a great revulsion in the North against the law. The abolitionists instantly recognized its propaganda value and focused all the energies of the antislavery organization upon the fugitive question. From press, pulpit, and rostrum, a storm of denunciation burst forth. The abolitionists, with their New England Puritan antecedents, were the heirs of a long tradition of rich pulpit invective against evil. Generations spent in excoriating the "Whore of Babylon" had given this style of discourse a deep Jehovah-like tone, which was now turned with full force not only upon the Fugitive Slave Act but upon everyone who supported it. Webster was called a "monster," "indescribably base and wicked," the "personification of all that is vile," a "fallen angel" who would receive the curses of posterity upon his grave, an "infamous New Hampshire renegade." Garrison's *Liberator,* in a lapse from this lofty wrath, accused him of keeping a harem of "big black wenches as ugly and vulgar as Webster himself." As for Fillmore, it were "better that he had never been born." George T. Curtis, who had accepted an appointment as commissioner, was "a Nero, a Torquemada." The act itself was to Theodore Parker "a hateful statute of kidnappers," to Emerson "a filthy law," to the Quincy, Illinois, *Whig* "an outrage to humanity." Anyone who obeyed it was "devoid of humanity" and ought to be "marked and treated as a moral leper"; anyone who had "even dreamed of obeying it" should "repent before God and ask His forgiveness." It was the duty of every citizen "to trample the law in the dust" and to see that it was "resisted, disobeyed at all hazards."[21] While the abolitionists vied to outdo

20. May, *Fugitive Slave Law,* describes some sixty cases of kidnapping of free Negroes.

21. For the denunciation of the Fugitive Slave Law, see Campbell, *Slave Catchers,* pp. 49–54; Craven, *Growth,* pp. 146–149; Nevins, *Ordeal,* I, 380–387; Ralph Volney Harlow, *Gerrit Smith, Philanthropist and Reformer* (New York, 1939), pp. 289–305; Claude M. Fuess, "Daniel Webster and The Abolitionists," Massachusetts Historical Society *Proceedings,* LXIV (1930–32), 28–42. Wendell Phillips, who had already raised defamation to an art, now exceeded himself: Irving H. Bartlett, "Wendell Phillips and the Eloquence of Abuse," *American Quarterly,* XI (1959), 509–520. But even Phillips did not match Theodore Parker in the grandiosity of his vituperation: Henry Steele

one another in this vein, it was perhaps even more ominous that moderate leaders like Edward Everett and Robert Rantoul expressed strong opposition to the law, as well as a conviction that it could not be enforced.[22]

It was no wonder, then, that when southern slaveowners sent their agents north to reclaim fugitives, trouble ensued. Within a month after the law was enacted, Negroes were claimed as slaves at New York, Philadelphia, Harrisburg, Detroit, and elsewhere. This caused panic among the blacks in many northern communities. Fugitives who feared capture and legally free Negroes who feared kidnapping wanted desperately to get beyond the reach of this new act, and as a result several thousand eventually fled over the northern border to Canada. Many of these refugees later returned to the United States, but to this day, Ontario has a small Negro population descended from the émigrés of 1850.[23]

Meanwhile, the abolitionists resolved that the enforcement of the law should not be permitted. In Boston—a city from which it was boasted that no fugitive had ever been returned[24]—violation was open and organized, led by Theodore Parker and other members of the city's elite. As early as October 1850, Parker's standing vigilance committee smuggled away two undoubted slaves—William and Ellen Craft—whom a jailer from Macon, Georgia, had come to claim. Also they intimidated the jailer himself so thoroughly that he fled from the city. Four months later a crowd of Negroes seized another fugitive, Shadrach, from the deputy marshal and spirited him away to Canada. Finally, in April 1851, the federal authorities succeeded in enforcing the law in Boston, when they secured the return of the slave Thomas Sims to his master. But this result was accomplished only at a cost of $5,000 and by making a vigorous

---

Commager, *Theodore Parker* (Boston, 1936), pp. 197–247; Octavius Brooks Frothingham, *Theodore Parker* (Boston, 1874), pp. 399–434.

22. Everett to Robert C. Winthrop, March 21, 1850, quoted in Nevins, *Ordeal*, I, 380–381; Luther Hamilton (ed.), *Memoirs, Speeches and Writings of Robert Rantoul, Jr.* (Boston, 1854), p. 744.

23. Fred Landon, "The Negro Migration to Canada after the Passing of the Fugitive Slave Act," *JNH*, V (1920), 22–36; eight other articles by Landon on this topic are cited in Robin W. Winks, *Canada and the United States: The Civil War Years* (Baltimore, 1960), pp. 8–10, footnotes; Benjamin Drew, *The Refugee, or The Narratives of Fugitive Slaves in Canada*, sometimes titled *A Northside View of Slavery* (Boston, 1856); Samuel Gridley Howe, *The Refugees from Slavery in Canada West* (Boston, 1864).

24. John Weiss, *Life and Correspondence of Theodore Parker* (2 vols.; New York, 1864), II, 107.

show of force and sending Sims out of the city at four o'clock in the morning. After that, the law was never but once again enforced in Boston, when Anthony Burns was sent back to Georgia in 1854.[25] Meanwhile in other cities, similar acts of vigilantism and violation occurred. In Detroit, it required military force to prevent the rescue of an alleged fugitive by a mob in October 1850. In September 1851, at Christiana, Pennsylvania, a slaveowner was killed in a shooting affray with a crowd of Negroes who were determined to prevent him from capturing a fugitive. In October, at Syracuse, New York, a mob of more than two thousand people broke into the courthouse and forcibly took a fugitive, Jerry McHenry, away from officers who had him in custody. In 1854, sympathizers broke down a jail door at Milwaukee and rescued Joshua Glover, an alleged fugitive.[26]

Along with these dramatic episodes of public resistance to the law, it appears that there was also an increase in the organized activity of private individuals who were helping fugitives to escape. There is no doubt that, as early as the late eighteenth century,

25. Campbell, *Slave Catchers*, pp. 117–121, 124–132, 148–151; Commager, *Theodore Parker*, pp. 214–247; Leonard W. Levy, "The 'Abolition Riot': Boston's First Slave Rescue," *NEQ*, XXV (1952), 85–92 (a case in 1836); Levy, "Sims' Case: The Fugitive Slave Case in Boston, 1851," *JNH*, XXXV (1950), 39–74; Levy, *The Law of the Commonwealth and Chief Justice Shaw* (Cambridge, Mass., 1957), pp. 72–108; Harold Schwartz, "Fugitive Slave Days in Boston," *NEQ*, XXVII (1954), 191–212; Schwartz, *Samuel Gridley Howe, Social Reformer, 1801–1876* (Cambridge, Mass., 1956), pp. 177–194; Samuel Shapiro, "The Rendition of Anthony Burns," *JNH*, XLIV (1959), 34–51; Shapiro, *Richard Henry Dana, Jr.* (East Lansing, Mich., 1961), pp. 58–66, 84–93. These authorities will refer the reader to many significant titles, especially of a memoir nature, by or about Parker, Howe, Dana, Bronson Alcott, Austin Bearse, Henry Ingersoll Bowditch, Lydia Maria Child, James Freeman Clarke, Moncure D. Conway, Ezra Stiles Gannett, William Lloyd Garrison, Thomas Wentworth Higginson, Julia Ward Howe, James Russell Lowell, Henry W. Longfellow, Horace Mann, Samuel J. May, Wendell Phillips, Franklin B. Sanborn, John Greenleaf Whittier, Elizir Wright, and others. A vivid, somewhat popular account is in Lawrence Lader, *The Bold Brahmins: New England's War Against Slavery, 1830–1863* (New York, 1961), pp. 155–185. Larry Gara, *The Liberty Line: The Legend of the Underground Railroad* (Lexington, Ky., 1961), pp. 106–109, and Louis Filler, *The Crusade Against Slavery, 1830–1860* (New York, 1960), offer illuminating insights.

26. Campbell, *Slave Catchers*, pp. 151–161; William Uhler Hensel, *The Christiana Riot and the Treason Trials of 1851* (Lancaster, Pa., 1911); W. Freeman Galpin, "The Jerry Rescue," *NYH*, XIII (1945), 19–34; Joseph Schafer, "Stormy Days in Court—The Booth Case," *Wisconsin Magazine of History*, XX (1936), 89–110; Vroman Mason, "The Fugitive Slave Law in Wisconsin, with Reference to Nullification Sentiment," State Historical Society of Wisconsin *Proceedings*, 1895, pp. 117–144. The special difficulty of enforcement in Ohio is shown in William C. Cochran, *The Western Reserve and the Fugitive Slave Law* (Cleveland, 1920).

Quakers in Pennsylvania had protected fugitives, and after the adoption of the Fugitive Slave Act of 1793 there was always a certain amount of assistance to refugees. How extensive this aid was, and how systematically organized, it is difficult to say. Without question, there were families ready to assist fugitives by lodging them, feeding them, concealing them if need arose, and directing or even escorting them on their way to other families who would do likewise. Such families sometimes became known to Negroes mostly through the grapevine system of intelligence. As a result, some fugitives escaped by a planned journey from one point of assistance to another, over strategic routes leading to safety in the North or in Canada.

Sometime prior to 1842, this apparatus became known as the "underground railroad," and under this name it took its place as part of what a recent writer has called "the anti-slavery myth." As the "myth" grew, the underground railroad came to be remembered as a vast and highly articulated network, with a "president," Levi Coffin, who had himself allegedly helped rescue two thousand slaves; a hierarchy of managers, conductors, stationkeepers, and agents; a complex of well-defined routes and "switch connections"; and an elaborate system of mysterious disguises, stratagems, and concealments. Through this apparatus, it was claimed, some three thousand operators aided in the escape of more than fifty thousand slaves in the period from 1830 to 1860.[27]

No accurate information existed to curb romantic imaginations, and partisans on both sides felt impelled to exaggerate the underground operation—the slaveholders to magnify their loss of property, the abolitionists to magnify their effectiveness in combating slavery. Consequently, extravagant estimates of the number of fugi-

27. Wilbur H. Siebert, *The Underground Railroad from Slavery to Freedom* (New York, 1898), presents the traditional history. Outstanding among the many writings which built up the tradition were Levi Coffin, *Reminiscences of Levi Coffin, the Reputed President of the Underground Railroad* (Cincinnati, 1876), and William Still, Negro secretary of the Philadelphia Vigilance Committee, *The Underground Railroad* (Philadelphia, 1886). Two recent works which uncritically follow the traditional version are William Breyfogle, *Make Free: The Story of the Underground Railroad* (Philadelphia, 1958), and Henrietta Buckmaster, *Let My People Go: The Story of the Underground Railroad and the Growth of the Abolition Movement* (New York, 1941). Siebert, pp. 346, 351, 403–439, estimated 3,000 operators, and though he gave no overall estimate of the number of fugitives who received aid, he did estimate 40,000 for Ohio alone and 9,000 for Philadelphia alone. Albert Bushnell Hart, *Slavery and Abolition, 1831–1841* (New York, 1906), p. 230, stated that for thirty years, about 2,000 slaves escaped annually, of whom perhaps 10 percent remained in the South and 10 percent went to Canada.

tives issued from both sides. Governor Quitman of Mississippi asserted that 100,000 of the South's slaves had been abducted in a forty-year period. At the other end of the axis, Josiah Henson, himself a fugitive slave and a runner of fugitives, reported in 1852 that there were 50,000 fugitives in the free states, while the Anti-Slavery Society of Canada, in the same year, estimated 30,000 fugitives north of the border.[28]

But comparison of these estimates with the census reports produces some puzzling anomalies. One of them arises directly from the census data on the number of slaves who ran away annually, reported, for instance, as 1,011 for 1850 and 803 for 1860. These figures included many runaways who did not reach the North and did not receive aid from the underground. But a more difficult question arises from the census returns of Negroes in the free states. Instead of increasing rapidly, as their numbers should have done if their ranks were being swelled by vast accretions of fugitives, free Negroes in the free states increased at a lower rate than the white population or the Negro slave population of the country as a whole, and scarcely more rapidly than the free Negro population in the slave states. It is also curious that, although the fugitive slaves were overwhelmingly males and a heavy influx of fugitives should have produced a preponderance of males in the population, the census figures for New York, for instance, showed more female than male Negroes. In Canada, likewise, there are glaring discrepancies between the claims of the abolitionists and the reports of the census. Antislavery men asserted that between 15,000 and 20,000 Negro fugitives fled to Canada, and almost wholly to Ontario, between 1850 and 1860. But the Canadian census indicated 5,469 Negroes in Upper Canada in 1848, increasing to about 8,000 in 1852, and to 11,223 in 1860. Even if there had been no natural increase at all, this would suggest either that the fugitives soon returned or that the migration was less than 6,000.[29]

28. Claiborne, *Life of Quitman*, II, 28; *The Life of Josiah Henson, Formerly a Slave, as Narrated by Himself* (Boston, 1849), p. 97; *Anti-Slavery Bugle*, April 10, 1852.

29. Between 1820 and 1850, the number of whites in the United States increased 148 percent; slaves, 108 percent; free Negroes, 86 percent (98 percent in the free states and 78 percent in the slave states). The increase of free Negroes in New England and the middle states was 10 percent and 70 percent respectively. Census data compiled from *Compendium of the Seventh Census* (Washington, 1854), p. 65; U.S. Census Office, *Preliminary Report of the Eighth Census, 1860* (Washington, 1862), pp. 11, 12, 131; *A Century of Population Growth* (Washington, 1909), pp. 222–223, showing free Negro population by states, 1820–1850; *First Report of the Secretary of* [Canada]

These data raise a fundamental question. Was the underground railroad really a large-scale organization, actually operating to facilitate the mass escape of fugitive slaves, or was it not rather a gigantic propaganda device, more significant psychologically than as an institution? Certainly it arose in a propaganda context. Originally designed to dramatize the fugitive slave issue and to compensate emotionally for the lack of channels of action through which the abolitionists could implement their strong feelings against slavery, it later became a legend to glorify not only a few men who had incurred danger to resist slavery but also a good many more who later wished that they had done so, and who translated their wishes into an epic of heroic adventure.

The historian must not be too impatient with the popular yearning to find drama in the past and to fabricate it where it is lacking, but he may well regret some of the side effects. One of the regrettable aspects of the legend of the underground railroad is that, while exalting the role of abolitionists, who seldom risked a great deal, it has drawn attention away from the heroism of the fugitives themselves, who often staked their lives against incredible odds, with nothing to aid them but their own nerve and the North Star. If anyone helped them, the evidence indicates that it was more likely to be another Negro, slave or free, who chose to take heavy risks, than a benevolent abolitionist with secret passages, sliding panels, and other stage properties of organized escape.[30]

Another unfortunate side effect has been the overlooking of sub-

---

Board of Registration and Statistics of the Census of the Canadas for 1851–52 (Quebec, 1853), pp. 36–37, 317; *Census of the Canadas, 1860–1861*, (Quebec, 1863), I, 79; *Census* [of Canada] *of 1871* (Quebec, 1873), I, 332; IV, 169; V, 18. The Canadian statistics give certain apparently discrepant figures, but none of them indicate the heavy influx of Negroes which has been claimed by writers on the fugitive slave question. Also, see Winks, *Canada and the United States*, pp. 7–11. For the ratio of males to females in the northern Negro population and its implications, see Gara, *Liberty Line*, pp. 38–39, and for additional discussion of census data, pp. 37–40.

30. Although Channing, Nevins, and other historians have discounted the claims of a vast underground organization, Gara, *The Liberty Line*, is the first study to explore the underground railroad fully as legend and folklore. For a notable discussion, see C. Vann Woodward, "The Antislavery Myth," *American Scholar*, XXXI (1962), 312–318. Prior to Gara, J. C. Furnas had taken a fresh look at the fugitive question in *Goodbye to Uncle Tom* (New York, 1956). An important branch of abolitionist literature, the many narratives purporting to be written by escaped slaves, is analyzed in Charles H. Nichols, *Many Thousand Gone: The Ex-Slaves' Account of Their Bondage and Freedom* (Leiden, 1963); also Marion W. Starling, "The Slave Narrative: Its Place in American Literary History" (Ph.D. dissertation, New York University, 1946); Margaret Y. Jackson, "An Investigation of Biographies and Autobiographies of American Slaves Published Between 1840 and 1860" (Ph.D. dissertation, Cornell University, 1954).

stantial evidence that a considerable segment of northern opinion was willing, for the sake of the Compromise, to accept even the Fugitive Slave Act. This is not to say that the North usually complied with the law, for it must be remembered that it was impossible to secure convictions of members of mobs who had taken fugitives from custody, that slaves were often recovered only at great public expense and with major display of official force, and that the law was defied primarily by spiriting slaves away before officers found them, rather than by resisting officers directly. Yet, the picture of overwhelming northern defiance must be qualified. There were, after all, in the first six years of the law, only three cases of forcible and successful rescue. During the same time, it is estimated that two hundred Negroes were arrested. Perhaps a third of these were taken back to the South without trial, while of the remaining two-thirds, eight were released, twelve were rescued, and the rest were remanded to slavery. In February 1851, Henry Clay asserted that the law was being enforced without any uproar in Indiana, Ohio, Pennsylvania, New York City, and everywhere except at Boston.[31]

While conservatives complained of the extent to which the law was resisted, antislavery men deplored the extent of public acquiescence. Their literature teemed with protests against the apathy with which the people tolerated brutality and indecent haste in the enforcement of the law. Although many clergymen condemned the measure, one prominent antislavery clergyman complained bitterly that, among the thirty thousand ministers of all denominations in the United States, not one in a hundred spoke out against it. Among the conservative and propertied class, in New York, Boston, and elsewhere, a vigorous sentiment of support for the law itself and for Daniel Webster as its sponsor was strongly in evidence.[32]

Conservative feeling showed up especially in the reaction of northern state legislatures. Long before passage of the Fugitive Slave Law of 1850, the Supreme Court had ruled in the case of *Prigg v. Pennsylvania* (1842) that the obligation of enforcement of the fugitive-slave clause of the Constitution was essentially federal, and

31. See the running debate in the Senate, Feb. 21–24, 1851, *Congressional Globe*, 31 Cong., 2 sess., appendix, pp. 292–326. On enforcement of the Act generally, see Campbell, *Slave Catchers*, pp. 110–147, 199–207.

32. Samuel J. May, *Some Recollections of Our Anti-Slavery Conflict* (Boston, 1869), pp. 349–373; David D. Van Tassel, "Gentlemen of Property and Standing: Compromise Sentiment in Boston in 1850," *NEQ*, XXIII (1950), 307–319; Nevins, *Ordeal*, I, 396–404; Campbell, *Slave Catchers*, pp. 63–79.

that the states need not devote their law-enforcing apparatus to this function. As a result, several states had enacted measures which became known as "personal liberty laws," by which they either forbade state officials to participate in the enforcement of the law or prohibited the use of their jails in fugitive slave cases. These laws had been one of the impelling causes which prompted the South to demand the enactment of a new federal law in 1850 to replace the earlier act of 1793, and the new law had carefully avoided any attempt to employ state officials in its enforcement. The question now arose whether the states would adopt new personal liberty laws to obstruct the enforcement of the new act. Ultimately nine northern states did adopt a new series of personal liberty laws, but it is a matter of considerable significance that only one of them did so within the first four years after the enactment of the "hated" law of 1850. Vermont in 1850 guaranteed jury trial to alleged fugitives, but other states waited until after the Kansas-Nebraska Act had reopened the political warfare between the sections in 1854.[33] In short, although personal liberty laws were a recognized antislavery device in the North, only one state adopted such a law during the furor that followed the Fugitive Slave Act of 1850. This does not mean that the northern public approved of the Fugitive Slave Act, nor that it did not entertain strong antislavery feelings, for at this time the state legislatures of Ohio, Massachusetts, and New York sent three new members to the Senate who opposed slavery most vigorously: Benjamin F. Wade, Charles Sumner, and Hamilton Fish. But it does mean that the public was not willing to tamper with the Compromise, as Sumner soon learned. During his first session he forced a vote in the Senate on the repeal of the Fugitive Act. It received only four votes in favor—those of Hale, Wade, Chase, and Sumner himself—while even such militant foes of slavery as Seward and Fish would not support it.[34] There is no convincing evidence that a preponderant majority in the North were prepared to violate or nullify the law. If, in concrete instances, they would shield a fugitive from his pursuers, this was more a matter of pity than of policy; it was something that also happened occasionally even in the South.[35]

33. *Ibid.*, pp. 87–88, 170–186; Norman L. Rosenberg, "Personal Liberty Laws and Sectional Crisis, 1850–1861," *CWH*, XVII (1971), 25–44.

34. *Congressional Globe*, 32 Cong., 1 sess., appendix, pp. 1113–1125; David Donald, *Charles Sumner and the Coming of the Civil War* (New York, 1960), pp. 224–237.

35. Gara, *Liberty Line*, pp. 55–57.

The real significance of the Fugitive Slave Act showed itself less in the philippics of Theodore Parker, and in the spectacular rescues of Shadrach, Jerry McHenry, and Joshua Glover, than in the public reaction to a fictional story of slavery which began to appear serially on June 5, 1851, in the *National Era,* Gamaliel Bailey's abolitionist journal in Washington, D.C. Before she began writing these weekly installments, the author, Harriet Beecher Stowe, had published only some amateurish stories in the "annuals" then in vogue. But *Uncle Tom's Cabin or Life Among the Lowly* was different. Announced to continue for three months, the serial ran away with both its author and its readers for a total of ten. Then it appeared as a book in March 1852 and quickly took the country by storm. In its first year, eight power presses, running simultaneously, turned out more than 300,000 copies to meet the public demand. In August, the story of Uncle Tom began its endless career as America's most popular play. Ultimately the book sold almost 3,000,000 copies in the United States and another 3,500,000 in other parts of the world, thus probably outselling any other single American work.[36]

In almost every respect, *Uncle Tom's Cabin* lacked the standard qualifications for such great literary success. It may plausibly be argued that Mrs. Stowe's characters were impossible and her Negroes were blackface stereotypes, that her plot was sentimental, her dialect absurd, her literary technique crude, and her overall picture of the conditions of slavery distorted. But without any of the vituperation in which the abolitionists were so fluent, and with a sincere though unappreciated effort to avoid blaming the South, she made vivid the plight of the slave as a human being held in bondage. It was perhaps because of the steadiness with which she held this focus that Lord Palmerston, a man notable for his cynicism, admired the book not only for "its story but for the statesmanship of it." History cannot evaluate with precision the influence of a novel upon public opinion, but the northern attitude toward slavery was never quite the same after *Uncle Tom's Cabin.* Men who had remained unmoved by real fugitives wept for Tom under the lash and cheered for Eliza with the bloodhounds on her track.[37]

36. On the popularity of *Uncle Tom's Cabin* see Frank Luther Mott, *Golden Multitudes: The Story of Best Sellers in the United States* (New York, 1947), pp. 114–122.

37. The literature on Harriet Beecher Stowe is copious, but see especially: Edmund Wilson, *Patriotic Gore: Studies in the Literature of the American Civil War* (New York,

Meanwhile, the administration of Millard Fillmore was steadily running its course, and as it did so, public attention shifted toward the next presidential election. The Democrats approached this contest with a confidence born of the fact that they had gained 140 congressional seats out of a total of 233 in the off-year elections of 1850, and they had a number of vigorous contenders for the nomination. From the Northwest, the veteran Lewis Cass of Michigan wanted another chance to retrieve his defeat in 1848. But Cass's northwestern support was contested by a relative newcomer, Stephen A. Douglas of Illinois, only thirty-nine years old but already an experienced and forceful leader. New York's William L. Marcy, formerly secretary of war under Polk and perhaps as talented and qualified as any candidate, was handicapped by the chronic factional feuds among New York Democrats. The South, having no major candidate of its own, despite the aspirations of Sam Houston of Texas and William O. Butler of Kentucky, threw most of its support to James Buchanan of Pennsylvania, Polk's secretary of state. Buchanan, as a "Northern man with Southern principles," had fully earned this backing.[38]

In the Democratic convention at Baltimore in May 1852, Cass, Douglas, and Buchanan successively held the lead in balloting that continued for forty-nine roll calls. But none could gain a majority,

---

1962), pp. 3–58; Charles Edward Stowe, *Life of Harriet Beecher Stowe* (Boston, 1889); Forrest Wilson, *Crusader in Crinoline: The Life of Harriet Beecher Stowe* (Philadelphia, 1941); Charles Howell Foster, *The Rungless Ladder: Harriet Beecher Stowe and New England Puritanism* (Durham, N.C., 1954); Philip van Doren Stern, *Uncle Tom's Cabin, an Annotated Edition* (New York, 1964); Chester E. Jorgenson (ed.), *Uncle Tom's Cabin as Book and Legend* (Detroit, 1952); Furnas, *Goodbye to Uncle Tom.* For southern reception of and replies to the book, see Craven, *Growth,* pp. 150–157; Jeannette Reid Tandy, "Pro-Slavery Propaganda in American Fiction of the Fifties," *SAQ,* XXI (1922), 41–50, 170–178. Because of her disapproval of the theater, Mrs. Stowe objected to having *Uncle Tom* made into a drama, and there were no bloodhounds in the novel.

38. The best account of the election of 1852 from the Democratic standpoint is Roy F. Nichols, *The Democratic Machine, 1850–1854* (New York, 1923), pp. 15–168. For the individual candidacies, see Frank B. Woodford, *Lewis Cass* (New Brunswick, N.J., 1950), pp. 292–294; Philip Shriver Klein, *President James Buchanan* (University Park, Pa., 1962), pp. 215–220; George Fort Milton, *The Eve of Conflict: Stephen A. Douglas and the Needless War* (Boston, 1934), pp. 79–96; Ivor Debenham Spencer, *The Victor and the Spoils: A Life of William L. Marcy* (Providence, R.I., 1959), pp. 175–183. On the campaign generally: Roy and Jeannette Nichols, "Election of 1852," in Arthur M. Schlesinger, Jr., *et al.* (eds.), *History of American Presidential Elections, 1789–1968* (4 vols.; New York, 1971), II, 921–950. Also Nevins, *Ordeal,* II, 3–39.

much less the two-thirds required for nomination. Douglas suffered from the hostility of the "old fogies," whom his "Young America" supporters had tactlessly assailed. His critics also said that he drank too heavily, lived too freely, and associated too much with corruptionists and looters. Both he and Cass were handicapped in the South by their identification with popular sovereignty, and Marcy was even more suspect because some Barnburners were supporting him. At the same time, the adherents of these three candidates were grimly resolved to prevent Buchanan from profiting by his role as the South's insurance against all the other candidates. Thus the convention turned at last to Franklin Pierce of New Hampshire, a dark horse who was known to the public only as a handsome, affable figure and a brigadier in the Mexican War. Like most nominations of this kind, Pierce's was not as impulsive as it appeared, but had been carefully planned by his friends in New Hampshire and by southerners who knew he would be sympathetic to southern views. The Democratic platform pledged the party to "abide by and adhere to a faithful execution of the acts known as the Compromise measures . . . the act for reclaiming fugitives included" and to forestall any renewal of the slavery agitation. To this platform, Democrats of all shades rallied with surprising unity. Not only did southern extremists show enthusiasm for Pierce, but also most of the Free Soilers of 1848 followed Martin Van Buren back into the ranks of the Democracy.[39] As a result, the remnant of the Free Soil party polled only 155,000 votes for John P. Hale in 1852, compared with 291,000 for Van Buren in 1848.[40]

No such harmony blessed the Whigs. Tainted with nativism and weakened by the feud between Fillmore and Seward factions in New York, they burdened themselves with a candidate unpopular in one section and a platform unpopular in the other. The nomination of Winfield Scott instead of the incumbent Fillmore, who had signed the Compromise measures, was a victory for northern delegates at the Whig convention, and it produced wholesale defections in the deep South. General Scott, inept and pompous, proved unable to

39. Roy Franklin Nichols, *Franklin Pierce, Young Hickory of the Granite Hills* (rev. ed.; Philadelphia, 1958), pp. 189–215; *Proceedings of the Democratic National Convention, 1852* (n.p., 1856).

40. Richard H. Sewell, *John P. Hale and the Politics of Abolition* (Cambridge, Mass., 1965), pp. 144–150.

save a party that had begun to disintegrate.[41]

The election results in November surprised no one. Pierce won 254 electoral votes to Scott's 42, carrying 27 of the 31 states in the most one-sided victory since the Era of Good Feelings. Yet, since the Democrats failed to get a majority of the popular vote in the North, it is by no means clear that the outcome constituted a bisectional endorsement of the Compromise.[42]

If the Democratic victory was not as sweeping, at least in the North, as it first appears, still it did make Pierce president for four years. Pledged to the finality of the Compromise and to keeping the slavery question out of politics, he held large Senate and House majorities. Antislavery men were profoundly discouraged,[43] and outward appearances all indicated that the national yearning for harmony would banish the slavery issue from politics. But beneath the surface, there were many indications that the sectional rapprochement of 1852 did not rest on broad or deep foundations. Times were changing. Between the nominations and the election, Henry Clay and Daniel Webster had both followed Calhoun to the grave. The antislavery bloc in Congress, strengthened by militant recruits like Sumner and Wade, was no longer a little handful of isolated men. In 1852, for every four votes that Franklin Pierce received in the free states, one copy of *Uncle Tom's Cabin* was sold.[44]

The cause of Union, to be sure, had won a victory and survived a crisis. But the strains of the crisis had weakened the basis of Union. The South, while deciding against secession, had accepted the doctrine that secession was a valid constitutional remedy, applicable in appropriate circumstances. Meanwhile, the North had refused to make a national issue of slavery, as the abolitionists desired, but it had accepted their doctrine that slavery was morally intolerable. Without embracing secession, the South had committed itself to the

41. For a more extensive analysis of the effect of the campaign and election on the Whig party, see below, pp. 232–247.

42. Nevins, *Ordeal*, II, 38–39. Statistics in Nichols, "Election of 1852," p. 1003.

43. According to Martin B. Duberman, *Charles Francis Adams, 1807–1886* (Boston, 1961), p. 179, the winter and spring of 1851–52 "marked the lowest point yet reached in the antislavery crusade," and a year later, after the election of Pierce, there was little sign of improvement.

44. Pierce received 1,153,097 votes in the free states; Mrs. Stowe's book sold 305,000 copies in its first year. "That number (adjusted by eliminating Southern population, which included almost no customers) is equivalent to a sale of more than 3,000,000 copies in the United States of 1947." James D. Hart, *The Popular Book: A History of America's Literary Taste* (Berkeley, 1961), p. 112.

principle of secessionism; without embracing abolition, the North had committed itself to the principle of abolitionism. Against these forces, the cause of Union had Franklin Pierce as a leader and the shibboleth of "finality" as a slogan. As events would soon show, they were not enough.

CHAPTER 7

A Railroad Promotion and Its Sequel

HINDSIGHT, the historian's chief asset and his main liability, has enabled all historical writers to know that the decade of the fifties terminated in a great civil war. Knowing it, they have consistently treated the decade not as a segment of time with a character of its own, but as a prelude to something else. By the very term "antebellum" they have diagnosed a whole period in the light of what came after. Even the titles of their books *The Coming of the Civil War, The Irrepressible Conflict, Ordeal of the Union, The Eve of Conflict, Prologue to Conflict*—are pregnant with the struggle which lay at the end.

Seen this way, the decade of the fifties becomes a kind of vortex, whirling the country in ever narrower circles and more rapid revolutions into the pit of war. Because of the need for theme and focus in any history, this is probably inevitable. But for the sake of realism, it should be remembered that most human beings during these years went about their daily lives, preoccupied with their personal affairs, with no sense of impending disaster nor any fixation on the issue of slavery. It is also realistic to recognize that for many people there were other public issues seeming more important than slavery. Questions of tariff policy, of banking policy, of public land policy, of subsidy to railroads—all loomed large and engendered strong feelings. Such questions were not necessarily sectional, and on their face they seemed unrelated to slavery, but they tended to get translated into terms of sectional conflict, with slavery somehow involved. The tariff issue had been so translated in the crisis of 1832.

Similarly, the question of expansion, which seemed only partially related to slavery in 1844, had become almost wholly an aspect of the slavery question by 1846.

One of the foremost issues of the early fifties was that of communication with the Pacific coast, by railroad or by trans-isthmus routes across Central America. By the time when Franklin Pierce's first Congress met in December 1853, this question had been gaining a steadily increasing share of public attention for almost a decade. In 1844 a New York merchant who was also a dreamer of great dreams, Asa Whitney, had published a proposal to build a railroad from Milwaukee to the Columbia River, if Congress would sell to him, for sixteen cents an acre, a strip of land sixty miles wide along the route. With the proceeds from the sale of this land, he would build the railroad for the government, constructing the line as the sales were made.[1] Whitney's vision was not to be fulfilled in those terms. The choice of a northerly route; the timetable of twenty-five, or, as he later said, fifteen years; the scheme of having the project executed by one man, without the sale of stock or the flotation of bonds; the dream of government ownership—all were to go by the board. But three features of Whitney's plan became articles of faith for many Americans of his generation. There must be a railroad to the Pacific; it must be financed by grants of public lands along the route; and it must be built by private interests which received these grants.

For different classes of people, the Pacific railroad scheme had different implications. For ordinary, civic-minded Americans, it meant the binding of a loose-jointed transcontinental republic into a closer unity.[2] But for many local citizens in communities throughout the Mississippi Valley, it was like a giant lottery in which a whole community might win the rich prize of becoming a great metropolitan terminal for all the vast traffic with the Pacific coast. For aspiring

1. The best authority on the movement for a transcontinental railroad is Robert R. Russel, *Improvement of Communication with the Pacific Coast as an Issue in American Politics, 1783–1864* (Cedar Rapids, Iowa, 1948). But also see Nelson H. Loomis, "Asa Whitney, Father of the Pacific Railroads," MVHA *Proceedings*, VI (1912), 166–175; Robert S. Cotterill, "Early Agitation for a Pacific Railroad, 1845–1850," *MVHR*, V (1919), 396–414; Margaret L. Brown, "Asa Whitney and His Pacific Railroad Publicity Campaign," *MVHR*, XX (1933), 209–224. For Whitney's own statements, Asa Whitney, *A Project for a Railroad to the Pacific* (New York, 1849); *Senatorial Executive Documents*, 29 Cong., 1 sess., No. 161 (Serial 473).

2. See chap. 1, note 16, above.

capitalists, it meant a chance to build and own the longest railroad on earth without paying for it themselves. In short, the project appealed to many motives and aroused a vast amount of excitement and promotional activity. For fifteen years the Mississippi Valley seethed with the rivalries of competing towns and with the intrigues of competing groups of promoters, each of which aspired to control the road—or rather the assets which would be granted for the building of it. Meanwhile, secondary and tertiary promoters speculated in local real estate whose value was contingent upon the ultimate route.

Less than a year after Whitney published his plan, Stephen A. Douglas, then a thirty-two-year-old freshman congressman from Illinois, wrote an open letter proposing an alternative to Whitney's scheme. Instead of running the road from Milwaukee to the Columbia, he would run it from Chicago. The war with Mexico had not yet begun, but he would put the western terminal at San Francisco Bay, "if that country could be annexed in time." Instead of depending upon the advance of settlers as a means by which to sell land and gradually construct the road as Whitney proposed, Douglas wanted to push the road ahead rapidly as a means of attracting settlers. To facilitate his plan he would organize the region west of Iowa as a new territory, to bear the Indian name Nebraska. Concurrently with his letter, he introduced a bill in the House for the organization of this territory.[3] The bill was not adopted, but it had marked the advent of Douglas as one of the first, the most persistent, and ultimately the most influential advocates of a Pacific railroad.

The remainder of the Polk administration and all of the Taylor-Fillmore administration saw a steady growth of interest in the Pacific railroad question and a rising intensification of the rivalries among various cities that hoped to gain by the location of the route. Much of the political energy of the times poured into this struggle. No fewer than eighteen state legislatures voted resolutions in favor of Whitney's plan.[4] Not a single session of Congress met without the

3. Douglas to Whitney, Oct. 15, 1845, in Robert W. Johannsen (ed.), *The Letters of Stephen A. Douglas* (Urbana, Ill., 1961), pp. 127–133; see also Frank Heywood Hodder, "Genesis of the Kansas Nebraska Act," State Historical Society of Wisconsin *Proceedings*, 1912, pp. 69–86. Douglas introduction of Nebraska bill in *Congressional Globe*, 28 Cong., 2 sess., p. 41.

4. Whitney, *A Project for a Railroad*, pp. 89–107; *House Reports*, 31 Cong., 1 sess., No. 140 (Serial 583).

introduction of a number of bills relating to the transcontinental project.

As the movement developed, certain alignments and patterns began to emerge. Broadly speaking, the strongest drive for a railroad came from the states along the Mississippi and beyond it, while most of the opposition centered in the East and South. Conservative advocates of strict economy and honesty in government stood aghast at the scale of the proposals to give away government land, and they were shocked by the greed of the potential spoilsmen. Old-fashioned strict constructionists of the Jeffersonian persuasion clung to the view that internal improvement measures of this kind were unconstitutional, and they deplored the consolidation of federal power which would result from such a grandiose project. But the West brushed aside such prudent considerations and acted almost as a unit in demanding that there should be a road.

Within the West, however, various localities competed against one another to secure the eastern terminal of the railroad. The principal rivals at first were Chicago and St. Louis, with New Orleans slow to enter the competition partly because her leaders realized that if a railroad within the United States should prove impossible from either an engineering or a financial standpoint, their city would be a logical port for traffic to the Pacific coast by way of Panama, Nicaragua, or the Isthmus of Tehuantepec. Active plans were developed for railroads or ship canals at each of the three points, and no less than five treaties were negotiated (though not all were submitted to the Senate) during the Fillmore and Pierce administrations to protect American interests at these strategic positions.[5] But while New Orleans continued to look south, Chicago, as early as 1849, was already pushing railroads west to Rock Island and to Galena to serve as "takeoff" points from which a Pacific line might be extended; and St. Louis was holding an immense railroad convention with more than a thousand delegates. Other cities also dreamed of greatness: Quincy, Illinois, was assured by an enthusiastic promoter that she might "rival Carthage in her pride of power,"

5. For an able general treatment of the Isthmian projects, see Russel, *Improvement of Communication*, pp. 54–94. See also John Haskell Kemble, *The Panama Route, 1848–1869* (Berkeley, 1943); Mary Wilhelmine Williams, *Anglo-American Isthmian Diplomacy, 1815–1915* (Washington, 1916); Paul Neff Garber, *The Gadsden Treaty* (Philadelphia, 1923), pp. 43–108; J. Fred Rippy, *The United States and Mexico* (New York, 1926), pp. 47–67.

while Memphis, Tennessee, was promised that she might receive "the untold wealth of the gorgeous east."[6] Where ambitions mounted so high, political leaders hastened to put themselves forward as champions of the projects upon which their communities' hopes were pinned—and also to gain a position where they might share in the profits. In New Orleans, both Judah P. Benjamin and Pierre Soulé were deeply involved in the Tehuantepec Railroad Company.[7] In Memphis, John C. Calhoun, at a convention in 1845, had almost abandoned strict construction in his advocacy of a railroad for the South.[8] In New York, President Polk's former secretary of the treasury, Robert J. Walker, headed a grandiose enterprise, the Atlantic and Pacific Railroad Company, which exceeded all others, if not in its accomplishments, at least in the government largess that it coveted.[9] In Missouri, Senator Thomas Hart Benton was proclaiming his readiness to be the Peter the Hermit of this railroad crusade in which he implied, though he did not state, that St. Louis was the Holy City.[10] Meanwhile, in Illinois, the proponents of Chicago as a terminus knew that their interests were safe in the hands of their "Little Giant," Stephen A. Douglas.[11]

As the Fillmore administration drew near an end, it appeared that the Pacific Railroad question was approaching a climax and that one or another of the rival cities soon must be awarded the prize. Thus the atmosphere was charged with expectancy when the short session of Congress met in December 1852. It had for some time been evident that a bill providing for any one specific route would be defeated by the combined proponents of all other routes. Therefore, Senator William M. Gwin of California sought a broad basis of support with a bill promising a terminal for everyone. His mea-

6. Russel, *Improvement of Communication*, pp. 20–25, 34–53; Robert S. Cotterill, "The National Railroad Convention in St. Louis, 1849," *MHR*, XII (1918), 203–215; Cotterill, "Memphis Railroad Convention, 1849," *Tennessee Historical Magazine*, IV (1918), 83–94; St. George L. Sioussat, "Memphis as a Gateway to the West," *ibid.*, III (1917), 1–27, 77–114.

7. Russel, *Improvement of Communication*, pp. 89, 102.

8. Charles M. Wiltse, *John C. Calhoun, Sectionalist* (Indianapolis, 1951), pp. 235–242.

9. Russel, *Improvement of Communication*, pp. 96–98, 126–129; James P. Shenton, *Robert John Walker: A Politician from Jackson to Lincoln* (New York, 1961), pp. 129–133.

10. William Nisbet Chambers, *Old Bullion Benton, Senator from the New West: Thomas Hart Benton, 1782–1858* (Boston, 1956), pp. 352–353, 397–398.

11. Hodder, "Genesis," and Hodder, "The Railroad Background of the Kansas-Nebraska Act," *MVHR*, XII (1925), 3–22, were basic in showing Douglas's railroad interests and their relation to the Kansas-Nebraska Act.

sure provided for a main trunk across New Mexico and North Texas, but with branches in the west to San Francisco and to Puget Sound, and in the east to Council Bluffs, Iowa, to Kansas City, and to the Gulf of Mexico. All in all, this would require the construction of 5,115 miles of railroad, for which 97,536,000 acres of public land grants would be needed. But in trying to please everyone, including all the spoilsmen, Gwin had overshot the mark. Too many senators agreed with Lewis Cass, who said the project "is entirely too magnificent for me." Gwin's plan failed, and the advocates of a railroad recognized that they would have to be satisfied with legislation for but a single line.[12]

Since it was still axiomatic that no bill which specified a given route could command a majority, Senator Thomas J. Rusk of Texas, as chairman of a special committee, brought in a measure which left the choice of route and of terminals to the president, and which left the award of contract to be determined by competitive bidding instead of naming the contractors in the bill. In these and other provisions, the Rusk Bill was adroitly drawn, and in February certain preliminary votes, testing the alignment of forces, indicated that it would certainly pass.[13] At the last moment, however, Senator Cass complained bitterly that no federal measure ought to subsidize the construction of a railroad within the limits of a state, which was constitutionally different from construction within a territory. His protest caused Senator James Shields of Illinois to consult with Douglas, who had by this time become a senator, and with Henry S. Geyer of Missouri, after which he offered an amendment that no portion of the $20 million provided in the bill should be used for "a road within the limits of any existing state of the Union." This appeared to be only an abstract constitutional restriction, but in effect it was a well-aimed blow at any southern route, for no road could extend west from New Orleans, or Vicksburg, or even Memphis, without running for hundreds of miles through the state of Texas. The realism with which senators grasped this functional implication is indicated by the fact that nine members from states to the north of Tennessee and Arkansas voted for the amendment, although they, as northerners, were usually less sensitive about

12. *Congressional Globe,* 32 Cong., 2 sess., pp. 280–285; Russel, *Improvement of Communication,* pp. 97–98.

13. *Congressional Globe,* 32 Cong., 2 sess., pp. 469–470; Russel, *Improvement of Communication,* pp. 98–102.

constitutional limitations than the southerners, who, in this case, showed no constitutional scruples and voted against the amendment. But though the northern bloc now had modified the bill in such a way as to necessitate the choice of a northern route, the triumph was short-lived. Southern senators withdrew their support altogether, and friends of the measure found themselves unable to bring it to a vote.[14] Thus, a sustained effort to secure railroad legislation had ended in a sectional deadlock, and Congress did nothing more during the session except authorize the secretary of war to spend $150,000 on surveys of possible railroad routes.[15]

Meanwhile, congressmen from Iowa, Missouri, and Illinois had introduced into the House a bill to organize the region west of Missouri and Iowa as the Nebraska Territory, and to extinguish the Indian titles there. In advocating it, they clearly stated that their purpose was to facilitate the building of a railroad westward through this region. The author of the bill in fact said, "Why, everybody is talking about a railroad to the Pacific Ocean. In the name of God, how is the railroad to be made if you will never let people live on the lands through which the road passes?" This measure passed easily in the House by a vote of 107 to 49, although Texans, who had never before shown any tenderness toward the Indians, now expressed great solicitude about the sanctity of Indian titles in the Nebraska area.[16] In the Senate, the measure came under the wing of the chairman of the Committee on Territories, Senator Douglas, who had himself twice already introduced bills for the organization of the Nebraska territory. But Douglas encountered a crowded calendar, which was always a hazard at the short session, and he could not get the bill to the floor until two days before adjournment in early March 1853. At that stage, David Atchison of Missouri stated that he would support the bill, but he expressed his firm opposition to organizing a territory west of Missouri from which slaveholders would be excluded. Douglas's bill did not exclude them, but they were excluded because Nebraska lay within the Louisiana Purchase

14. *Congressional Globe,* 32 Cong., 2 sess., Cass statement, p. 711; Shields amendment, and vote on it, pp. 714, 744; Russel, *Improvement of Communication,* pp. 103–106.

15. *Congressional Globe,* 32 Cong., 2 sess., pp. 798–799, 814–823, 837–841, 996–998; Russel, *Improvement of Communication,* pp. 107–108; Allan Nevins, *Ordeal of the Union* (2 vols.; New York, 1947), II, 84–85.

16. *Congressional Globe,* 32 Cong., 2 sess., pp. 7, 47, 474–475; 542–544, 556–565; Russel, *Improvement of Communication,* pp. 156–159.

north of 36°30′ and was, therefore, under the Missouri Compromise of 1820, closed to slavery. Atchison had looked to the possibility of repealing the Missouri Act, he said, but had found it impossible to rally the necessary support, and therefore he would, with misgivings, accept the measure as it stood. But while Atchison was only reluctantly willing, other slave state senators were not willing at all. Senators from Texas, Arkansas, Mississippi, and Tennessee made it plain that they would filibuster if the bill were pressed. In these circumstances, on the last day of the session, the proposal was tabled by a vote of 23 to 17. Every senator voting from every state south of Missouri voted to table. The southerners thus gave their tit for tat to the supporters of the Shields amendment. If one faction had effectively checkmated a railroad through Texas, the other, by their veto of the Nebraska bill, had checkmated a road west from St. Louis or Chicago.[17]

This defeat brought the career of Stephen A. Douglas to a crucial point. During the months that followed, before the first Congress under President Pierce assembled in the subsequent winter, Douglas had a great deal to ponder. He was deeply committed in every way to the cause of a railroad, or even two railroads running from the Northwest to the Pacific Coast. Personally, he had invested heavily in real estate at Chicago and at Superior City, Michigan, and he would gain by the construction of roads either along a central route from Council Bluffs westward through the South Pass or along a northern route from the head of Lake Superior to Puget Sound. But even more vitally, he was a recognized champion of the Northwest's interests and an eloquent herald of its future greatness. It was he who had proclaimed in the Senate: "There is a power in this nation greater than either the North or the South—a growing, increasing, swelling power that will be able to speak the law to this nation. . . . That power is the country known as the Great West—the Valley of the Mississippi, one and indivisible from the Gulf to the Great Lakes, and stretching . . . from the Alleghanies to the Rocky Mountains. There, sir, is the hope of this nation—the resting place of the power that is not only to control, but to save, the Union." It was he who, as chairman of the Senate Committee on Territories, had presided at the organization of governments for the

17. *Congressional Globe,* 32 Cong., 2 sess., pp. 1020, 1111–1117; Russel, *Improvement of Communication,* pp. 159–160.

western area extending from Texas (whose statehood bill he introduced in the House) to Minnesota (whose territorial bill he had sponsored in the Senate). It was he who had sought legislation for Nebraska ever since 1844. His talent, resourcefulness, and drive had made him the recognized champion of the Democracy in the Northwest.[18]

Yet while the demands for a northern route continued to mount, the actual prospects for the northern route were being thrown into serious jeopardy. To begin with, Franklin Pierce had fallen under the influence of the old party regulars and the southerners in the Democratic organization, and Douglas had been snubbed in the distribution of the patronage; clearly he possessed little influence with the new administration. But among those who did have influence was Jefferson Davis. As secretary of war, Davis was in charge of the railroad surveys. He had sent out surveyors to make reconnaissances on three transcontinental routes: one, northerly, from St. Paul, via the Great Bend of the Missouri; another between the 38th and 39th parallels from the source of the Arkansas through Cochetopa Pass (in southern Colorado) and by way of Salt Lake; and a third from Fort Smith, Arkansas, via Albuquerque. Davis explained the omission of any survey for Douglas's favored central route by way of the South Pass on the ground that it had already been sufficiently explored. But this assurance did not inspire confi-

18. The explanation of the Kansas-Nebraska Act in terms of Douglas's desire for a Pacific Railroad was first advanced by Frank H. Hodder, as cited in notes 3 and 11 above. The importance of the railroad factor was denied by P. Orman Ray, *The Repeal of the Missouri Compromise* (Cleveland, 1909), and "The Genesis of the Kansas-Nebraska Act," AHA *Annual Report,* 1914, I, 259–280. More recently, Nevins, *Ordeal,* II, 102–107, has also discounted this factor. But George Fort Milton, *The Eve of Conflict: Stephen A. Douglas and the Needless War* (Boston, 1934), pp. 8, 9, 97–107, introduced new evidence, including evidence on Douglas's personal investments, in support of the railroad explanation. Also, the first direct statement by Douglas of the need for a railroad as a reason for territorial organization was found by James C. Malin, in "The Motives of Stephen A. Douglas in the Organization of the Nebraska Territory: A Letter Dated December 17, 1853," *Kansas Historical Quarterly,* XIX (1951), 351–352. Malin gives a more general discussion of the importance of the railroad factor in *The Nebraska Question, 1852–1854* (Lawrence, Kan., 1953), pp. 123–153 and *passim.* Douglas's speech on the destiny of the West: *Congressional Globe,* 31 Cong., 1 sess., appendix, p. 365; for his dominant legislative role as an organizer of territories, see Roy F. Nichols, *Blueprints for Leviathan: American Style* (New York, 1963), pp. 286–287; also John Bell's comment in the Senate that Douglas deserved "ten civic crowns" because of his "passion . . . for the organization of new territories." *Congressional Globe,* 33 Cong., 1 sess., appendix, pp. 407–415, cited in Albert J. Beveridge, *Abraham Lincoln, 1809–1858* (4 vols.; Boston, 1928), III, 206.

dence, for Davis was known to be sympathetic to a southern route, and his chief of the Corps of Topographical Engineers, William H. Emory, was even more clearly committed to such a route. Emory had publicly expressed enthusiasm for the Gila River route before 1850; he owned real estate in San Diego; and his brother-in-law, Robert J. Walker, was head of the Atlantic and Pacific Railroad Company, which had been chartered in New York in 1853 and had already sent private surveyors to the Gila River country.[19] What was even more ominous, Pierce had appointed James Gadsden, a South Carolina railroad promoter, as minister to Mexico, and had sent him there with instructions to negotiate for the purchase of an area south of the Gila, which would be strategic in the construction of a railroad by the southern route.[20] At this point, Memphis, Vicksburg, New Orleans, and Texas were about to make their supreme effort to win the Pacific railroad for their region and seemed likely to succeed.

Douglas returned to Washington in December 1853, still hoping to organize the Nebraska Territory. But with this continuity there was also much discontinuity. After the death of his wife less than two months before the end of the previous session, he had gone to Europe for an extended trip just as Franklin Pierce came into the presidency. On his return, he found that Pierce had failed to exercise any effective initiative and was surrounded by partisan southerners, while Gadsden, long the advocate of a railroad by the southern route, was on a mission to Mexico. Worse still, he quickly learned that he could no longer count on support from Senator Atchison for his Nebraska Bill, and in fact he soon came under heavy pressure from Atchison.

During the congressional recess, Atchison had entered the first phase of a campaign for reelection in Missouri, in which he was opposed by Thomas Hart Benton. The campaign was a grudge fight, for the Atchison forces had unseated Benton in 1851 after

19. For critical discussion of the role of Davis and Emory and their purpose to secure a southern route, see William H. Goetzmann, *Army Exploration in the American West, 1803–1863* (New Haven, 1959), pp. 262–278; Russel, *Improvement of Communication*, pp. 168–173, 183–186; Shenton, *Walker*, pp. 129–133; John Muldowny, "The Administration of Jefferson Davis as Secretary of War" (Ph.D. dissertation, Yale, 1959); for Douglas's lack of influence in the Pierce administration, Milton, *Eve of Conflict*, pp. 94–97; Roy Franklin Nichols, *Franklin Pierce, Young Hickory of the Granite Hills* (rev. ed.; Philadelphia, 1958), pp. 228–229.

20. Garber, *The Gadsden Treaty*, pp. 77–108.

thirty years in the Senate. The Old Roman wanted revenge, and knew how to get it, for he was a popular, dangerous, and often unscrupulous adversary. Probing for Atchison's vulnerable spots, Benton had hit on his support for the Nebraska Bill, which would make Nebraska free soil, and which was therefore objectionable to Atchison's proslavery supporters. Benton had, in fact, impaled Atchison on a dilemma: if Atchison supported the bill, he was betraying Missouri's slavery interests; if he opposed it, he was betraying Missouri's railroad interests.

After a severe mauling by Benton, Atchison had begun to say that he would see Nebraska "sink in hell" before he would hand it over to the free-soilers. Also, he had perceived a way to reverse the dilemma and impale Benton: take the Missouri Compromise feature out of the Nebraska Bill, and confront Benton with a choice of accepting it, which would antagonize his free-soil supporters, or opposing it, which would antagonize the Missouri railroad interests that supported him.

Exactly what kind of pressure Atchison put upon Douglas has been a matter of prolonged and rather unnecessary controversy. Atchison apparently declared later that he had originated the idea of repealing the Missouri Compromise and had forced Douglas to carry out his plan by threatening to take over the chairmanship of the Committee on Territories and bring in a bill himself if Douglas would not do so.[21] Whether Atchison in fact made such a threat scarcely matters. Without doubt he let Douglas know that he had changed his mind and would not again support a Nebraska Bill with the Missouri Compromise still intact. Douglas knew that Atchison was a powerful senator—indeed the senior member of the Senate in point of service—and a messmate of James M. Mason and Robert M. T. Hunter of Virginia and Andrew P. Butler of South Carolina—as powerful a trio as there was in the Senate. He knew, too, that his bill had not passed in 1853 even with Atchison's support, and if that was true, it certainly could not pass in 1854 without Atchison's support. More fundamentally, Douglas recognized that with

21. On the impact of political and other conditions in Missouri, see Ray, as cited in note 18; Malin, *The Nebraska Question*, pp. 123–153, 416–443, and *passim*—in fact Malin's is the first adequate history of the Nebraska question as a western question, involving the interests and the activities of local elements in Iowa, Missouri, Nebraska, and Kansas; William E. Parrish, *David Rice Atchison of Missouri, Border Politician* (Columbia, Mo., 1961), p. 143.

Eastern antirailroad interests opposed, the bill could not possibly pass if it incurred heavy opposition from southern senators also. Yet southerners simply had no incentive to vote for a measure which would create another free territory and would also help Chicago or St. Louis to snatch the Pacific railroad away from the southern cities at a time when their prospects were brighter than ever before. The Nebraska proposition therefore must be framed with concessions to win some southern support, or it would fail. To a mind like Douglas's this must have seemed both axiomatic and controlling.[22]

During January and February, Douglas made concessions in a series of steps. Their rationale can only be inferred, but some of the inferences are very strong. To begin with, the Missouri Compromise was a major obstacle to Douglas because as long as it remained, he could not hope to induce southern senators to vote for the organization of Nebraska. But the obstacle seemed too formidable to be removed directly by outright repeal. For too long a time, too many people had regarded the Act of 1820 as a "solemn compact" not subject to repeal. Douglas himself had spoken of it with veneration as late as 1849, and his Nebraska Bill of the preceding session had tacitly accepted the Missouri restriction. When Atchison said on the Senate floor that he would like to remove the restriction, but that he saw no hope of doing so, the silence of Douglas and other senators showed that they then shared Atchison's understanding that the Missouri Compromise still applied.[23]

Yet despite the sanction which this Compromise had once enjoyed, Congress had repeatedly refused between 1846 and 1850, as Douglas well knew, to extend it to the Mexican Cession, and had

22. Frank H. Hodder and P. Orman Ray (notes 3, 11, 18) conducted a long and contentious debate as to whether Douglas or Atchison was "author" of the repeal of the Missouri Compromise. In this controversy, somehow, the two failed to see how complementary their findings were. Hodder showed why Douglas needed the territorial organization, but not why this need led him unwillingly to undertake the repeal. Ray showed the nature of the forces that were exerting pressure for repeal, but not the urgency of the need that made Douglas vulnerable to them. Douglas was subject to generalized southern pressures, and it is useless to argue the relative importance of one southerner (Atchison) or another (Dixon).

23. Douglas in 1849 said that the Missouri Compromise had become "canonized in the hearts of the American people as a sacred thing which no ruthless hand would ever be reckless enough to disturb." Speech at Springfield, Oct. 23, 1849, as quoted in John G. Nicolay and John Hay, *Abraham Lincoln: A History* (10 vols.; New York, 1890), I, 335; Atchison's statement in *Congressional Globe*, 32 Cong., 2 sess., pp. 1111, 1113.

finally applied the principle of popular sovereignty as a new basis of settlement. As a matter of logic, this raised a question: If in 1850 Congress had taken the position that the people of a territory should handle the slavery question locally, and that Congress should abstain, did this not conflict with the kind of regulation imposed by Congress in the Act of 1820? And if it did conflict, had not the later basis of settlement replaced the earlier one? In short, had not the Missouri Compromise been "superseded" by the legislation of 1850? To a person who needed desperately to get away from the restrictions of the Missouri Compromise, but who feared to move against it directly, this seemed a rewarding line of inquiry.

As pure syllogistic reasoning, this argument was not without merit,[24] but two points had to be whisked out of sight. First was the fact that Congress was not necessarily committed to a single "principle" for determining the status of slavery in the territories. Historically, it had not adhered to any one principle, but had used the principle of exclusion in the Northwest Territory, the principle of geographical division in the Louisiana Territory, and the principle of popular sovereignty in the Mexican Cession. In the pragmatic tradition of Anglo-American politics, there was no reason why the adoption of one of these principles for one region should interfere with the application of another principle for another region. Geographical partition applied to one area; popular sovereignty to another; and there was no reason why the twain should meet.[25] Second was the palpable historic reality that no one in 1850 had supposed that the Utah and New Mexico legislation had any bearing upon the Missouri Compromise. Proslavery men had not claimed that it did;

24. A number of students of Douglas have emphasized the sincerity of his belief, even before 1854, in popular sovereignty as a basis for territorial policy. E.g., Robert W. Johannsen, "The Kansas-Nebraska Act and the Pacific Northwest Frontier," *Pacific Historical Review,* XXII (1953), 129–141; Harry V. Jaffa, *Crisis of the House Divided: An Interpretation of the Issues in the Lincoln-Douglas Debates* (New York, 1959), pp. 133–146; Gerald M. Capers, *Stephen A. Douglas, Defender of the Union* (Boston, 1959), pp. 43–44. But Jaffa is almost alone in accepting the contention that "long before 1854" Douglas had regarded the Missouri Compromise as "superseded" by popular sovereignty. And even Jaffa qualifies his position by saying that Douglas expected to apply popular sovereignty in Kansas-Nebraska without repealing the Missouri Compromise (p. 107).

25. The intentional diversity of the series of settlements which determined the status of slavery throughout the Union and the absence of any single controlling principle of settlement were stated with conclusive effectiveness by Thomas Hart Benton in *Congressional Globe,* 33 Cong., 1 sess., appendix, pp. 557–558.

antislavery men, who would have raised the roof at the hint of such an idea, had not protested that it did.[26] Thus, the desire of Douglas to by-pass the Missouri Compromise rather than face the task of removing it led him into a remarkable fiction. Instead of merely claiming that the Missouri settlement had been inadvertently removed four years previously by men who did not know what they were doing—which would itself have strained credulity—he made, in effect, the even more stunning claim that this crucial political action had been taken knowingly and yet without a contest by men who did not even bother to discuss what they were doing.

Douglas did not come willingly to this tour de force of logic. Inexorable circumstances drove him to it, step by step. He made the move from which there was no turning back on January 4, when he brought in a new bill for Nebraska. This bill simply provided for the organization of Nebraska Territory, and it specified in the exact language which had been used in the Utah and New Mexico acts of 1850 that "when admitted as a State or States, the said territory, or any portion of the same, shall be received into the Union, with or without slavery, as their constitution may prescribe at the time of their admission." It further provided that the new territory should include not merely the area west of Iowa and Missouri, as the previous bill had done, but that it should extend to the Canadian border and embrace all of the area of the Louisiana Purchase which remained unorganized.[27]

This bill of January 4 said nothing about the Missouri Compromise or about the status of slavery in the territory. Whether Douglas intended it to be a silent repeal of the Act of 1820, as many historians have assumed, or a subtle device to placate the southerners by making them think he had abandoned the Act of 1820 without actually abandoning it, as has been contended, is not entirely clear. It is, on the other hand, quite clear that he was offering the least concession which, he hoped, might win southern support.

He quickly discovered that this minimum was not enough, for the

26. On the lack of any intent in 1850 to repeal the Missouri Compromise, see Nevins, *Ordeal,* II, 100.

27. *Senate Reports,* 33 Cong., 1 sess., No. 15 (Serial 706); *Congressional Globe,* 33 Cong., 1 sess., p. 115. Technically this bill was in the form of an amendment to a bill which Senator Augustus Caesar Dodge of Iowa had introduced at the beginning of the session, and which was an exact copy of Douglas's Nebraska Bill of the previous session.

southerners pointed out, quite correctly, that the Act of 1820 still applied: the bill only allowed the people of a territory to adopt a proslavery constitution when they were admitted to statehood; while they were a territory, the Act of 1820 would still remain in force. In short, Douglas's bill would create a situation under which, at the time of admission to statehood, slaveholders might vote for a proslavery constitution, but also under which no such slaveholders could establish themselves in the territory prior to this vote.[28] Atchison and others apparently applied strong, and perhaps even harsh, pressure on this point. Douglas tried to meet it in a curious way, for on January 10 the Washington *Union* printed an additional section of the bill which it stated had been omitted from the original published version through "clerical error." This addition specified that "all questions pertaining to slavery in the Territories, and in the new states to be formed therefrom are to be left to the people residing therein, through their appropriate representatives."[29] But the South's more astute legal minds were still not satisfied even with this second step, for unless the Act of 1820 were repealed outright, it would still exclude slaves until the territorial government arrived at the decision to let them in—which such a government could never be expected to do if no slave interest had been permitted to establish itself in the first place. Representative Philip Phillips of Alabama perceived this point and aroused other southern Democrats to its significance.[30] They were privately urging Douglas to make a fur-

28. Lincoln expressed this point in 1854 when he said, "Keep it [slavery] out [of a territory] until a vote is taken, and a vote in favor of it cannot be got in any population of forty thousand on earth," Roy P. Basler (ed.), *The Collected Works of Abraham Lincoln* (8 vols.; New Brunswick, N.J., 1953), II, 262–263.

29. On the manuscript of the bill, the additional section was added separately, which suggests that the "omission" in the first printing may have been more than a "clerical error." Allen Johnson, *Stephen A. Douglas* (New York, 1908), p. 233.

30. For the southern pressure on Douglas, see Henry Barrett Learned, "The Relation of Philip Phillips to the Repeal of the Missouri Compromise in 1854," *MVHR*, VIII (1922), 303–317; Henry S. Foote, *Casket of Reminiscences* (Washington, 1874), p. 93; Samuel S. Cox, *Three Decades of Federal Legislation* (Providence, R.I., 1888), p. 49; Milton, *Eve of Conflict*, p. 112; Ray, *Repeal of Missouri Compromise*, pp. 209–219; Nevins, *Ordeal*, II, 95.

After Douglas accepted the principle of repeal, he could no longer admit that he did not wholly favor repeal, or that the previous versions of his bill might not have effectuated it. The suspicions of southerners that Douglas was trying to gain their support by seeming to offer repeal while not actually doing so were, therefore, soon forgotten. More than a century later, however, Harry V. Jaffa, in *Crisis of the House Divided*, p. 175 and *passim*, came independently to the view to which Phillips had come in 1854—Douglas's first proposal, he believes, would have eliminated the Missouri

ther concession when Senator Archibald Dixon, a Kentucky Whig, arrived independently at the conclusion that repeal by inference was not enough. On January 16, Dixon offered an amendment for the explicit repeal of as much of the Act of 1820 as prohibited slavery north of 36°30′.[31] This amendment at last brought the question of repeal of the Compromise fully into the open. The week that followed was a busy one for a number of people. Dixon offered his amendment on a Monday. On Wednesday, Douglas went for a carriage ride with Dixon, who explained his views as to the necessity of explicit repeal. Douglas showed that he was reluctant to accept Dixon's plan, but he responded to Dixon's logic, and after considerable discussion he at last exclaimed impulsively, "By God, Sir, you are right. I will incorporate it in my bill, though I know it will raise a hell of a storm."[32] Between then and the end of the week, the territorial committees of the Senate and House worked on new drafts of the bill. They agreed to take the drastic third step—to include a specific repeal of the Missouri Compromise—and also to organize two territories—Kansas west of Missouri and Nebraska west of Iowa and Minnesota. This was apparently done to give the widely separated settlements in the Platte Valley and in the valley of the Kansas an equal chance to develop without one being dependent upon the other, or to equalize the chances for the proponents of a northern railroad route and a central route, but it was widely construed to intend that one territory should become a slave state

---

Compromise at the time of admission to statehood, but would have left it in force during the territorial stage, thus effectually barring slaveholders by keeping them out until after an election was held to determine whether they might come in. Jaffa's treatment has an importance which historians have regrettably failed to recognize. But it is ironical that Jaffa should apparently regard this equivocal conduct by Douglas as statesmanship, while Phillips regarded it as mere duplicity. Douglas's circumstances forced him to take an equivocal position, and the result was that antislavery men suspected him of covertly selling out to proslavery while proslavery men suspected him of covertly selling out to antislavery. Jaffa's suggestion that he was gulling his southern associates by maintaining the Missouri Compromise while seeming not to is scarcely more creditable to him than Nevins's conviction that he was gulling his northern associates by scuttling the Missouri Compromise while seeming not to.

31. *Congressional Globe,* 33 Cong., 1 sess., p. 175; Mrs. Archibald Dixon, *True History of the Missouri Compromise and Its Repeal* (Cincinnati, 1898), p. 445. The role of Douglas has confused historians because he was publicly spokesman and manager for forces which privately he could not control. In his Committee on Territories, he had to make concessions to Sam Houston (Nichols, *Blueprints,* p. 95); in the Democratic caucus, he was forced unwillingly to accept phrasing which, on the floor, he vigorously defended as if it were his own.

32. Dixon, *Missouri Compromise,* p. 445.

and the other a free state. This implication, of course, further antagonized the antislavery forces.[33]

At this late stage, Franklin Pierce, who thus far had assumed no initiative whatever, now sought to have a voice in the matter. Both Senator Cass and Secretary William L. Marcy had warned the president that the repeal might involve his administration in serious difficulties. The cabinet discussed the problem on Saturday, and it appears that all of the members except James C. Dobbin of North Carolina and Davis of Mississippi disapproved of the pending action. Further, it appears that the president and cabinet actually drafted and sent to Douglas an alternative proposal, which would have sought a judicial determination of the question of the constitutionality of the Compromise. But with Atchison, Mason, Butler, and the other southern leaders all working to hold Douglas in line, he rejected the cabinet plan.

Late on Saturday night, the Committee decided to report on the following Monday, but they recognized that before they brought in a measure as important and as controversial as this one promised to be, they must commit the president to its support. They knew that Pierce disliked to transact business on Sunday, and therefore, instead of approaching him directly, they went to Jefferson Davis and asked him to arrange an interview. Davis, Douglas, Atchison, Mason, Hunter, John C. Breckinridge, and Phillips proceeded together to the White House, and Davis went directly to the president. Pierce received them all in the library, and it appears that his manner was distant and unenthusiastic, as it might well have been in view of their rejection of his proposal of the previous day, and his own misgivings as to their plans. But Pierce could no more resist the power of the southern senatorial junto than could Douglas. After some discussion, he agreed that the administration would support their plan. Pierce had by this time been in office long enough for knowledgeable congressmen to be aware that he had a way of giving impulsive commitments which he was afterward reluctant to keep, so Douglas

33. Capers, *Douglas*, pp. 94–97, has an especially lucid account of the steps by which the original bill was modified. After Douglas had accepted the idea of repeal, he was forced, in a final concession, to change the statement that the Missouri Compromise was "inoperative" to a statement that it was "void." On the two territories, see Johnson, *Douglas*, pp. 226–227, 238–239; Milton, *Eve of Conflict*, pp. 148–149; Douglas, in Senate, *Congressional Globe*, 33 Cong., 1 sess., pp. 221–222; appendix, p. 382.

prudently contrived to have him write out in his own hand the fatal statement that the Missouri Compromise "was superseded by the principles of the legislation of 1850, commonly called the compromise measures and is hereby declared inoperative and void." Pierce had erred in letting this meeting take place without any of his own advisers except Davis in attendance, and in a final, half-hearted effort to keep one possible escape hatch open, he asked the conferees to consult Secretary Marcy also. But they had already got what they came for, and they later used the fact that Marcy had not been at home when they called as an excuse for not consulting him at all.[34]

Douglas, Pierce, and the senatorial junto were not the only people in Washington who were busy that Sunday. In another part of the city, Salmon P. Chase and Charles Sumner from the Senate, and Gerrit Smith, Joshua R. Giddings, and two other members from the House, were meeting to complete their plans for protesting, as antislavery men, against the reopening to slavery of an area which had been formally dedicated to freedom. In fact, antislavery spokesmen had voiced their disapproval instantly, from the time earlier in the month when Douglas had brought in his bill in its first form, and Sumner had offered an amendment reaffirming the Missouri Compromise. But events had moved so rapidly that up to this time, no major resistance had developed. Now, however, the six antislavery congressmen were preparing to launch an organized opposition. They took the rough draft of a statement written by Giddings primarily for use in Ohio, and Chase prepared a new version in broader terms. Sumner gave it a final literary polish, and then these six "Independent Democrats," as they called themselves, sent their composition in to the editor of the *National Era,* an antislavery weekly published in Washington.[35]

34. On the policy of the administration and the Sunday interview, see Jefferson Davis to Mrs. Dixon, Sept. 27, 1879, in Dixon, *Missouri Compromise,* pp. 457–460; New York *Herald,* Jan. 23, 24, 1854; letter of Philip Phillips, Aug. 24, 1860, in Washington *Constitution,* Aug. 25, 1860; Learned, "Relation of Philip Phillips to the Repeal"; Nichols, *Pierce,* pp. 321–324; Milton, *Eve of Conflict,* pp. 115–117; Claude M. Fuess, *The Life of Caleb Cushing* (2 vols.; New York, 1923), II, 146–147. On the ignoring of Marcy, statement by Reuben Fenton in Henry Wilson, *History of the Rise and Fall of the Slave Power in America* (3 vols.; Boston, 1872–77), II, 382–383; Ivor Debenham Spencer, *The Victor and the Spoils: A Life of William L. Marcy* (Providence, R.I., 1959), pp. 278–279.

35. On the authorship of the "Appeal of the Independent Democrats," George W. Julian, *Life of Joshua R. Giddings* (Chicago, 1892), p. 311; Chase to J. T. Trowbridge,

On Monday morning, Douglas reported the new bill from the committee. On Tuesday, he proposed to take it up for debate. Chase, with affected casualness, asked for a delay in order that he might study the bill. Douglas agreed to this request. But before the day was over, the *National Era* came off the press. In it was the "Appeal of the Independent Democrats." The Appeal clearly foreshadowed a bitter fight by the free-soil Democrats against the bill and against the administration, for it attacked the measure "as a gross violation of a sacred pledge, as a criminal betrayal of precious rights, as part and parcel of an atrocious plot" to make free Nebraska a "dreary region of despotism, inhabited by masters and slaves." It denounced Douglas personally, accusing him of sacrificing the tranquillity of the nation to satisfy his presidential ambitions, and flaying the "servile demagogues" who served the "slavery despotism." It was the first cannonade in what is, perhaps to this day, America's fiercest congressional battle.[36]

It would be a matter of nice discrimination to say when the Nebraska issue ceased to be primarily a railroad question and became primarily a slavery question. But certainly it was never a railroad question again after the January 24 issue of the *National Era* left the press.

The Appeal of the Independent Democrats was significant for its highly effective use of an antislavery tactic which had been used already by the abolitionists and which is perhaps always used in any situation of angry controversy, but which reached a supreme level of effectiveness in 1854, and the years following. This was the tactic of attacking the defenders of slavery not on the merits or demerits of their position, but on the grounds that they were vicious, dishonest, and evil. Ironically, this accusation, which was in many cases not true, proved much more effective for publicity or propaganda purposes than the accusation that they were supporting a pernicious

---

Jan. 19, 1854, in Robert B. Warden, *An Account of the Private Life and Public Services of Salmon Portland Chase* (Cincinnati, 1874), p. 338; J. W. Schuckers, *Life and Public Services of Salmon Portland Chase* (New York, 1874), pp. 140, 160–161; Chase to E. L. Pierce, Aug. 8, 1854, in *Diary and Correspondence of Salmon P. Chase*, AHA *Annual Report*, 1902, II, 263. An admirable concise statement on the authorship in Nichols, *Blueprints*, pp. 290–291. See also James A. Rawley, *Race and Politics: "Bleeding Kansas" and the Coming of the Civil War* (Philadelphia, 1969), pp. 44–45.

36. Introduction of bill and Chase request in *Congressional Globe*, 33 Cong., 1 sess., pp. 221, 239–240; text of Appeal of Independent Democrats, *ibid.*, pp. 281–282; critical note on the date of the Appeal in Nichols, *Blueprints*, p. 291.

system, which was true. Thus, it was not sufficient for the "Independent Democrats" to base their attack on the ground that Douglas's measure disturbed the peace, tampered with a settled law, and gave an advantage to slavery, and was wrong and irresponsible. Instead they had to picture him, like Webster, as a lost soul. Douglas had turned traitor, they said, in return for slaveholder support for the presidency.

This publicity relied heavily on moral absolutes: the Missouri Compromise was not just an act of Congress; it was a sacred pledge. The repeal was not just a political maneuver; it was the result of an atrocious plot. Douglas was not, conceivably, trying to find a way to keep Nebraska free and also to get it organized; he was a Judas, a Benedict Arnold, selling Nebraska into slavery. What he did was not a mistake; it was a criminal betrayal. These denunciations were couched in language of the most sanctimonious indignation, leaving the impression that the complainants were, as Charles Sumner described himself, "slaves of principle" who would never stoop to politics. Yet the fact was that the antislavery congressmen were usually political free lances who lacked a normal basis of political support through party organization, and they found in this type of propaganda a strategic way to compensate for the weakness of their organizational position. The fact was also that they were, for the most part, quite astute politically, and capable of extremely sharp practices. For instance, at the very time they were consigning Douglas to outer darkness because of his sacrilege in playing politics with the Missouri Compromise, William H. Seward, the foremost antislavery leader, was making political medicine for the Whig party by encouraging the southern Whigs to demand the outright repeal of the Compromise so that Douglas would be forced to propose it also in order to match the Whig bid for southern support.[37] Historians for more than a century have denigrated Douglas for his moral callousness, yet the record reveals no act of political trickery on the part of Douglas comparable to this. The point here, however, is not that the antislavery leaders were both more adroit in their politics and less moral in their practice than they appeared to be. It is that, increasingly after 1854, they had a strength which derived not only

37. Montgomery Blair to Gideon Welles, May 17, 1873, *The Galaxy*, XVI (1873), 691–692; Milton, *Eve of Conflict*, pp. 151–152; Nichols, *Blueprints*, p. 290, citing Seward to Thurlow Weed, Jan. 7, 8, 1854, in Thurlow Weed Papers.

from the righteousness of their cause but also from the technical skill of a distinctive style of publicity, which discredited their opponents as not only wrong on principle but also morally depraved and personally odious.

The next three and a half months witnessed a struggle of unprecedented intensity. The chorus of free-soil response to the Appeal of the Independent Democrats rose to a roar which stunned the supporters of Kansas-Nebraska and must have filled them with fear. Only the coolest of political warriors could have kept his head in the midst of such a furor. But the senatorial junto consisted of seasoned veterans, and behind the scenes they maintained a steady pressure to hold the lines of party regularity. Meanwhile, on the floor of the Senate, Douglas waged a battle such as the oldest members had never seen. A natural fighter, he performed best when under attack, and no member of an American Congress had ever been assailed both in and out of Congress as he was now. As he himself later said, he could have traveled to Chicago by the light of his own burning effigies. In the face of this assault, he defended himself with astonishing resources—an accurate memory for the minor details of political transactions for thirty years back, a slashing directness and cogency in rebuttal, a sustained power suggested by the volume and pertinence of the evidence which he could bring to bear on almost any point, and a supreme virtuosity in the give-and-take of debate. For five weeks he dominated the Senate, and then finally, in a five-and-a-half-hour address, beginning near midnight on March 3, he cornered his adversaries and forced them to admit flaws in their arguments. After this conclusion of the debate, the Kansas-Nebraska bill was passed 37 to 14.[38]

In the House the revolt of the free-soilers was far more extensive than it had been in the Senate. On March 21 the House refused to refer the bill to the Committee on Territories but sent it instead to the Committee of the Whole, where it was buried beneath fifty other bills. This action showed that the

38. The vote and Douglas's speech, *Congressional Globe*, 33 Cong., 1 sess. p. 532; appendix, pp. 325–338; 14 free-state senators and 23 slave-state senators in affirmative; 12 free-state senators and 2 slave-state senators (Bell of Tennessee and Houston of Texas) in negative. Excellent accounts of the debates in Milton, *Eve of Conflict*, pp. 118–141; Beveridge, *Lincoln*, III, 198–217 (both favorable to Douglas); Nevins, *Ordeal*, II, 113–145 (adverse to Douglas).

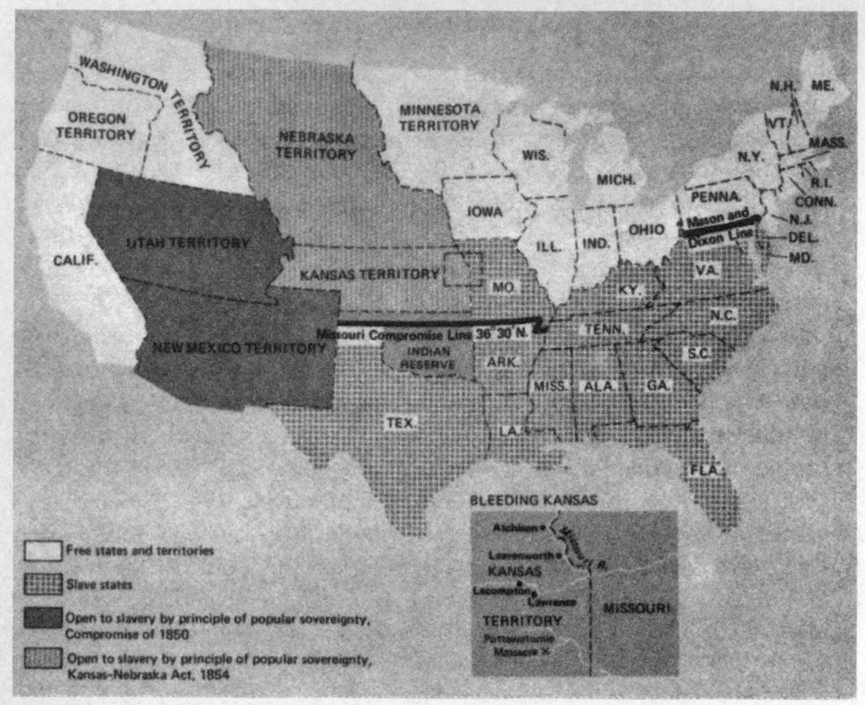

The Kansas-Nebraska Act, 1854

Democratic party was split and that a majority of House members opposed the measure. But administration leaders began at once to wield the patronage, to mend fences, and to apply pressure. Gradually, during March and April, they amassed votes, until in May, Douglas, who was managing the campaign, believed that a majority had been gathered, and he determined on a strategy of tabling all bills ahead of the Kansas-Nebraska measure. Accordingly, with roll call after roll call, the House laid aside the other bills and came at last to the measure that had been buried. The contest lasted for fifteen days, and tension mounted as the final vote drew nearer. Amid scenes of wild excitement, weapons were drawn, and bloodshed seemed imminent. Alexander H. Stephens of Georgia, less commanding than Douglas, but no less resourceful in bringing the details of many years of political history to bear in his argument, and no less tenacious in his purpose, acted as floor manager for the administration. On May 22, applying "whip, and spur," he succeeded in bringing the bill to its third reading, and it was passed by a vote of 113 to 100. Pierce signed it eight days later.[39]

Thus the Kansas-Nebraska bill became law and produced a number of results, foreseen and unforeseen, desired and undesired. After ten years of effort, the land beyond the Missouri River now had a territorial government, to the gratification of certain enterprising types at Wyandot City and at Council Bluffs who cared very little, one way or the other, about the moral issues which had convulsed Congress. A preliminary condition had been met for building a Pacific railroad west from Chicago or from St. Louis, but in fact no track would be laid until the sons of the northern and the southern Democrats who had stood together to pass the bill were killing one another on the battlefields of Virginia. Of more immediate bearing was the fact that the Democratic party had been split into opposing factions, ending the uneasy truce of 1850. Franklin Pierce might have come in as a symbol of the finality of the Compromise, but the main concern of his administration was to be the turmoil of "Bleeding Kansas."

Few events have swung American history away from its charted course so suddenly or so sharply as the Kansas-Nebraska Act. Because of this deflection, it remains something of an enigma. Why did the Pierce administration, triumphantly elected on the "finality" platform, sponsor a measure so certain, as Douglas recognized, to raise "a hell of a storm"? Why did a young and ambitious free-state senator, who held no brief for slavery, introduce a measure which served the "slave power" and blasted his own career? Douglas's own shifting course as he made concessions step by step, his own equivocation—not unlike the equivocation which had surrounded the territorial question ever since the Nicholson letter—and his silence concerning his railroading goals all contributed to the mystery. But the greatest problem for the historian confronting the developments of 1854 is not to penetrate what is hidden so much

39. Incomparably the best account of the previously neglected story of enactment in the House is in Nichols, *Blueprints*, pp. 104–120. The division in the House showed 44 free-state and 69 slave-state members in affirmative (all Democrats save 12 southern Whigs), 91 free-state and 9 slave-state members in negative (of these 51 were Whigs—44 northern, 7 southern—and 3 were free-soil). From this, it will appear that the northern Democrats were evenly divided at 44 to 44.

as to clear away the propaganda smokescreen employed with great effectiveness by the free-soilers in their campaign against the Kansas-Nebraska Act. To say this is certainly not to deny that they were right in opposing it or to question that they were justifiably outraged by the repeal of the Missouri Compromise. If the Compromise had not stood throughout eternity as they suggested, it had stood for thirty years, and it deserved, as Douglas knew, to be treated with respect.

But for the psychological purposes of waging a grim battle, it was not enough for the free-soilers that diverse forces, represented by various men, had converged to produce the repeal of the Missouri Compromise. They needed one villain who was solely responsible, and their villain was Douglas. It was not enough, in their version, that he should pursue a mistaken or an unwise policy; he must do his evil deed for a sordid and evil purpose. And further, it was not enough that what he did might work to the advantage of slavery; he must appear as one of a group of proslavery conspirators. Hence the free-soilers made Douglas the sole architect of the Kansas-Nebraska Act; they made ambition for the presidency his motive; and they made the repeal of the Act of 1820 a deliberate effort to convert the Nebraska region into a stronghold of slavery. The combatants themselves should not be blamed too much, for these are recognized tactics of political conflict. But leading historians of the late nineteenth century adopted the view that Douglas alone had engineered the Kansas-Nebraska Act as part of a scheme to attain the presidency. When the accusations of partisan warfare became the conclusions of historical scholarship, attention was diverted from certain important questions: To what extent was Douglas really in command of the situation? What factors other than presidential ambition motivated the measure of repeal? Did Douglas intend that Kansas-Nebraska should be added to the domain of slavery?

If the evidence proves anything, it proves that Douglas was not really in command of the situation. To be sure, he showed himself a tremendously able tactician. He directed the entire legislative campaign, with the powerful members of the senatorial clique and even the president serving as his lieutenants. He forged and wielded the majorities which snatched victory from defeat. But in a larger sense, he had lost the initiative, for he had not been able to prevent his political allies from defining his objective and his political opponents from defining the issue on which he won. Beginning with a straightforward bill, to organize a territory, for a straightforward purpose, to facilitate a Pacific railroad, he had run aground on the Missouri Compromise, and in an effort to get afloat again, he had been tempted into tactics of indirection, which his critics called shiftiness or deceit. He was a man whose strength usually lay in his directness, and his equivocations at this juncture showed what straits he was in. The very fact that he hung back from asserting a direct purpose to repeal the Missouri Compromise was his own inverted tribute to the measure he was undoing. It indicated the extent to which Atchison and others were forcing his hand. Roy F. Nichols has well said, "In the midst of the cataclysm, one sees Douglas crashing and hurtling about, caught like a rock in a gush of lava."[40]

But why was he so vulnerable to these pressures which played upon him? The oldest and flimsiest reason ever adduced, but one still embalmed in history, is that he was trying to procure southern support for his candidacy for the presidency. This accusation, however, is one that can always be used to explain the motivation of any act at any time by any person who has been seriously mentioned as a presidential candidate. But in specific terms it does not fit the circumstances very closely. Douglas already enjoyed some favor among southern Democrats, and there is no evidence that he had to offer them any special inducements to win their support or that the repeal of the Missouri Compromise was a favor they were asking for. In fact one of the tragic aspects of the repeal is that while it was offered to the South as a kind of bait, southerners had not been pressing for it and were, in fact, decidedly unresponsive to the idea when Douglas first introduced it. Only after the antislavery leaders pilloried Douglas as a tool of the slave power did the proslavery men rally to his support. In this ironical but very real sense, Salmon P. Chase contributed a great deal to the creation of Douglas's proslavery support, simply by accusing him of deserving it.[41]

Another explanation sometimes offered is that Douglas was in-

40. Roy F. Nichols, "The Kansas-Nebraska Act: A Century of Historiography," *MVHR*, XLIII (1956), 187–212, at 212. This is a masterful review of the complex history of how historians have handled Kansas-Nebraska, and an admirable interpretation of the events.

41. Avery O. Craven, *The Growth of Southern Nationalism, 1848–1861* (Baton Rouge, 1953), pp. 192–205, offers extensive evidence of the initial southern lack of responsiveness to the Kansas-Nebraska Act; also, Malin, *The Nebraska Question*, pp. 314–315, 320–324, 328–330.

fluenced by Atchison. If this means that Douglas voluntarily sponsored what he knew would be an unpopular measure, in order to help Atchison, one can only say that it is the kind of thing which would be hard to prove, there is no evidence for it, and it is not plausible. No doubt Douglas disliked Benton and hoped that Atchison would annihilate him in Missouri, but it does not necessarily follow that Douglas was prepared to jeopardize his own career in order to influence political events in a neighboring state. If it means that Douglas acted involuntarily in response to Atchison's political leverage, probably this is true, but it in turn raises the question why the Nebraska Bill was so vital to him that he would put his neck in the noose in order to get it enacted.[42] Here one is brought back to the question of a Pacific railroad. It is an indisputable fact that Douglas had been deeply interested in the Pacific railroad project, both personally and politically, ever since 1844. If he did not himself proclaim his intention to use Kansas-Nebraska as a stepping-stone toward the realization of a railroad by the central or northern route, it was because he could not do so without admitting that he had baited his bill with the meat of repeal of the Missouri Compromise in order to entice the southerners into supporting his scheme to win the Pacific Railroad for his own region. To have stated his purpose would have been to defeat it. But it is a significant fact, often ignored by historians, that in the next session of Congress after the Kansas-Nebraska bill, Douglas's main activity was the sponsorship of a Pacific railroad bill. This bill provided for the construction of three railroads, running westward from Texas, from Missouri or Iowa, and from Minnesota. No such proposal would have been practicable until after the region included in the Kansas and Nebraska territories had been organized. In the House, this bill was reduced by amendment to a proposal for a single road, westward from Iowa or Missouri. If it had passed the House in this form, which was perhaps what Douglas intended, it might then have been brought to the Senate and passed there also, in which case the Kansas-Nebraska Act would have borne immediate fruit in the form of a Pacific railroad by the central route. But after first being passed in the House, it was then reconsidered and defeated by a vote of 105 to 91. Subsequently, Douglas secured the adoption of his original three-road bill in the Senate, and it remained the only Pacific railroad bill

42. See note 22, above.

that passed either chamber of Congress prior to the Civil War. Perhaps the unrecognized tragedy in the career of Stephen A. Douglas was the fact that after initial success in both houses, he was defeated on a vote to reconsider in the lower house, and by this narrow margin lost his great objective, the prerequisite for which had cost him such effort and such sacrifice in 1854.[43] In any event, that he failed to secure the adoption of a Pacific railroad bill certainly should not blind historians to the fact that it long remained one of his major objectives.[44]

While Douglas has suffered historically from imputations of sordid motives, he has probably suffered even more from the implicit assumption that his measure amounted to a simple surrender of Nebraska to slavery. This assumption is often made without even a query as to what Douglas actually thought the effect of the repeal of the Missouri Compromise might be. Certainly the repeal removed an exclusion of slavery, and certainly this was objectionable to antislavery men. But the repeal was by no means a proposal to give slavery a franchise in Nebraska. Rather, it was a proposal to remove one form of control, namely congressional exclusion, which was widely believed to be unconstitutional in any case and which was in fact declared unconstitutional by the Supreme Court three years later, and to substitute another form of control, namely local regulation through a democratic process. Such control meant, in Douglas's opinion, that the people of Nebraska would make the territory free. At the time of the Compromise of 1850, he had declared: "We have a vast territory, stretching from the Mississippi to the Pacific, which is rapidly filling up with a hardy, enterprising, and industrious population, large enough to form at least seventeen new free states, one-half of which we may expect to be represented in this body during our day. . . . I think I am safe in assuming that each of these will be free territories and free states, whether Congress shall prohibit slavery or not." In 1854 and throughout his career, he continued to believe that the application of the principle of popular sovereignty would make all the territories free. But in addition to

43. See note 18, above. *Congressional Globe,* 33 Cong., 2 sess., pp. 210, 224, 251, 814 (action in Senate), 281–356 (action in House).

44. Very able overall discussions of Douglas's motives are in Milton, *Eve of Conflict,* pp. 144–154, Nevins, *Ordeal,* II, 102–107 (discounting the railroad motive, but published before Malin's discovery, mentioned in note 18); Capers, *Douglas,* pp. 97–103.

this sanction, he also believed, in the case of Nebraska, that the territory was further protected by the fact that climate and physiographical conditions would serve as barriers to slavery. During the Nebraska debate, he wrote to a New Hampshire newspaper that every intelligent person knew that the proposed repeal was a "matter of no practical importance," for, he said, "all candid men who understand the subject admit that the laws of climate, and of production, and of physical geography have excluded slavery from that country."[45]

Functionally, Douglas was an antislavery man in the sense that he did not want slavery to go into the territories. But he was not an antislavery man in the same way as Chase and Seward. He disagreed with them partly because he was willing to rely on physical conditions to keep slavery out and they were not; he was willing to rely on local decisions to keep slavery out and they were not. But it is not always recognized that their difference lay deeper than a mere disagreement as to what effects physical conditions and local decisions might have upon slavery. They also disagreed as to whether the United States should be committed to a national policy toward the institution of slavery. In a sense, the exclusion of slavery north of 36°30′ (as well as in the Northwest Territory under the Ordinance of 1787) had embodied such a policy. It had implicitly recognized a necessity for conceding some areas to the South, but had established a kind of national preference for freedom over slavery. The repeal of the exclusion might or might not lead to the introduction of slaves into Nebraska, but it would certainly mark an abandonment of the national policy. Thus, even if the Kansas-Nebraska Act would not have abetted the spread of slavery at all, it was still objectionable to antislavery men because it abrogated a policy of disapproval of slavery and established the policy that slavery was a local issue, not a subject of any national preference one way or the other.

Douglas was a vigorous believer in the democratic principle of

45. Speech of Douglas, *Congressional Globe,* 31 Cong., 1 sess., appendix, p. 371; account of private conversation, Jan. 1854, in which Douglas called slavery "a curse beyond computation to both white and black," in George Murray McConnel, "Recollections of Stephen A. Douglas," ISHS *Transactions,* 1901, pp. 48–49; letter of Douglas to editor of the *State Capitol Reporter* (Concord, New Hampshire), Feb. 16, 1854, in Johannsen (ed.), *Letters of Douglas,* pp. 284–290. For the view that Douglas was callously indifferent to the spread of slavery, see evidence in Nevins, *Ordeal,* II, 107–109.

local autonomy, but his opponents were equally vigorous believers in the moral primacy of freedom. Douglas apparently thought that slavery could be and should be kept out of new territories, and that it could be done by local action. The antislavery men thought that slavery violated national values and that it was too important to oppose by limited or local means. Douglas was satisfied to contain slavery without condemning it, but in the eyes of the free-soilers, the man who would not condemn could not be trusted to contain. Douglas cared more about the Union than about the eradication of slavery and would never push the slavery issue to a point where it imposed too much strain upon the Union. Many antislavery men thought the Union hardly worth preserving so long as it had slavery in it. Such were the barriers that always arose, at all crucial moments, between the antislavery of Douglas and the antislavery of the free-soilers.

In an era of many futile measures, the Kansas-Nebraska Act approached the apex of futility. No matter how measured, it seems barren of positive results. Even at the level of a mere political combination, it did not fulfill anyone's expectations, for though it combined the votes of northerners who hoped to gain a transcontinental railroad and southerners who hoped or were induced to hope for the extension of slavery, the ensuing railroad bill failed to pass, and despite years of turmoil, Kansas never had slavery except in a nominal sense. Yet the nation paid an inordinate price for passage of the bill.

One of its most damaging effects was to contaminate the doctrine of popular sovereignty, by employing it as a device for opening free territory to slavery. When Douglas used it in this way, the effect was to make people forget (and even to make subsequent historians forget) that, prior to 1854, popular sovereignty held an important potentiality as a possible means of blocking the extension of slavery without causing political convulsions. Of course it did not guarantee freedom, but Douglas believed, as Polk and Webster and others had believed, that the kind of people who would be settling the territories would not want slavery. Thus popular sovereignty could be viewed as a less troublesome way of accomplishing all that the Wilmot Proviso would have accomplished. Antislavery men who wanted to "contain" slavery by moderate methods, and to achieve freedom without bringing on a crisis—men, for instance, like Lincoln—had every reason to take this idea seriously. Before 1854, popular sovereignty may have been perhaps the country's best hope for keeping the territories free and at the same time avoiding sectional disruption.

But when Douglas, with a broad wink to the southerners in Congress, invited them to vote for popular sovereignty as a device for overthrowing the guarantees of freedom north of 36°30′, he permanently discredited his own doctrine in the eyes of any potential antislavery supporters. Their revulsion against his stratagem strengthened the position of the militants in the antislavery ranks —men like Sumner and Chase. Antislavery men never trusted Douglas again, even when he was later fighting the battle for freedom in Kansas. No other event of this period—not even the Dred Scott decision—did more to stimulate antislavery elements to such steps as attempting to nullify the Fugitive Slave Act by adverse state legislation and sanctioning the use of violence by John Brown. Thus the Kansas-Nebraska Act went far to close off moderate means of action which the antislavery forces might have followed and to forfeit the very real influence which, up to 1854, Douglas might still have exercised in shaping antislavery policy.

These consequences were apparent only in the perspective of time, but it quickly became evident that Kansas-Nebraska had destroyed the ascendancy of the Democratic party in the free states and had also upset the bisectional balance within the Democratic party. In any analysis of the dynamics of Unionism, it is important to recognize that ever since the time of Andrew Jackson, the Democrats had been the dominant party both in the North and in the South. The strength of each geographical wing within its own area had given it a strong voice in the affairs of the party and had made it a valuable ally for the other wing. In short, the Democratic party had been a powerful force for nationalism because neither of its sectional wings was weak enough to be subordinate to the other, while each was powerful enough to exercise a tempering influence on the other. Thus, Jackson in 1828 and 1832, Van Buren in 1836, Polk in 1844, and Pierce in 1852 had carried a majority of both the slave states and the free states. In Congress, the two geographical wings of the party had been rather evenly balanced. In the House of Representatives in 1847–1849, there were 108 Democrats (as opposed to 115 Whigs), of whom 54 were from free states and 54 from slave states. In the first Congress of the Taylor-Fillmore administration the Democrats, with 116 members, held a plurality in the House; 55 of these were from free states, 61 from slave. In the second Congress of the Taylor-Fillmore regime, they had increased their total to 141 members, divided between the free and slave states at 81 and 60 respectively. In the Congress which adopted the Kansas-Nebraska Act the total stood at a new high of 162 with free-state Democrats outnumbering slave-state Democrats 91 to 67.[46]

But the reaction to Kansas-Nebraska was a bitter one, and in the congressional elections of 1854 and 1855, the northern Democrats sustained smashing defeats. These setbacks involved the sudden rise of a new political organization, the Know-Nothing party, as well as the demise of the Whigs, and the beginnings of the Republican party, with complexities that will require discussion in a later chapter. Because of these complexities, Douglas always maintained, in his best partisan manner, that it was Know-Nothingism, and not opposition to Kansas-Nebraska, which caused the Democratic defeat of 1854–1855. But though there may be room for controversy as to what caused the result, there is no doubt that it was sweeping. While the Democrats retained all but four of the 67 slave-state seats which they had held, they were able to save only 25 of the 91 free-state seats which they had won in 1852.[47] This loss of 66 of their 91 free-state seats was never recovered. The Democratic party did, to be sure, win another presidential election in 1856, and it gained control of the House in the same election. But Buchanan carried only five of the sixteen free states in 1856 and he owed his election mostly to southern support. Northern Democrats were never again half as numerous as the southern Democrats in the House until long after the Civil War. Douglas remained the outstanding Democrat in the country, but he was identified with a minority faction within his own party. The Democratic party remained bisectional in the sense that it was active in both sections and sought policies that would be acceptable to both, but the equilibrium within the party did not survive the repeal of the Missouri Compromise. The effect was cumulative: the susceptibility of the Democratic party to southern control weakened it in the North, and

46. Compiled from the lists of members, with party affiliations shown, in the various issues of the *Congressional Globe*, the *Whig Almanac*, and the *Tribune Almanac*.

47. Only seven of the forty-four Northern Democrats who had voted for Kansas-Nebraska survived the election. But only fifteen of the forty-eight who voted against it or did not vote were reelected.

its weakness in the North increased its susceptibility to southern control.

Thus the organization of two new territories on the Northern Plains accomplished nothing that anyone intended and a great deal that no one intended. It did not lead on to the railroad which was Douglas's objective, or to the extension of slavery which was the southerners' objective. But it stimulated a reaction against slavery even greater than that produced by the Fugitive Slave Act. It infected the doctrine of popular sovereignty with a fatal proslavery taint. And it undermined the structure of the Democratic party, which was the strongest national organization that still sustained the Union.

CHAPTER 8

The Ebb Tide of Manifest Destiny

HISTORIANS customarily write about past events as if each one occurred in isolation, neatly encapsulated in a sealed container, or chapter, which keeps it from being mixed up with other events in their own containers. This practice is based on a sound assumption that both the writer and the reader can best do one thing at a time. The alternative would be to make history as chaotic as a dictionary of dates. Yet a realist will always want to remember that this neat historical order is only a convenient fiction, and sometimes a deceptive one, and that diverse events constantly impinge upon and modify one another.

Some events which intersected the Kansas-Nebraska bill illustrate this. The bill did not have a long legislative history when compared with many major congressional enactments. Douglas brought in his original measure on January 4, 1854. It passed the Senate two months later. It passed the House on May 22, and on May 30, Pierce signed it. The entire time interval was a little less than five months.

What else was happening during this time? Any answer by the historian involves an arbitrary selection, but here are seven other things that took place:

1. On January 18, the same day when Douglas agreed with Dixon to make the repeal of the Missouri Compromise explicit, William Walker, self-styled president of the sovereign Republic of Lower California, issued a decree annexing Sonora to his Republic and changing its name to the Republic of Sonora. With fewer than three hundred men, Walker was challenging all of Mexico. The delicious

absurdity of his pretensions led the San Francisco *Alta California* to observe, "Santa Anna must feel obliged to the new president that he has not annexed any more of his territory than Sonora. It would have been just as cheap and easy to have annexed the whole of Mexico at once, and would have saved the trouble of making future proclamations."[1]

2. On February 10, President Pierce submitted to the Senate a treaty with Mexico which the American minister, James Gadsden, had recently negotiated with Santa Anna, for the purchase of land south of the Gila River, in what is now southern New Mexico and Arizona. This area was vital for the construction of a Pacific railroad by the southern route and was, therefore, almost as strategic in the railroad rivalries of the time as the Kansas-Nebraska bill.

3. On the evening before Washington's birthday, George N. Sanders, the American consul in London, held a dinner party at which the guests included seven revolutionists—Mazzini, Garibaldi, and Orsini of Italy, Kossuth of Hungary, Arnold Ruge of Germany, Ledru-Rollin of France, Alexander Herzen of Russia—and one non-revolutionist, James Buchanan of Pennsylvania. Sanders was one of the most aggressive members of the Young America movement, whose members believed emphatically in both the world mission and the territorial growth of the United States. He and his guests drank toasts to "a future alliance of America with a federation of the free peoples of Europe."[2]

4. On February 28, the Marques de la Pezuela, governor-general of Cuba, confiscated in Havana harbor an American merchant vessel, the *Black Warrior*, and placed her captain, James D. Bulloch, an American naval officer, under arrest. He did so on the ground that the ship's manifest misrepresented what was in the cargo, but this was a mere pretext, for the ship was only touching at Havana, was neither discharging nor taking on cargo, and had repeatedly made out her manifest in this way by agreement with the port authorities.[3]

5. On April 3, Secretary of State William L. Marcy sent to Pierre

1. *Alta California*, Jan. 30, 1854, quoted in William O. Scroggs, *Filibusters and Financiers: The Story of William Walker and His Associates* (New York, 1916), p. 42.

2. Merle Curti, "Young America," *AHR*, XXXII (1926), 34–55, has an account of the Sanders dinner.

3. Henry L. Janes, "The Black Warrior Affair," *AHR*, XII (1907), 280–298; Basil Rauch, *American Interest in Cuba, 1848–1855* (New York, 1948), pp. 279–281, 284–285; Amos Aschbach Ettinger, *The Mission to Spain of Pierre Soulé, 1853–1855* (New Haven, 1932), pp. 252–290, 484–488.

Soulé, American minister at Madrid, instructions to negotiate for the purchase of Cuba for a maximum price of $130 million. If this could not be arranged, Marcy continued, far more cryptically, "You will then direct your efforts to the next most desirable object, which is to detach that island from the Spanish dominion and from all dependence on any European power."[4]

6. On May 16, Solon Borland of Arkansas, the American minister to Nicaragua, was injured by a broken bottle, thrown by someone in a hostile crowd which he was trying to address, at Greytown (San Juan del Norte) on the Mosquito Coast. Six years previously, the British had set Greytown up as an independent, sovereign "free city," for though it was a miserable little cluster of huts, it was also the logical eastern terminus for a route across an isthmus which Britain did not want the United States to control. An American corporation, the Accessory Transit Company, operated ships on the San Juan River, and antagonism had developed between the company and the natives. It was in this situation that, on May 16, the American captain of one of the company's boats shot a native member of his crew. When Greytown Negroes attempted to arrest the captain, Borland intervened, holding them off with a gun. This in turn led to the arrest of Borland himself. It was while protesting against this arrest that he was injured. He was released two days later, and resumed his journey to Washington, where he laid his case before Secretary Marcy. The government responded by sending Commander George H. Hollins, with the United States man-of-war *Cyane*, to Greytown to demand satisfaction. The instructions to Hollins directed him to teach the offenders at Greytown that the United States would not tolerate such outrages, and at the same time to avoid destruction of property or loss of life, but they left him broad latitude. Accordingly, demands were twice made upon the authorities of Greytown for an apology and for promises to respect American rights in the future. When these demands were ignored, Hollins, on July 13, after giving twenty-four hours' notice and providing facilities for the inhabitants to leave, began a bombardment of the town. By the end of the day he had destroyed it, but without loss of life.[5]

4. Marcy to Soulé, April 3, 1854, in William R. Manning (ed.), *Diplomatic Correspondence of the United States: Inter-American Affairs, 1831–1860* (12 vols.; Washington, 1932–39), XI, 175–178.

5. Scroggs, *Filibusters and Financiers*, pp. 72–78; Mary Wilhelmine Williams, *Anglo-*

7. On May 31, the day after signing the Kansas-Nebraska Act, President Pierce issued a proclamation warning that the administration would prosecute all violations of the Neutrality laws.[6] It was generally understood that this proclamation was aimed at John A. Quitman, known to be heading a gigantic filibustering project against Cuba, with a million dollars and fifty thousand men reportedly at his disposal.[7] Quitman had previously assumed that he also had the backing of the administration.

In retrospect, it is easy to recognize that these events marked the climax and also the end of an aggressive, aggrandizing foreign policy which bears the ironic label "Manifest Destiny." Identified as it was with a program of territorial acquisitions and with what now seems a naïve, parochial, and self-righteous belief in the "regenerating" effect of American values upon societies which were then bluntly called "backward," rather than "underdeveloped," Manifest Destiny runs counter to prevailing twentieth-century attitudes and is now in almost complete disrepute, even among Americans who do not propose to retrocede California to Mexico. But in 1854, the aspirations of Manifest Destiny still had much luster. As Franklin Pierce saw it, he was the bearer of a glorious Democratic tradition of extending the sway of democracy, under the American flag, in the Western Hemisphere. Jefferson had acquired Louisiana from Napoleon, with no questions asked about Napoleon's right to sell what had been Spanish territory. Polk, coming to office after eight years of delay in the annexation of Texas, had again released the forces of expansion and had pushed the national boundaries to the Pacific. Except for Whig obstruction, he probably would have annexed a large part of Mexico south of the Rio Grande. Polk had also tried to purchase Cuba.

After Polk, the Whigs had again let Manifest Destiny subside. Clayton, as secretary of state under Taylor, had negotiated the Clayton-Bulwer treaty with Britain, by which the United States renounced claim to exclusive control of any future isthmian canal, and he had also permitted British occupation of the Mosquito Coast.[8]

American Isthmian Diplomacy, 1815–1915 (Washington, 1916), pp. 171–183; Richard W. Van Alstyne (ed.), "Anglo-American Relations, 1853–1857," *AHR*, XLII (1937), 491–500; Ivor Debenham Spencer, *The Victor and the Spoils: A Life of William L. Marcy* (Providence, R.I., 1959), pp. 309–317.

6. James D. Richardson (ed.), *A Compilation of the Messages and Papers of the Presidents* (11 vols.; New York, 1907), V, 272–273.

7. *National Intelligencer*, June 22, 1854; see note 20, below.

8. Richard W. Van Alstyne, "British Diplomacy and the Clayton-Bulwer Treaty,

Webster, as secretary of state under Fillmore, had failed to press hard for the most favorable line when the Mexican boundary was being run.[9] Fillmore himself had made no protest in 1851 when the Cuban government put to death the American members of a defeated filibustering expedition led by the Cuban revolutionist Narciso López (the dead included the nephew of Attorney General Crittenden).[10] Nor had Fillmore offered any resistance in 1852 when the captain-general of Cuba refused to let the United States Mail Steamship Company's *Cresent City* land at Havana so long as a certain purser who had written disparagingly of the captain-general was aboard.[11] But Fillmore had paid dearly for his policy in loss of popularity, and the assistant editor of the Washington *Union* wrote, "It is general opinion here that Cuba has killed Fillmore."[12]

The Democrats of 1852 meant to resume where Polk had left off, and to repudiate the Whig program as failure to uphold the national honor. They made no concealment of their expansionism. On the contrary, they gloried in it. Pierce was elected on a platform which praised the Mexican War as "just and necessary" and "congratulate[d] the American people on the results of that war."[13] His inaugural address, devoted mostly to generalities and to negative promises such as a pledge of "rigid economy," included two positive affirmations: that the measures of 1850 were to be "unhesitatingly carried into effect," and that "my administration will not be controlled by any timid forebodings of evil from expansion. Indeed,

---

1850–1860," *Journal of Modern History,* XI (1939), 149–183. See also Williams, *Anglo-American Isthmian Diplomacy;* [David] Hunter Miller, *Treaties and Other International Acts of the United States of America* (8 vols.; Washington, 1931–48), V, 671–802.

9. J. Fred Rippy, *The United States and Mexico* (New York, 1926), pp. 106–125; William H. Goetzmann, *Army Exploration in the American West, 1803–1863* (New Haven, 1959), pp. 153–208; Odie B. Faulk, *Too Far North, Too Far South* (Los Angeles, 1967).

10. Robert Granville Caldwell, *The López Expeditions to Cuba, 1848–1851* (Princeton, 1915); Rauch, *American Interest in Cuba,* pp. 121–180; Edward S. Wallace, *Destiny and Glory* (New York, 1957), pp. 56–96; Philip S. Foner, *A History of Cuba and Its Relations with the United States* (2 vols.; New York, 1962–63), II, 41–65; Herminio Portell Vilá, *Narciso López y su Epoca* (Havana, 1930); Vilá, *Historia de Cuba en sus Relaciones con los Estados Unidos y España* (4 vols.; Havana, 1938–41), I, 347–483; C. Stanley Urban, "New Orleans and the Cuban Question during the Lôpez Expeditions of 1849–1851," *LHQ,* XXII (1939), 1157–1165.

11. Rauch, *American Interest in Cuba,* pp. 231–235; Foner, *History of Cuba,* II, 68–70.

12. Charles Eames to William L. Marcy, Sept. 14, 1851, as quoted in Rauch, *American Interest in Cuba,* p. 163.

13. Kirk H. Porter and Donald Bruce Johnson (eds.), *National Party Platforms, 1840–1960* (Urbana, Ill., 1961), pp. 17–18.

it is not to be disguised that our attitude as a nation and our position on the globe render the acquisition of certain possessions not within our jurisdiction eminently important for our protection."[14]

Pierce's appointments reflected his commitment to this policy. To head the State Department, he chose William L. Marcy of New York, who had served as secretary of war under Polk. For foreign posts, he appointed a disproportionate number of southerners and members of what has been called the "Mexico gang." Some of the minor appointments were especially revealing: they included Edwin de Leon, who had originated the name "Young America," and John L. O'Sullivan, editor of the *Democratic Review* and creator of the term "Manifest Destiny"—a man who had twice been indicted for violating the neutrality laws. James Buchanan, as minister to London, was the one northerner named to a major post. To France, Pierce sent John Y. Mason of Virginia; to Spain, Pierre Soulé of Louisiana; to Mexico, James Gadsden of South Carolina.[15]

Gadsden was both a southerner and a railroad president, and he had been a tireless worker for a transcontinental railroad along the southern route. He was sent to Mexico to acquire the territory needed for running a railroad through the Gila River region, and his instructions authorized him to negotiate for a cession. If possible he was to acquire the northern part of Tamaulipas, Nuevo Leon, Coahuila, Chihuahua, and Sonora and all of Lower California, for which he might offer up to $50 million. At the very least, he must seek a line that would provide the necessary railroad route, and also a port on the Gulf of California. Despite his most vigorous urging, however, Gadsden could not persuade Santa Anna to cede anything more than the Gila River region, without even the port. He accepted this minimum, and in January 1854 brought back a treaty. Pierce, though disappointed that Gadsden had not acquired more, nevertheless submitted the treaty to the Senate.[16]

14. Richardson (ed.), *Messages and Papers*, V, 198. The Crimean War (1854–56), preoccupying Britain and France, left the United States freer to indulge in expansionist adventures during the Pierce administration.

15. Spencer, *Victor and the Spoils*, pp. 228–231; Roy Franklin Nichols, *Franklin Pierce, Young Hickory of the Granite Hills* (rev. ed.; Philadelphia, 1958), pp. 255–257, 287–288; Henry B. Learned, "William Learned Marcy," in Samuel Flagg Bemis (ed.), *The American Secretaries of State and Their Diplomacy* (10 vols.; New York, 1927–29), VI, 174–182; Rauch, *American Interest in Cuba*, pp. 218–219. Among the minor advocates of expansionism was George Francis Train, who supported Cuban annexation by arguing that Cuba was a deposit of alluvium from the Mississippi River. "What God has joined together let no man put asunder," said Train.

16. Paul Neff Garber, *The Gadsden Treaty* (Philadelphia, 1923), especially pp. 64–

The Senate's action quickly showed that the reluctance of Mexico to yield territory was no longer the main obstacle to American expansion, for the senators dealt harshly with Gadsden's small triumph. A few opposed the treaty because it did not acquire enough land; others did so because it accepted American responsibility for claims by parties who had held Mexican franchises, now repudiated by Mexico, for the building of a railroad across the Isthmus of Tehuantepec (this factor was complicated by the rivalry of two separate groups of Tehuantepec claimants); but the major opposition came from free-state senators, who simply did not want to acquire more territory that might extend the area of slavery. At one time, the opposition actually defeated ratification, when twelve antislavery senators joined with three advocates of more extensive acquisition and three defenders of Tehuantepec claimants. Twenty-seven affirmative votes did not provide the necessary two-thirds against these eighteen, and it appeared that the treaty had been killed. Its friends were able to revive it, but they accomplished this only by accepting an amendment which cut 9,000 square miles from the area that Gadsden had obtained. Even then, twelve northern foes of the treaty balanced twelve northern proponents, and only the support of a solid bloc of twenty-one southern senators prevented defeat. The southern route for a Pacific railroad had squeaked through, but for the first time in history, the Senate had refused to accept land ceded to the United States.[17]

If Pierce fared badly in his effort to acquire Mexican territory, he fared even worse with his second major objective, which was Cuba. Here, the recklessness of his tactics was shown at the outset by his appointment of Pierre Soulé of Louisiana as minister to Spain. In general, Soulé was disqualified by a melodramatic temperament and a tendency to excess in all that he did. A native of France and an exile from Europe because of his revolutionary activities, Soulé strangely combined a red republican identification with revolutionary causes in Europe on the one hand, and an ardent support of

---

108; Rippy, *United States and Mexico,* pp. 126–147; correspondence between the State Department, Gadsden, and Mexican officials concerning the treaty in Manning (ed.), *Diplomatic Correspondence of the United States,* IX, 134–169, 600–696; texts of the treaty and valuable history of negotiations and ratification in Miller, *Treaties of the United States,* VI, 293–437.

17. Garber, *Gadsden Treaty,* pp. 109–145; Rippy, *United States and Mexico,* pp. 148–167; Rippy argues that the importance of the sectional factor has been exaggerated; *Journal of the Executive Proceedings of the Senate of the United States of America* (Washington, 1828–87), IX, 238–240, 260–315.

slavery in America on the other. Thus, European conservatives and American liberals alike had reason to distrust him. In particular, Madrid was the worst place in the world to send him, for he had distinguished himself in the Senate by his flowery eulogies of López and the Cuban filibusters, and also by his assertion that military conquest of Cuba was consonant with the spirit of Young America. The night before he sailed for Spain, this incredible envoy was serenaded by the Cuban junta in New York. He stood on a hotel balcony and listened complacently while a spokesman for the junta, in the presence of several thousand people, enjoined him to bring back "a new star" to "shine in the sky of Young America." In his response, he said that America would speak "tremendous truths to the tyrants of the old continent." On the following day he set sail for the court of one of the tyrants.[18]

Soulé was in Spain for fifteen months, during which there were few dull moments. With his hyperthyroid personality and his penchant for pyrotechnics, he kept Madrid in an uproar much of the time. Within two months after his arrival he wounded the French ambassador in a duel because someone—not the ambassador—had commented too freely on Mrs. Soulé's décolletage. Two months later, when Secretary Marcy instructed him to protest the seizure at Havana of the American steamship *Black Warrior,* Soulé improved upon his instructions by delivering a forty-eight-hour ultimatum; the Spanish foreign minister shrewdly suspected that this demand had not been authorized in Washington, and he refused to yield to it. After another two months, Spanish republicans attempted a revolution; Soulé was already in touch with them, and it was believed that he had subsidized them; he publicly hailed the revolution "with all the fervor of holy enthusiasm"; and he reported to Marcy that he had obtained from the revolutionists a promise to relinquish Cuba to the United States if Marcy would pay them $300,000. Before the end of his mission, he had become involved with an international network of revolutionists, including, perhaps, one regicide, and had been temporarily barred from France.[19]

18. Ettinger, *Mission of Soulé,* is the standard authority. For Soulé's Senate speech of Jan. 25, 1853, see *Congressional Globe,* 32 Cong., 2 sess., appendix, pp. 118–123; for his speech to the junta, New York *Herald,* Aug. 6, 1853; *National Intelligencer,* Aug. 9, 1853; New York *Evening Post,* Aug. 6, 1853.

19. Ettinger, *Mission of Soulé,* pp. 190–338; on the *Black Warrior,* see note 3, above. The ardent expansionists had struggled to prevent Marcy's appointment, and a

The extravagance of Soulé's conduct has diverted attention from the more important question of Washington's actual intentions with regard to Cuba. Soulé acted from the outset as though the sole purpose of his mission was to acquire Cuba by hook or by crook, but in fact, his instructions specified that the United States must not disturb Spanish sovereignty in Cuba and that he must refrain from any purchase negotiations. He himself had stated that purchase was "obsolete." The administration hoped, Soulé was told, that Cuba would "release itself or be released" from Spanish control. This phrase had a well-understood meaning at the time as referring euphemistically to the kind of internal revolution, supported by aid from outside, which had taken place in Texas. Such a revolution was, in fact, being prepared, with John A. Quitman of Mississippi as its publicly recognized leader. Quitman's filibustering project enjoyed financial and political support in many places, particularly in New York, New Orleans, and Kentucky. He had friends in the cabinet, especially Jefferson Davis and Caleb Cushing. He visited Washington in July 1853, consulted with these friends, and apparently obtained assurances that the administration would not intervene to obstruct his invasion plans. In August 1853, Quitman signed a formal agreement with the leaders of the Cuban junta in New York, by which he became "the civil and military chief of the revolution," with complete control of all funds, the power to issue bonds and military commissions, the power to raise troops and charter vessels, and all the prerogatives of a dictator. Quitman was to devote these powers to the creation of an independent government in Cuba which would retain slavery; he was to receive $1 million if and when Cuba became free.[20]

---

group of them including A. Dudley Mann of Virginia, assistant secretary of state; John L. O'Sullivan, minister to Portugal; George N. Sanders, consul at London; and Daniel E. Sickles, secretary of the London legation, sought to by-pass Marcy and work directly with Pierce. Exactly how far they succeeded, it is difficult to say, but certainly enough to cause Marcy to worry about whether he was really in charge of foreign policy. Also, it was believed in Europe that Soulé was by-passing Pierce and that he received and disbursed large sums provided by the Cuban expansionists to subsidize revolution in Spain in order to bring to power a government which would sell Cuba. Nichols, *Pierce*, p. 358; Spencer, *Victor and the Spoils*, pp. 326–328; Ettinger, pp. 300, 304–306, 316–338, 342, 349–355.

20. Rauch, *American Interest in Cuba*, pp. 262–264, makes the important point, overlooked by Ettinger, that initially, the administration looked to filibuster or revolution and not to purchase as the means of obtaining Cuba. Using the Quitman papers, Rauch, pp. 265–301, also provides what is probably the best critical account

As if to meet the challenge of Quitman's widely publicized project, the Spanish government, in September 1853, took an extraordinary step. It appointed the Marqués de la Pezuela as captain general of Cuba, and Pezuela soon launched a program for Cuba's Negro population which was unlike anything ever heard of in the island up to that time. Before his appointment Cuba had remained one of the few places where the African slave trade still flourished on a large scale. Neither the Cuban planters nor the reactionary rulers of Spain had felt any humanitarian concern about slavery. It came, therefore, like a bolt from the blue when Pezuela decreed harsh measures to suppress the slave trade and announced that all slaves brought into the island since 1835 should be freed. A large part of Cuba's Negro population had in fact arrived after 1835, and so this was tantamount to a proclamation of emancipation. Further, he encouraged racial intermarriage and organized freed Negroes into a militia, at the same time forbidding the whites to bear arms. Coming as it did from a government which made no pretense toward liberal or reformist purposes, this policy of "Africanization" had a meaning

---

of Quitman's filibuster project. Other important treatments include Portell Vilá, *Historia de Cuba*, II, 9–134; John F. H. Claiborne, *Life and Correspondence of John A. Quitman* (2 vols.; New York, 1860), II, 195–209, 346–366, 379–392; C. Stanley Urban, "The Ideology of Southern Imperialism: New Orleans and the Caribbean, 1845–1860," *LHQ*, XXXIX (1956), 48–73; Urban, "The Abortive Quitman Filibustering Expedition, 1853–1855," *Journal of Mississippi History*, XVIII (1956), 175–196. For the text of Quitman's formal agreement with the junta see Claiborne, II, 389–390. The most important and also the most obscure question about Quitman is exactly what understanding he had with the administration. Claiborne, II, 195, speaking of a visit of Quitman to Washington, says, "His designs were frankly communicated to distinguished persons at the seat of government, and he left there with the distinct impression upon his mind not only that he had their sympathies, but that there could be no pretext for the intervention of the federal authorities." But who were these distinguished persons? Did they include Pierce himself? A number of historians have believed that Jefferson Davis, secretary of war, and Caleb Cushing, attorney general, dominated Pierce and used their influence with him to support Quitman, who was a friend of Davis. But if they initially supported him, the evidence of this support was always carefully covered up, and was at least partially withdrawn after May 1854. William E. Dodd, *Jefferson Davis* (Philadelphia, 1907), pp. 132–141, and Claude M. Fuess, *The Life of Caleb Cushing* (2 vols.; New York, 1923), II, 137–177. Though Claiborne, II, 209, speaks of Quitman's "notes of what transpired at Washington," the evidence remains almost wholly circumstantial. For the belief of third parties that the administration was supporting Quitman, see W. H. Holderness to Lord Palmerston, Sept. 22, 1854, in Gavin B. Henderson (ed.), "Southern Designs on Cuba, 1854–1857 and Some European Opinions," *JSH*, V (1939), 375; Alexander H. Stephens to J. W. Duncan, May 26, 1854, in Ulrich Bonnell Phillips (ed.), *The Correspondence of Robert Toombs, Alexander H. Stephens, and Howell Cobb*, AHA *Annual Report*, 1911, II, 345.

which was paradoxical, and at the same time perfectly clear. The Cuban government was preparing to use Negro troops against any would-be filibusters and against any Cuban planters in sympathy with the filibusters.[21] It was an ironic counterpart to Soulé's own policy, for where Soulé sought to create an alliance between the revolutionary republicans of Europe and the slaveholding planters of the South and of Cuba against the government of Spain, Pezuela, with an even more daring opportunism, sought to make Cuba's enslaved black masses a bulwark of support for Spanish absolutism against American and Cuban republicanism. Pezuela's policy was risky as well as imaginative, for it aroused the people of the southern states to a strong sense of the need to move quickly before the "Africanization" program went into effect. But while increasing the risk of American intervention, it also gave to would-be filibusters a sobering awareness that invasion of Cuba might involve grim fighting against embattled slaves defending their new freedom. Pezuela further underscored his readiness to defy the Americans when he seized the *Black Warrior* and refused to treat with the American consul for her release.

Thus, at the height of the Kansas-Nebraska crisis, there was also a Cuba crisis. With the Louisiana legislature calling for "decisive and energetic measures," with Pierce informing Congress that the seizure of the *Black Warrior* was a "wanton injury" for which he had demanded "immediate indemnity," with Senator Slidell of Louisiana pressing for a repeal of the neutrality laws that restricted the activities of the filibusters, and with Caleb Cushing, in the cabinet, urging a blockade of Cuba, it appeared that some action must be imminent.[22] If John A. Quitman had chosen to move at precisely this time, he might have forced the administration to support him. But Quitman was too prudent to be a successful filibuster, and he continued his endless preparations. On April 16, he informed the junta

21. C. Stanley Urban, "The Africanization of Cuba Scare, 1853–1855," *Hispanic American Historical Review*, XXXVII (1957), 29–45; Portell Vilá, *Historia de Cuba*, II, 9–134; Rauch, *American Interest in Cuba*, pp. 276–281.

22. Resolution of the Louisiana legislature in *Senate Miscellaneous Documents*, 33 Cong., 1 sess., No. 63 (Serial 705); Pierce message, March 15, 1854, in Richardson (ed.), *Messages and Papers*, V, 234–235; Fuess, *Cushing*, II, 163. On May 1, John Slidell secured passage of a resolution in the Senate instructing the Committee on Foreign Relations to inquire into the possibility of suspending the neutrality laws (*Senate Journal*, 33 Cong., 1 sess., p. 354). Such action was not eliminated as a possibility until Pierce's proclamation in May (discussed below).

that he would move as soon as he had three thousand men, one armed steamer, and $220,000 at his disposal.[23]

But while Quitman was waiting, the indulgent attitude of the administration began to cool, and a shift in policy took place. The government decided to rely on purchase rather than a filibuster as a means of acquiring Cuba. The reason for this change is not entirely clear. Apparently some ardent expansionists genuinely believed that purchase could be easily accomplished, that filibustering endangered it, and that Quitman ought therefore to be suppressed. But also, the change was in part a retreat from expansionism altogether, resulting from the fact that the Pierce administration, already shaken by the Kansas-Nebraska affair, began to see what overwhelming criticism would result from an aggressive expansionist policy, and especially from the support of a proslavery invasion of Cuba.[24]

Hence the move in the direction of purchase. The turning point in this change of policy came on April 3, when Secretary Marcy sent Soulé entirely new instructions, revoking his previous injunction to abstain from purchase negotiations, authorizing him to offer as much as $130 million, and adding the cryptic statement, already quoted, that if this offer should fail, "you will then direct your effort to the next desirable object, which is to detach that island from the Spanish dominion."[25] Eight weeks later the administration completed its shift when Pierce issued a proclamation warning that the government would prosecute anyone who violated the neutrality laws. This came the day after he signed the Kansas-Nebraska bill, and he may have acted partly because he did not have the stomach for another such fight. At the very time when the Senate Foreign Relations Committee was about to clear the way for the filibusters by reporting favorably on Slidell's bill to repeal the neutrality laws, Pierce not only stopped them but also positively reaffirmed these laws.[26] This was perhaps the most decisive step that Pierce person-

23. Quitman to the junta, in Claiborne, *Quitman,* II, 391.

24. Urban, "Africanization of Cuba Scare," p. 41, shows that in Aug. 1854, A. Dudley Mann, an ardent southern annexationist, was advising that Quitman be curbed so that purchase might be consummated, but for influence of Kansas-Nebraska issue, see below, p. 198, and Fuess, *Cushing,* II, 163–164.

25. See footnote 4.

26. See footnote 6; also Nichols, *Pierce,* pp. 341–343, on Pierce's role in restraining the supporters of Slidell's resolution. While taking public steps against the filibusters, Pierce was, at the same time, trying to placate them by sending a message through Marcy and the district attorney at New Orleans assuring them of his inten-

ally took during the four years of his presidency.

Quitman, of course, protested. Through a spokesman in Washington, he complained with good reason that Pierce's proclamation violated his understanding with the administration. Also, he continued to busy himself with preparations for an expedition. But after a grand jury in New Orleans required him to enter into a recognizance in the sum of $3,000 to observe the neutrality laws for a period of nine months, he postponed his expedition until 1855. The delay worsened his prospects. In January the governor general arrested more than a hundred Cuban supporters of the filibuster, and some of them were put to death. Quitman had always hoped to go to the support of a revolution in Cuba rather than simply to invade the island, and this was a blow to his hopes. Later in the winter, Pierce apparently called Quitman to Washington, where he was shown convincing evidence that the island was strongly defended. Finally, in April 1855, after almost two years of postponements, Quitman gave back to the junta the powers that it had conferred upon him.[27]

Meanwhile, the Cuban crisis had been subsiding. Spain not only refused to sell Cuba; it would not even give Soulé an opening to offer to buy it. But the *Black Warrior* was returned to its owners, after the exaction of a fine of $6,000, against which the United States continued to protest. Pezuela grew somewhat less draconian in his measures of "Africanization" and in September 1854 he returned to Spain.[28] The whole Cuban affair seemed about to end with a whimper from Quitman, when Soulé, characteristically, contrived to terminate it with one more bang that brought his ministry to an end.

Soulé had never lost sight of that cryptic passage in Marcy's instructions—"the next desirable object, which is to detach that island from the Spanish dominion." Possibly Marcy himself had forgotten it. At any rate, Marcy, perhaps under pressure from Pierce, permitted himself to be persuaded that it would be a good idea for the three major American ministers in Europe—Buchanan at London, Mason at Paris, and Soulé—to meet privately for "a full and free

---

tion to gain Cuba by purchase. This message was sent after a conference with Pierce on May 30, at which Jefferson Davis, James M. Mason, Slidell, and Douglas were present. Five months earlier, three of these men—Davis, Mason, and Douglas—had been present when Pierce committed himself to Kansas-Nebraska. Significantly, Marcy, though secretary of state, was not included in either conference.

27. Foner, *History of Cuba*, II, 86–95; Claiborne, *Quitman*, II, 195–209, 391–392; Rauch, *American Interest in Cuba*, pp. 286–300.

28. *Ibid.*, p. 285; Ettinger, *Mission of Soulé*, pp. 272–290.

interchange of views" concerning Cuba. In August 1854 he authorized Soulé to arrange such a meeting. Buchanan, at this time, was arguing persuasively that pressure to sell Cuba might be brought upon the Spanish government by the holders of Spanish bonds, and it is likely that Marcy expected Buchanan to substitute this more subtle policy for Soulé's histrionic methods. In any case, he must have hoped that Buchanan's well-known caution would have a restraining effect upon Soulé. But again he had reckoned without Soulé's talent for turning every transaction into a melodrama. Great damage was done before the conferees even met, for the secrecy of their meeting had been announced in such stage whispers that every diplomat in Europe knew something singular was afoot. Then, when the three men gathered, instead of Buchanan's imposing his views upon Soulé, Soulé somehow imposed his views upon Buchanan.

The conferees met at Ostend in October 1854, adjourned to Aix-la-Chapelle, and after three days of discussion put their names to a statement which Marcy had intended to be a memorandum for the State Department but which suddenly assumed the character of a pronouncement to the world. In this Ostend Manifesto, as it came to be called, the three envoys recited their shared belief that "Cuba is as necessary to the North American republic as any of its present members, and that it belongs naturally to that great family of states of which the Union is the Providential Nursery," and also that the United States should make an "immediate and earnest effort" to buy Cuba "at any price for which it can be obtained," so long as the price did not exceed $120 million. With overblown rhetoric, they then pictured the prosperity which the purchase price would bring to Spain, as that country "would speedily become what a bountiful Providence intended she should be, one of the first Nations of Continental Europe—rich, powerful, and contented."

So far this was only one more specimen of a prose style which Manifest Destiny had already made familiar to most Americans, if not to Europeans. But the sting of the Ostend statement was in its tail. If Spain should refuse to sell, and if Spain's possession of Cuba "should seriously endanger our internal peace"—perhaps by an Africanization program—then "by every law, human and Divine, we shall be justified in wresting it from Spain if we possess the power."[29]

29. *Ibid.*, pp. 339–412, provides an excellent account of every aspect of the Ostend Manifesto. For Buchanan's plan to work through the Spanish bondholders, see

What ever induced James Buchanan to put his signature to this statement remains a matter for speculation. Perhaps, it has been suggested, he was mesmerized by Soulé. But Buchanan was not easily enticed into steps that would operate to his own disadvantage, and it is quite possible that he saw an opportunity to embarrass William L. Marcy, his most serious future rival for the presidential nomination. Marcy's old maneuvering to "detach that island" left him vulnerable. He could not wholly disavow the Ostend statement, yet it would place him in a very awkward position, and it would make Buchanan popular with expansionists. Perhaps this explanation attributes too much Machiavellian skill to a tired, elderly alumnus of the Pennsylvania school of politics, but in any event, Buchanan signed.[30]

On the same day in November 1854 that Marcy received the ministers' statement, he also learned that not one of the nine New York congressmen who voted for the Kansas-Nebraska Act had survived the election.[31] It would be difficult to say which news was worse. But worst of all was the fact that within two weeks, the New York *Herald* got wind of what had happened and published the content of the ministers' recommendations. This aroused such insistent demands for the administration to abandon its secrecy that in the following March, after months of prodding, Pierce had to send the correspondence to Congress—with a little editing. Marcy's "detach that island" was omitted, but the insistence of Soulé and the expansionists made it impossible to suppress anything else. Marcy forced Soulé's resignation by coldly repudiating the whole thing, but the damage had been done. For months the administration was

---

Buchanan to Pierce, Dec. 11, 1852; Buchanan to Marcy, July 11, 1854, in John Bassett Moore (ed.), *The Works of James Buchanan* (12 vols.; Philadelphia, 1908–11), VIII, 493–499; IX, 211–213; Nichols, *Pierce,* pp. 357–358; Rauch, *American Interest in Cuba,* pp. 258–259; Spencer, *Victor and the Spoils,* p. 325; for Marcy's instructions, Aug. 16, 1854, to Soulé, to arrange the conference of ministers, see Manning (ed.), *Diplomatic Correspondence of the United States,* XI, 193–194; for text of the Ostend statement, see *ibid.*, VII, 579–585. For a facsimile rough draft in the hand of James Buchanan, see Nichols, *Pierce,* following p. 596.

30. Ettinger, *Mission of Soulé,* pp. 364–365, suggests that Soulé had "hypnotized" and "beguiled" Buchanan; Spencer, *Victor and the Spoils,* p. 331, accepts the New York *Herald*'s argument that Buchanan had cast in his lot with the extremists and was working to undermine Marcy's position as a presidential candidate (Dec. 29, 1854); Philip Shriver Klein, *President James Buchanan* (University Park, Pa., 1962), pp. 237–241, stresses Buchanan's reluctance to participate in the ministers' meeting, and his insistence on the point that the Manifesto did not speak unconditionally of wresting Cuba from Spain.

31. Ettinger, *Mission of Soulé,* pp. 376–378.

held up to the country and to the world as the advocate of a policy of "shame and dishonor," the supporter of a "bucaneering document," a "highwayman's plea." American diplomacy, said the London *Times,* was given to "the habitual pursuit of dishonorable objects by clandestine means."[32]

The Ostend Manifesto and the Kansas-Nebraska Act were the two great calamities of Franklin Pierce's presidency. This is true in the obvious sense that both brought down an avalanche of public criticism upon the administration. But it is also true in the deeper sense that each permanently discredited an administration doctrine which, up to that time, had been regarded as quite respectable. The doctrine of popular sovereignty was respectable until the repeal of the Missouri Compromise linked it with the goals of slavery extension. The doctrine of Manifest Destiny, with its purpose of spreading American democratic institutions under the American flag, was widely regarded as respectable until the Ostend Manifesto linked it with naked aggression. Douglas and Soulé, between them, therefore, had spiked two of the Democratic party's best weapons in what would today be called the battle for men's minds.

Both the Act and the Manifesto resulted from pressures that had been brought to bear by the advocates of slavery. Both measures cost the administration a fearful loss of political support. On the balance sheet of politics, such a lavish expenditure of political strength can be justified only by solid and important gains. One must ask, therefore, what the slaveholding interest gained in return for this squandering of the power won in 1852. By this criterion, their policies in 1854 were the work of folly. They paid more in unpopularity for an empty right to take slaves where few intended to take them than they might have paid to carve a new slave state out of Texas. They incurred as much opprobrium for indulging in a piece of fancy rhetoric at Ostend as they might have incurred for supporting John A. Quitman with guns and money in an operation to bring Cuba along the path of Texas and California.

For practical purposes, the Ostend Manifesto gave the *coup de*

32. *Ibid.*, pp. 391–407; Marcy to Soulé, Nov. 13, 1854, in Manning (ed.), *Diplomatic Correspondence of the United States,* XI, 196–201; Foner, *History of Cuba,* II, 101–102. Marcy stated that he did not understand the ministers to mean that the United States should confront Spain with the alternative of cession or seizure. Marcy later wrote a semipublic letter to L. B. Shepherd, a leading New York Democrat, April 15, 1855, "The robber doctrine, I abhor." Quoted in Ettinger, *Mission of Soulé,* p. 393.

grâce to expansionism—at least until 1898, when slavery had been dead for thirty years. The fact that it was a turning point is clearer in retrospect than it was at the time,[33] for the Democratic administrations of Pierce and Buchanan continued to support expansionism, and Lincoln was still afraid of it in 1861.[34] This presidential support resulted in at least two steps which momentarily had the appearance of victories for Manifest Destiny. The first of these came in May 1856, when the Pierce administration accorded diplomatic recognition to the government of William Walker as president of Nicaragua. Walker was an unimpressive-looking, somewhat inarticulate little man from Tennessee, but he did not lack powers of decision. After moving to California, he became a filibuster, convinced of his destiny to "regenerate"—and rule—in Latin America. In 1853 he invaded Lower California unsuccessfully and after proclaiming his Republic of Lower California (and later of Sonora) was forced to retreat to San Diego, where he surrendered to American authorities who placed him under arrest. A San Francisco jury acquitted him after being out for eight minutes. This vindication encouraged him to try again, and in May 1855 the "grey-eyed man of destiny" sailed with about sixty followers ("the immortals") to participate in a civil war in Nicaragua. Within half a year he was in control of the country. Within little more than a year, he made himself president, and Franklin Pierce recognized his government. But he was not recognized by Cornelius Vanderbilt, for he had recklessly revoked a franchise for a Vanderbilt-controlled steamship company in Nicaragua, and this proved his undoing. Vanderbilt was able to cut off his support, and within another year his adversaries in Nicaragua had overwhelmed him and permitted him to flee from the country on an American naval vessel. But he had filibustering in his blood, and in 1860 he returned to Central America, where he met his death before a firing squad.[35]

33. At the time, the antislavery men were pessimistically convinced that slavery expansionism was completely triumphant. Wendell Phillips wrote to Mrs. Elizabeth Pease Nicol, Aug. 7, 1854, "The *government* has fallen into the hands of the slave power *completely*. So far as *national* politics are concerned, we are beaten—there's no hope. We shall have Cuba in a year or two, Mexico in five." Wendell P. and Francis J. Garrison (eds.), *William Lloyd Garrison, 1805–1879: The Story of His Life Told by His Children* (4 vols.; New York, 1885–89), III, 411.

34. Benjamin P. Thomas, *Abraham Lincoln* (New York, 1952), p. 230.

35. The standard scholarly treatment of Walker is Scroggs, *Filibusters and Financiers*, but see also Wallace, *Destiny and Glory*, pp. 142–240; Laurence Greene, *The Filibuster:*

Walker's career furnishes an interesting contrast to that of Quitman, for Walker was a true filibuster, while Quitman turned out to be only an expansionist politician who dreamed of being a filibuster. Walker understood that he must not wait for timid men to agree to bold measures. Instead, by confronting them with *faits accompli,* he would make it easy for them to accept what they desired but did not dare. Quitman, on the other hand, waited for a consensus in support of his invasion of Cuba, and it never materialized. It is perhaps going beyond the scope of this history to add that of course Quitman died in bed.

Walker's experience also offers an insight into the relationship between filibustering and slavery. The Man of Destiny was, of course, from a slave state, and he accepted slavery as a matter of course, but there is no evidence that he was dedicated to the expansion of slavery, and the impulse of some historians to picture him as a minion of the "slave power" reflects a failure to recognize that Walker may have been exploiting the proslavery elements, instead of their exploiting him.[36] In September 1856, with defeat staring him in the face, Walker revoked the decrees of the former Federation of Central American States which had abolished slavery in Nicaragua, and in 1860, in his book *The War in Nicaragua,* he pictured his republic as a potential field for the expansion of slavery. But in both cases, it is clear, he was trying to win desperately needed support for his own personal rule in Nicaragua.[37] Until this need arose, his history had been simply an adventure story, a drama of daring and conquest to fulfill the glorious destiny of a superman rather than to serve the interests of a section. As such, it had appealed immensely to the romantic imagination of Americans who were at that time uninhibited by notions of international responsibility, and Walker had seemed something

---

The Career of William Walker (Indianapolis, 1937); Spencer, *Victor and the Spoils,* pp. 353–364. Walker's own account, *The War in Nicaragua* (Mobile, 1860), is a detailed and important source. For Pierce's recognition of Walker, see Marcy to John H. Wheeler, June 3, 1856, in Manning (ed.), *Diplomatic Correspondence of the United States,* IV, 85–86.

36. Dodd, *Jefferson Davis,* p. 136, equates Walker with Quitman, calling both of them "propagandists" of the "slave states."

37. Scroggs, *Filibusters and Financiers,* pp. 6–8, 49–51, 67–69, 210–212, and Greene, *The Filibuster,* pp. 311–314, both offer convincing refutation of the idea that Walker's activity was a manifestation of slavery expansionism, and Greene regards Walker as being, in some ways, an idealist. He had opposed the Kansas-Nebraska Act.

of a hero to the American public, northern as well as southern.

The second seeming gain for expansionism came in January 1860, when President Buchanan submitted to the Senate a treaty negotiated with the Juárez government, which was at the time trying to overthrow the government in Mexico City. The treaty came after a long period of turmoil in Mexico had led to the default of payment on obligations due American citizens, to conditions that endangered the lives of Americans in Mexico, and to a feebleness and a desperate need for money on the part of the Mexican government. Some sharp-eyed Americans had seen in Mexico's weakness an opportunity to gain valuable property rights and to extend American control. Under the Pierce administration, the American minister, John Forsyth, had negotiated a treaty for a $15 million loan to Mexico, which he said would constitute a kind of "floating mortgage upon the territory of a poor neighbor" which she could not pay off, and which "could only be paid by a peaceable foreclosure with her consent." Thus, "finding it impossible to acquire territory immediately," Forsyth had sought "to pave the way for the acquisition hereafter."[38] But President Pierce, still smarting from the burns of the Ostend Manifesto, had not submitted Forsyth's treaty to the Senate and had left the Mexican problem for Buchanan to solve. Buchanan likewise refrained from submitting Forsyth's treaty, but he spoke up vigorously for expansion, raising the subject repeatedly in his messages to Congress, with reference both to Mexico and to Cuba. He also instructed his envoys to seek territorial acquisitions and recommended to Congress that it grant him the "necessary power to take possession of a sufficient portion of the remote and unsettled territory of Mexico, to be held in pledge" for the payment of American claims. Further, he proposed that the United States assume a temporary protectorate over the northern parts of Sonora and Chihuahua. Congress ignored the proposals, but a year later Buchanan asked for authority to send "a sufficient military force to enter Mexico" for the purpose of obtaining "indemnity for the past and security for the future." Meanwhile, he had authorized his minister to Mexico, Robert M. McLane, to negotiate with Juárez. The Juárez regime, involved as it was in civil war, desperately needed American cooperation, and it gave McLane a treaty containing ex-

38. Forsyth to Lewis Cass, April 4, 1857, in Manning (ed.), *Diplomatic Correspondence of the United States,* IX, 902–909.

traordinary concessions. For the sum of $4 million, Mexico would grant to the United States two perpetual rights-of-way from the Atlantic to the Pacific, one across the Isthmus of Tehuantepec, another from a point on the lower Rio Grande to the port of Mazatlán. The United States was also authorized to protect these routes by unilateral military action and to intervene with the use of force to maintain the rights and the security of American citizens in Mexico. Few treaties submitted to the Senate have ever granted so much to the United States as this one, and it is easy to imagine the welcome with which it would have been received during the Polk administration. But the Senate rejected it, 18 to 27. Fourteen southerners and four northerners voted for it; four southerners and twenty-three northerners, against it.[39]

The 1850s may have marked, as some historians suggest, the high tide of Manifest Destiny, but when all the dust of manifestoes, filibusters, annexation treaties, and spread-eagle speeches had settled, the only territory that had changed hands during this decade was the strip of land obtained in the Gadsden Purchase.

Expansionism had seemingly won a popular mandate in 1852, and Fillmore's popularity had suffered severely because he failed to take an expansionist position. Yet by 1855, its force was virtually spent. The explanation for this precipitate decline must lie in the fact that it had lost its national quality and had become a sectional issue. Polk himself had presumably won his election in 1844 by linking the "re-occupation" of Oregon with the "re-annexation" of Texas, and thus by transcending the sectional limitations of the Texas Question. But his failure to "re-occupy" all of Oregon, after

39. The best general accounts of relations with Mexico between 1854 and 1861 are in Rippy, *The United States and Mexico*, pp. 212–229, and in James Morton Callahan, *American Foreign Policy in Mexican Relations* (New York, 1932), pp. 244–275. But see also Rippy, "Diplomacy of the United States and Mexico Regarding the Isthmus of Tehuantepec, 1848–1860," *MVHR*, VI (1919–20), 503–531; Howard L. Wilson, "President Buchanan's Proposed Intervention in Mexico," *AHR*, V (1900), 687–701; Callahan, "The Mexican Policy of Southern Leaders under Buchanan's Administration," AHA *Annual Report*, 1910, pp. 133–151. For Buchanan's messages repeatedly urging Congress to support an expansionist policy in Mexico and to purchase Cuba, see Richardson (ed.), *Messages and Papers*, V, 510–511, 536, 561, 642 (concerning Cuba), and 514, 568, 578–579 (concerning Mexico). For the diplomatic correspondence between the State Department, McLane, and Mexican officials, see Manning (ed.), *Diplomatic Correspondence of the United States*, IX, 260–293, 1037–1234. For the defeat of the McLane-Ocampo treaty in the Senate, see W. Stull Holt, *Treaties Defeated by the Senate* (Baltimore, 1933), pp. 92–96; *Journal of Executive Proceedings of Senate*, XI, 115–199.

the "re-annexing" of all of Texas, had shown how difficult it was to preserve the bisectional balance of expansionism. The expansionist "Young America" movement of the fifties, with its bumptious republicanism, its noisy scorn for "decadent monarchies," and its shrill insistence upon the regenerative mission of America, represented another effort to make expansionism again a national program. This was why George Sanders and the slavery expansionists consorted with Mazzini, Kossuth, and the firebrands of European revolution.[40] But expansionism meant expansionism southward, and expansion southward meant the extension of slavery. Therefore, expansion became more and more a southern goal, and thus a sectional issue.[41] In the later fifties, two principal agencies of expansionism were *De Bow's Review,* an ardently prosouthern periodical published at New Orleans, whose editor, James D. B. De Bow, wanted to make New Orleans the commercial center of a rich tropical empire;[42] and the Knights of the Golden Circle, a secret society of southerners who aspired to extend slavery and the power of the South all around the circle of tropical and semitropical golden lands bordering the Gulf of Mexico. In 1860, the Knights, with an imperial program of expansion, claimed a membership of 65,000, including all but three of the governors of the slave states, and several members of President Buchanan's cabinet.[43]

By the time the southern states seceded, Manifest Destiny had reached a supreme paradox: northern unionists who believed in American nationalism resisted most proposals for further territorial growth of the nation, while states' rights southerners who denied that the Union was a nation sought to extend the national domain from pole to pole. The expansionists were not nationalists, and the nationalists were not expansionists. Thus, many of the southerners

40. Merle Curti, "Young America," pp. 34–55; Curti, "George N. Sanders, American Patriot of the Fifties," *South Atlantic Quarterly,* XXVII (1928), 79–87; Julius W. Pratt, "John L. O'Sullivan and Manifest Destiny," *NYH,* XXXI (1933), 213–234; Rauch, *American Interest in Cuba,* pp. 213–297. On expansionism generally: Albert K. Weinberg, *Manifest Destiny* (Baltimore, 1935); Frederick Merk, *Manifest Destiny and Mission in American History* (New York, 1963).

41. Urban, "The Ideology of Southern Imperialism."

42. Robert F. Durden, "James D. B. De Bow: Convolutions of a Slavery Expansionist," *JSH,* XVII (1951), 441–461; Rollin G. Osterweis, *Romanticism and Nationalism in the Old South* (New Haven, 1949), pp. 155–185.

43. Ollinger Crenshaw, "The Knights of the Golden Circle: The Career of George Bickley," *AHR,* XLVII (1941), 23–50; C. A. Bridges, "The Knights of the Golden Circle: A Filibustering Fantasy," *SWHQ,* XLIV (1941), 287–302.

who were most grandiose in their dreams of bringing distant and exotic lands under the American flag—who were most extravagant in their claims for the mission of America in foreign parts—were also most jealous in denying the supremacy of the American government on the domestic scene. For many, there was but a short interval between their last efforts to bring new potential states into the Union and their decisions to take their own states out.[44]

But this is not the only irony of Manifest Destiny. The supreme irony, it may be argued, takes us back to the Kansas-Nebraska Act and was suggested by William L. Marcy only a few weeks after that disastrous measure had been adopted and before most southerners realized that their hopes for Cuba were already a lost cause. "The Nebraska Question," said Marcy, "has sadly shattered our party in all the free states and deprived it of that strength which was needed and could have been much more profitably used for the acquisition of Cuba."[45] In the calculations of *realpolitik,* the party and the southern leaders in the party had enough strength to force through one thoroughly unpopular measure, but not two. They could use this strength to create a dubious opportunity for slavery in Kansas, or they might use it to annex Cuba. But they could not use it for both. Without ever recognizing the necessity for a choice, they had followed a policy which, in effect, sacrificed the Cuban substance for a Kansan shadow. Many intelligent southerners, even at this juncture, realized the emptiness of their victory in Kansas, but it is not likely that even a handful of them saw just how empty it really was.

44. Thomas Corwin of Ohio clearly defined this irony in a speech in the House, Jan. 21, 1861, addressing the southerners: "You say you must acquire other territory, and you gravely sit down here in the halls of legislation, in the only successful republic that has yet appeared, in our form, on the face of the earth, and distribute among yourselves the dominions of neighboring states while you are about to break in pieces your own government. . . . You are looking toward Mexico, Nicaragua, and Brazil to determine what you will do with all their territory when you get it, while you are not sure you will have a government to which these could be ceded." *Congressional Globe,* 36 Cong., 2 sess., appendix, p. 74. There was still, of course, some northern expansionist sentiment looking toward Canada. See Donald F. Warner, *The Idea of Continental Union: Agitation for the Annexation of Canada to the United States, 1849–1893* (Lexington, Ky., 1960).

45. Marcy to Mason, July 23, 1854, quoted in Spencer, *Victor and the Spoils,* p. 324.

CHAPTER 9

Two Wars in Kansas

ON May 25, 1854, shortly after the Kansas-Nebraska bill had passed the House, William H. Seward made a fighting speech in the Senate. "Come on then, Gentlemen of the Slave States," he said, "since there is no escaping your challenge, I accept it in behalf of the cause of freedom. We will engage in competition for the virgin soil of Kansas, and God give the victory to the side which is stronger in numbers as it is in right."[1]

Whether Seward meant this literally or not, it was in fact a singularly accurate forecast for territorial Kansas. Instead of settling a controversy, the adoption of the act transplanted the controversy from the halls of Congress to the plains of Kansas. The forces which had fought one another so fiercely in Washington continued to fight beyond the wide Missouri.

Each side later accused the other of taking the initiative in starting this contest, but apparently the first planned effort to organize migration to Kansas in a way that would bear upon the slavery question was made by Eli Thayer of Massachusetts. Aroused very early in the course of the battle in Congress, Thayer had moved fast—so fast indeed that a month before passage of the Kansas-Nebraska Act, he had obtained a charter from the Massachusetts legislature incorporating the Massachusetts Emigrant Aid Company, with a capital stock not to exceed $5 million, "for the pur-

1. *Congressional Globe,* 33 Cong., 1 sess., appendix, p. 769.

pose of assisting emigrants to settle in the West."[2]

Thayer's grand design was, of course, a hope rather than a reality, and he did not have any part of $5 million. In fact, when his project finally got started it did not operate under this original charter. His irresponsible way of treating dreams as if they were facts was later to distress his friends even more than it alarmed his enemies. But when men in western Missouri read in Horace Greeley's New York *Tribune* an account of Thayer's "Plan of Operations," it conjured up in their minds the picture of a vast, wealthy, and overpowering abolitionist organization ready to hurl 20,000 hirelings upon their borders.[3] Their reaction was reflected by a correspondent of Senator Atchison, who wrote in alarm, "we are threatened . . . [with] being made the unwilling receptacle of the filth, scum, and offscourings of the East . . . to pollute our fair land . . . to preach abolitionism, and dig underground Rail-roads."[4] Missourians were not the kind of men to submit meekly to such an incursion, and on July 29, at Weston, Missouri, an excited gathering organized the "Platte County Self-Defensive Association," asserting their readiness to go to Kansas "to assist in removing any and all emigrants who go there under the auspices of Northern Emigrant Aid Societies."[5] Thus, while antislavery men were first to organize migration as a means of continuing the contest over slavery, Missourians were first openly to invoke the use of force. Soon, the Missourians began to perceive the advantages of operating without publicity, whereupon they organized secret societies, including the "Blue

2. *Private and Special Statutes of the Commonwealth of Massachusetts* (Boston, 1861) X, 204 (Act of April 26, 1854), 282–283 (Act of Feb. 21, 1855); *Resolutions and Private Acts of the General Assembly of the State of Connecticut, May session, 1854* (New Haven, 1854), pp. 118–119; Samuel A. Johnson, *The Battle Cry of Freedom: The New England Emigrant Aid Company in the Kansas Crusade* (Lawrence, Kan., 1954), pp. 16–17.

3. New York *Tribune*, May 29, 30, 31, June 22, 1854; Johnson, *Battle Cry*, pp. 94–95; Mary J. Klem, "Missouri in the Kansas Struggle," MVHA *Proceedings*, IX (1917–18), 400; *House Reports*, 34 Cong., 1 sess., No. 200 (Serial 869), titled *Report of the Special Committee Appointed to Investigate the Troubles in the Territory of Kansas* (cited hereafter as *Howard Committee Report*), pp. 356, 838.

4. William Wyandot to Atchison, July 11, 1854, in William E. Parrish, *David Rice Atchison of Missouri: Border Politician* (Columbia, Mo., 1961), p. 161.

5. Johnson, *Battle Cry*, p. 96; Klem, "Missouri in the Kansas Struggle"; Elmer LeRoy Craik, "Southern Interest in Territorial Kansas, 1854–1858," KSHS *Collections*, XV (1919–22), 334–450; James A. Rawley, *Race and Politics: "Bleeding Kansas" and the Coming of the Civil War* (Philadelphia, 1969), pp. 85–86; James C. Malin, "The Proslavery Background of the Kansas Struggle" *MVHR*, X (1923), 285–305.

Lodges," and "Platte County Regulators."[6]

Thus the lines were already drawn before Andrew Reeder arrived as the first governor of the territory in October. Reeder ordered a census and scheduled an election of a territorial legislature. The census showed a population of 8,601, of whom 2,905 were eligible to vote. The election was bound to be irregular in any case, for the loosely phrased law permitted any "resident" to vote, however recently he had arrived. This encouraged both sides to deploy all the last-minute "residents" that they could muster. Thus the Emigrant Aid Company hastened the departure of emigrant parties in the hope that they might reach Kansas in time for the election. Missourians responded energetically to this threat. A large body of them, led by Senator Atchison, came across the border, full of whiskey and resentment toward abolitionist "invaders," and voted in the election on March 30, 1855, thereby producing a topheavy proslavery majority in the total vote of 6,307.[7]

For most of its length, the Missouri-Kansas border was only a surveyor's line, and as the excitable Missourians saw it, they were defending their own homeland against an invasion by Yankee mercenaries. But what they had done was to steal an election. Aside from the moral wrong of this transaction, it was also tactically wrong, for they would have won on a fair ballot. The concentration of free-soil settlers in one area around Lawrence, the trumpetings of the free-soil newspaper, the *Herald of Freedom*, and the exaggerated claims of the New York *Tribune* all made the free-state contingent appear formidable out of proportion to its actual size, but in fact the number of settlers sent out by the several emigrant aid societies now in operation was small. Thayer's society—the first and the most important—assisted about 650 emigrants in 1854 and about 1,000 in 1855, but it is not likely that many of the latter had arrived by March 30. Cool-headed people like Alexander H. Stephens knew when they studied the census that emigrants from the slave states outnumbered emigrants from the free states 1,670 to 1,018, and that the proslavery forces stood to win. The Missourians'

6. Johnson, *Battle Cry*, pp. 97–98; J. N. Holloway, *History of Kansas* (Lafayette, Ind., 1868), pp. 122–124; William M. Paxton, *Annals of Platte County, Missouri* (Kansas City, Mo., 1897); *Howard Committee Report*, pp. 896, 902–903.

7. For census, see *ibid.*, pp. 72–100, 934; for the statute permitting vote by "actual residents," p. 866; for copious testimony on voting by recently arrived Emigrant Aid groups and by Missourians, pp. 101–523, 834–872, 894–900; for the election returns, pp. 30–33. Also, James R. McClure, "Taking the Census and Other Incidents in 1855," KSHS *Transactions*, VIII (1903–04), 227–250.

act of aggression was, therefore, also an act of supererogation by which they compromised their own victory.[8]

If the Missourians made a mockery of popular sovereignty by casting these fraudulent votes, Governor Reeder destroyed the prospects of normal local adjustment when he allowed the result to stand. Reeder permitted himself to be governed by the formal consideration that, in many districts, the vote was not challenged. He threw out the returns in certain districts where the result was contested, but he felt that where there was no challenge, he could not interfere. His self-restraint left a fraudulently elected group in control, and this led to reactions that soon brought anarchy to Kansas.[9]

The great anomaly of "Bleeding Kansas" is that the slavery issue reached a condition of intolerable tension and violence for the first time in an area where a majority of the inhabitants apparently did not care very much one way or the other about slavery. The evidence is clear that an overwhelming proportion of the settlers were far more concerned about land titles than they were about any other public question. Most of them were land-hungry westerners, engaged in the hallowed democratic practice of squatting on new lands in order to stake a claim. Most were conspicuously indifferent to law, either in its "higher" form or in the ordinary statutory version. The most natural reason for strife among them lay in the fact that, at the time of their arrival, the government had done a poor job of making land with clear title available. A number of Indian tribes still held titles, and when the territory was opened to settlement on May 30, 1854, the land had not even been surveyed. Six months later, not a single acre was legally available, either for preemption or for cash purchase, and the first surveyor's plats did not reach the land office until January.[10] During the first great influx

8. On the numbers of migrants sent by the Aid Society, see Johnson, *Battle Cry*, p. 75; Louise Barry, "The Emigrant Aid Company Parties of 1854," and "The New England Emigrant Aid Company Parties of 1855," *Kansas Historical Quarterly*, XII (1943), 115–155, 227–268; *Howard Committee Report*, pp. 873–893 (with lists of emigrants). On Stephens's analysis of the relative strength of the parties, *Congressional Globe*, 34 Cong., 1 sess., appendix, pp. 1070–1076. On the geographical origins of the population of Kansas at about this time, see James C. Malin, *John Brown and the Legend of Fifty-Six* (Philadelphia, 1942), pp. 511–515.

9. Testimony of Reeder, in *Howard Committee Report*, pp. 935–936. For results of the reelection in districts where Reeder threw out the vote, see *ibid.*, pp. 36, 524–546. The proslavery vote was 560; antislavery, 802.

10. The definitive study of the public land situation in territorial Kansas and its relation to the slavery contest is Paul Wallace Gates, *Fifty Million Acres: Conflicts over*

of emigrants, therefore, no one held title to the land he occupied, and contention between rival claimants became chronic. Accordingly, the ensuing quarrels fell into a kind of pattern. The Missouri claimants, thinking of Kansas as their own neighborhood, regarded the immigrant Yankees as invaders, while the Yankees hated the Missourians for grabbing the best land without honestly settling on it, and stigmatized them as half-savage "Pukes." Such friction was not unusual in frontier situations, and it often led to controversy, lawlessness, and even violence.

Much of the friction in Kansas in the fifties began as this kind of diffused contention over land claims. There is little evidence of any deep ideological divisions on the questions of slavery or the Negro, although one faction did want to bring in Negroes as slaves, while the other did not want them to come in either slave or free. At a later time, when the "free-state" faction had set up a governmental organization of their own, they adopted severely discriminatory laws, prohibiting the entry of Negroes into Kansas and excluding them from the franchise. One free-soil clergyman explained his position by saying, "I kem to Kansas to live in a free state and I don't want niggers a-trampin' over my grave."[11]

Thus, the issue of slavery was perhaps not the basic source of division between the "proslavery" and the "antislavery" parties in Kansas. But if it was not crucial in producing friction, it was certainly crucial in structuring and intensifying the friction. When one group of land claimants came with the aid of money furnished by antislavery organizations and the other group looked for leadership to Senator Atchison, who talked fiercely of his yearning "to kill every God-damned abolitionist in the district";[12] when one group gravitated toward the town of Lawrence, subsidized by Aid Society money, and the other concentrated around Leavenworth; when each squatter found himself applauded in his aggressive actions by a vast sectional claque, and found himself opposed not merely by another squatter like himself but by an organized adversary group, the effect was to polarize and organize all the diffused and random

Kansas Land Policy, 1854–1890 (Ithaca, N.Y., 1954), esp. pp. 19–22, 48–71. Also, Malin, *Brown and the Legend of Fifty-Six*, pp. 498–508.

11. William A. Phillips, *The Conquest of Kansas by Missouri and Her Allies* (Boston, 1856), pp. 127–140; *Howard Committee Report*, pp. 54, 713–756; Malin, *Brown and the Legend of Fifty-Six*, pp. 509–536; Malin, "The Topeka Statehood Movement Reconsidered: Origins," in *Territorial Kansas: Studies Commemorating the Centennial* (Lawrence, Kan., 1954), pp. 33–69.

12. Testimony of Dr. G. A. Cutler, *Howard Committee Report*, p. 357.

antagonisms, which might otherwise have remained merely individual and local.

The effect of the slavery issue in thus aligning many of the personal drives and frustrations of the Kansas pioneers would have been serious enough in itself, but it was rendered far more acute by the political sequel to the election of 1855. When the fraudulently elected legislature met, it acted in the most bigoted and despotic way. Over the governor's veto, it adopted a uniquely repressive set of statutes for the protection of slavery, making it a capital offense to give aid to a fugitive slave and a felony to question the right to hold slaves in Kansas. Also, the proslavery majority expelled the handful of antislavery legislators who had been elected in the districts where Governor Reeder had ordered reelections.[13] Although less spectacular than the incredible statutes, this expulsion proved far more serious in its consequences, for it impelled the free-state men to deny the validity of the territorial government and to set up a rival government of their own. During the summer and fall of 1855, they prepared the way for a convention at Topeka, which drafted a state constitution for Kansas. In December an election in which only free-state men participated "ratified" this constitution, and in January 1856 free-state voters elected a "governor" and members of a "legislature." In March the legislature convened at Topeka to take the steps preparatory to statehood—adopting "statutes" and even naming United States senators.[14]

This procedure had a highly ambiguous quality about it, for in one sense it was a recognized right of territorial populations to set up organizations which might be called "shadow governments" in preparation for statehood. On the other hand, there was no legal sanction of any kind for defying the authority of the officially recognized territorial government—no matter how outrageous its legislation might be. For a time, it was not clear which the organizers of the Topeka government were doing, and in fact they disagreed among themselves. The more sober and prudent leaders, like Amos

13. *Statutes of the Territory of Kansas, Passed at the First Session of the Legislative Assembly, 1855* (Shawnee M. L. School, 1855), p. 715.

14. On the Topeka movement: election of convention and text of Topeka Constitution in *Howard Committee Report*, pp. 661–712, 607–649; Malin, "Topeka Movement Reconsidered"; Charles Robinson, "Topeka and Her Constitution," KSHS *Transactions*, VI (1897–1900), 291–305; "The Topeka Movement: Record of the Executive Committee of Kansas Territory," KSHS *Collections*, XIII (1913–14), 125–249; New York *Tribune*, Nov. 10–18, 1855 (proceedings of Topeka Convention).

A. Lawrence, chief backer of the New England Emigrant Aid Society, cautioned that the free-staters could not set up an operative state government "without coming in collision directly with the United States." Later, Andrew Reeder, who had been dismissed from the territorial governorship and had then joined the free-state group, warned his associates that, by "putting a set of laws in operation in opposition to the territorial government," they would place themselves, "so far as legality is concerned, in the wrong."[15]

Most of the antislavery men heeded this advice, and in their legislature they voted that the laws which they were framing should go into effect only upon the admission of Kansas to statehood. At their Big Springs convention, the moderates brought in a resolution declaring that it would be "untimely and inexpedient" to try to set up a free-state government which would conflict with the territorial government.[16]

But the more militant elements in the free-state faction grew impatient under this kind of forbearance, and Reeder, who had not yet arrived at his later moderate position, declared in ringing tones, "We owe no allegiance or obedience to the tyrannical enactments of this spurious legislature." The more impetuous supporters of the free-state "government" cheered Reeder's pronouncement and adopted a resolution endorsing forcible measures when peaceful methods proved unsuccessful. The most violent figure in this wing of the party was Jim Lane, who did not hesitate to raise troops and to menace the territorial officials with military force.[17]

At times, the division between hotheads and men of caution caused deep internal tensions within the free-state ranks. But ultimately, instead of weakening the antislavery position, this division seemed actually to strengthen it. It served the same purpose which "nonviolent resistance" was to serve for later protest groups, by giving them the dramatic and psychological advantage of defying the authority which they opposed, while making their adversaries

15. Letter of Lawrence, Aug. 16, 1855, and other statements by him in Malin, *Brown and the Legend of Fifty-Six*, pp. 521, 525, 526; letter of Reeder, Feb. 12, 1856, in *Howard Committee Report*, p. 1136.

16. Resolutions at the Big Springs Convention, in Daniel W. Wilder, *The Annals of Kansas* (Topeka, 1886), pp. 75–77; R. G. Elliot, "The Big Springs Convention," KSHS *Transactions*, VIII (1903–04), 362–377.

17. Wilder, *Annals*, p. 77. On Lane, Leverett W. Spring, "The Career of a Kansas Politician," *AHR*, IV (1898), 80–104; Wendell Holmes Stephenson, "Political Career of General James H. Lane," KSHS *Publications*, III (1930), 41–95.

appear to be the aggressors. In Kansas, overt acts of resistance to the territorial authorities were often threatened but seldom committed. The result was a form of brinksmanship that threw the proslavery faction off balance and more than once goaded them into acts of repression which discredited their own cause.

The formation of a rival government also had a decisive effect in consolidating and perpetuating the division of the people of Kansas into hostile camps. In a situation where there were almost no slaves, the slavery issue, by itself, could hardly have been so deeply divisive if it had not been reinforced by organizational divisions which could not be bridged. But when the "proslavery" faction supported a government at Lecompton which the other faction regarded as fraudulent or "bogus," and when the "antislavery" faction supported a government at Topeka which the other faction regarded as illegal and revolutionary, the demarcation between "proslavery" and "antislavery" became far sharper than attitudes toward the peculiar institution, as such, would have made it. Structurally, this was the opposite of a democratic situation, in which the gravitational forces draw rival groups toward a middle ground of accommodation. In Kansas, the situation caused the gravitational forces to pull toward the extremes. If one government was valid, the other was spurious, either morally or legally, as the case might be. If the acts of one were binding upon the citizens, then submission to the authority of the other by, for instance, paying its taxes or serving in its militia would constitute sedition, or even treason.[18]

Rhetorically, the two factions were by this time at war, and as befitted belligerents, they soon began gathering their armaments. Only three days after the election of the territorial legislature, Charles Robinson, agent of the Emigrant Aid Society in Kansas and later "governor" in the Topeka movement, sent to Boston an urgent plea for two hundred Sharps rifles and two field guns. In May, Amos A. Lawrence and other antislavery men in Massachusetts sent a hundred rifles in response to this plea; later, additional shipments by the same group raised the total to 325 rifles. Also, at a later time, rifles were provided for the northern emigrants at their departure, and other groups began to send weapons. The Reverend Henry Ward Beecher, for instance, became especially identified with this activity, because the rifles which his congregation provided were

18. Malin, *Brown and the Legend of Fifty-Six*, pp. 509–536.

spoken of as "Beecher's Bibles," but there were many others who busied themselves in shipping firearms to Kansas.[19] On the proslavery side, there was no comparable effort, perhaps because weapons were part of the normal costume of the adult male Missourian to begin with. But a certain Colonel Jefferson Buford of Alabama was organizing an "expedition" with about three hundred able-bodied young southerners who would not hesitate to fight, and he was spending $20,000 of his own money for this project. Throughout the South, there were efforts to raise funds for Kansas and to stimulate migration thither.[20]

As the year 1855 drew toward a close, the polarization of forces in Kansas was almost complete. The population was divided into two groups, each group armed to the teeth and organized into secret military units. A clash was by this time perhaps unavoidable, and it came in a series of episodes between November 1855 and May 1856. The sequence began when a proslavery man, Coleman, killed a free-soil man, Dow, in a quarrel over land claims and then pleaded self-defense. Coleman was not arrested, and free-state men retaliated by threatening the lives of Coleman and two of his witnesses and by burning their cabins. The sheriff of Douglas County, Samuel Jones, thereupon arrested one of the free-staters who had made the threats, but before he could take his prisoner to jail, he was intercepted by Samuel N. Wood and a band of armed free-staters who rescued the arrested man by force. The sheriff thereupon determined to collect a posse of three thousand men to arrest Wood and his followers. But to the consternation of the governor, Wilson Shannon, who had succeeded Reeder, it soon developed that Jones

19. W. H. Isely, "The Sharps Rifle Episode in Kansas History," *AHR*, XII (1907), 546–566. Johnson, *Battle Cry*, pp. 104–165, with additional sources, also reviews this topic and largely confirms Isely's findings. The Aid Society caused great controversy by denying that it sent weapons. This denial was based on the technical point that the persons involved were acting as individuals and not as officers of the company, but this distinction was merely an evasion, for the purchase, shipment, and distribution of weapons was handled through the aid company office, the raising of funds and the ordering of rifles was managed by the officers of the company, and "the guns were consigned to and distributed by the Company's agents in Kansas." *Ibid.*, p. 126.

20. On southern (as distinguished from Missourian) aid to the proslavery faction, see Walter L. Fleming, "The Buford Expedition to Kansas," *AHR*, VI (1900), 38–48; Edward Channing, *A History of the United States* (6 vols.; New York, 1905–25), VI, 163–166; Craik, "Southern Interest in Territorial Kansas"; Parrish, *Atchison*, pp. 183–192, describing the Missouri senator's campaign to organize southern migration to Kansas; Allan Nevins, *Ordeal of the Union* (2 vols.; New York, 1947), II, 428–430.

had accepted the service of an army of invading Missourians to enforce "law and order in Kansas." For the first time since the election of the "bogus" legislature, the Border Ruffians had returned in force, and their mobilization outside the town of Lawrence looked so ominous that it has been recorded in history as the Wakarusa War. But in fact, the governor contrived to avert hostilities. He called on President Pierce to give him the support of federal troops which were stationed in Kansas, and though Pierce weakly failed him in this request, he was able, by a combination of authority and persuasion, to convince Atchison and his lieutenants that only a few free-staters had resisted the sheriff, and that there ought not to be any wholesale action against all free-state men indiscriminately. Atchison's army therefore disbanded, albeit very reluctantly, and in the months that followed, the rigors of one of the severest winters in the history of Kansas served to keep the peace.[21]

But in the spring of 1856 Sheriff Jones went back to Lawrence. There he was twice forcibly prevented from making arrests. A few nights later he was wounded by gunshot from an unknown source.[22] Simultaneously with this development, the grand jury of Douglas County met at Lecompton before Samuel D. Lecompte, chief justice of the territorial supreme court. It heard his instructions that the laws of the territory had been defied, that insurgent military forces were organizing, equipping, and drilling, and that such acts were treasonable. In response, the grand jury returned indictments against three free-state leaders, against two newspapers at Lawrence—the *Herald of Freedom* and the *Kansas Free State*—and against the Free State Hotel at Lawrence, which, it said, was in fact a fortress, "regularly parapeted and port-holed for use of small cannon and arms." Armed with these indictments and with warrants of arrest, the federal marshal, rather than the sheriff, moved on the

21. Testimony of Coleman, of Governor Shannon, and of other witnesses to the events leading from the death of Dow to the Wakarusa War, in *Howard Committee Report*, pp. 1040–1116; O. N. Merrill, *A True History of the Kansas Wars* (Cincinnati, 1856), pp. 1–16; G. Douglas Brewerton [New York *Herald* correspondent], *The War in Kansas: A Rough Trip to the Border* (New York, 1856), pp. 149–231, 293–298; Phillips [New York *Tribune* correspondent], *Conquest of Kansas*, pp. 152–223. Among secondary accounts, Alice Nichols, *Bleeding Kansas* (New York, 1954), pp. 47–70, combines a lively narrative with judicious separation of propaganda from eyewitness testimony.

22. *Ibid.*, p. 278, has a concise, able comment on the sources for the shooting of Jones and the efforts of the antislavery party to disclaim or minimize their responsibility.

offending town. Supported by a posse in which, again, volunteer Missourians were numerous, he meant to keep these followers under control. He entered the town accompanied only by an escort of federal troops and a handful of leaders of the posse while the main body of men remained outside. He arrested a few minor figures, after discovering that the free-state leaders had all fled, and then told his posse that they were dismissed.[23] This was the second time the Missourians had been thwarted in their purpose to come to grips with the "abolitionists," and they would very probably have mutinied then and there if Sheriff Jones, by this time recovering from his wound, had not offered them an immediate alternative. He instantly enlisted them as a sheriff's posse and took them into Lawrence.

Jones and his men entered the town with flying banners, as if they were a conquering army. They threw two printing presses in the river, "liberated" as much whiskey as they could discover, and trained their five cannon on the Free State Hotel. (Later, when all this was history, free-state men alternated between protesting that this structure was never intended as a fortification, and boasting that it had been built so impregnably that five volleys from five cannon could do no more than scar the walls.) Ultimately, Jones's men fired the building. They also burned "Governor" Robinson's house and made off with a certain amount of movable property. Free-state spokesmen called it "the sack of Lawrence," but despite looting and riotous uproar, no lives were lost except for one slave-state man, struck by a falling piece of wall from the Free State Hotel.[24]

The sack of Lawrence took place on May 21. On May 22, in Washington, Representative Preston Brooks of South Carolina made a visit to the Senate chamber when the Senate was not in session. He was looking for Senator Charles Sumner, because Sumner had two days previously delivered a philippic entitled "The Crime Against Kansas." Coming to the Senate in 1851, Sumner had

23. Wilson Shannon to Franklin Pierce, May 31, 1856, in KSHS *Collections*, IV (1886–88), 414–418; Phillips, *Conquest of Kansas*, pp. 288–309; Charles Robinson, *The Kansas Conflict* (Lawrence, Kan., 1898), pp. 251–256; relevant documents, such as indictments, marshal's proclamation, letters of the Committee of Safety in Lawrence, etc., in Holloway, *History of Kansas*, pp. 314–319, and KSHS *Transactions*, V (1891–96), 393–403; James C. Malin, "Judge LeCompte and the 'Sack of Lawrence,' May 21, 1856," *Kansas Historical Quarterly*, XX (1953), 465–494; Johnson, *Battle Cry*, pp. 155–160; Nichols, *Bleeding Kansas*, pp. 106–109; Nevins, *Ordeal*, II, 433–437.

24. Johnson, *Battle Cry*, p. 315, n. 49, has an excellent summary on the Free State Hotel as a fortification; also see Nichols, *Bleeding Kansas*, p. 280.

compensated for a lack of legislative aptitude by using the Senate as a sounding board from which to arouse public opinion by delivering a series of carefully planned and remarkably vituperative speeches against slavery. "The Crime Against Kansas"—florid, polished, and vitriolic—was the most abusive of these somewhat theatrical productions. Alternating between pompous rectitude and studied vilification, Sumner had assured Senator Douglas that "against him is God" and had characterized Senator Andrew P. Butler of South Carolina as a "Don Quixote who had chosen a mistress to whom he has made his vows, and who . . . though polluted in the sight of the world is chaste in his sight—I mean the harlot, slavery." For good measure, Sumner had sneered at "the loose expectoration" of Senator Butler's speech, alluding to the imperfect labial control of an old man. Senators had found the oration almost uniquely offensive, but none of them had taken it quite as seriously as Representative Brooks, who was related to Butler and who felt the obligation of the southern code to retaliate for an insult to his elderly kinsman. Knowing that Sumner would not accept a challenge, Brooks had hesitated as to the course he should follow, but his decision was now formed. Armed with a gutta-percha cane, and finding Sumner seated at his Senate desk, he first accosted the Massachusetts senator, saying that his speech was a libel upon South Carolina and upon Butler, and then he began to rain blows upon Sumner's head with the cane. Sumner, struggling to get to his feet, wrenched loose his desk, which was screwed to the floor. Brooks continued to strike, although the cane, which was a light one, broke after the first five or six blows. After an interval that was much shorter than it must have seemed, someone—apparently Representative Ambrose S. Murray—seized Brooks to restrain him. Sumner had collapsed with a bloody head on the Senate floor, and there was controversy afterward as to whether Brooks continued to hit him after he was down.[25]

Sumner did not come back to his seat in the Senate for the next two

25. David Donald, *Charles Sumner and the Coming of the Civil War* (New York, 1960), pp. 278–311, provides a definitive scholarly account. For the text of Sumner's speech and the subsequent rebuke of Sumner by other senators, see *Congressional Globe*, 34 Cong., 1 sess., appendix, pp. 529–547. The most important source material concerning the attack by Brooks is in *House Reports*, 34 Cong., 1 sess., No. 182 (Serial 868), titled: *Alleged Assault upon Senator Sumner*. Also, Robert L. Meriwether (ed.), "Preston S. Brooks on the Caning of Charles Sumner," *South Carolina Historical and Genealogical Magazine*, LII (1951), 1–4; "Statement by Preston S. Brooks," Massachusetts Historical Society *Publications*, LXI (1927–28), 221–223.

and a half years. His enemies said he was shamming; his friends said that he had suffered disabling physical injuries; we now know that neither was literally correct, but that in fact Sumner had experienced severe psychosomatic shock. It is well to remember, however, that all that anyone knew at the time was that Brooks had assaulted Sumner and had injured him, and that after these injuries, he appeared disabled and did not return to the Senate.[26]

The assault on Sumner had repercussions which will be examined later, but first it is necessary to mention a third event that followed the day after the assault, just as the assault had followed the day after the raid on Lawrence. This event involved a Kansas emigrant named John Brown. In May 1856, Brown was fifty-six years old. Born in Torrington, Connecticut, he had lived a life of vicissitudes, involving no less than twenty distinct business ventures in six different states. A number of these enterprises had ended badly; Brown ran into bankruptcy and was frequently a defendant in litigation. Yet despite his failures and his record of unreliability, he was able to impress men of influence and standing and even to inspire their loyalty. As early as 1834, he became an ardent sympathizer with the Negroes, and he was vitally interested both in rearing a Negro youth in his own family and in offering guidance to a colony of Negroes on the farm of the wealthy abolitionist Gerrit Smith at North Elba, New York. It is by no means certain that the slavery issue was uppermost in his mind when he followed five of his sons to Kansas in October 1855. After his arrival, however, the strife between free-state and proslavery men preyed on his mind, and he soon grew contemptuous of the moderate free-staters because they hesitated to violate the laws of the territorial government. In May 1856 he went with one of the free-state volunteer companies, the Pottawatomie Rifles, to protect the town of Lawrence, but before they arrived they learned that the town had been "sacked," that United States troops were now in charge, and that there was no need to go on.[27]

26. Donald, *Sumner*, pp. 312–342, contains an exhaustive and judicious analysis of Sumner's disability.

27. Oswald Garrison Villard, *John Brown, 1800–1859* (Boston, 1910), long regarded as the best full-length biography, has been superseded by Stephen B. Oates, *To Purge This Land with Blood: A Biography of John Brown* (New York, 1970). Another modern study, written for general readers, is Jules Abels, *Man on Fire: John Brown and the Cause of Liberty* (New York, 1971). On the Kansas phase of Brown's career, Malin, *Brown and the Legend of Fifty-Six*, is exhaustive, scholarly, and decidedly anti-Brown. Many source materials on Brown have been assembled in Louis Ruchames (ed.), *A John Brown Reader* (London, 1959), with a revised, paperback edition titled *John*

On the following day John Brown persuaded seven members of this company to leave their unit and go with him. These seven included four of his sons and a son-in-law. He armed his party with broadswords honed to razor sharpness and set out southward toward Pottawatomie Creek.

At about eleven o'clock on the night of the twenty-fourth, Brown and his men went to the cabin of a settler named James Doyle. When Doyle answered their knock, they forced their way in, ordered him to surrender in the name of the Army of the North, and, leaving two of their number to stand guard, took him outside. A few minutes later they returned and took Doyle's two eldest sons, though they left the youngest, aged sixteen, when his mother pleaded for his life. They shot the father dead, split the skulls of the two sons with their broadswords, and hacked the bodies of all three. About an hour later, they visited the cabin of Allen Wilkinson, a member of the territorial legislature, and, despite the entreaties of his wife, hacked open his skull and pierced his side. From there they went to the home of James Harris, where they took a house guest, William Sherman, but left Harris and another guest. Sherman, too, had his skull split open and his side pierced, and in addition a hand was severed from his body. Brown and his men drove off a number of horses that belonged to the men they had killed, and then they rode back to rejoin the Pottawatomie Rifle Company. These killings have been known in history as the Pottawatomie Massacre.[28]

It was never entirely clear why Brown had chosen these particular victims. Perhaps the one thing they had in common, besides a loose general identification with the proslavery party, was the fact that all of them except one, a minor, were connected with the territorial district court for the Osawatomie area—one was a grand juror, one a bailiff, one a district attorney pro tem, and the fourth an owner of the house at which the court met. Less than a month before the "Massacre," John Brown, Jr., as captain of the Pottawatomie Rifles, had come into the court and had demanded to know whether the

---

Brown: The Making of a Revolutionary (New York, 1969). For additional titles, see footnotes to chap. 14. For a historiographical critique, see Stephen B. Oates, "John Brown and His Judges: A Critique of the Historical Literature," *CWH*, XVII (1971), 5–24.

28. Malin, *Brown and the Legend of Fifty-Six*, is the fullest treatment of the massacre, but see also Oates, *To Purge This Land*, pp. 126–135, 383–388, for a more concise account and different interpretation.

territorial laws would be enforced. When the court ignored this inquiry, the company had adopted resolutions pledging forcible resistance to any attempt to compel obedience to the territorial authority, and a committee, which may or may not have been escorted by fifty armed men, had delivered these resolutions to the court. By this act the members of the Pottawatomie Rifles had laid themselves open to charges of treason. Two days after the "Massacre" the attorney general asserted that the victims had been killed to prevent them from testifying to the treasonable conduct of the men who killed them.[29] But even if this explanation for a very controversial matter should be accepted, it would still not account for the hacked skulls, the severed hand, or the stolen horses. In any case, many writers have seen Brown primarily as a man who believed himself to be the agent of Jehovah's wrath, and at least one has seen him primarily as a horse thief.[30] Whether a man who sincerely believed himself to be an agent of Jehovah could stoop to steal horses, and whether a man with his mind set on running off horses could sincerely believe this to be Jehovah's work, are difficult questions. But whatever the motivation, it left the bodies of five dead men to be discovered by their neighbors along the Pottawatomie the next morning.

There had been a great deal of gunplay in Kansas, and some fatalities, nearly all of which had resulted from fights between parties who were both armed, but up to this time the murder of defenseless captives had not been part of the pattern. The Pottawatomie massacre, combined as it was with the sack of Lawrence, brought both sides in Kansas to the belief that civil war was upon them and that they must kill their adversaries or be killed by them. On the proslavery side, the Border Ruffians of Missouri came back organized as an army, and in the antislavery camp leadership passed into the hands of Jim Lane, a violent, brawling, political adventurer who put himself at the head of an army of several hundred men and appealed to the blood lust of his followers by threatening to exterminate the entire proslavery population of Kansas. Throughout the summer and early fall of 1856, armies marched and countermarched, threatening one another with blood-curdling threats, ter-

29. This is Malin's explanation, *Brown and the Legend of Fifty-Six*, pp. 509–592, but Oates questions it, *To Purge This Land*, pp. 387–388.

30. Hill Peebles Wilson, *John Brown, Soldier of Fortune: A Critique* (Lawrence, Kan., 1913).

rorizing peaceably inclined settlers, committing depredations upon those who could not defend themselves, and killing with enough frequency to give validity to the term "Bleeding Kansas."[31]

In the autumn a new governor, John W. Geary, came to the territory and succeeded in restoring order.[32] He was able to do this by persuading each side that he would protect it against violence from the other. Fundamentally, personal safety for themselves and their families was what most men on each side had wanted all along. But when the men at Lawrence saw their neighborhood invaded by profane and violent ruffians from Missouri, with the acquiescence of the governmental authorities, they took up arms. When men at Leavenworth learned of the arrival of subsidized immigrants armed with Sharps rifles and encouraged by the northern press to defy the local authority, they prepared to fight. Each side constantly threatened the other with wholesale slaughter, and it seemed necessary for the leaders to make such threats in order to keep up the fighting spirit of their volunteer forces. But it was no mere good luck that when the hostile armies faced each other, they always avoided pitched battle. In reality, both sides wanted peace, and prepared to fight only because they felt threatened by a frightening adversary. Both knew that after the fighting was over, they would have to become neighbors and fellow-citizens again. Therefore each side was probably secretly relieved to submit to Governor Geary's vigor-

31. Malin, *Brown and the Legend of Fifty-Six,* pp. 593–628, based on very extensive sources; Nichols, *Bleeding Kansas,* pp. 120–150; Nevins, *Ordeal,* II, 476–486; Johnson, *Battle Cry,* pp. 181–230.

32. Andrew Reeder, the first governor, and Wilson Shannon, the second governor, had both been forced out of office. Both had begun on terms of cordiality with the proslavery faction but had antagonized this group by their attempts to be impartial. Both men did things that gave some ground for their removal, as Roy F. Nichols demonstrates in *Franklin Pierce, Young Hickory of the Granite Hills* (rev. ed.; Philadelphia, 1958), pp. 407–418, 435–436, 444, 473–475, 478–479. Reeder speculated in Indian lands and called the first legislature to meet at Pawnee, on the open prairie where his land holdings were located. Shannon embarrassed the administration by using federal troops, instead of civil authority, to prevent the meeting of the Topeka legislature. See *Howard Committee Report,* pp. 933–949; *National Intelligencer,* June 20, 1855; *Senate Reports,* 34 Cong., 1 sess., No. 34 (Serial 836); Robinson, *Kansas Conflict,* pp. 202–203; "Documentary History of Kansas," containing executive minutes, correspondence, speeches, resolutions, etc., concerning Reeder and Shannon administrations, in KSHS *Transactions,* III (1883–85), 226–337; IV (1886–88), 385–403; V (1891–96), 163–264; Nichols, *Bleeding Kansas,* pp. 31–36, 130–139.

ous measures of pacification, although each was careful to make a great show of reluctance at first, and afterward to claim a complete attainment of all its objectives.[33]

At the end of the Pierce administration, the question of Kansas still remained a major problem. In the territory itself, belligerent armies were no longer marching and countermarching, but the proslavery leaders in the judiciary and in the legislature were still using their control to hold free-state leaders in imprisonment and to rig the adoption of a proslavery constitution for statehood without submitting it to the voters. They made life so difficult and even unsafe for Geary that he resigned his governorship on the day that Pierce's term ended.[34] Meanwhile, in Washington, the disruptive effect of the Kansas issue was making itself felt in Congress in debates as prolonged and bitter as those over the Kansas-Nebraska Act. Senator Seward had introduced a bill to admit Kansas as a free state under the Topeka Constitution, despite the fact that the Topeka Convention was neither legal nor representative of the people of Kansas. This bill stood no chance of being enacted, but it was useful for keeping the emotions of the public at a high pitch. Almost the only attempt on either side to attain a constructive solution was a bill by Robert Toombs to hold a new registration of voters in Kansas, under the supervision of federal commissioners, and an election of delegates to a convention which would frame a constitution for statehood. Toombs's bill finally passed the Senate early in July 1856 by a vote of 33 to 12, but it was too impartial for the House, and there it received scant consideration. Douglas indignantly asserted that disturbances in Kansas were a vital source of political advantage for the antislavery people, and that they did not want pacification in the territory until after the presidential election.[35]

33. On Geary's pacification and subsequent experience as governor, John H. Gihon (Geary's secretary), *Geary and Kansas* (Philadelphia, 1857), with text of Geary's proclamations, etc. Geary used federal troops to force the "armies" on both sides to disband; he ordered the disbanding of all existing "militia"; and he ordered all adult males to enroll in a new militia. Johnson, *Battle Cry*, pp. 231–234; Nichols, *Bleeding Kansas*, pp. 145–185; Allan Nevins, *The Emergence of Lincoln* (2 vols.; New York, 1950), I, 133–140; "Documentary History of Kansas," containing executive minutes, correspondence, etc. of the Geary governorship, KSHS *Transactions*, IV, (1886–88), 520–742; V (1891–96), 264–289.

34. Nevins, *Emergence*, I, 133–139.

35. Glyndon G. Van Deusen, *William Henry Seward* (New York, 1967), pp. 168–169; Nevins, *Ordeal*, II, 419–428, 471–472; Ulrich Bonnell Phillips, *The Life of Robert Toombs*

But if controversy still raged, at least the era of organized violence was over, and in some respects it looked as if peace might come to Kansas. From the outset, most of the pioneers had been motivated primarily by a purpose to exploit the economic resources of the territory, and Geary's pacification had created a situation in which, for the first time, such motives could come into play. Men of both factions had responded to these new circumstances with alacrity. At once, opportunities for speculation took the center of the stage, with results that would have seemed incredible a few months earlier. A writer for the *Missouri Republican* wrote: "We find Stringfellow, Atchison, and Abell [all militant pro-slavery Missourians] and the notorious Lane lying down together, 'hail fellows well met' and partners in trade; growing fat in their purses and persons by speculations in town sites; eating roasted turkies and drinking champagne with the very money sent there from Missouri and elsewhere to make Kansas a *slave state;* and refusing to render an account, although demanded, as to how they have disbursed their funds."

At about the same time, the New York *Tribune* reported that "the love of the almighty dollar had melted away the iron of bitterness and Anti-Slavery and Pro-Slavery men were standing together as a unit on their rights as squatters." Samuel C. Pomeroy, an antislavery leader whose sanctimonious manner covered an insatiable craving for boodle, and who was later to serve as the prototype for Mark Twain's Colonel Mulberry Sellers in *The Gilded Age,* wrote to the head of the Emigrant Company that everyone's attention was now turning to real estate and "we don't think or care now whether the laws are 'bogus' or not." Very soon Pomeroy would be associated with Benjamin Stringfellow, previously one of the fiercest of the Border Ruffians, in amassing a fortune by manipulations of land grants and railroad charters that culminated later in the forming of the Atchison, Topeka, and Santa Fe Railroad.

One more comment on the new departure came from John W. Whitfield, who had at one time been elected by the proslavery legislature as territorial delegate to Congress. Whitfield wrote from Leavenworth, "All the world and the rest of mankind are here. Speculations run high. Politics seldom named, *money* now seems to be the question. Stringfellow and Lane good chums, and don't be

(New York, 1913), pp. 125–128; *Congressional Globe,* 34 Cong. 1 sess., appendix, pp. 749–805, 844.

alarmed when I tell you I live in the same town with *Jim Lane*. Thank God I have a little too much self-respect to make him an associated [sic]. . . . What will Greely [sic] do, now that Kansas has ceased to bleed?"[36]

Whitfield's final question reflected only a partial awareness of the real significance of what had happened in Kansas. He perceived that, for the antislavery forces on the national scene, what mattered about Kansas was its propaganda value. The nominal status of slavery in Kansas—even the presence or absence of a negligible number of slaves—was far less important than the nationwide response to the territorial melodrama. What Whitfield perhaps failed to see was that this response had already been determined by the handling of the Kansas story in the northern press. For Kansas, locally, the war was a kind of bushwhacking contest between rival factions for the control of land claims, political jobs, and local economic opportunities, as well as a struggle over slavery. At the end of the Pierce administration, the result of this contest was still in doubt. But for the United States, the war was a propaganda war (or, alternatively, a struggle for the minds of men), and by 1857 the South and the administration had lost it decisively. The Kansas crusade in particular and the antislavery crusade in general, like most moral crusades in democratic societies, represented a struggle for ideals. But the crusaders, like most crusaders, were publicists as well as idealists and not so wholly idealistic as to suppose that they could rely simply upon the attractiveness of their ideals. Rationally, a case against the proslavery party might have rested simply upon the fact that it sought to legalize slavery in the territory. But this was not enough. To arouse public opinion against the proslavery party, a drama was necessary, in which there would be heroes and villains embodying good and evil. Once this conception was put into effect, it worked to distort much of the evidence available to the historian. And yet, for purposes of understanding what took place in the nation, it is possibly less important to know what happened in Kansas than to know what the American public thought was happening in Kansas.

36. This discussion is based on Gates, *Fifty Million Acres*, pp. 106–108, with citations to *Missouri Republican*, Aug. 12, 1857, New York *Tribune*, Dec. 15, 1856, and letters of Pomeroy, Dec. 19, 1856, and Whitfield, May 9, 1857, as quoted. On June 12, 1857, David Atchison wrote to the mayor of Columbia, South Carolina, "Some of our friends . . . are turning their attention to speculation and money making. I therefore would suggest that no more money be raised [for the proslavery cause] in South Carolina." Parrish, *Atchison*, p. 208.

What the public learned about Kansas came largely through the antislavery press and was, in a sense, the manufactured product of a remarkable propaganda operation. Superficially, the abolitionists appeared to be badly handicapped for purposes of conducting a great campaign to win public opinion. They were never popular personally—never overcame an adverse public image as cranks and fanatics—and they never possessed more than negligible financial resources. Yet, seizing upon a succession of issues—the gag rules, the Mexican War, the Wilmot Proviso, the Fugitive Slave Act, the Ostend Manifesto, and later the Dred Scott decision and the martyrdom of John Brown—they kept up a constant and tremendously effective barrage of publicity. After the Kansas-Nebraska Act, they focused on the Kansas territory, and "Bleeding Kansas" became the supreme achievement of their publicity. Here they attained some of their most striking effects; here, also, they practiced some of their most palpable and most successful distortions of the evidence.

The information about Kansas which reached the American public came, of course, through specific channels. First of all, there were the newspapers of Kansas itself. There were several proslavery papers, all with limited newsgathering facilities and with purely local circulation. There were at least three antislavery papers, but the most important of these, and the first newspaper in Kansas, was the *Herald of Freedom,* originally published in Pennsylvania. It is a notable fact that long before the first Sharps rifles were sent west, the New England Emigrant Aid Society had financed the removal of this paper to Kansas and had acquired ownership of its press. The Society also served as a distributing agent, circulating the *Herald* widely throughout New England, so that it became the only Kansas paper with more than a local audience.[37] Second, there were the eastern newspapers, such as the *National Intelligencer,* at Washington, and leading New York papers, including particularly the *Times,* the *Herald,* and the *Tribune.* But these papers were by no means alike in the way they handled the news from Kansas. The *Intelligencer,* for instance, carried dispatches from the territory only when disturbances were more acute than usual, and then it relied on exchanges and

37. Malin, *Brown and the Legend of Fifty-Six,* is the classic exploration of the propaganda war, which also receives pointed treatment in Nichols, *Bleeding Kansas,* especially in many of the notes pp. 265–296. Also see Nichols, *Pierce,* pp. 473–480. On the *Herald of Freedom,* Malin, pp. 32–33, 63–68; Johnson, *Battle Cry,* pp. 89–91. See also Ralph Volney Harlow, "The Rise and Fall of the Kansas Aid Movement," *AHR,* XLI (1935), 1–25.

telegraphic dispatches, rather than on correspondents.[38] By far the most active newspaper in reporting affairs in Kansas was the New York *Tribune*, edited by Horace Greeley, who proved a true field marshal in the propaganda war and clearly stated his strategy as follows: "We cannot, I fear, admit Reeder [as a delegate in Congress]; we cannot admit Kansas as a state; we can only make issues on which to go to the people at the Presidential election." Accordingly, Greeley kept one of his best correspondents, William Phillips, in Kansas, where he produced a steady and dependable supply of antislavery news. Phillips was a good antislavery man, but perhaps not quite as good as the correspondent of the *National Era*, John H. Kagi, who proved his zeal by shooting a proslavery territorial judge.[39] A third major source of information about Kansas came from speeches in Congress, for the *Congressional Globe* was circulated throughout the country. Sumner's speech "The Crime Against Kansas" was the most conspicuous example of legislative oratory by which antislavery congressmen kept the Kansas issue before the public. Also, the House of Representatives appointed a committee, with two Republicans—William Howard of Michigan and John Sherman of Ohio—and one Democrat—Mordecai Oliver of Missouri—to go to Kansas and investigate conditions there. The Howard Committee produced a report containing the testimony of 323 witnesses, and running to more than 1,300 pages.[40]

With antislavery elements tending to monopolize the "manufacture of Kansas news," proslavery men in the territory were systematically placed in the worst possible light. During intervals when the Missourians had not committed any offending act, they could still be castigated for their profane speech, their uncouth manners, and their whiskey-guzzling ways. In fact the extent to which they were denounced for these features is a kind of inverted tribute to the fact that their bark was very much worse than their bite. The term "ruffian" was fixed upon them so firmly that they fell to using it themselves, and Senator Atchison was reduced to proclaiming the

38. Malin, *Brown and the Legend of Fifty-Six*, p. 34.

39. *Ibid.*, pp. 89–92 (containing quotation from Greeley), 228–229 (on Kagi), 231–238; Jeter Allen Isely, *Horace Greeley and the Republican Party, 1853–1861: A Study of the New York Tribune* (Princeton, 1947), pp. 130–142, 173–184; Bernard A. Weisberger, *Reporters for the Union* (Boston, 1953), pp. 23–41.

40. Donald, *Sumner*, p. 302, estimates that "perhaps a million copies of Sumner's Crime against Kansas' speech were distributed"; on the partisan attitude of the Howard Committee: Malin, *Brown and the Legend of Fifty-Six*, pp. 50, 59; Nichols, *Bleeding Kansas*, p. 118.

virtues of a true Border Ruffian.[41] When the Missourians did resort to violence, which was not seldom, their acts were described in a rhetoric borrowed from accounts of the persecutions of the early Christians.

A specific example of what happened to the news as it was filtered through these media may be found in the treatment of the events of May 22–24, 1856—the "sack" of Lawrence, the assault on Sumner, and the "massacre" at Pottawatomie.

When Sheriff Jones marched into Lawrence with a large "posse" he was returning to a town where he had twice been resisted in making an arrest and where he had once been shot. (Antislavery papers reported that he had not been shot at all; that the shooting had been done by a proslavery man; and that, although shot by an antislavery man, he had deliberately sat in a lighted tent making a target of himself.)[42] Jones was, of course, vulnerable to criticism for going into Lawrence at all, especially after the federal marshal had just been there, and vulnerable also for permitting riotous conduct and looting. But these were not the offenses with which he was charged in the press. Instead the sack of Lawrence was depicted as an orgy of bloodshed. The New York *Tribune* introduced its account with shrieking headlines: "Startling news from Kansas—The War Actually Begun—Triumph of the Border Ruffians—Lawrence in Ruins—Several Persons Slaughtered—Freedom Bloodily Subdued." The *New York Times* also headlined its first story with wholesale slaughter. A few days later, in less conspicuous type, both papers got around to indicating that Lawrence had been sacked with scarcely anyone getting hurt, but the melodramatic headlines of the initial stories had done their work.[43]

The antislavery press reported the assault upon Sumner quite accurately, for the truth was damaging enough. But again, it made

41. Malin, "Proslavery Background of the Kansas Struggle," p. 301, quotes a manuscript report by William Hutchinson of a speech by Atchison, Feb. 4, 1856, "I would not advise you to burn houses. I would not advise you to shoot a man. If you burn a house, you turn a family out of doors; if you shoot a man, you shoot a father, a husband. Do nothing dishonorable. No man is worthy of a border ruffian who would do a dishonorable act."

42. Malin, *Brown and the Legend of Fifty-Six,* pp. 93–94, 73, citing New York *Tribune,* May 31, June 5 (Jones not shot), May 8 (shot by a proslavery man), and May 15 (at fault for exposing himself to being shot by an antislavery man)—the last also in *Herald of Freedom,* April 26, 1856.

43. New York *Tribune,* May 28, 1856; Isely, *Greeley and Republican Party,* pp. 130–142; Malin, *Brown and the Legend of Fifty-Six,* pp. 92–94; Weisberger, *Reporters,* pp. 33–34.

the fullest use of the attack for propaganda purposes. Personal assaults were relatively commonplace at that time in most parts of the Union, but for one member of Congress to beat another in the chamber of the United States Senate was something new, and for Brooks to strike Sumner when he was seated was a gross violation even of the code of men who regarded personal assault as a proper way of responding to personal insult. Thus Brooks's deed stood as a kind of travesty of the chivalry which the South claimed to represent; accordingly, the "civilization" of the South was denounced generally in the antislavery press. When many southerners fell into the error of defending what they would not have done themselves, simply because it had been done to a man they hated, their defense gave confirmation to the most serious part of the North's accusation —that the spirit of Brooks was the spirit of the South.[44]

But the treatment of the Pottawatomie killings offers perhaps the most revealing insight into a highly developed propaganda technique.

It was known in Kansas that John Brown and his men were the Pottawatomie killers. John Brown, Jr., understood this so clearly that he experienced an acute mental breakdown within less than forty-eight hours after he learned what had happened on Pottawatomie Creek. The members of the Pottawatomie Rifles understood it so plainly that they forced John Brown, Jr., to resign his captaincy. Antislavery men in the area were so deeply distressed by the deed that many of them joined with proslavery men in holding a meeting at which they condemned the killings very strongly and pledged themselves to lay aside all sectional and party feeling and to "act together to ferret out and hand over to the criminal authorities the perpetrators for punishment." Brown's role in the "massacre" was mentioned in newspapers in New York and Chicago several times within a month after the event.[45]

44. The House appointed a committee of investigation, the majority of which recommended that Brooks be expelled; this proposal was voted 121 to 95 (every southern congressman but one in this minority), but failed of the necessary two-thirds; Brooks nevertheless resigned, ran for reelection, and won. *Congressional Globe*, 34 Cong., 1 sess., p. 1628. For southern and northern reactions to Brooks's assault, see Nevins, *Ordeal*, II, 446–448; Donald, *Sumner*, pp. 297–311; Avery O. Craven, *The Growth of Southern Nationalism, 1848–1861* (Baton Rouge, 1953), pp. 228–236 (at variance with Donald in concluding that the South was not united in approval of the assault).

45. On June 6, 1856, James Harris, who was present in the house from which the Brown party took William Sherman outside to kill him, testified that he "recognized" two of the party—"a Mr. Brown, whose given name I do not remember, commonly

Soon the eastern antislavery press took hold, however, and began to deny Brown's participation, to impugn the character of the victims, to suggest that perhaps no killings had really occurred,[46] and to fabricate stories that the killings had been committed in self-defense.[47] The *New York Times* first printed the story of the massacre, in eleven inconspicuous lines, as coming from the St. Louis *Republican*, and discounted it as being "quite as improbable as many other [stories] that have appeared in that journal."[48] The New York *Tribune*'s correspondent, with real dexterity, used the savage features of Brown's crime to exonerate Brown, by arguing that the mutilation of Henry Sherman's body showed that he had been killed by the Comanche Indians and that the proslavery men had tried to pin this atrocity upon the free-staters.[49] The *Tribune*'s man also blackened the character of each of the victims, described the killings as the result of a fight between armed and evenly matched proslavery and antislavery groups, and added, piously, "Terrible stories have floated through the newspapers, distorted and misrepresented by those whose interest it was to misrepresent them."[50]

---

known by the appellation of 'Old Man Brown' and his son, Owen Brown." Affidavit in *Howard Committee Report*, p. 1178. John Doyle also *described* the leader of the murder party but, not having known him before, could not *identify* him as Brown, and Mrs. Wilkinson identified "one of Captain Brown's sons" (pp. 1175–1181). Malin, *Brown and the Legend of Fifty-Six*, pp. 568–577, shows that four witnesses swore affidavits against Brown and that a warrant for his arrest on a charge of murder was issued May 28, 1856. The Chicago *Democratic Press*, June 5, and the New York *Tribune*, June 17 (quoted in Malin, pp. 111, 100), carried reports naming Brown as the killer. In short, Brown had been identified, formally accused, and named in the press as leader of the murder band within three weeks after the massacre. Despite these facts, the free-state propagandists continued for more than twenty years to maintain the public impression that he had been falsely accused.

46. *New York Times*, June 5, 12, 1856; Sara T. D. Robinson, *Kansas, Its Interior and Exterior Life* (Boston, 1856), p. 318, spoke of the Pottawatomie killings, but mentioned them as rumors along with other rumors which had been proved false, thus effectively suggesting, without asserting, that the Pottawatomie reports were false.

47. New York *Tribune*, June 4, 5, 6, 9, 10, 12, 1856, and discussion in Malin, *Brown and the Legend of Fifty-Six*, pp. 95–111, also quoting Chicago *Tribune*, June 3, and Springfield *Republican*, June 4, 1856.

48. *New York Times*, May 31, 1856.

49. Phillips, *Conquest of Kansas*, p. 317.

50. *Ibid.*, pp. 316–317. There were many reports that a group of antislavery men surprised a group of proslavery men in the act of doing violence to another antislavery man, that a fight started, and the proslavery men were killed in the fight. The proslavery men were described as engaged in robbery or other predatory or aggres-

At the time of the Pottawatomie killings, Congressmen Howard, Sherman, and Oliver were in Westport, Missouri, conducting their fact-finding investigation of conditions in Kansas. They called in James Harris, from whose house Henry Sherman had been summoned to his death, and Harris started to tell some facts. A majority of the committee, consisting of the two antislavery members, stopped him on the grounds that "no testimony in regard to acts of violence committed since the resolutions organizing this commission" will be received. (Later, they did not adhere to this rule very strictly, and they found space for two nonfatal acts of violence committed by proslavery men.) However, the minority member, who was proslavery, introduced affidavits to incorporate the testimony which had been suppressed. These included a statement in which Harris named "the man called Old Man Brown" as the leader of the eight-man Army of the North. They also showed the testimony of the widows of James Doyle and Allen Wilkinson. John Doyle, aged sixteen, had a statement telling how he had found the bodies of his father and his two brothers the next day. Representative Oliver got all this evidence published in the voluminous transcript of testimony. But Howard and Sherman kept it out of the majority report, which attained wide circulation and became an arsenal of material for the northern press.[51]

John Brown himself was a man who always fearlessly admitted anything which could be incontrovertibly proved. In this case, proof would not have been easy, and he chose to let his friends deny his participation, while he himself said nothing except to remark almost incidentally, in one letter to his family: "We were immediately after this accused of murdering five men at Pottawatomie,"[52] as if the accusation were not worthy of a denial. His companions continued to deny his part until James Townsley, one of the band, at last decided to speak out, affirming in a formal statement that he had guided Brown and his men, and "that John Brown, Sr., did command the party and did order the killing of Wilkerson [*sic*], Doyle

---

sive acts, and were even disparaged as illiterate. Any recognition that unarmed men not engaged in any offensive act were taken from their cabins and murdered was carefully avoided.

51. Minority report by Mordecai Oliver, pp. 104–107 (not to be confused with pp. 104–107 of testimony) in *Howard Committee Report.* Also pp. 1175–1181 of testimony.

52. John Brown to his wife and children, June 1856, in F. B. Sanborn (ed.), *The Life and Letters of John Brown* (Boston, 1891), pp. 236–241.

and his two sons and William Sherman."[53]

But that was not until 1879. In the twenty years intervening, much water had flowed along the Pottawatomie. Also, the Southern Confederacy had come and gone. By the time Townsley said what he had to say, it was not of much interest to anyone except historians. Bleeding Kansas had long since ceased to bleed, and John Whitfield had long since had an answer to his question of what Horace Greeley would do when that happened. The answer was that Kansas had bled long enough to serve Horace Greeley's purposes.

In the great struggle which raged throughout the decade of the fifties, the South labored under the insuperable handicap that, in the Western world in the middle of the nineteenth century, it was trying to defend a vast system of human slavery. This handicap was probably inescapable for a slave-based society. But at the same time, southerners suffered further and even more damaging handicaps by fighting battles for nonessential objectives. In these gratuitous conflicts, the South won a series of victories which cost more than they were worth. The Fugitive Slave Act had been such a victory. The Ostend Manifesto was another. The Kansas-Nebraska Act was a third. But none proved more barren for the proslavery forces than winning control of the first territorial government in Kansas. When Pierce left office, men in Kansas were still contesting for control of the territory, but the battle of symbols had already ended, and "Bleeding Kansas" had been awarded to the antislavery cause by public opinion as one of the most decisive victories ever won in a propaganda war.

53. Malin, *Brown and the Legend of Fifty-Six*, pp. 363–364, 385–387. The Townsley confession, first published in Lawrence *Daily Journal*, Dec. 10, 1879.

CHAPTER 10

The Political Parties in Metamorphosis

IN his famous last speech read to the Senate during the crisis of 1850, John C. Calhoun made a striking analysis of the "cords" that held the Union together. These cords, he said, were many and various, and some of them had already snapped under tension as the sections drifted apart. For instance, the national church organizations of Methodists, of Baptists, and of Presbyterians had already parted under the strain. But other cords continued to hold.[1]

Of these remaining ties, none were more generally recognized as strong unifying agencies than the two national political parties of Whigs and of Democrats. These two remarkable organisms performed a unique function in America's federal system. By the nature of this system, each state had, at the state level, separate political issues and political organizations all its own, but each was also a common participant in national affairs and, as such, needed a coordinating mechanism to bring its own political impulses into working relationship with the political life of the other states. The two national parties had met this important political need.[2] As loosely articulated structures, they were able to function as opportunistic

1. Above, chap. 5, note 18.

2. On June 10, 1852, David Outlaw of North Carolina said in the House of Representatives, "Party ties are among the strongest associations which bind men together. . . . the very name of party has a talismanic power on the passions and prejudices of the people." *Congressional Globe,* 32 Cong., 1 sess., appendix, p. 678. See Allen Johnson, "The Nationalizing Influence of Party," *Yale Review,* XV (1907), 283–292.

coalitions of diverse state organizations. Yet, at the same time a certain like-mindedness among the Democrats and also among the Whigs gave a measure of philosophical cohesiveness to each group. The Democrats had a generalized and mildly populistic orientation; the Whigs an equally mild orientation toward property values. These differences gave some real meaning to party distinctions. Yet, in the Anglo-American political tradition, the parties expressed themselves diffusely, in attitudes and tonal qualities rather than in doctrine or dogma. Representing interests more than ideologies, they displayed the easy-going, accommodative, somewhat cynical, and anti-intellectual tendencies that coalitions of interest groups are apt to show. This lack of sharp rationalization of objectives was conducive to a certain looseness which enabled both parties to hold a mixed bag of diverse state organizations in combination.

Relatively unencumbered by ideological mission, the two parties did not have enough intellectual focus to offer voters clear-cut alternatives. Thus they failed in one of the classic functions theoretically ascribed to political parties. But if they defaulted in this way, they performed admirably another equally important if less orthodox function: they promoted consensus rather than divisiveness. By encouraging men to seek a broad basis of popular support, they nourished cohesiveness within the community and avoided sharpening the cutting edge of disagreement to dangerous keenness. Without ideological agreement as a basis for cohesiveness, the parties could still cultivate unity, based upon the esprit that men develop by working together, or upon the practical need that diverse groups may have for one another's support.

The national political parties in America overemphasized these consensual elements to an extreme degree. Far from pressing issues to logical conclusions, they often practiced the arts of evasion and ambiguity in order to gain the broadest possible base of support. They substituted the ties of personal loyalty to a leader—a Jackson or a Clay—for shared beliefs in policy objectives. They relied heavily on the sentimental bonds which develop among men who have worked as a team in victory and defeat, and on the pragmatic importance of winning for the sake of gaining office or exercising power.

This combination of esprit and interest proved a powerful cement, even in the absence of any real agreement on policy. Thus, party unity seemed capable of surviving basic differences of opinion, and party regulars valued party harmony above party policy. When the slavery question began to take shape as a public issue, both parties, sensing its divisive potential, vigorously resisted its introduction into politics.

The ideologues, both of southernism and of antislavery, had found themselves repeatedly confounded by this resistance. Calhoun, for instance, failed again and again to achieve a united front of southerners because southern Whigs did not trust him politically, and southern Democrats did not want to form any combinations that would separate them from their northern Democratic allies. In 1848, when Calhoun seemed about to gain bipartisan southern support for his Address to the People of the Southern States, the southern Whigs veered off at the last moment because, having just elected Taylor to the presidency, they did not want to jeopardize the results of their victory before their candidate took office. To the Georgia Whig Robert Toombs, Calhoun's project was simply "a bold stroke to disorganize the Southern Whigs."[3] Among the opponents of slavery, also, the gravitational pull of party loyalties sometimes took a priority over antislavery ideals. Thus, when antislavery dissenters within both of the old parties combined to form the Free Soil party in 1848, running the former Democrat Van Buren and the former Whig Charles Francis Adams for the presidency and vice-presidency, many earnest antislavery men decided to stay with their traditional parties. Although Thomas Hart Benton had begun to thunder against the proslavery group within the Democratic party, he chose to give at least nominal support to Lewis Cass rather than to join the Free Soilers. Similarly, among the Whigs, even such antislavery men as William H. Seward, Horace Greeley, and Benjamin F. Wade—not to mention Abraham Lincoln—gave their backing to the Louisiana slaveholder Zachary Taylor rather than to the Free Soil ticket.[4]

3. Toombs to John J. Crittenden, Jan. 3, 1849, in Ulrich Bonnell Phillips (ed.), *The Correspondence of Robert Toombs, Alexander H. Stephens, and Howell Cobb*, AHA *Annual Report*, 1911, II, 139. On Jan. 22, Toombs wrote to Crittenden, "We have completely foiled Calhoun in his miserable attempt to form a Southern party. . . . I told him that the Union of the South was neither possible nor desirable until we were ready to dissolve the Union." *Ibid.*, p. 141.

4. William Nisbet Chambers, *Old Bullion Benton: Senator from the New West: Thomas Hart Benton, 1782–1858* (Boston, 1956), pp. 332–337; Glyndon G. Van Deusen, *William Henry Seward* (New York, 1967), pp. 107–110; Jeter Allen Isely, *Horace Greeley and the Republican Party, 1853–1861: A Study of the New York Tribune* (Princeton, 1947), p. 36; Hans L. Trefousse, *Benjamin Franklin Wade, Radical Republican from Ohio* (New York, 1963), pp. 56–59; Reinhard H. Luthin, "Abraham Lincoln and the Massachusetts Whigs in 1848," *NEQ*, XIV (1941), 619–632.

Not until after the election of 1848 did the divisive effect of the slavery question begin to make itself deeply felt within the two great bisectional organizations. As late as 1844, slavery had not yet become a dominant issue, and in 1848 a large part of the antislavery solvent was drained off into the Free Soil party. For organizers of a third party, waging their first campaign, the Free Soilers made a remarkable showing in 1848. They carried 14.4 percent of the popular vote cast in the free states, and ran ahead of the Democrats in New York, Massachusetts, and Vermont.[5] Their rise seemed almost meteoric, and many enthusiastic antislavery men hoped that the new party would become the dominant political organization in the North.

If the Free Soil combination had held together, the force of antislavery would have operated upon the older parties primarily from the outside, drawing antislavery men away from both the Whig and the Democratic organizations. But in fact the Free Soilers could not effectively capitalize on antislavery sentiment in the North because too many antislavery men preferred to fight the battle of slavery restriction within the framework of the traditional parties. Also, the Free Soil movement of 1848 had enjoyed the support of the Barnburners of New York, who were more interested in whipping a rival faction in state politics than they were in serving the cause of antislavery.[6] In 1849, John Van Buren led most of his father's Barnburner followers back into the Democratic fold.[7] Since 43 percent of the Free Soil vote had been concentrated in the Empire State, this move by itself dealt irreparable damage to the Free Soilers, and by 1852, the third party had almost collapsed. In the 1852 election it carried only 6.6 percent of the vote cast in the free states. Many antislavery men were profoundly discouraged. Even before the election, Charles Francis Adams lamented: "The moral tone of the Free States never was more thoroughly broken." Another Free Soiler wrote to Charles

5. Compiled from data in W. Dean Burnham, *Presidential Ballots, 1836–1892* (Baltimore, 1955).

6. Stewart Mitchell, *Horatio Seymour of New York* (Cambridge, Mass., 1938), p. 111, remarks that Barnburners, who later supported Pierce in 1852 and Buchanan in 1856, could not have been very deeply committed to antislavery in 1848. "It seems clear," he says, "that in 1848, the Barnburners tricked the Free Soilers into supplying the disguise of reform for their own political revenge on Polk and Cass."

7. *Ibid.*, pp. 112–114; Walter L. Ferree, "The New York Democracy: Division and Reunion, 1847–1852" (Ph.D. dissertation, University of Pennsylvania, 1953).

Sumner that "the *morale* of our party is *chloroformed.*"[8]

The Free Soil movement mitigated the strain on the old parties by removing the strongest antislavery pressures within them. This diversion left control in the hands of men who wanted party solidarity and cultivated equivocation on slavery as an expedient for preserving bisectional harmony. But disintegration of the Free Soil party brought new hazards to the two old parties by confronting both with the same dilemma. Northern Democrats and northern Whigs needed part of the former Free Soil vote to win in state elections, and they also needed the support of southern party allies to win in national elections; but they could not cultivate the one without antagonizing the other. Insofar as they strengthened their state organizations, they weakened their national organization, and vice versa.

Of the two, local strength was more essential than national strength. A political party might prove hardy despite defeat at the national level, but it could not endure without strength at the state level. For this reason, as well as because most of them felt antipathy to slavery, northern Whigs and northern Democrats between 1848 and 1852 frequently found themselves in the position of bidding against one another for the Free Soil support which constituted a balance of power between them. Political events in Ohio and Massachusetts after the election of 1848 serve as good examples.

In Ohio the Free Soilers held the balance of power between Whigs and Democrats in the legislature. Their strategy in an intricate situation was to favor whichever party would support an ardent antislavery man for the Senate—either Salmon P. Chase on the part of the Democrats or Joshua R. Giddings on the part of the Whigs. The Democrats accepted this overture, while many of the Whigs did not, and a Democratic–Free Soil coalition elected Chase in 1849.[9] The lesson of Chase's election was by no means wasted on the Whigs. In 1850–1851, when another senator was to be chosen, they no longer persisted unrealistically in supporting an old-line Whig

8. Theodore Clarke Smith, *The Liberty and Free Soil Parties in the Northwest,* (New York, 1897), pp. 160–161, 176; Martin B. Duberman, *Charles Francis Adams, 1807–1886* (Boston, 1961), pp. 160–161, 174–179; see also David Donald, *Charles Sumner and the Coming of the Civil War* (New York, 1960), p. 249.

9. Smith, *Liberty and Free Soil Parties,* pp. 164–175; Reinhard H. Luthin, "Salmon P. Chase's Political Career before the Civil War," *MVHR,* XXIX (1943), 517–540; J. W. Schuckers, *The Life and Public Services of Salmon Portland Chase* (New York, 1874), pp. 91–96.

like Thomas Ewing. Instead, they offered the Free Soilers a number of possible candidates, including Benjamin F. Wade—a flaming crusader against slavery. When the Free Soilers responded favorably to Wade, he was elected with solid Whig support.[10]

The Free Soil party had "collapsed" in Ohio. While it was doing so, the Democrats prevented the Whigs from electing one senator, and the Whigs prevented the Democrats from electing another. They achieved these ends by giving the state's Senate seats to Chase and Wade, two of the most pronounced antislavery men in public life. The "defunct" Free Soil party was the only real victor, and the relationship between Ohio Whigs and southern Whigs, Ohio Democrats and southern Democrats, was no longer an illustration of bisectional harmony.

In Massachusetts the Free Soilers were somewhat less opportunistic than they were in Ohio, for the dominant faction, led by Henry Wilson and Charles Sumner, preferred to ally with the Democrats rather than the Whigs. But here, as in Ohio, they intended that the Democrats should reward them in the impending senatorial contest by supporting a Free Soil candidate.

The Democrats, in their state convention in 1849, paved the way for this rapprochement with resolutions opposing slavery "in every form and color." After the state election of 1850, in which the Whigs gained a plurality, the Democrats and the Free Soilers formed a distinct coalition, despite strong opposition from prosouthern elements among the Democrats and from Whiggish elements among the Free Soilers. In 1850 this coalition elected George S. Boutwell, a Democrat, to the governorship, and in 1851 it sent Charles Sumner to the Senate.[11]

With both party organizations torn between the need for Free Soil support in the North and for proslavery support in the South, the era of the truly bisectional parties had already passed by 1852. But as Calhoun said, the cords of Union could not be destroyed at a single stroke, and in 1852 the bisectional principle gained its last major triumph. Franklin Pierce, candidate of the national Democratic party, carried fourteen free states and twelve slave states, thus defeating Winfield Scott, candidate of the national Whig party, who

10. Trefousse, *Wade*, pp. 64–67.

11. The best and fullest account of the extremely complex maneuvers leading to Sumner's election is in Donald, *Sumner*, pp. 164–202, but see also Duberman, *Charles Francis Adams*, pp. 158–174; Ernest A. McKay, "Henry Wilson and the Coalition of 1851," *NEQ*, XXXVI (1963), 338–357.

carried only two free states and only two slave states, but 43.6 percent of the popular vote in the free states and 44.2 percent in the slave states. The Free Soilers, as has been mentioned, carried only 6.1 percent of the popular vote in the free states.[12]

Pierce's victory as a winner both north and south of the Mason-Dixon line seemed to mark the triumphant reaffirmation of a bisectional pattern that had prevailed in every presidential election except four. In 1796, 1800, 1824, and 1828, one section had predominantly supported the winner and the other section had predominantly opposed him. But in the other twelve presidential elections held up to 1852, the person elected had been victorious both in the free states and in the slave states: one section had not imposed its choice upon the other.[13]

Close observers in 1852 may well have noted a flaw in this picture, for Pierce's heavy electoral vote concealed some significant weaknesses. His margins of victory in state after state were extremely narrow, and he did not, in fact, carry a majority of the total popular vote cast in the free states. But even so, when this fact was recognized, his victory still seemed impressive, and no one could have dreamed how long it would be before North and South again gave the preponderance of their electoral vote to the same candidate. Woodrow Wilson would achieve such a victory in 1912, but even then, he would not win a majority of the popular votes outside the South.[14] Not until 1932 would anyone come closer than Pierce to carrying both the North and the South, as victorious candidates had customarily done during the first six decades of the republic.[15]

If the election of Pierce represented a final manifestation of bisectional harmony, powerful divisive forces were also at work. The

12. Compiled from data in Burnham, *Presidential Ballots;* on the election generally, see Roy F. and Jeannette Nichols, "Election of 1852," in Arthur M. Schlesinger, Jr., *et al.* (eds.), *History of American Presidential Elections, 1789–1968* (4 vols.; New York, 1971), II, 921–950.

13. This statement based on analysis of electoral vote by states for slave states and free states. See U.S. Bureau of the Census, *Historical Statistics of the United States, Colonial Times to 1957* (Washington, 1960), p. 685. In 1836, Van Buren was elected without a clear electoral majority in the slave states, but his vote there was 61, while the Whigs' was 54, and South Carolina's 11 votes went to an anti-Jackson ticket.

14. This statement is subject to the exception that in 1868, Grant carried more of the former slave states (seven) than Seymour (five) (52 electoral votes to 37), but he did not carry a majority, for three states with 26 votes had not been reconstructed in time to cast their votes; and in 1872, the Democrats carried six former slave states and lost seven (63 to 53) while two, with 12 electoral votes, did not have their votes counted. *Ibid,* pp. 688–689.

15. *Ibid.*, pp. 686–687.

Whigs were badly divided in 1852, and their bisectional organization did not survive its defeat in the election. As for the Democrats, they appeared much better united, but they were soon to suffer a traumatic destruction of the bisectional balance within their party as a result of the Kansas-Nebraska crisis. Though these developments have already been treated, it may still be worthwhile to examine their impact upon the party structures more closely.

The election of 1852 seemed to have a unifying effect upon the Democrats. In nominating Franklin Pierce, they rejected Lewis Cass and James Buchanan, both of whom enjoyed support that was concentrated in one section or the other.[16] In asserting the "finality" of the compromise, they shrewdly avoided making an issue of its merits and took the ground that controversies which had been peaceably settled ought not to be reopened. Northern and southern Democrats alike had smelled victory and, with a lively anticipation of the patronage to come, had worked together with such effect that they carried all but four states and won topheavy congressional majorities.

For the Whigs, however, the strains of sectionalism proved far more devastating. When the party convention met at Baltimore in June 1852, the southern delegates were still bitter because the slaveholder whom they had worked so hard to elect in 1848 had turned out to be a political protégé of William H. Seward. This memory haunted them because Seward was again supporting a southerner, a military hero—and a possible protégé—in the person of Winfield Scott of Virginia.[17] Their distrust of the northern Whigs was further activated by the refusal of northern Whig congressmen to support

16. On the first ballot in the convention, Cass received 116 votes out of a total of 288; of these, 72 came from the free states; 38 from the upper South (Delaware, Maryland, Kentucky, Tennessee, Missouri), and only 6 from the lower South (Louisiana). Buchanan reached 101 votes on the 25th ballot (his maximum was 104); of these 27 were from his home state of Pennsylvania, only 20 from other free states, and 54 from the slave states. The ultimate source for convention votes is *Official Report of the Proceedings of the Democratic National Convention held at Baltimore, May 22, 1848* (n.p., 1848), but the voting data are conveniently assembled in Richard C. Bain, *Convention Decisions and Voting Records* (Washington, D.C., 1960).

17. On the extent to which Scott was politically a protégé of Seward, see Van Deusen, *Seward*, pp. 141–142; Charles Winslow Elliott, *Winfield Scott, the Soldier and the Man* (New York, 1937), pp. 607–613; Arthur Charles Cole, *The Whig Party in the South* (Washington, 1913), pp. 224–226, 258–261; Harry J. Carman and Reinhard H. Luthin, "The Seward-Fillmore Feud and the Disruption of the Whig Party," *NYH*, XLI (1943), 335–357.

resolutions affirming the finality of the Compromise.[18] They were determined that the convention should endorse the Compromise and should nominate Millard Fillmore in recognition of his leadership in securing its adoption.

With the Whigs thus divided both as to platform and as to candidate, the party convention of 1852 was little more than a stage for acting out the drama of division. Evidences of distrust between northern and southern delegates were conspicuous from the outset. The delegates could not even agree to let a "reverend gentleman" open the proceedings with prayer, and the secretary was accused of reading the results of a roll call in the wrong tone of voice. Southerners successfully insisted that the framing of the platform should precede the nomination of a candidate, and they secured the adoption of a resolution endorsing the Compromise, including the Fugitive Slave Act, as a final settlement, "in principle and in substance." Any satisfaction this vote might have given them was much diminished by two facts. First, every one of the 66 votes against the resolution came from a Scott supporter in the free states; second, Henry J. Raymond, a delegate and also the editor of the *New York Times,* stated in the *Times* that there had been an understanding by which the North would yield on the platform and the South would accept Scott as the candidate. Southern delegates, who were not at all reconciled to the nomination of Scott, read this to mean that they had been betrayed by a deal which would give them only an abstract paper resolution, while it gave the North control of the ticket and power to ignore the resolution. Fearful that what Raymond said was true, they denounced him as a liar and demanded his expulsion from the convention.

Sectionalism governed the balloting. On the first ballot, Fillmore received 133 votes, of which all but 18 came from the slave states; Scott received 132, of which all but 4 came from the free states. Daniel Webster received 29. The deadlock between Fillmore and Scott continued for fifty-two ballots—a number exceeded in American party history only in the Democratic conventions of 1860 and 1924. Scott's victory with 159 votes on the fifty-third ballot still included only 17 from the slave states.[19]

18. Above, p. 122.

19. George Ticknor Curtis, *Life of Daniel Webster* (2 vols.; New York, 1870), II, 621, gives the vote for each of the 53 ballots.

After the nomination, Scott was completely unable to heal the breach in the party. During the convention John Minor Botts of Virginia had produced a letter from Scott stating that if he were nominated, he would fully endorse the Compromise resolutions in his letter of acceptance. Later, southern Whigs pressed him for fulfillment of this promise. Under strong free-soil pressure, however, he issued only the grudging statement that he accepted the nomination "with the resolutions annexed."[20] This was Scott's only comment, throughout the campaign, on the leading issue of the election. Six days later, nine southern Whig congressmen, correctly inferring that the nominee dared not endorse the platform, announced that they would not support him. They included Alexander H. Stephens and Robert Toombs of Georgia, and others from Alabama, Mississippi, Virginia, and Tennessee. During the campaign, such prominent Whigs as Kenneth Rayner of North Carolina, Waddy Thompson of South Carolina, and William G. Brownlow of Tennessee also went into opposition.[21]

As the campaign continued, Scott proved unable to stem this defection in the lower South. It is easy to criticize his tactical incompetence, but in fact, he faced a terrible dilemma. He could regain lost ground in the South only by endorsing the Compromise, which would have resulted in more than equivalent losses in the North. His dilemma was worse than that of the Democrats, because the northern Whigs, being on the whole more strongly opposed to slavery than the northern Democrats, would not make the kind of concessions that the northern Democrats were willing to make for the sake of southern support. Therefore, he could not avoid taking heavy losses in one section or the other, and he chose, in effect, to take them in the South.[22]

The sweeping consequences of this choice appeared in the election results. The six states of the lower South gave Scott only 35

20. *National Intelligencer*, June 29, 1852.

21. Letter of southern Whigs, *ibid.*, July 5, 1852. On the divisions among the southern Whigs, Cole, *Whig Party in the South*, pp. 257–276; Ulrich Bonnell Phillips, *The Life of Robert Toombs* (New York, 1913), pp. 110–115; Horace Montgomery, *Cracker Parties* (Baton Rouge, 1950), pp. 72–116; Joseph Howard Parks, *John Bell of Tennessee* (Baton Rouge, 1950), pp. 271–282; Albert D. Kirwan, *John J. Crittenden: The Struggle for the Union* (Lexington, Ky., 1962), pp. 265–288.

22. The hopelessness of Scott's dilemma is suggested by the fact that Isely, *Greeley and Republican Party*, p. 40, says Scott "publicly pledged his support of the platform . . . thus tying his candidacy into a knot" (he was too proslavery), while Cole, *Whig Party in the South*, p. 275, says "Seward and his allies had been fatal to Scott's success" (he was too antislavery).

percent of their vote. The same states had given 49.8 percent to Taylor in 1848. Any reader who knows how stable party strength usually remains from one election to the next, even amid so-called landslides and debacles, will recognize that the southern Whigs had suffered one of the sharpest losses in American political history.

It is an exaggeration to speak, as some writers have done, of the "destruction" of the Whig party in the South. Although Scott's candidacy was very weak in the Gulf states, he retained substantial strength in the upper South, carrying Tennessee and Kentucky and making a strong showing in the other border states. As a consequence, Scott carried 44.2 percent of the total popular vote in the slave states, which was more than his percentage in the free states. Although many prominent Whigs of the lower South had abandoned him, others fought hard in his cause.[23] The state Whig organizations continued to hold together and to send senators and representatives to Congress, and in 1856, Millard Fillmore, running as a dual Whig and American candidate, won 43.9 percent of the popular vote in the slave states, and 41 percent of the lower South —6 percent more than Scott had carried. But this total was gained only in a combination with the American party, and at a time when the candidate was not bidding seriously for northern votes and could therefore make a sectional appeal for southern support. Scott's experience had demonstrated that a strong bisectional combination of Whigs would no longer hold together.

The strain imposed by the slavery issue bore harder upon the Whigs than upon the Democrats for a number of reasons. For one thing, the Whig coalition had been from the beginning exceptionally loose; it never achieved the cohesion which the Democrats had attained under Jackson, Van Buren, and Polk. The two presidential campaigns that the Whigs won in 1840 and 1848 were both conducted without a party platform. The Whigs had found a basis for unity in their loyalty to two inspiring leaders—Henry Clay and Daniel Webster—both of whom died between the time of Scott's nomination and his defeat, but not before Webster, embittered by his own failure to win the nomination and antagonized by the sectionalism of the northern Whigs, had refused to support the Whig ticket.[24]

Second, the ideological disagreement over slavery struck the

23. Cole, *Whig Party in the South*, pp. 259–276.
24. Curtis, *Webster*, II, 626–627, 688–689, 693.

Whigs at a moment of strategic weakness when, as the minority party, they lacked the material assets of victory which will sometimes hold a party together even in the absence of agreement on principles. The simple attraction of winning will often lead the factions to avoid pressing the issues on which they are ideologically incompatible. Democratic unity after the election of 1852 was sustained by just such expedient considerations. But the northern and southern Whigs disagreed on principle and, at the same time, distrusted one another as political liabilities whose support cost more than it was worth. In this view, both were substantially correct. As northern Whigs saw it, the southern Whigs had demanded an obnoxious platform as the price of southern support and had then had the bad grace to desert the party in droves. Southern Whigs, on the contrary, resented the imposition of Scott as nominee by a northern faction which, even with a candidate of its own choosing, still lost all of the North except the perennial Whig states of Massachusetts and Vermont. At a time when both Whigs and Democrats were divided along sectional lines, it was strategically decisive that the Democrats gained a victory which impelled them to subordinate their divisions, while the Whigs suffered a reverse which inflamed theirs.

Apart from these weaknesses, which were inherent in the historical looseness of their organization and in their circumstances as a defeated party, the Whigs were also probably more susceptible to sectional disruptions than the Democrats because of a difference in their degree of responsiveness to the slavery question. Although there was much antislavery feeling in the northern wings of both parties, it appears that the Whigs reacted against slavery much more strongly.[25]

From the time of Jefferson and Jackson, the Democratic party, with its southern leadership, had drawn support in the North from elements more or less congenial to the South, but it had never had much appeal to the hard-core Yankees of New England. The political opposition to the Democrats had centered in New England, first

25. The relatively greater strength of antislavery feeling among northern Whigs as compared with northern Democrats is clearly indicated by two votes in the House of Representatives. On the Fugitive Slave Act (1850), northern Democrats voted 26 in favor, 16 opposed; northern Whigs, 3 in favor, 50 opposed. On the Kansas-Nebraska Act (1854), northern Democrats voted 44 in favor, 44 opposed; northern Whigs, 0 in favor, 44 opposed.

under the Federalists and later under the Whigs. Thus, the only two states that never voted for Jackson and remained Whig in every presidential election from 1836 through 1852 were Massachusetts and Vermont.[26]

To say that the Whig party contained a high proportion of Yankees is to say that it contained a high proportion of men with Puritan attitudes. And to say this is to say that it had more than its share of potential antislavery men.[27] It would, of course, be a gross oversimplification to suppose that all Puritans were abolitionists, for Puritanism was far too complex for any such easy generalization. Puritans believed in moral stewardship—in being their brothers' keepers—and this or some other aspect of their belief led to a propensity toward antislavery. But they also believed in property and respectability, and many of them were repelled by the violent language of abolitionism, its denunciation of the Union as a league with slaveholders, and its reckless contempt for the status quo and for legal property rights.

The more conservative exponents of the New England tradition, therefore, resisted the introduction of the slavery issue, and men like Webster, Edward Everett, Rufus Choate, Robert C. Winthrop, Millard Fillmore, and Thomas Ewing sought to maintain a cordial relationship with the slaveholding Whigs of the South. In Massachusetts, their group was known as "Cotton" Whigs, because of the alleged economic alliance between the cotton planters of the South and the cotton textile manufacturers of New England.[28] But if these features complicated the response of the Whigs to slavery, the fact remained that the "conscience" elements—those that responded

26. Maine and New Hampshire were Democratic, but prior to 1852, Connecticut and Rhode Island had voted for the Jacksonian party only once between 1824 and 1848. Lee Benson, *The Concept of Jacksonian Democracy: New York as a Test Case* (Princeton, 1961), pp. 177–179, says that the Whiggish solidarity of Yankee stock has been exaggerated, but that New York voters of New England descent were slightly Whiggish. See below, p. 244. See also Wilfred E. Binkley, *American Political Parties: Their Natural History* (2nd ed.; New York, 1945), pp. 163–165.

27. On the correlation of Whiggism and Puritanism, see Benson, *Concept of Jacksonian Democracy*, pp. 198–207. On the relationship between Puritanism and reform movements, see Clifford S. Griffin, *Their Brothers' Keepers: Moral Stewardship in the United States, 1800–1865* (New Brunswick, N.J., 1960); David Donald, "Toward a Reconsideration of Abolitionists," in his *Lincoln Reconsidered* (New York, 1956), pp. 27, 29.

28. On the Cotton Whigs, see Thomas H. O'Connor, *Lords of the Loom: The Cotton Whigs and the Coming of the Civil War* (New York, 1968).

primarily to the moral issue of slavery—had gained a preponderance in the northern wing of the Whig party by 1852. So complete was this ascendancy that it became common in the 1852 campaign to say, "We accept the candidate, but we spit on the platform."[29]

Thus, by 1852, the tensions over slavery, which severely strained both Democratic and Whig bisectionalism, had at last disrupted the Whig organization as a national party. This fact is well known. But what has not been understood, or even adequately recognized as a separate problem, is why the Whig party in the North also broke down at almost the same time that the sectional wings were splitting apart. Because the two events coincided in time, historians holding their coroners' inquests over the Whig party have often assumed that the two processes were one, and have equated them by suggesting that the loss of sectional balance inevitably caused the decline of the party in the North, on the theory, apparently, that a party, like a bird, cannot fly with only one wing. But plausible though this may seem, the evidence shows that sectional parties can be vigorous and successful. The Republican party, successor to the Whigs, sprang up as a sectional party and flourished for a century without developing any appreciable strength in the South. Also, it is evident that the Democrats in 1854 suffered a loss of sectional equilibrium comparable to that of the Whigs. Yet the Democrats survived.

The extent of the Democrats' loss of sectional balance and their capacity to go on despite this loss are worth scrutinizing carefully. The loss itself resulted from the Kansas-Nebraska Act. When this measure came to the floor of the House in 1854, the Democrats held a triumphant majority, with ninety-one free-state and sixty-seven slave-state members. Presumably, each group was large enough to command the respect of the other and to insist that all major policies should be based on consensus. Sectional equilibrium thus seemed assured. But when Douglas and the administration decided to force the passage of Kansas-Nebraska, they weakened the northern wing, first by causing some northern members to quit the party, and second by exposing those who followed the party mandate to decimation by northern voters. The bitter parliamentary battle tore the party badly, and the southern Democrats, in effect, overrode their northern allies by casting 57 votes in favor of the bill and only

29. A. G. Riddle, *The Life of Benjamin F. Wade* (Cleveland, 1888), p. 219; Elliott, *Scott*, p. 627.

2 votes against it, while 88 free-state Democrats were dividing 44 to 44. The whiphand tactics of Douglas in the Senate and of Alexander H. Stephens in the House left some deep scars. But more serious in the long run was the fact that the northern Democrats were so badly defeated in the subsequent congressional election that they could no longer hold their own against southern Democrats in the party caucuses. Having once lost this equality in the counsels of the party, they remained a minority for the next eighty years. In the elections of 1854, northern Democratic representation fell at one stroke, as has already been shown, from 91 to 25, while the southern representation slipped only slightly, from 67 to 58.

This meant that after 1854 the southern wing could dictate party decisions as it had never been able to do in previous Congresses. Up to this time, the Whig accusation that the party was dominated by its southern elements had been a partisan allegation, never more than partially true. But the elections of 1854, in a sense, made it true. The northern Democrats never again reached a parity with southern Democrats in the House of Representatives until the days of the New Deal (except for a period during the Civil War and Reconstruction), and for much of the time they were a small minority. In 1856 they rallied from their crushing post-Nebraska defeat, capturing 53 seats instead of 25, but in the counsels of the party, they were still outnumbered by 75 southerners, and in the Senate they were outnumbered, within their own party, 25 to 12. In 1858 their strength declined again, and the Democrats in the House numbered 34 from the North and 68 from the South, while those in the Senate numbered 10 from the North and 27 from the South.[30] This was the last Congress in which southerners sat until after Appomattox, which means that after 1854, and until the Civil War, the bisectional balance in the Democratic party had been destroyed. The party, in contrast to the Republicans, still attempted to maintain its strength in both sections, but at the same time it was in the grasp as never before of its southern wing.

The full meaning of this imbalance was to become evident in 1858 when Douglas rallied the northern Democrats against the proslav-

30. All these counts of sectional distribution of Democratic strength in Congress are compiled from the lists of members, with party affiliations shown, in the relevant issues of the *Congressional Globe,* the *Whig Almanac,* and the *Tribune Almanac.* For purposes of this tally, all slave states (including Delaware) are regarded as "southern" and all free states (including California and Oregon) as "northern."

ery Lecompton Constitution for Kansas. If he had done something like this before 1854, he might have carried almost half the party with him. But in 1858 the Southern bloc, controlling both the administration and the party organization in Congress, was able to treat him as a deviationist and to bring all the machinery of party discipline to bear against him. The only place where he could fight on terms of equality was in the party's quadrennial convention, because there the northern states were fully represented, whether they had elected Democrats to office or not.[31]

Therefore, both Whigs and Democrats suffered a loss of sectional balance. The Whigs sustained smashing blows to their southern wing in 1852; the Democrats, to their northern wing in 1854. But while the losses of the Whig party in the South seemed to pave the way for a collapse of the party in the North also, the losses of the Democrats in the North seemed actually to make the Democrats stronger in the South, as the southerners gained control and made the party increasingly subservient to southern interests and therefore increasingly attractive to section-minded southern Whigs. Thus, while the Whig party collapsed less than two years after losing its bisectional balance, the Democratic party endured and was still electing its candidates to the presidency more than a century later. If the Democratic party grew stronger in the South as it grew weaker in the North, why—the question clamors to be answered—did not the Whig party grow stronger in the North as it was growing weaker in the South?

Even in the massive defeat of 1852, there was some evidence of such a tendency. Scott had carried only two northern states, but in nine of the fourteen free states he had polled a larger vote than Taylor polled in 1848. In Rhode Island, New York, Illinois, Indiana, Michigan, Wisconsin, and Iowa, Scott received more votes than any other Whig candidate ever received.[32] While the slavery conflict was weakening the ties between the northern Whigs and their southern allies, the gravitation of the northern group toward an antislavery position seemed to be strengthening the party in the North. Seward

31. See below, pp. 325–326. For a comparison of the relative strength of divisive forces within the Whig and the Democratic parties in 1852, see Don E. Fehrenbacher, *Prelude to Greatness: Lincoln in the 1850's* (Stanford, 1962), p. 26.

32. See significant percentage tables in Svend Petersen, *A Statistical History of the American Presidential Elections* (New York, 1963). Taylor won 45.5 percent of the popular vote in the free states, and Scott won 43.6 percent, but since Scott received much of the former Free Soil vote, his percentage loss was more disappointing than the figures alone would indicate.

was already developing this potential of the Whig party as an antislavery party in New York, and Abraham Lincoln was about to try doing so in Illinois.[33] Yet the potential was not fulfilled, and why not remains one of the great unrecognized riddles of this era in American history.

It has remained unrecognized, perhaps, because of the excessive preoccupation of historians with the slavery issue as the only key to the events of the fifties. Yet it should be clear that whatever destroyed the Whig party in the North, it was not solely the "disruptive effect of the slavery issue." There was, however, one wholly different development which did do the party serious injury. This was the rising tension in American society between immigrant groups, which were predominantly Catholic, and native elements, which were overwhelmingly Protestant.

To appreciate the disruptive impact of this antagonism around the middle of the nineteenth century, it is necessary to recognize two factors which are now hard to appreciate. One of these is the sheer magnitude of the wave of immigration that suddenly hit the country in the late forties; the other is the degree of frank, unconcealed antagonism then existing between Protestants and Catholics.

It is widely known, of course, that migration to America at the time of the Irish famine was very heavy. But it is seldom realized that, proportionately, this was the heaviest influx of immigrants in American history. The total of 2,939,000 immigrants in the decade between 1845 and 1854 was less than one-third of the number in the decade before the First World War, but the total population was also much smaller, and in fact the immigrants of 1845–1854 amounted to 14.5 percent of the population in 1845, whereas the 9,000,000 new arrivals of 1905–1914 were but 10.8 percent of the population of 1905. Moreover, this riptide of immigration between 1845 and 1854 struck with severe shock in a society with a very small proportion of foreign-born members. Total immigration had never reached 100,000 before 1842, nor 200,000 before 1847, but it exceeded 400,000 three times in the four years between 1851 and 1855.[34]

33. Frederic Bancroft, *The Life of William Henry Seward* (2 vols.; New York, 1900), I, 365–368; Albert J. Beveridge, *Abraham Lincoln, 1809–1858* (4 vols.; Boston, 1928), III, 218–361; Fehrenbacher, *Prelude*, pp. 19–47; Reinhard H. Luthin, "Abraham Lincoln Becomes a Republican," *Political Science Quarterly*, LIX (1944), 420–438.

34. Data computed from tables in *Historical Statistics of the United States;* Maldwyn Allen Jones, *American Immigration* (Chicago, 1960), pp. 92–116.

In addition to the general fact that immigration was extremely heavy, there was a further, more specific feature: no less than 1,200,000 of the immigrants of 1845–1854 came from a single country—Ireland. In 1851 alone, a total of 221,000 recorded immigrants arrived from Ireland, which means that Irish immigrants, alone, in one year totaled more than 1 percent of the population. By contrast, the massive migrations of 1905–1914 never showed an influx from one country in one year even half as great proportionately.[35]

Such a heavy influx of strangers, and especially of often impoverished strangers, might have produced tensions under the best of circumstances. But in this case the antagonism was rendered far more acute by the fact that only a small proportion of the newcomers were Protestants from Ulster, while the vast majority, coming from the western and southern counties of Ireland, were Roman Catholics. Many Americans in this era were hostile to Catholicism, partly because they identified it with monarchism and reaction in a world where a republic was still somewhat lonely, and even more because of the Puritan heritage of antagonism to "popery"—an antagonism dating back to Bloody Mary, the Armada, the Gunpowder Plot, and the Revolution of 1688, when a Catholic king had been driven from the English throne. Because of this heritage, copies of Foxe's *Book of Martyrs* were still preserved in many American homes, and the observance of an annual Pope Day, as an occasion for anti-Catholic demonstrations, had been sanctioned in Boston as late as 1775. In the mid-nineteenth century, Catholics still treated Protestants harshly in countries where they were in the ascendancy, and Protestants still imposed disabilities on Catholics. American Protestants and Catholics did tolerate one another, but their "toleration" was in the literal sense, and not in the modern sense of according respect to one another's beliefs. Clergymen and church periodicals of the Protestant churches frequently denounced Catholicism as popery, idolatry, or "the beast," and even such respected and influential figures as the Reverend Lyman Beecher shared in this baiting

35. *Ibid.* On the Irish migration see Marcus Lee Hansen, *The Atlantic Migration, 1607–1860* (Cambridge, Mass., 1940), pp. 242–306; Hansen, *The Immigrant in American History* (Cambridge, Mass., 1940), pp. 154–174; Cecil Woodham-Smith, *The Great Hunger: Ireland, 1845–1849* (New York, 1962). Oliver MacDonagh, "Irish Emigration to the United States of America and the British Colonies during the Famine," in R. Dudley Edwards and T. Desmond Williams (eds.), *The Great Famine* (New York, 1957).

of Catholics. Catholic priests and Catholic periodicals proved quite capable of retorting in kind.[36]

In the 1850s, religious toleration was regarded more as an arrangement among the Protestant sects than as a universal principle. Given this background of religious antagonism, a certain amount of ethnocentrism on both sides, and a measure of economic rivalry between natives and immigrants in the competition for jobs, friction between native Protestants and immigrant Catholics became almost inevitable. With a high degree of social separation—perhaps even segregation—between them, they viewed one another from a distance with distrust and hostility. Many natives regarded the Irish as intruders and treated them as inferiors. The Irish, in turn, resented the discrimination and even persecution which they encountered at the hands of the Yankees. Ill will led to hostile acts, which, of course, reinforced the ill will in a vicious circle. Though it is largely forgotten today, and has consistently been minimized in American history, it is nevertheless true that for a considerable part of the nineteenth century the Catholic church was chronically under fire. Its beliefs were denounced; its leaders were assailed; its convents were slandered, and its property was threatened or even attacked. With both the Protestant press and the secular press keeping up a constant barrage of abuse, mob action sometimes resulted. Between 1834 and the end of the fifties, serious riots, with loss of lives, occurred in Charlestown, Massachusetts, in Philadelphia, in Louisville, and elsewhere. Convents were attacked, and one in Charlestown was burned to the ground, while probably as many as twenty Catholic churches were burned in cities or towns from Maine to Texas.[37]

It was in the midst of serious tensions, therefore, that the immigrant Irish began their participation in American political life. One of the earliest steps in this participation was choosing between the Whigs and the Democrats. If this choice had been purely an intellectual one, involving a decision between the formalistic "principles" of the two parties, the Irish response might have been fairly evenly

36. On hostility toward Catholics in the first half of the nineteenth century, the primary study is Ray Allen Billington, *The Protestant Crusade, 1800–1860: A Study of the Origins of American Nativism* (New York, 1938, and Chicago, 1964—citations are to the Chicago edition). Also Sister Mary Augustina Ray, *American Opinion of Roman Catholicism in the Eighteenth Century* (New York, 1936). Other studies of nativism are cited in notes 38, 45.

37. Billington, *Protestant Crusade*, pp. 53–90, 196–198, 220–237, 302–314.

divided. But in fact, the traditions of the two parties made it almost inevitable that the Irish should prefer the Democratic party. From its Jeffersonian beginnings, the Democratic party had been somewhat more cosmopolitan, less sectarian, and more concerned about the welfare of ordinary people than the opposition party. The Federalists, and after them the Whigs, on the other hand, were eminently the bearers of a political tradition which reflected the conservative Puritanism of eighteenth-century New England. They tended to believe in a "church and state" establishment, dominated by the spiritually and temporally elect. Compared to the Democrats, they were aristocratic in tone, deferential toward property, tenaciously faithful to Puritan values. They had hated Jefferson for his deism, his Gallicism, and his sympathy for revolution, and many of them were narrowly Protestant, suspicious of anything exotic, and intolerant of any deviation from accepted Yankee values.

Of course, all Puritans were not Whigs; much less were all Whigs Puritans. But there was a correlation, and it was quite strong enough to be clearly visible to Irish voters. According to Lee Benson in a study of New York state politics, voters of New England descent tended to vote Whig in a ratio of about 55:45. Immigrants from England, Benson believes, went Whig by a ratio of 75:25, while the ratio for those from Scotland, Ulster, and Wales stood at about 90:10. German immigrants, in contrast, were usually Democrats by a ratio of 80:20.[38]

The Irish, one imagines, took one look, saw the British and the

38. Benson, *Concept of Jacksonian Democracy*, pp. 278–287. The role of the Irish in this phase of American history is the subject of an extensive literature. See especially Oscar Handlin, *Boston's Immigrants, 1790–1880: A Study in Acculturation* (Cambridge, Mass., 1941; rev. ed., 1959); Hansen, *Immigrant in American History*, pp. 97–128; William G. Bean, "Puritan versus Celt, 1850–1860," *NEQ*, VII (1934), 70–89; Philip D. Jordan, "The Stranger Looks at the Yankee," in Henry Steele Commager (ed.), *Immigration and American History: Essays in Honor of Theodore C. Blegen* (Minneapolis, 1961), pp. 55–78; Thomas N. Brown, "The Origins and Character of Irish-American Nationalism," *Review of Politics*, XVIII (1956), 327–358; Thomas T. McAvoy, "The Formation of the Catholic Minority in the United States, 1820–1860," *Review of Politics*, X (1948), 13–34; Robert Ernst, *Immigrant Life in New York City, 1825–1863* (New York, 1949); Max Berger, "The Irish Emigrant and American Nativism as Seen by British Visitors, 1836–1860," *PMHB*, LXX (1946), 146–160; Florence E. Gibson, *The Attitudes of the New York Irish toward State and National Affairs, 1848–1892* (New York, 1951), pp. 59–110; Carl Wittke, *The Irish in America* (Baton Rouge, 1956), pp. 125–134; George W. Potter, *To the Golden Door: The Story of the Irish in Ireland and America* (Boston, 1960), pp. 371–386; William G. Bean, "An Aspect of Know Nothingism—the Immigrant and Slavery," *SAQ*, XXIII (1924), 319–334.

Puritans on one side, and knew they must belong on the other. Whatever caused them to do so, they went overwhelmingly with the Democrats. Benson maintains that by 1844, the Catholic Irish of New York were Democratic by a ratio of 95 to 5. And once the first Irish immigrants joined the Democratic party, the process became a self-reinforcing one. New Irish immigrants, guided by earlier Irish forerunners, accepted it as a fact of life that the Democratic party was the Irishman's party. Democrats in general, glad to have new allies, welcomed the Irish as friends, while the Whigs took a censorious view of these Democratic reinforcements and began to talk sourly of requiring a residence of twenty-one years for naturalization.

Some Whigs, of course, saw the need to compete against the Democrats for Irish support. For instance, William H. Seward, as Whig governor of New York, had advocated that public funds should be appropriated for the support of Catholic schools.[39] But few Whigs advocated specific measures attractive to the Irish, and most of them did little more than to try to cajole Irish voters. A clear instance appeared in 1852, when Winfield Scott wooed these voters with elephantine clumsiness. Planting Irish supporters in his audiences, he would greet their prearranged interruptions with assurances that he "loved to hear that rich Irish brogue." But his stratagems were as unsuccessful as they were transparent, and after the election Tom Corwin wrote, in an extremely despondent vein, "we know they *all* voted the other ticket."[40]

The millstone around the neck of the northern Whigs in 1852 was not the loss of the southern wing of their party; it was a volume of immigration which in four years exceeded Scott's total popular vote. The Whigs knew that this reservoir of potential new votes would soon overwhelm them. If this ominous factor had not cast a pall over the future of the Whig organization, it is entirely likely that antislavery Whigs would have launched a drive to make the Whig party an

39. Vincent P. Lannie, *Public Money and Parochial Education: Bishop Hughes, Governor Seward, and the New York School Controversy* (Cleveland, 1968); Bancroft, *Seward*, I, 96–101; Billington, *Protestant Crusade*, pp. 142–165.

40. Elliott, *Scott*, pp. 625–646; Corwin to James A. Pearce, Oct. 20, 1854, in Bernard C. Steiner, "Some Letters from the Correspondence of James Alfred Pearce," *Maryland Historical Magazine*, XVII (1922), 40–41. Kirwan, *Crittenden*, pp. 293–296, shows copious evidence of the Whigs' conviction that the foreign vote was the deciding factor in their defeat in 1844 and also in 1852; Binkley, *American Political Parties*, pp. 162–163, 187–190; Nichols, "Election of 1852," pp. 947–948.

antislavery party and nativist Whigs would have set to work to make it a nativist party. But with circumstances what they were, both antislavery men and nativists had reason to doubt that they could gain their objectives inside the party as well as they could outside it. The antislavery men were held in the thrall of an embarrassing affiliation with cotton Whigs and were thus separated from antislavery men in the Democratic and Free Soil parties who were their natural allies. The nativists were tied to a party containing many men who either condemned nativism strongly, like William H. Seward,[41] or at least avoided it because they felt it discreditable. Thus the nativists found themselves separated from those Democrats who shared their hostility toward immigrants or Catholics. Many antislavery men wanted to be rid of the nativist issue, and many nativists wanted to get away from the slavery issue and to stress the idea of Union. The Whig party, with its strange bedfellows, its equivocations, and its record of defeat, frustrated all of these impulses without offering any significant political advantages to compensate for the frustrations.

The impulse toward a clear-cut antislavery party had shown itself as far back as 1840 with the Liberty party, and had reached major proportions with the Free Soil movement of 1848. Nativist parties also began to emerge at the local level in the 1830s, but none developed into a national organization. At the same time, a number of nonpolitical organizations like the American Bible Society and the American Tract Society were becoming increasingly anti-Catholic in purpose and therefore nativist in influence. Of particular interest is the Order of the Star-Spangled Banner, a secret society apparently founded in New York in 1849 but rising to prominence as a political force after the election of Pierce, when it acquired the "Know Nothing" label.[42]

Thus, after 1852, many nativist Whigs and many antislavery Whigs were alike ready to leave their party for new political alliances if they could be arranged. The Kansas-Nebraska Act triggered these impulses by causing many anti-Nebraska Democrats to bolt the Democratic party and thereby make themselves available as potential allies. Thus, a proslavery measure (the Kansas-Nebraska Act)

41. Seward's opposition to the Know-Nothings was so strong and outspoken that it probably later cost him the Republican nomination for the presidency. See his speech of July 12, 1854, in *Congressional Globe,* 33 Cong., 1 sess., pp. 1708–1709.

42. Billington, *Protestant Crusade,* pp. 166–288, 337–338; 380–407; W. Darrell Overdyke, *The Know-Nothing Party in the South* (Baton Rouge, 1950), pp. 34–44.

not only did serious injury to the proslavery party which sponsored it, but also it injured the rival Whig organization even more, for it offered antislavery Whigs a new set of potential allies whose support they might win by leaving the Whig party.

When both antislavery Whigs and nativist Whigs saw the anti-Nebraska men breaking away from the Democratic party, they reacted at once with steps to form new political organizations. The spring and summer of 1854 witnessed rapid change and intense political activity. The exertions of the antislavery men were, of course, relatively more conspicuous than those of the nativists, for the reaction to the Nebraska bill was an antislavery reaction. In every free state, this reaction had a significant impact on the structure of parties. In some states, such as Michigan and Wisconsin, Indiana and Maine, where the Whigs had been weak to begin with, it was possible, with relative ease and rapidity, to form a fully integrated new party, based essentially on antislavery. But the tempo of the transition was not the same in any two states. Where the Whig organization had been strong, as in New York, or where contiguity to slave state neighbors had induced an indulgent attitude toward slavery, as in Pennsylvania and New Jersey, the Whig organizations proved tenacious, and antislavery men had to be content with loose coalitions. In New York, the sagacious Whig boss, Thurlow Weed, knew that he must carry the Whigs into a new organization. But he meant to do it in his own way and at his own time, and only after he had seen to the safe election of William H. Seward for another six years in the Senate. In Illinois, a group of men whose antislavery zeal amounted almost to abolitionism announced plans to launch a "Republican" party and sought to place Abraham Lincoln on their central committee. He refused this overture and chose to run for the state legislature on the Whig ticket instead. Lincoln was not indifferent to the antislavery cause, but in 1854 he hoped to make the Whig party the antislavery party, at least in Illinois.

With this wide difference from state to state in the process by which antislavery Whigs gravitated toward a new antislavery organization, party patterns presented a bewildering variety, and changes took place under a confusing diversity of labels. The term "Republican" was proposed at Ripon, Wisconsin, in February 1854; was endorsed by a meeting of thirty congressmen in Washington in May; and was adopted by a state convention at Jackson, Michigan, on July 6. But antislavery men rallied to the banner of a People's party in

Ohio and Iowa, and they used the terms Fusion party and Anti-Nebraska party in some states. Terminology counted for less than what was happening. Within a few weeks after the enactment of Kansas-Nebraska, new antislavery combinations had formed in Vermont, Massachusetts, Ohio, Indiana, Michigan, Iowa, and Wisconsin, and the ferment was at work to bring about the creation of similar parties in every northern state.[43]

The architects of the new antislavery party brought their activity to its climax on July 13, the anniversary of the Northwest Ordinance, when they held conventions simultaneously in Vermont, Ohio, and Indiana, only one week after the Michigan convention had adopted the Republican name.[44] But the antislavery Whigs were by no means alone in striving for a new party. The nativists were also in the field. Four days after the three simultaneous antislavery meetings, the Order of the Star-Spangled Banner held a convention in New York of delegates from thirteen states to set up a national organization. They established a Grand Council for the Order, with a hierarchy of subordinate state and local councils; they fixed a secret ritual, with many alluring devices, such as an arrangement that meetings would be called by the distribution of heart-shaped bits of paper—in normal times, white paper, but if danger threatened, red paper; and they adopted a pledge by which members would promise to renounce all party allegiance and never to vote for any foreign-born or Roman Catholic candidate for office. Members were also pledged to keep all information about the Order secret, and when questioned, to say, "I know nothing." Politically, they designated themselves as the American party, but, inevitably, they were called Know-Nothings.[45]

43. On the origins of the Republican party, see Andrew Wallace Crandall, *The Early History of the Republican Party, 1854–1856* (Boston, 1930); Eric Foner, *Free Soil, Free Labor, Free Men: The Ideology of the Republican Party before the Civil War* (New York, 1970); Isely, *Greeley and Republican Party*, pp. 86–170; Bancroft, *Seward*, I, 363–424; Beveridge, *Lincoln*, III, 263–296; George H. Mayer, *The Republican Party, 1854–1964* (New York, 1964), pp. 23–47; Fehrenbacher, *Prelude*, pp. 19–47; Binkley, *American Political Parties*, pp. 206–221; Hans L. Trefousse, *The Radical Republicans: Lincoln's Vanguard for Racial Justice* (New York, 1969), pp. 66–102; Michael Fitzgibbon Holt, *Forging a Majority: The Formation of the Republican Party in Pittsburgh, 1848–1860* (New Haven, 1969); Morton M. Rosenberg, *Iowa on the Eve of the Civil War: A Decade of Frontier Politics* (Norman, Okla., 1972).

44. New York *Tribune*, and *New York Times*, July 14, 16, 1855.

45. On political nativism and the Know Nothing party, see Billington, *Protestant Crusade*, pp. 380–436; Overdyke, *Know-Nothing Party in South*, pp. 36–155; Harry J.

In the general political ferment of the fifties, antislavery and nativism were not the only new forces in evidence. There was also a powerful temperance movement which had achieved the adoption in Maine in 1851 of a law against the sale of liquor.[46] By 1854, the advocates of temperance had organized politically in other states and were supporting candidates in many elections. Voters in 1854, therefore, faced a stunning array of parties and factions. Along with the old familiar Democrats, Whigs, and Free Soilers, there were also Republicans, People's party men, Anti-Nebraskaites, Fusionists, Know-Nothings, Know-Somethings (antislavery nativists), Maine Lawites, Temperance men, Rum Democrats, Silver Gray Whigs, Hindoos, Hard Shell Democrats, Soft Shells, Half Shells, Adopted Citizens, and assorted others.

To historians writing after the dust had settled, it appeared that the emergence of the Republican party was the central development in all this complicated process of political disintegration and reintegration. But in 1854, the results of the elections, although in some

---

Carman and Reinhard H. Luthin, "Some Aspects of the Know-Nothing Movement Reconsidered," *SAQ,* XXXIX (1940), 213–234; John Higham, "Another Look at Nativism," *Catholic Historical Review,* XLIV (1958), 147–158; Thomas J. Curran, "Seward and the Know-Nothings," New York Historical Society *Quarterly,* LI (1967), 141–159; Ira M. Leonard, "The Rise and Fall of the American Republican Party in New York City, 1843–1845," *ibid.,* L (1966), 151–192; Louis Dow Scisco, *Political Nativism in New York State* (New York, 1901); Mary St. Patrick McConville, *Political Nativism in the State of Maryland, 1830–1860* (Washington, 1928); Laurence Frederick Schmeckebier, *The History of the Know Nothing Party in Maryland* (Baltimore, 1899); Philip Morrison Rice, "The Know-Nothing Party in Virginia, 1854–1856," *Virginia Magazine of History and Biography,* LV (1947), 61–75, 159–167; James H. Broussard, "Some Determinants of Know-Nothing Electoral Strength in the South, 1856," *Louisiana History,* VII (1966), 5–20; Leon Cyprian Soulé, *The Know-Nothing Party in New Orleans* (Baton Rouge, 1962); Agnes Geraldine McGann, *Nativism in Kentucky in 1860* (Washington, 1944); Carl Fremont Brand, "The History of the Know Nothing Party in Indiana," *IMH,* XVIII (1922), 47–81, 177–206, 266–306; John P. Senning, "The Know-Nothing Movement in Illinois, 1854–1856," ISHS *Journal,* VII (1914), 7–33; Joseph Schafer, "Know-Nothingism in Wisconsin," *Wisconsin Magazine of History,* VIII (1924), 3–21; Ralph A. Wooster, "An Analysis of the Texas Know-Nothings," *SWHQ,* LXX (1967), 414–423; Peyton Hurt, "The Rise and Fall of the 'Know Nothings' in California," *California Historical Society Quarterly,* IX (1930), 16–49, 99–128.

46. On the early temperance movement, see John Allen Krout, *The Origins of Prohibition* (New York, 1925), pp. 176–177, 217–220, 262–304; Griffin, *Their Brothers' Keepers,* pp. 112–151, 217–241; Joseph R. Gusfield, *Symbolic Crusade: Status Politics and the American Temperance Movement* (Urbana, Ill., 1963), pp. 36–60; Alice Felt Tyler, *Freedom's Ferment: Phases of American Social History to 1860* (Minneapolis, 1944), pp. 308–350; Frank L. Byrne, *Prophet of Prohibition: Neal Dow and His Crusade* (Madison, Wis., 1961).

respects ambiguous, seemed to indicate the possible triumph of Know-Nothingism rather than of antislavery. At that juncture, there seemed to be a likelihood that the Catholic or immigrant question might replace the slavery question as the focal issue in American political life.[47]

In May 1854, even before the Know-Nothings had set up their national organization, they demonstrated their unsuspected strength by capturing the mayoralty of Philadelphia with a majority of over 8,000 votes. By July, it was evident that nativism might become the dominant national issue, and Stephen A. Douglas, although still bleeding politically from wounds inflicted by the antislavery men during the Kansas-Nebraska debate, began assailing the Know-Nothings, rather than the antislavery groups, as the principal danger to the Democratic party.[48]

In various state and municipal elections during the summer and autumn, the Know-Nothings scored astonishing successes, often with secret candidates whose names had not even been printed on a ballot. Finally, in the November elections, they gained some stunning victories. In Massachusetts, especially, they swept everything, polling 63 percent of the vote, and electing all of the state senators and all but two of the 378 representatives. They cast more than 40 percent of the vote in Pennsylvania and 25 percent in New York, despite the continued strength of the Whigs in that state. A sizable minority of the men elected to the national House of Representatives were Know-Nothings. These victories, it may be added, were followed in 1855 by further triumphs in three other New England states and in New York, Pennsylvania, California, and the South. In these circumstances, it seemed entirely plausible for the New York *Herald* to predict that the Know-Nothings would win the presidency in 1856.[49]

47. Binkley, *American Political Parties,* p. 195.

48. Douglas declared that the Know-Nothings, not the Republicans, had won the election of 1854; the anti-Nebraska movement became, he said, "a crucible into which they poured Abolitionism, Maine liquor-lawism, and what there was left of Northern Whiggism, and then the Protestant feeling against the Catholic, and the native feeling against the foreigner." *Congressional Globe,* 33 Cong., 2 sess., appendix, pp. 216–230.

49. Billington, *Protestant Crusade,* p. 389, quoting New York *Herald* and Boston *Pilot.* On the magnitude of Know-Nothing victories, *ibid.,* pp. 387–406; Allan Nevins, *Ordeal of the Union* (2 vols.; New York, 1947), II, 326–328, 341–346; Overdyke, *Know-Nothing Party in South,* pp. 57–90.

The antislavery and nativist groups frequently avoided a contest with one another for the good reason that both appealed to the same elements in the population. It may seem paradoxical in the late twentieth century to say that the same people who opposed the oppression of a racial minority also favored discrimination against a religious minority, but history is frequently illogical and the fact is that much of the rural, Protestant, Puritan-oriented population of the North was sympathetic to antislavery and temperance and nativism and unsympathetic to the hard-drinking Irish Catholics. Politicians of course realized that it might be possible to join the support of Republicans, Know-Nothings, and Temperance groups to form a winning political combination. Thus it happened that nativism and antislavery operated in conjunction in 1854 more often than in opposition. When the new Congress was elected, there were about 121 members who had been chosen with Know-Nothing support and about 115 who had been elected as Anti-Nebraska men, with antislavery support. About 23 were antislavery but not nativist; about 29 were nativist but not antislavery (most of these were Southerners); but some 92 were both antislavery and associated with nativism. This situation meant that most of the nativists were antislavery and most of the antislavery members were in some degree nativists. Confusing though it may be, it was possible to say that the anti-Nebraska men held a majority in the House and also that the Know-Nothings held a majority in the House.[50] At that juncture, it seemed clear that antislavery would be strongly linked with nativism, and the only question, apparently, was which of these forces would be predominant in the coalition.

American historians have been slow to recognize the relation between Know-Nothingism and Republicanism in 1854. Perhaps this is partly because they have been confused by a complicated situation, almost unique in American history, in which two different parties could both legitimately claim to have won an election. But

50. For a list of Americans (Know-Nothings) in the 34th Congress, see speech by Representative S. A. Smith of Tennessee, April 4, 1856, in *Congressional Globe*, 34 Cong., 1 sess., appendix, p. 352. For a list of antislavery or anti-Nebraska men in the same Congress, see *The Whig Almanac and United States Register for 1855* and *The Tribune Almanac and Political Register for 1856*. A comparison of these lists shows the distributions indicated above. William A. Richardson, Illinois Democrat, addressing the opposition, said, "In either view of your principles you are in the majority. As Republicans, you have a majority. As Americans, you have also a majority." *Congressional Globe*, 34 Cong., 1 sess., p. 314.

also, it has been psychologically difficult, because of their predominantly liberal orientation, for them to cope with the fact that antislavery, which they tend to idealize, and nativism, which they scorn, should have operated in partnership.

The affinities of nativism and antislavery are striking, even apart from the fact that both drew their strength from the same religious and social constituencies.[51] Both, for instance, reflected psychologically a highly dramatized fear of a powerful force which sought by conspiratorial means to subvert the values of the republic: in one case this was the slavocracy, with its "lords of the lash," in the other, the Church of Rome with its crafty priests and subtle Jesuits. Both reflected in their propaganda a prurient fascination with the alleged sexual excesses of slaveholders and priests. In an age when sexual repression was widespread and sex as a theme in most branches of literature was taboo, the "exposure" of evil provided a sanction for the salacious description of sexual transgressions. In the lurid and sensational literature of the two movements, the lecheries of the priests and the miscegenation of the slaveholders were favorite themes. Endangered chastity—whether of lovely octoroon girls or of virginal nuns—was a vital part of the message of reform. If the escape of a mulatto girl was the high point of *Uncle Tom's Cabin,* the escape of a nun from the convent was the high point of *The Awful Disclosures of Maria Monk.* If *Uncle Tom* outsold *Maria, Maria* outsold everything else, and was called, with perhaps more significance than was intended, "the *Uncle Tom's Cabin* of Know-Nothingism." If Wendell Phillips said that the slaveholders had made the entire South "one great brothel," the *American Protestant Vindicator* said that an unmarried priesthood had converted whole nations into "one vast brothel."[52]

51. On the relationship between nativism and Republicanism, see Foner, *Free Soil,* pp. 226–260.

52. On the use of sex sensationalism in abolitionist literature (as distinguished from serious attention to the sexual exploitation of slave women) see Arthur Young Lloyd, *The Slavery Controversy, 1831–1860* (Chapel Hill, 1939), pp. 83–92, 97–99. On the same theme in anti-Catholic literature, see Billington, *Protestant Crusade,* pp. 99–108, 361–367. On "one vast brothel," *ibid.*, p. 167, quoting *American Protestant Vindicator,* Dec., 1, 1841; and speech of Wendell Phillips, Jan. 27, 1853, in Phillips, *Speeches, Lectures and Letters* (Boston, 1864), p. 108; on "the Uncle Tom's Cabin of Know-Nothingism," Billington, p. 108. Note the spectacular effects and results obtained by the Reverend Henry Ward Beecher in holding mock auctions of nubile Negro girls at the Plymouth Church in Brooklyn, Paxton Hibben, *Henry Ward Beecher* (New York, 1927), pp. 136, 150. David Brion Davis, "Some Themes of Counter-Subversion: An Analysis of Anti-Masonic, Anti-Catholic, and Anti-Mormon Litera-

Recognition of these parallels should not obscure the difference in principle between antislavery and nativism—a difference Lincoln pointed out when he said, "How can anyone who abhors the oppression of negroes be in favor of degrading classes of white people? . . . As a nation, we began by declaring that 'all men are created equal.' We now practically read it 'all men are created equal, except negroes.' When the Know-Nothings get control, it will read 'All men are created equal, except Negroes and foreigners and Catholics.' "[53] But although the rational appeals of nativism and antislavery may have been wholly dissimilar, the irrational appeals of the two, especially to men with high levels of fear or anxiety, were somewhat the same.[54]

In these circumstances, it seemed likely that the two movements would remain mutually supportive. Where the component of irrationality was strong, there was no assurance that the more rational crusade would be enduring and the less rational one transitory. In some ways, the anti-Catholic impulse seemed to have more psychological voltage than the antislavery impulse. The number of dead and wounded in the anti-Catholic riots of Louisville's "Bloody Monday" in 1855 far exceeded the casualties resulting from John Brown's raid.[55] Certainly, both issues had enough power to make a coalition of antislavery men and nativists highly expedient, and even Abraham Lincoln kept very silent in public about his disapproval of Know-Nothingism.[56] If nativism did not crowd antislavery off the track altogether, the antislavery party, it appeared, would at least have to accept nativist planks in its platforms and nativist candidates on its tickets. But to gain this nativist support, it would have to accept the stigma of nativist intolerance.

Such was clearly the situation when the Congress met in Decem-

---

ture," *MVHR*, XLVII (1960), 205–224, shows the similarities in the "projective fantasies" that were used against the three groups named, despite their actual dissimilarity. The antislavery propaganda of the time presents obvious parallels and used the same kind of "projective fantasies," but it is not included in Davis's analysis. See also, however, his *The Slave Power Conspiracy and the Paranoid Style* (Baton Rouge, 1969), in which he concludes (p. 85) that the "Slave Power" image was irrational but put to such good use that it was "almost providential."

53. Lincoln to Joshua Speed, Aug. 24, 1855, Roy P. Basler (ed.), *The Collected Works of Abraham Lincoln* (8 vols.; New Brunswick, N.J., 1953), II, 320–323.

54. See Nevins, *Ordeal*, II, 329–331, for the "natural sympathy . . . among temperance men, anti-slavery men, Whigs, and Northern Know-Nothings."

55. Billington, *Protestant Crusade*, p. 421; Kirwan, *Crittenden*, p. 300.

56. Charles Granville Hamilton, *Lincoln and the Know-Nothing Movement* (Washington, 1954), p. 9.

ber 1855. Yet six months later when the presidential candidates of 1856 had been named, there was not an antislavery nativist on any ticket, and the northern Know-Nothings were giving their support to John C. Frémont, a man who had never been in a Know-Nothing lodge and whose marriage to the daughter of Senator Benton had been performed by a Catholic priest.[57] The story of how this came about is one of the obscure and neglected aspects of American political history.

To begin with, the nativists learned in June 1855 that, as a bisectional organization, they enjoyed no more immunity from the disruptive effects of the slavery question than their predecessors, the Whigs. They had sought to exalt nationalism in the creed of the order, as a bulwark against sectional forces, and had even inaugurated a "Union degree" in their ritual. As many as 1,500,000 members, pledging themselves to stand together against sectional forces from either North or South, are estimated to have taken this degree. But once the Order embarked on national politics, it had to take a position on Kansas-Nebraska, and on this question northern and southern nativists found that the secret rituals they shared did not help them to agree. When the National Council met at Philadelphia in June 1855, the southern delegates took advantage of an opportunity to force through a resolution, known as the Twelfth Section, declaring that existing laws must be maintained as a final settlement of the slavery question.[58] This indirect endorsement of the Kansas-Nebraska Act caused the entire delegations of all the free states except New York, Pennsylvania, New Jersey, and California to withdraw from the meeting and gather in a separate conclave to voice their protest.[59]

This separation was not final, and need not have been fatal. Northern state councils did not intend to withdraw from the order, but to renew the fight at another session of the National Council just

57. On the marriage, compare Allan Nevins, *Frémont, Pathmarker of the West* (New York, 1939), p. 69, with his *Ordeal of the Union,* II, 496.

58. On the Union degree, Billington, *Protestant Crusade,* p. 423; on the Twelfth Section, *New York Times,* June 7–14, 1855; Henry Wilson, *History of the Rise and Fall of the Slave Power in America* (3 vols.; Boston, 1872–77), II, 423–432; Scisco, *Political Nativism in New York,* pp. 144–147; Overdyke, *Know-Nothing Party in South,* pp. 127–133; Speech of Ethelbert Barksdale, July 23, 1856, in *Congressional Globe,* 34 Cong., 1 sess., appendix, p. 1178; Kirwan, *Crittenden,* pp. 298–299.

59. *New York Times,* June 15, 1855; *Harper's New Monthly Magazine,* XI (Aug 1855), 399; Wilson, *Rise and Fall of Slave Power,* II, 431–433.

before the national convention in February 1856. At this second Council meeting, the Twelfth Section was rescinded, but the harmony thus restored did not last a week.[60] The national convention became a scene of strife where "members ran about the hall as if they were mad, and roared like bulls." On the third day, the southern delegates, with the aid of New York, voted down a resolution in favor of the restoration of the Missouri Compromise (that is, the repeal of Kansas-Nebraska). This vote precipitated action by fifty northern delegates from eight states. They walked out of the convention, reassembled in a separate meeting that evening, and issued a call for a separate convention of northern Know-Nothings in June. The remaining delegates then nominated Millard Fillmore for president, with Andrew J. Donelson as his running mate.[61] From this time forward, northern and southern Know-Nothings were completely divided, and the nomenclature of political parties was enriched by two new terms—North Americans and South Americans.

This sectional split signalized Know-Nothing failure in efforts to establish a national party, but it left the North Americans competing strongly with the Republicans for the role of major opposition party in the free states. What complicated the competition was the fact that nativist and anti-Nebraska sentiments were so often united in the same man. Also, there was as yet little unity at the national level in anti-Nebraska ranks; the birth of the Republican party was still in process. The disarray of partisan politics became evident when the Thirty-fourth Congress convened in December 1855. Viewing this body of varied and often multiple allegiances, the editor of the *Congressional Globe* set aside his practice of designating members' party affiliation. The Democrats, in the reaction against the Kansas-Nebraska Act, had lost control of the House, but the opposition could not be consolidated into a majority under the Know-Nothing, anti-Nebraska, or any other label. The result was a fierce two-month contest for the speakership ending in

60. *New York Times,* and New York *Herald,* Feb. 19, 20, 21, 22, 1855. Antislavery men wanted the quarrel in the June 1855 meeting of the Council to result in a definite schism, and therefore described it as such. But in fact the state councils of the North did not break with the national Council; they did, however, endorse demands for a second meeting of the Council to reconsider the Twelfth Section. Delegations from nine northern states, meeting at Cincinnati in Nov. 1855, adopted resolutions leading to this second meeting. *New York Times,* Nov. 24, 1855.

61. *New York Times* and New York *Herald,* Feb. 23, 24, 25, 26, 27, 1856.

the election of Nathaniel P. Banks of Massachusetts.[62]

Banks, formerly a Democrat and recently a Know-Nothing, but now plainly a Republican, had only gradually accumulated enough anti-Nebraska support to win. It was a sectional, antislavery victory and a significant one in several ways. Banks personified the link between nativism and antislavery, but also the greater appeal of antislavery. His election meant that the large bloc of congressmen with both nativist and anti-Nebraska associations had, with few exceptions, given its primary allegiance to antislavery and thus to the emerging Republican party.[63] In the words of one editor on the scene, "Some who came here more 'American' than Republican are now more Republican than American."[64] The nativist-antislavery alliance had been made to work with a minimum of nativism and a maximum of antislavery. At the same time, the speakership contest compelled the loose anti-Nebraska coalition in Congress to take a long step toward unity and permanent organization.[65]

Four months later, the Republicans completed the process of neutralizing nativism and capturing control of the Know-Nothing movement in the North. This time they maneuvered the North Americans into accepting a presidential candidate who was not even a nativist. This was done with great adroitness and amid the most difficult circumstances, for the North Americans, at the time of their breach with the southern nativists in February, had shown much

62. Fred Harvey Harrington, "'The First Northern Victory,'" *JSH,* V (1939), 186–205; Harrington, *Fighting Politician: Major General N. P. Banks* (Philadelphia, 1948), pp. 28–31. The long contest ended only when the House voted (as in 1849–50) to elect by a plurality. Banks won on the 133rd ballot with 103 votes to 100 for William Aiken of South Carolina and 11 scattered. For comments of participants, see *Congressional Globe,* 34 Cong., 1 sess., pp. 86, 174, 231, 242–245, 306, 308, 313, 315, 326, 1043; Temple R. Hollcroft (ed.), "A Congressman's [Edwin B. Morgan] Letters on the Speaker Election in the Thirty-Fourth Congress," *MVHR,* XLIII (1956), 444–458.

63. Apparently a good many nominal Know-Nothings had joined the Order as a means of advancing the antislavery cause and were in fact subversives within the organization. Henry Wilson is a prime example. See his *Rise and Fall of Slave Power,* II, 417–419; also, Ernest A. McKay, "Henry Wilson: Unprincipled Know-Nothing," *Mid-America,* XLVI (1964), 29–37; and Isely, *Greeley and Republican Party,* p. 165, quoting Greeley on "bogus" Know-Nothings.

64. Gamaliel Bailey to Charles Francis Adams, Jan. 20, 1856, quoted in Foner, *Free Soil,* p. 247.

65. Harrington, "First Northern Victory," pp. 204–205, quoting Joshua R. Giddings: "We have got our party formed, consolidated, and established"; and Thurlow Weed: "The Republican party is now inaugurated." Twenty days later, on Feb. 22, the Republican party held its first national meeting at Pittsburgh.

resourcefulness in calling a convention to nominate a presidential candidate and scheduling it to meet five days before the Republican nominating convention.[66] Such a convention at such a time presented the Republicans with a dilemma: If the North Americans nominated a different candidate from the Republicans, it would split the antislavery vote; if they nominated the same candidate, the fact that their nomination came first would make it appear that the candidate was primarily a North American also endorsed secondarily by the Republicans.[67] The only escape from this dilemma would be to persuade the North Americans to defer their nomination, which was impossible, or to maneuver them into nominating a stalking horse, who would withdraw in favor of the Republican nominee at the strategic moment. They induced Nathaniel P. Banks to accept this dubious role,[68] and after that, matters moved like clockwork. The Know-Nothings nominated Banks for president and William P. Johnston for vice-president on June 16; the Republicans nominated John C. Frémont, uncontaminated by a prior Know-Nothing endorsement, on June 18. The Know-Nothings made desperate efforts to persuade the Republicans to accept a joint ticket of Frémont and Johnston, so that the nativists could save face. But the Republicans, realizing that they were in the saddle at last, refused all overtures;[69] on June 19 the North Americans reassembled and, knowing that Banks would not accept their nomination, capitulated completely, though with bitter protests. Banks was withdrawn, Frémont was nominated, and they permitted themselves only the gesture of nominating Johnston for vice-president.[70] This charade ended in August when Johnston had an interview with Frémont, who may have promised him patronage, after which he too withdrew from candidacy.[71]

66. Harrington, *Fighting Politician,* p. 36.

67. Horace Greeley declared, "Our real trouble is the K.N. convention on the 12th." Quoted in Nevins, *Frémont, Pathmarker,* p. 430.

68. Harrington, *Fighting Politician,* pp. 36–38; *New York Times* and New York *Herald,* June 13, 14, 15, 16, 17, 1856.

69. *New York Times,* June 17, 19, 20. 1856; New York *Herald,* June 17, 19, 1856; William H. Seward to his family, July 7, 1856, in Frederick W. Seward, *Seward at Washington as Senator and Secretary of State, 1846–1861* (New York, 1891), p. 283.

70. *New York Times* and New York *Herald,* June 20, 21, 1856.

71. Frémont at first secured the consent of the North Americans to his own nomination by promising to secure the withdrawal of the Republican vice-presidential nominee and to accept Johnston as his running mate, but this promise was not fulfilled. See letter of Z. K. Pangborn to Banks, June 25, 1856, in Fred Harvey Harrington, "Fremont and the North Americans," *AHR,* XLIV (1939), 847. Also

In 1855, it had appeared that the Know-Nothings might win the presidency next year. But when nativists went to the polls in 1856, their choice lay between Fillmore, whom the South Americans had nominated, but whose membership in the Know-Nothing order was questionable,[72] and Frémont, nominated by the Republicans, endorsed by the North Americans, widely though incorrectly believed to be a Catholic, and certainly not a Know-Nothing. Such was the anticlimax of the great Know-Nothing excitement.

The antislavery party had now clearly, and somewhat unexpectedly, emerged as the dominant party in the North. Thus, as it appears in retrospect, the Know-Nothing phase of American politics had served as a kind of intermediate stage in the transition from Whiggism to Republicanism. To historians of a generation now gone, with their teleological view of history, all this seemed preordained: as one of them wrote, the Know-Nothing party had "performed its historical mission" when it had "prepared the way for the Republicans," and "the only task the party now had was to die."[73] Today, a more skeptical critic might agree that nativism was intrinsically a transient phenomenon, and that the slavery issue was a more lasting one. But he will question the providential role of Know-Nothingism, and will ponder uneasily what the relationship of nativism to antislavery might have been if the Know-Nothings had been more clever or the Republicans less so in their political tactics, and if Nathaniel P. Banks, the Bobbin Boy of Massachusetts, had not

---

letters of Francis Ruggles and Lucius Peck, of the North American National Committee, to Republican headquarters, June 30, Aug. 4, 1856, quoted by Roy Franklin Nichols, "Some Problems of the First Republican Presidential Campaign," *AHR*, XXVIII (1923), 493–494.

72. Robert J. Rayback, *Millard Fillmore* (Buffalo, 1959), pp. 386–414, discusses Fillmore's relationship with the Know-Nothings. In 1855, Fillmore was privately initiated into the Order but apparently never attended a meeting. Fillmore's principal lieutenants had carefully gone about gaining political control of the Know-Nothing organization in New York, and they used it to promote his fortunes. But Fillmore was not anti-Catholic; his daughter had been educated by nuns; and he scarcely touched upon the nativist issue during the campaign. During the campaign, Fillmore's membership was both affirmed (*New York Times*, March 3, 1856) and denied (New York *Herald*, Feb. 27, 1856).

73. H. E. von Holst, *The Constitutional and Political History of the United States* (8 vols.; Chicago, 1876–92), V, 198; Schuckers, *Chase*, p. 161, speaks of the Know Nothing party as "a stepping stone" for voters who were on the way to becoming Republicans. Nevins, *Ordeal*, II, 331, says that "the voters were sure to fall away from" Know-Nothingism, because "national and religious tolerance are absolutely basic elements in American life." But, in fact, discriminatory attitudes toward immigrants and minority groups did not disappear at all; they merely ceased to be expressed in overt political form.

decided to turn his remarkably flexible talents to politics.

After 1856, the Know-Nothing party was dead, and it has been easy to assume that nativism was dead also. But the political equivalent of the law of the conservation of matter should remind us that nativism did not merely evaporate. Though deprived of a formal organization of its own, it remained as a powerful force, and in fact, it went into the Republican party, just as it had come out of the Whig party. No event in the history of the Republican party was more crucial or more fortunate than this sub rosa union. By it, the Republican party received a permanent endowment of nativist support which probably elected Lincoln in 1860 and which strengthened the party in every election for more than a century to come. But this support was gained without any formal concessions that would have forfeited the immigrant support also vital to political success. The Republicans were able to eat the cake of nativist support and to have too the cake of religious and ethnic tolerance.

The process of party realignment reached completion in the presidential election of 1856. In the campaign of that year, the first and the last candidate nominated was Millard Fillmore—first by the South Americans in February, after the second secession of the North Americans from the Know Nothing party, and last by the residue of the Whig party in September, at a gathering conspicuous for its high proportion of elderly men.[74] The next nomination came from the Democrats who met at Cincinnati in early June. The southern delegations there favored first Pierce and then Douglas, in both cases because of the vital part they had played in fostering the Kansas-Nebraska Act, but the northern delegations opposed them for the same reason. On the seventeenth ballot, however, both sides agreed on a candidate who was known to be sympathetic to southern views but who, happily, had been American minister to Britain and therefore out of the country at the time of Kansas-Nebraska.[75] This was James Buchanan of Pennsylvania, a sixty-four-year-old

74. *New York Times* and New York *Herald,* Feb. 22–27, 1856, for the Know-Nothing nomination; Fillmore, who was abroad, did not accept the nomination until May 21 —letter of acceptance in *New York Times,* June 16, 1856; *ibid.*, Sept. 18, 1856, for the Whig nomination of Fillmore by a convention at Baltimore of 150 delegates, representing 21 states; Cole, *Whig Party in the South,* pp. 322–326; Overdyke, *Know-Nothing Party in the South,* pp. 73–155; Llerena Friend, *Sam Houston, the Great Designer* (Austin, 1954), p. 294; Kirwan, *Crittenden,* pp. 302–304, on Whig efforts to persuade the Know-Nothings to defer their nomination; Rayback, *Fillmore,* pp. 403–405.

75. On the Democratic convention, Roy F. Nichols, *The Disruption of American Democracy* (New York, 1948), pp. 2–18; Philip Shriver Klein, *President James Buchanan* (University Park, Pa., 1962), pp. 245–260; Roy F. Nichols, *Franklin Pierce, Young*

veteran of the political wars, with a decade in the House, another decade in the Senate, diplomatic service at St. Petersburg and London, and four years as secretary of state under Polk. The "Old Public Functionary," as he called himself, was a seasoned politician—a man of some ability and of much experience, but an organization man, unlikely ever to do anything unorthodox. The third nomination, as has already been recounted, was that of John C. Frémont, by the Republicans primarily, and secondly by the North Americans. This choice was something of an anomaly, for Frémont, although a celebrated explorer, had no credentials either as a Republican or as a political leader, and the Republican managers, including especially Thurlow Weed, would not have nominated him if they had thought they had any real chance of winning the election. But they expected to lose, and Weed wanted to save his own candidate, William H. Seward, for 1860. Consequently, the Republicans nominated the "Pathfinder," a man of youthful vigor, handsome appearance, and popular appeal. His wife Jessie, Thomas Hart Benton's daughter, whom he had married after a romantic elopement, was, as Abraham Lincoln later ruefully said, "quite a female politician," and she took a much greater part in planning the campaign than her husband, who was politically incompetent.[76]

The campaign that followed, between three candidates who held five different nominations, was in several ways paradoxical. Frémont, nominated by the Republicans and the North Americans, was scarcely a Republican and certainly not a Know-Nothing. Fillmore, nominated by the South Americans and the Whigs, had so tenuous a relation with the Know-Nothings that there was dispute as to

---

Hickory of the Granite Hills (rev. ed.; Philadelphia, 1958), pp. 450–469; George Fort Milton, *The Eve of Conflict: Stephen A. Douglas and the Needless War* (Boston, 1934), pp. 211–229; *Official Proceedings of the Democratic National Convention Held in Cincinnati in 1856* (Cincinnati, 1856). On the campaign generally, see Roy F. Nichols and Philip S. Klein, "Election of 1856," in Schlesinger, *et al.* (eds.), *Presidential Elections*, II, 1007–1033.

76. Nevins, *Frémont, Pathmarker*, pp. 421–458; Nevins, *Ordeal*, II, 462–464; Ruhl Jacob Bartlett, *John C. Frémont and the Republican Party* (Columbus, Ohio, 1930); Glyndon G. Van Deusen, *Thurlow Weed, Wizard of the Lobby* (Boston, 1947), pp. 208–211; Van Deusen, *Seward*, pp. 174–178; Crandall, *Republican Party*, pp. 154–288; William Ernest Smith, *The Francis Preston Blair Family in Politics* (2 vols.; New York, 1933), I, 299–379; Beveridge, *Lincoln*, IV, 29–81; Isely, *Greeley and Republican Party*, pp. 151–195; George W. Julian, "The First Republican National Convention," *AHR*, IV (1899), 313–322; Francis P. Weisenburger, *The Life of John McLean: A Politician on the United States Supreme Court* (Columbus, Ohio, 1937), pp. 146–152; William B. Hesseltine and Rex G. Fisher (eds.), *Trimmers, Trucklers and Temporizers: Notes of Murat Halstead from the Political Conventions of 1856* (Madison, 1961).

whether he had ever joined the Order and agreement that he had never attended a meeting. But in the election, Fillmore suffered some of the disfavor that was beginning to attach to nativism, while Frémont escaped it completely. This difference arose partly from the fact that Frémont had really not flirted with nativism as much as Fillmore, but it resulted even more from the fact that Frémont's Know-Nothing nomination had followed his Republican nomination, so that he was identified in the public mind as the Republican candidate initially, while Fillmore had the misfortune to receive his Know-Nothing nomination first, so that his endorsement by the Whigs seven months later did not substantially alter his image as the nativist candidate.

Three-cornered contests, on the rare occasions when they occur in American presidential politics, tend to produce curious electoral patterns, and the contest of 1856 was no exception. The basic peculiarity of this election lay in the fact that although it appeared superficially to present a triangular rivalry, there were actually two separate contests in progress at the same time—one between Buchanan and Frémont in the free states, the other between Buchanan and Fillmore in the slave states. Fillmore did not carry any free state, and the only one in which he even ran second was California. In New Jersey he received 24 percent of the vote; in his own state of New York 21 percent; in Pennsylvania 13 percent; in six other free states, between 5 percent and 15 percent; and in the remaining six, less than 5 percent. Altogether, Frémont and Buchanan divided between them 86 percent of the free-state vote. In the slave states, on the other hand, Frémont was not on the ticket except in Delaware, Maryland, Virginia, and Kentucky, and only in the first of these four did he receive as much as 1 percent of the vote.

In the slave states, the race was entirely between Fillmore and Buchanan. Fillmore carried only the one state of Maryland, but showed substantial strength and actually won a higher proportion of the popular vote than Scott had won in Missouri and Maryland and in every state of the lower South from Georgia to Texas. Fillmore received more than 40 percent of the vote in ten of the slave states. In Missouri and Texas, he ran stronger than any Whig had ever run.[77]

In one sense, the nature of the contest placed Buchanan at a disadvantage, for it forced him to take a position that would win

77. See data in Petersen, *Statistical History of American Presidential Elections*, pp. 33–35.

favor in both the North and the South, while Frémont could court the northern votes exclusively, and Fillmore could concentrate on winning southern support. But in the long run, this situation helped Buchanan, for it identified him as the only truly national candidate in the race—the only one whose victory would not be a clear-cut sectional victory. It also marked him as the only candidate who could possibly beat Frémont, and this consideration, although purely tactical, proved ruinous to Fillmore. Buchanan's role as a national candidate grew more and more important. For the first time since the Compromise of 1850, there was widespread fear for the safety of the Union that undoubtedly influenced many voters.

Ostensibly, the chief difference between the candidates was on the question of slavery in the territories. On this point, the Democrats took a stand for "non-interference by Congress with slavery in state or territory or in the District of Columbia." This clearly ruled out congressional exclusion and at the same time preserved a convenient ambiguity as to whether the territorial governments themselves could (as Douglas insisted) or could not (as the South insisted) exclude slavery within their jurisdictions. The American platform condemned the repeal of the Missouri Compromise, but did not promise to restore it, and like the Democratic statement offered the voters an ambiguous form of popular sovereignty. But the Republicans condemned slavery and polygamy as "twin relics of barbarism" and affirmed the right and duty of Congress to exclude both from all the territories.

The Republican position was that of the Wilmot Proviso and thus repudiated the terms of the Compromise of 1850. After the Compromise, several southern states had solemnly declared that they would secede from the Union if the settlement were violated, and as the possibility of Frémont's election loomed up, the question of disunion arose again in the South. Throughout the summer and early autumn of 1856, a succession of southern leaders announced that if Frémont were elected, they would advocate disunion. *De Bow's Review* took this position in June, almost before Frémont's nomination. In July, Robert Toombs wrote, "The election of Frémont would be the end of the Union, and ought to be." James M. Mason stated that the South's answer to a Republican victory must be "immediate, absolute, eternal separation." John Slidell, Jefferson Davis, Andrew P. Butler, and many others spoke in a similar vein. In September, Henry A. Wise, governor of Virginia, invited other southern governors to meet at Raleigh, North Carolina, to take counsel on the course which the South should follow. He received a mixed response, with negative replies from Maryland, Georgia, and Louisiana, and acceptances from North Carolina, South Carolina, Florida, and Alabama. Ultimately, only two other governors met with Wise, but he had probably accomplished his purpose by dramatizing the will of many southerners to secede if Frémont were elected.[78] Some spokesmen of the South, like Governor Herschel V. Johnson of Georgia, might deny that a majority of southerners entertained disunionist ideas, and some northerners might scoff, as did Henry Wilson, that the South could not be kicked out of the Union,[79] but a great many northern citizens became convinced that the election of Frémont meant disunion, and that the candidate to vote for was the one who could beat Frémont. Since Buchanan appeared to have the better chance, this tactical factor, as much as anything else, brought about his victory and Fillmore's ruin. In proportion as old-line Whigs feared Frémont's victory, they began to be drained away from Fillmore to Buchanan. Men who for all their lives had been enemies of the Democratic party—Rufus Choate, James A. Pearce of Maryland, and the sons of Henry Clay and Daniel Webster—now announced their support of Buchanan, and as each one did so, it strengthened the gravitational pull of the

78. For evidence of southern intent to secede in the event of Frémont's election, see Avery O. Craven, *The Growth of Southern Nationalism, 1848–1861* (Baton Rouge, 1953), pp. 243–244; Nevins, *Ordeal,* II, 497–500; Milton, *Eve of Conflict,* p. 240; Barton H. Wise, *The Life of Henry A. Wise of Virginia, 1806–1876* (New York, 1899), pp. 209–210; Wilson, *Rise and Fall of Slave Power,* II, 521 (on Henry A. Wise); Percy Scott Flippin, *Herschel V. Johnson of Georgia, State Rights Unionist* (Richmond, 1931), pp. 75–79; Laura A. White, *Robert Barnwell Rhett, Father of Secession* (New York, 1931), pp. 137–138; Parks, *John Bell,* pp. 310–311; Henry T. Shanks, *The Secession Movement in Virginia, 1847–1861* (Richmond, 1934), pp. 52–54; Joseph Carlyle Sitterson, *The Secession Movement in North Carolina* (Chapel Hill, 1939), pp. 132–135; Harold S. Schultz, *Nationalism and Sectionalism in South Carolina, 1852–1860* (Durham, N.C., 1950), p. 123; Percy Lee Rainwater, *Mississippi: Storm Center of Secession, 1856–1861* (Baton Rouge, 1938), pp. 37–38; Clement Eaton, "Henry A. Wise and the Virginia Fire Eaters of 1856," *MVHR,* XXI (1934–35), 495–512.

79. Wilson, in Senate, April 14, 1856, *Congressional Globe,* 34 Cong., 1 sess., appendix, p. 394; Edwin Troxell Freedley, *The Issue and Its Consequences* (n.p., n.d.—a campaign pamphlet, 1856), represented the southerners as saying, "Oh, blood and thunder, don't you know we are all cooperative disunionists? Don't you know we are all colonels? If you do [elect Frémont], on the fourth of March next a whole battalion of colonels will come up to Washington and seize the Federal archives, and seize the Federal treasury—particularly the treasury."

Buchanan candidacy and weakened that of Fillmore.[80]

Buchanan himself stated the central issue plainly and emphatically. In a private letter, he wrote, "I consider that all incidental questions are comparatively of little importance . . . when compared with the grand and appalling issue of Union or Disunion. . . . In this region, the battle is fought mainly on this issue."[81] Democrats knew that in the North they had no chance of carrying a majority, but they hoped to win enough electoral votes to make up a majority when combined with the electoral vote of the South. If they could carry Buchanan's own state of Pennsylvania, along with New Jersey, California, and virtually all of the southern states, they would win the election. If Fillmore could carry a few southern states, he might throw the election into the House of Representatives, where there would be a good chance of his being chosen.

Much depended upon Pennsylvania, and money and energy were spent upon that state accordingly. The contest there was fierce and desperate, and very much in doubt until the Democrats won the state election in October. Thereafter, Buchanan's election seemed assured.[82]

In November, Buchanan was elected to be the fifteenth president of the United States. He won the contest against Fillmore in the slave states, losing only Maryland. But he lost the contest in the free states against Frémont, carrying only the five states of Pennsylvania, New Jersey, Illinois, Indiana, and California, while the Pathfinder carried eleven others.[83] The completeness of sectionalization is sug-

80. On the importance of the stop-Frémont movement, and the strategic value to the Democrats of the public belief that Buchanan could stop him and Fillmore could not, see Kirwan, *Crittenden*, p. 306; Parks, *John Bell*, pp. 310–311; Montgomery, *Cracker Parties*, pp. 169–171; Rayback, *Fillmore*, pp. 409–413; Nevins, *Ordeal*, II, 491–492; Overdyke, *Know-Nothing Party in South*, pp. 146–151; Cole, *Whig Party in the South*, pp. 324–325; David M. Potter, "The Know-Nothing Party in the Presidential Election of 1856" (M.A. thesis, Yale University, 1933), pp. 98–103.

81. Buchanan to Nahum Capen, Aug. 27, 1856, in George Ticknor Curtis, *Life of James Buchanan* (2 vols.; New York, 1883), II, 180–181.

82. Nichols, *Disruption*, pp. 41–50; Klein, *Buchanan*, pp. 257–260; Nevins, *Ordeal*, II, 505–507.

83. Although Buchanan carried 178 electoral votes to Frémont's 114 and Fillmore's 8, he received only 45 percent of the popular vote to Frémont's 33 percent and Fillmore's 21 percent. In the free states, Frémont received 45.2 percent; Buchanan, 41.4 percent, and Fillmore, 13.4 percent; in the slave states, Buchanan, 56.1 percent; Fillmore, 43.9 percent; Frémont, .0005 percent. Rayback, *Fillmore*, pp. 413–414, shows that although Fillmore appeared to be overwhelmingly beaten, a percentage change of less than 3 percent of the popular vote (or of 8,016 votes) in Kentucky,

gested by the fact that, with the exception of Ohio, all of the eleven Frémont states were farther north than any of the twenty Buchanan states. Buchanan won overall because his preponderance in the South was heavier than Frémont's preponderance in the North. But the Republicans correctly diagnosed their loss as a "victorious defeat," for they knew that, in 1860, if they could add Pennsylvania and either Indiana or Illinois to the bloc of states already captured, they would win the election. The Democrats condemned the Republicans for their "sectionalism," but their own bisectional equilibrium was badly upset, for Buchanan was the first president since 1828 to win an election without carrying a majority of the free states as well as of the slave states. The fact that slave-state Democrats heavily outnumbered free-state Democrats in Congress underscored the imbalance.

If Calhoun had been alive to witness the result, he might have observed a further snapping of the cords of Union. The Whig cord had snapped between 1852 and 1856, and the Democratic cord was drawn very taut by the sectional distortion of the party's geographical equilibrium. It could not stand much more tension without snapping also. As for the Republican party, it claimed to be the only one that asserted nationalist principles, but it was totally sectional in its constituency, with no pretense to bisectionalism, and it could not be regarded as a cord of Union at all.

Tennessee, and Louisiana would have given Fillmore enough electoral votes to throw the contest into the House of Representatives.

Electoral and Popular Votes Cast for President: 1848

	Popular	Electoral
Democratic	1,222*	127
Whig	1,361	163
Totals	2,879†	290

Electoral and Popular Votes Cast for President: 1852

	Popular	Electoral
Democratic	1,601*	254
Whig	1,385	42
Totals	3,162†	296

Electoral and Popular Votes Cast for President: 1856

	Popular	Electoral
Democratic	1,833*	174
Republican	1,340	115
American	872	8
Totals	4,045	297

U.S. Bureau of the Census, *Historical Statistics of the U.S.: Colonial Times to 1957* (Washington, D.C., 1960).

*In thousands.

†The total vote exceeds the sum of Democratic and Whig votes because of votes cast for minor candidates.

CHAPTER 11

Dred Scott and the Law of the Land

TROUBLES have come to a number of American presidents soon after they took office, but James Buchanan fared worse in this respect than any, except Abraham Lincoln and Herbert Hoover. Two days after he was sworn in, the Supreme Court rendered its decision in the case of Dred Scott versus John F. A. Sanford.

This was a litigation which had been brewing for a long time. In 1834, an army surgeon named John Emerson reported for duty at Rock Island, Illinois, and with him went Dred Scott, a slave whom he had recently bought from the Peter Blow family in St. Louis. Emerson kept Scott with him in Illinois for two years, despite that state's laws forbidding slavery. In 1836, Emerson was sent to Fort Snelling, in the northern part of the Louisiana Purchase, in what was then Wisconsin Territory (now Minnesota). Again, he took his slave along, although the Missouri Compromise forbade slavery in the part of the Louisiana Purchase north of 36°30′. At Fort Snelling, Emerson bought a slave woman named Harriet, and Harriet and Dred were formally married. After several years, Emerson was again transferred, and the slaves returned to Missouri.[1]

Emerson died in 1843. He left his estate, including his slaves, for the lifetime use of his wife and ultimately as a bequest to his daugh-

1. On the early history of Dred Scott, see Vincent C. Hopkins, *Dred Scott's Case* (New York, 1951), pp. 1–8; Walter Ehrlich, "Was the Dred Scott Case Valid?" *JAH*, LV (1968), 256–265. But the fullest treatment is Ehrlich's Ph.D. dissertation (Washington University, 1950), titled: "History of the Dred Scott Case Through the Decision of 1857."

ter. At some time during the following years, Dred Scott apparently tried unsuccessfully to buy his freedom. Then, in April 1846, the year of the Wilmot Proviso, Dred Scott or the members of the Blow family who were supporting him[2] took a drastic step. He, or they in his name, brought suit against Mrs. Emerson in the Circuit Court of St. Louis County for his freedom, on the ground that his former residence in Illinois and in Wisconsin Territory had made him free. He lost this suit, but a retrial was granted, and in 1850—the year of the Compromise—the jury brought in a verdict in his favor. Apparently, Dred Scott was a free man.[3]

Mrs. Emerson, however, did not accept this decision. She appealed to the Missouri Supreme Court, which reversed the lower court in 1852 by a two-to-one decision. Admitting that in previous cases, the Missouri courts had freed Missouri slaves who had come under the provisions of emancipating laws in other states, the court observed that this had been done under the principle of comity between states, and this comity was optional, not mandatory: every state retained "the right of determining how far, in a spirit of comity, it will respect the laws of other states." Speaking frankly of the growing rancor which the slavery controversy had generated, the court now exercised its option to withhold the extension of comity which it had previously allowed. Scott was currently held as a slave under the law of Missouri, and the Missouri courts would not invoke the laws of another jurisdiction to set him free.[4]

If any further recourse remained, it was only in the Supreme Court of the United States, but a recent decision of that tribunal (*Strader* v. *Graham,* 1851) indicated that it would probably not accept jurisdiction of such a case on appeal from a state supreme court. Mrs. Emerson, meanwhile, had remarried and now lived in Massa-

2. Scott was befriended to the end of his life by the Blow family. See John A. Bryan, "The Blow Family and Their Slave Dred Scott," Missouri Historical Society *Bulletin,* IV (1948), 223–231, V (1949), 19–25.

3. Hopkins, *Dred Scott's Case,* pp. 10–18, 181–183; Ehrlich, "History of Dred Scott Case," pp. 51–97; John D. Lawson (ed.), *American State Trials* (17 vols.; St. Louis, 1921), XIII, 223–238.

4. Scott *v.* Emerson, 15 Missouri 413; on the political background in Missouri see Richard R. Stenberg, "Some Political Aspects of the Dred Scott Case," *MVHR,* XIX (1933), 571–577; Benjamin C. Merkel, "The Slavery Issue and the Political Decline of Thomas Hart Benton, 1846–1856," *MHR,* XXXVIII (1944), 388–407; Hopkins, *Dred Scott's Case,* pp. 18–22. On the precedents which influenced the Missouri decisions, see Helen T. Catterall, "Some Antecedents of the Dred Scott Case," *AHR,* XXX (1924), 56–71.

chusetts with her husband, Calvin C. Chaffee, who later would be elected to Congress as an antislavery nativist. She left Dred Scott and his family in St. Louis under the control of her brother, John Sanford, who had also moved east and become a citizen of New York while retaining business interests in Missouri. This situation made it possible for Scott's lawyer to initiate in 1853 a new suit against a new defendant in the federal circuit court at St. Louis, under the diverse citizenship clause of the Constitution.[5]

The circuit court ruled, over the protest of the defendant, that Scott might bring his case under the diversity of citizenship formula, but in 1854—the year of Kansas-Nebraska—it ruled against Scott on the merits, holding him to be still a slave.[6] His lawyers appealed the decision, on a writ of error, to the United States Supreme Court.

In February 1856 the Supreme Court heard the case argued by eminent counsel. A future member of Lincoln's cabinet, Montgomery Blair, spoke for the plaintiff; and for the defendant, a former member of Taylor's cabinet, Reverdy Johnson, together with an incumbent senator from Missouri, Henry S. Geyer. In the course of the pleading it began to be evident that the justices faced some difficult problems, not only in working out answers to the questions arising, but even more in deciding which questions had to be answered and which did not. There were two substantive questions involved in the case, but for technical reasons it seemed uncertain whether either had to be answered. First was the question of whether Dred Scott was a citizen of the state of Missouri in the sense that would make him eligible to bring a suit against a citizen of

5. The legal relationship of Sanford to Dred Scott has been one of the minor mysteries of the case. The agreed statement of facts used in the federal suit identified Sanford as Scott's owner. Yet it was the Chaffees who later acted the part of owners when Scott was manumitted. For many years, historians generally adopted the view that Mrs. Chaffee (formerly Mrs. Emerson) transferred Scott to her brother, Sanford, by a "fictitious sale" and then later recovered ownership. Hopkins, *Dred Scott's Case*, pp. 23–24, 29–30, 176, rejected this explanation, maintaining that Sanford merely acted for his sister, probably because he had been named an executor of Emerson's will. Ehrlich, "Was the Dred Scott Case Valid?" stressed a fact mentioned by Hopkins —that Sanford had never qualified as Emerson's executor—and suggested that the case was consequently "not properly before the Supreme Court" (p. 256). What must be added, however, is that a freedom suit could be brought against anyone *holding* a person as a slave, whether he claimed to be the legal owner or not. Since Sanford did admit to holding the Scotts as slaves, the legal relationship between them did not affect the validity of the suit.

6. Hopkins, *Dred Scott's Case*, pp. 23–25; Lawson, *American State Trials*, XIII, 242–255.

another state. If he were a slave, he was certainly not a citizen, and even if he were free, his citizenship was legally questionable. But the circuit court had ruled that Scott could sue, and Sanford, satisfied with the ultimate decision in his favor, had not appealed from this ruling; it was Scott who had appealed. Therefore it could be argued that this part of the decision did not come up on appeal, and all that did come up was the validity of Scott's claim.[7]

Second was the question of whether residence in Illinois or in Wisconsin Territory had made Scott free and whether the antislavery law of the latter was constitutional. But here, too, the Supreme Court would not necessarily rule directly on the question. It might refuse jurisdiction on the ground that Scott was not a citizen, or it might rule, as it had done in the case of *Strader* v. *Graham* (1851), involving Kentucky slaves who had been carried into Ohio, that the decision of the state courts was final in determining the slave or free status of a Negro who lived within the state at the time of the decision. In short, the Court had first to decide what questions it was going to decide.

Specifically this meant that nine justices, who regarded themselves as largely outside the orbit of politics, would have to make up their minds whether or not to confront a political question which the politicians had been dodging for years. Through all the interminable debates over the Wilmot Proviso, the Compromise of 1850, the Kansas-Nebraska Act, had run the question of whether Congress possessed power (and could delegate it to a territorial legislature) to regulate slavery in the territories. Congressional "agreements" had consistently been undercut by a hidden disagreement on this point. Thus, in 1850, southern and northern congressmen "agreed" that Utah and New Mexico territories might exercise as much control over slavery as Congress could constitutionally delegate, but they disagreed as to the extent of such control. In 1854 they agreed that the congressional restriction on slavery in Kansas and Nebraska should be removed, but they disagreed as to whether

7. Technically, Scott had claimed citizenship in the lower court; Sanford had contested this claim in a plea in abatement; Scott had demurred to the plea; the court had sustained the demurrer but had then awarded judgment to Sanford on the merits of the case. Scott had appealed, but of course had not appealed the ruling on his demurrer, since it was in his favor; Sanford, by pleading to the merits in the lower court, had, according to one view, waived his right to raise the jurisdictional question again.

this meant that a restriction could be enacted by the territorial legislature in place of it.[8] All the seeming agreement on popular sovereignty, which had held the Democratic party together, was just an acceptance "on principle" of the proposition that the territories should have as much power as Congress could give them, without any agreement on how much power that was.

Congress, while dodging this question, had repeatedly tried to pass the responsibility on to the courts. Thus, as early as 1848, Senator John M. Clayton's compromise, adopted by the Senate, had provided that the status of slavery in the California and New Mexico territories should not be determined by Congress, but that the introduction or prohibition of slavery should rest "on the Constitution, as the same shall be expounded by the [territorial] judges, with a right to appeal to the Supreme Court of the United States."

Clayton's measure had failed to pass the House, but two years later, when Congress adopted the Compromise of 1850, both the Texas and New Mexico Act and the Utah Act had borrowed the literal language of the Clayton bill, and had also reproduced its provision to make appeals from the territorial courts easy. It was this provision for a test by the courts which led Senator Thomas Corwin of Ohio to say that Congress had enacted not a law but a lawsuit. Nevertheless, four years later, the Kansas-Nebraska Act had once again incorporated the exact words of the Clayton bill in providing that "all cases involving title to slaves and 'questions of personal freedom' are referred to the adjudication of local tribunals, with the right of appeal to the Supreme Court of the United States."[9]

From the statutes of 1850 and 1854, it was evident that Congress would welcome a chance to rid itself of the vexing territorial issue. In fact, Congress had done all it could to foster a judicial resolution of the problem. But despite the protracted quarrel over slavery in the territory acquired from Mexico, and despite the virtual war in Kansas, no cases had come to the courts from those areas, perhaps because a case could not easily arise until either Congress or a territorial legislature sought to exclude slavery from a territory, and this had not happened in New Mexico, Utah, Kansas, or Nebraska.

8. This substructure of underlying congressional disagreement is admirably shown in Wallace Mendelson, "Dred Scott's Case—Reconsidered," *Minnesota Law Review*, XXXVIII (1953), 16–28.

9. *Ibid.; Congressional Globe*, 30 Cong., 1 sess., pp. 950, 1002; 9 *U.S. Statutes at Large*, 446, 453; 10 *ibid.*, 277.

Consequently, the case that ultimately brought this burning question before the Supreme Court did not arise in one of the areas of active controversy where the congressional acts had specifically invited a judicial settlement. Instead it arose from the residence almost twenty years previously of a slave in an area—the Louisiana region north of 36°30′—which had been declared free by a statute—the Act of 1820—that had been repealed almost three years before the case was decided.

The question in 1856 was whether nine justices, or rather a majority of nine justices, would decide to rush in where Congress had feared to tread. But for several months this question was obscured as the justices wrestled with the jurisdictional question of citizenship and the related technical problem as to whether the plea in abatement involving this jurisdictional question came up to the Supreme Court, since neither plaintiff nor defendant had appealed the circuit court's ruling on this plea. When the Court held consultations in April 1856, it developed that the justices were divided, four to four, on this latter point, with Justice Samuel Nelson of New York in doubt. Nelson moved for a reargument, which was ordered for the next term of court. The justices may have issued this order with some relief, for it would put the decision over until after the pending election and would give them a little more time before taking steps which they all recognized would be serious.[10]

After the reargument in December 1856, the justices did not consult on the case until February 14, three weeks before their final decision. At this late stage, the majority decided to follow *Strader* v. *Graham*, ruling that the case was governed by the law of Missouri as applied by the courts of that state, and that it did not properly come before the federal courts. Justice Samuel Nelson received the assignment to write an opinion of the Court along these lines.[11]

10. Nelson's role in bringing about the reargument was first established by Charles Warren, *The Supreme Court in United States History* (rev. ed., 2 vols.; Boston, 1926), II, 285, on the basis of a statement by John A. Campbell, Oct. 13, 1874, in 20 Wall x, xi. The fact that the decision might coincide with the presidential campaign was apparently gratifying to Justice McLean, who hoped for the Republican presidential nomination and would have welcomed a chance to dissent from a decision against Scott. Perhaps other justices did not want either to accommodate McLean or to decide the case in the midst of an election campaign. Edward S. Corwin, "The Dred Scott Decision in the Light of Contemporary Legal Doctrines," *AHR*, XVII (1911), 53. Francis P. Weisenburger, *The Life of John McLean* (Columbus, Ohio, 1937), pp. 197–198.

11. John A. Campbell, Nov. 24, 1870, and Samuel Nelson, May 13, 1871, to Samuel

This kind of "narrow" decision had been indicated from the beginning and was what many people had been expecting all along. As early as April 1856, Justice Benjamin R. Curtis had written to his uncle, George Ticknor, that "the Court will not decide the question of the Missouri Compromise line—a majority of the judges being of the opinion that it is not necessary to do so." At an earlier stage, the New York *Tribune* had denounced the Court for its "convenient evasion," had accused the majority of preventing the minority from upholding the constitutionality of the Missouri Compromise, and had complained, "The black gowns have come to be artful dodgers."[12]

But between February 14 and February 19, the Court suddenly changed its position and decided to rule on the Missouri Compromise. Probably there were several factors which entered into this fateful decision. For one thing, two northerners on the Court, John McLean of Ohio and Benjamin R. Curtis of Massachusetts, had both made it clear that they would write dissenting opinions, declaring Scott free under the terms of the Missouri Compromise, which they would pronounce constitutional. If the majority said nothing about this question, they would appear to let the argument go by default. As Justice John Catron expressed it, "a majority of my brethren will be forced up to this point by the two dissentients," and as Justice Robert C. Grier said, "Those who hold a different opinion from Messrs. McLean and Curtis on the powers of Congress and the validity of the Compromise Act feel compelled to express their opinions on the subject."[13] But apart from the situation created by Curtis and McLean, some of the justices felt that sectional conflict had fed for a decade upon the uncertainty about the constitutional question, and that it was their judicial responsibility to settle the question. Also, there was no doubt some desire among all five of the southern members of the Court—Chief Justice Roger B. Taney of Maryland, James M. Wayne of Georgia, John Catron of Tennessee,

---

Tyler, in Tyler's *Memoir of Roger Brooke Taney* (Baltimore, 1872), pp. 382–385; Robert C. Grier to James Buchanan, Feb. 23, 1857, in John Bassett Moore (ed.), *The Works of James Buchanan* (12 vols.; Philadelphia, 1908–11), X, 106–108.

12. Curtis to Ticknor, April 8, 1856, in Benjamin R. Curtis, Jr., *Memoir of Benjamin Robbins Curtis* (2 vols.; Boston, 1879), I, 180; New York *Tribune*, May 15, 1856.

13. Catron to Buchanan, Feb. 19, 1857; Grier to Buchanan, Feb. 23, 1857, in Moore (ed.), *Works of Buchanan*, X, 106–108. The centrality of the role of McLean and Curtis in forcing a broad decision was asserted with special vigor by Frank H. Hodder, "Some Phases of the Dred Scott Case," *MVHR*, XVI (1929), 3–22.

Peter V. Daniel of Virginia, and John Archibald Campbell of Alabama—to implement their belief that the act of 1820 was unconstitutional.

It was Justice Wayne, at a consultation not attended by Nelson, who moved that the chief justice be requested to prepare a broad decision. This motion apparently struck a responsive chord with others, like Justice Daniel, who had wanted such a decision all along.[14]

Wayne's motion carried, but he and his southern brethren doubtless sensed the reluctance of some of their colleagues. It would be embarrassing in the extreme if the five southern members alone declared against the Compromise, yet if Nelson and Grier of Pennsylvania held to a narrow decision that would be the result. Because of this anxiety, Justice Catron took a step which was certainly of doubtful propriety, though he had been tempted into it by the president-elect. Buchanan had written to him early in February asking simply when the case would be decided, and this now led Catron to reply, telling Buchanan that a broad decision was pending and expressing his anxiety lest Justice Grier might "take the smooth handle for the sake of repose." Catron then urged Buchanan to "drop Grier a line, saying how necessary it is, and how good the opportunity is to settle the agitation by an affirmative decision of the Supreme Court, the one way or the other." Buchanan sent Grier a prompt letter which is no longer extant, but the tenor of which can be inferred from Grier's reply that he and Taney and Wayne were committed to a broad decision and that they would try to enlist Daniel, Campbell, and Catron to support the same position.[15] By the time the Court issued its opinion, all of the justices except Nelson had decided to grapple with the constitutional question. With what varying degrees of foreboding or inner doubt each one of them had done this, it is impossible to say.

14. Letter of John A. Campbell, cited in note 11, and also letter of to George T. Curtis, Oct. 30, 1879, quoted in Allan Nevins, *The Emergence of Lincoln* (2 vols.; New York, 1950), II, 473, concerning the Court's intention on Feb. 14 and Wayne's subsequent motion. Nevins (pp. 473–477) argues that Taney, Daniel, and especially Wayne wanted a broad decision quite as much as McLean and more than Curtis. See also Alexander A. Lawrence, *James Moore Wayne, Southern Unionist* (Chapel Hill, 1943), pp. 153–156; Curtis, *Curtis*, I, 234–235.

15. On the Catron-Grier-Buchanan correspondence, see citations in note 13; on Buchanan's intervention, Philip G. Auchampaugh, "James Buchanan, the Court, and the Dred Scott Case," *Tennessee Historical Magazine*, IX (1926), 231–240.

Thus, all the signals were changed less than three weeks before the final decision. The opinion that Justice Nelson had written as the opinion of the Court he delivered as solely his own, and the seventy-nine-year-old chief justice delivered on March 6 a long opinion which he had prepared in less than three weeks.[16]

Taney began by taking up at length the question of Dred Scott's citizenship, and with an intricate argument he held that Scott was not a citizen and further that emancipated slaves or their descendants could not become citizens. Distinguishing between citizenship of the United States and citizenship of a particular state, he argued that a person could attain federal citizenship only by being born a citizen, which slaves were not, or by being naturalized, which slaves had not been and could not be. He argued further that Negroes had not been accorded citizenship by any state.

In the light of much subsequent analysis, it appears that Taney's argument on this point was certainly in error. Justice Curtis showed in his dissent that, historically, Negroes had been recognized as citizens and had exercised the functions of citizenship in several states. If state citizenship for Negroes existed, it would apparently qualify them to sue in a federal court under the diversity of citizenship clause, regardless of whether they held federal citizenship or not, and making irrelevant all of Taney's argument about the impossibility of federal citizenship. But instead of emphasizing such flaws, many critics of the Chief Justice bitterly assailed him for a provocative passage taken out of context. In reviewing the public attitude toward Negroes at the time of the framing of the Constitution, Taney declared: "They had for more than a century been regarded as beings of an inferior order . . . so far inferior that they had no rights which the white man was bound to respect." Antislavery denunciation of these words, though fully justified, gave the impression of misrepresenting the chief justice, and it also tended to divert attention from more fundamental weaknesses in his argument.[17]

Taney's opinion had begun by denying that Scott was eligible to sue even if free, but as he continued, he turned to the point that if Scott were not free he certainly could not sue. This second aspect

16. 19 Howard 393–454. In the record of the case John A. Sanford's name is misspelled as Sandford.

17. Cases of this alleged misrepresentation are cited in Warren, *Supreme Court,* II, 303–304, 389–390; Charles W. Smith, Jr., *Roger B. Taney, Jacksonian Jurist* (Chapel Hill, 1936), p. 174; Tyler, *Memoir of Taney,* p. 373.

of the question of jurisdiction therefore led Taney to consider whether residence in Illinois and Wisconsin had made Scott free. His own words made it clear that, in taking up this question, he was not turning to the merits of the case but rather continuing with the problem of jurisdiction. "Now," he said, "if the removal [to Illinois and Wisconsin] . . . did not give them [the Scott family] their freedom, then by his own admission he is still a slave; and whatever opinions may be entertained in favor of the citizenship of a free person of the African race, no one supposes that a slave is a citizen of the State or of the United States. If therefore, the acts done by his owner did not make them free persons, he is still a slave, and certainly incapable of suing in the character of a citizen."[18]

In this context, Taney took up, as the second major part of his decision, the question of whether Congress had been constitutionally able to exclude slavery from the territory north of 36°30′. In a lengthy discourse, he argued that the citizens of all states alike enjoyed a right to take their property into the territories, and that an act of Congress which excluded one type of property and not another was an impairment of this property right and was a violation of the guarantee in the Fifth Amendment that no person should "be deprived of life, liberty, or property without due process of law." Up to that time, "due process" had been generally regarded as a matter of procedure—involving jury trial, rights to cross-examine witnesses, and the like—but in the Dred Scott decision, Taney gave the clause the meaning it came to have in the twentieth century: a statute that encroaches upon the constitutionally protected rights of the individual, such as freedom of speech, of the press, or of religion is in itself a violation of due process—a substantive rather than a procedural violation. To Taney, the act of 1820 was a violation of due process because "an act of Congress which deprives a citizen of the United States of his liberty or property [including slave property] merely because he came himself or brought his property into a particular territory of the United States, and who had committed no offense against the laws, could hardly be dignified with the name of due process of law."[19]

Constitutionally, the argument was important as one of the first applications of a concept with a significant future history ahead of

18. 19 Howard 427. Also, pp. 394–395.

19. *Ibid.*, pp. 450–451; Edward S. Corwin, "The Doctrine of Due Process of Law Before the Civil War," *Harvard Law Review*, XXIV (1911), 366–385, 460–479.

it. In the context of the immediate case, however, Taney's contentions were again faulty, for he resorted to an extremely narrow view of federal powers and was forced to give a tortured construction to the constitutional clause providing that "Congress shall have power to dispose of and make all needful rules and regulations respecting the territory or other property belonging to the United States." By this construction he reached the conclusion that the Missouri Compromise had been unconstitutional and therefore had not freed Scott, who remained a slave and therefore was doubly lacking the citizenship which would have enabled him to bring his suit.

Each one of the other justices submitted a concurring or dissenting opinion, and beyond saying that Scott remained a slave, it is not easy to state with certainty on what points a majority of the Court agreed. Six of the justices took the view that Scott was not a citizen, without fully agreeing that a free Negro could not be a citizen.[20] In the same way, six also agreed that the Act of 1820 was unconstitutional, without agreeing as to why. Calhoun, of course, had proclaimed as early as 1848 that it was unconstitutional, on the ground that the territories were owned jointly by the states and that Congress was merely the agent for the joint owners, without power to discriminate between those recognizing slavery as a form of property and those not so recognizing it. Since this argument had become orthodox southern dogma, it was natural for critics to assume that the Court had accepted it, and to denounce the decision as pure Calhounism. In fact, however, no justice relied exclusively upon the Calhounist doctrine. Justice Campbell made some use of it.[21] Justice Catron argued that slavery in the Louisiana region was protected by the Louisiana Purchase Treaty of 1803. Taney, speaking for himself, for Justices Grier and Wayne, and to some extent for Justice Daniel (who did lean heavily toward Calhounism), based his argument primarily upon the due process clause.

Taney, Wayne, Catron, Grier, Daniel, and Campbell formed a straggling majority, but it must have been an embarrassment to

20. Taney, Wayne, and Daniel held that a Negro could not be a citizen; Campbell, Catron, and Grier held that Scott was a slave who had not been freed by his time in Illinois and Wisconsin Territory and therefore not a citizen; Nelson, that the question did not arise; Curtis and McLean, that Scott was a citizen.

21. It was ironical that Campbell should have followed the Calhoun line of argument in any degree for he had written a letter to Calhoun in 1848, arguing that Congress possessed full power to exclude slavery from the territories. Eugene I. McCormac, "Justice Campbell and the Dred Scott Decision," *MVHR*, XIX (1933), 565–571.

them that Grier was the only northerner among their number[22]—otherwise the majority was a southern majority. Nelson concurred in their decision in favor of the defendant, but not in their reasoning, and he did not take a position on the Missouri Compromise. The two other justices, McLean and Curtis, dissented—McLean with more emphasis than logic, but Curtis with a powerful and closely reasoned argument both that free Negroes were citizens and that the Act of 1820 was constitutionally valid and had made Scott a free Negro. Curtis also announced that he did not regard the opinion of the majority on the Act of 1820 as having judicial force: "I do not hold any opinion of this court or any court binding, when expressed on a question not legitimately before it. . . . The judgment of this court is that the case is to be dismissed for want of jurisdiction, because the plaintiff was not a citizen of Missouri. . . . Into that judgment, according to the settled course of this Court, nothing appearing after a plea to the merits can enter. A great question of Constitutional law, deeply affecting the peace and welfare of the country, is not, in my opinion, a fit subject to be thus reached."[23]

In coming to a decision not to "take the smooth handle for the sake of repose," the majority of the Court had grasped a nettle. Overestimating the power of the judiciary to settle a troubled political question, they had committed themselves on a point around which the legislators had built an elaborate structure of evasion. In one sense, they had met the standard of official responsibility far better than Congress, for they confronted a question which they could have dodged. For several years, both antislavery and proslavery advocates had been proclaiming that it was the duty of the courts to "settle" the uncertainty about the power to regulate slavery in the territories.[24] They had accepted this duty and,

22. Grier himself had been conscious of this point and had written to Buchanan on Feb. 23: "I am anxious that it should not appear that the line of latitude should mark the line of division in the court."

23. 19 Howard 589–590. Justice McLean also denied the judicial force of the Taney opinion: "Nothing that has been said by them [the majority] which has not a direct bearing on the jurisdiction of the Court, against which they decided, can be considered as authority. I shall certainly not regard it as such. . . . The question of jurisdiction, being before the Court, was decided by them authoritatively, but nothing beyond that question."

24. Mistaken though it proved to be, the opinion was widespread that the Court might quiet the dispute over slavery by rendering a clear-cut decision. Justice Wayne, in his concurring opinion, spoke of disputed constitutional questions "about which there had become such a difference of opinion, that the peace and harmony of the

as far as they could, had removed the uncertainty.

But both the decision that a free Negro was not a citizen and the decision that Congress could not exclude slavery from the territories were intensely repugnant to many people in the free states, and everything about the decision tended to accentuate the controversial aspect of it. Here was an opinion on a question widely regarded as sectional and political, decided in favor of the southern position by a majority consisting of five southerners and one northerner, with three other northerners actively dissenting or not concurring in the crucial parts of the decision. Here, for the first time since the framing of the Constitution, was the Supreme Court invalidating a major act of Congress. Here was a seventy-nine-year-old chief justice advancing arguments concerning citizenship and congressional power in the territories which did not stand up well under historical or logical analysis. But the quality of the reasoning counted for less than the fact that the Supreme Court was upholding slavery. It was not surprising that the decision met with a furious outburst of protest and that the Court came under concentrated attack.

What was, perhaps, surprising, however, was the form which this attack took. Despite the vulnerability of Taney's position on the questions of Negro citizenship and legislation for the territories, the debate did not rage primarily around these points. Instead, critics more often emphasized two other contentions which were far less well supported than the arguments against the Court's constitutional reasoning. First, it was said that the statements concerning the Act of 1820 were not a necessary part of the ruling of the Court on Scott's right to bring suit, and were therefore mere dicta without the force of judicial ruling.[25] Second, the decision was attacked as

---

country required the settlement of them by judicial decision" (19 Howard 454–456). Justice Curtis later declared that Wayne and Taney had "become convinced that it was practicable for the Court to quiet all agitation on the question of slavery in the territories by affirming that Congress had no constitutional power to prohibit its introduction" (Curtis, *Curtis*, I, 206, 234–236). Justice Catron advised Buchanan that he could safely say in his inaugural that "it is due to its [the Supreme Court's] high and independent character to suppose that it will decide and settle a controversy which has so long and seriously agitated the country, and which *must* ultimately be decided by the Supreme Court" (see note 13 above).

25. One weakness in this contention, generally ignored by historians, was pointed out by the Savannah, Georgia, *Republican*, March 24, 1857: "Of this point [obiter dictum] the Court is also the supreme judge, as much so as it is of the merits of the case, and from their decision there is no appeal under the constitution."

the product of a deep-laid conspiracy between the justices and other intriguers. In short, the attack which quickly developed was not an attack on the decision but an attack on the Court.

The groundwork for the attack had, in fact, been laid before the day of the decision. As early as April 1856, Republican newspapers had begun to learn, probably from Justice McLean, what alignments were developing among the justices, and knowing that the decision, whether broad or narrow, would not be in favor of Scott, they had begun to discount it in advance. The New York *Courier* declared, "The Court, in trying this case, is itself on trial." James E. Harvey, a friend of McLean, wrote in the *Tribune,* "The urgency of the slave power is great . . . our judicial decisions upon constitutional questions touching the subject of slavery are rapidly coming to be the enunciation of mere party dogmas."[26] Republicans called on the Court to face up to the territorial question, but alternated these demands with warnings that a decision against Scott would not be accepted. Two months before the decision, the *Tribune,* commenting on a rumor that the decision would deny the power of Congress to restrict slavery in the territories, said, "Judicial tyranny is hard enough to resist under any circumstances, for it comes in the guise of impartiality and with the prestige of fairness. If the Court is to take a political bias and to give a political decision, then let us, by all means, have it distinctly and now. The public mind is in a condition to receive it with the contempt it merits."[27]

Once the decision was rendered, a fierce onslaught was launched and sustained for days, with the New York *Tribune* taking the lead, and the *Independent,* the *Evening Post,* and other antislavery papers swelling the chorus. These assaults especially pilloried Taney. The

26. New York *Courier,* Dec. 18, 23, 1856, New York *Tribune,* Dec. 20, 1856, both as quoted in Warren, *Supreme Court,* II, 286, 289. On Justice McLean's political ambitions and his giving information on the status of the case while it was being argued, to James Harvey of the New York *Tribune,* see Weisenburger, *Life of John McLean,* pp. 196–197; Jeter Allen Isely, *Horace Greeley and the Republican Party, 1853–1861: A Study of the New York Tribune* (Princeton, 1947), p. 226; Hopkins, *Dred Scott's Case,* pp. 41–46; Auchampaugh, "Buchanan and Dred Scott Case," p. 234, quotes Catron to Buchanan, Feb. 6, 1857, "All our opinions were published in the N.Y. *Tribune* the next day after the opinions were expressed [in consultation]. This was, of course, a gross breach of confidence, as the information could only come from a judge who was present. This circumstance, I think, has made the chief more wary than usual."

27. New York *Tribune,* Jan. 5, 1857, and other quotations, in Warren, *Supreme Court,* II, 291–292.

chief justice, a member of an old Maryland Catholic family of planters, was a man of personal integrity and conscience. He had emancipated his own slaves many years previously. But the antislavery press, constantly speaking of him as a slaveholder, charged him with "gross historical falsehoods" and with a "jesuitical decision," while endlessly repeating that he had denied that the Negro had any rights which a white man was bound to respect. As for the decision itself, it was "a wicked and false judgment" . . . "an atrocious doctrine" . . . "a deliberate iniquity" . . . a "willful perversion" . . . "If the people obey this decision, they disobey God." One correspondent remarked, "If epithets and denunciations could sink a judicial body, the Supreme Court of the United States would never be heard of again."[28]

The epithets and denunciations served to rouse emotions, but as the argument against the decision developed, it took the form, above all, of an elaboration of the statement in Justice Curtis's dissent, that in dealing with the constitutionality of the Missouri Compromise, the Court had taken up a question which was not properly before it. The outcry against the dictum was accompanied by outraged protests against a "political decision." These accusations gained force from the assurance with which Curtis originally stated the point, and they were widely taken up. Thomas Hart Benton, almost on his deathbed, wrote a long and polemical attack on the decision, in which he said that Scott was "turned back from the door, for want of a right to enter the court room—debarred from suing for want of citizenship; after which it would seem to be a grave judicial solecism to proceed to try the man when he was not before the court." George Ticknor Curtis, brother of Justice Curtis

28. For copious evidence, from which the above quotations are taken, concerning the attack on the Court, see *ibid.*, II, 302–309. The five southern justices were constantly denounced as "slaveholders" (examples in Albert J. Beveridge, *Abraham Lincoln, 1809–1858* [4 vols.; Boston, 1928], IV, 127), and their personal relation to slavery is a matter of some interest: Wayne at one time owned a rice plantation with between thirty and forty slaves, and in 1856 he owned nine slaves; Daniel usually had four or five slaves, used as house servants; Catron held title to several slaves in Tennessee but permitted them to live as free Negroes, though this practice was illegal; Taney and Campbell had both emancipated their slaves. Lawrence, *James Moore Wayne,* p. 144; John P. Frank, *Justice Daniel Dissenting: A Biography of Peter V. Daniel, 1784–1860* (Cambridge, Mass., 1964), p. 58; J. Merton England, "The Free Negro in Ante-Bellum Tennessee," *JSH,* IX (1943), 46–47 (on Catron); Carl Brent Swisher, *Roger B. Taney* (New York, 1935), p. 94; Henry G. Connor, *John Archibald Campbell* (Boston, 1920), p. 71.

and counsel for Scott along with Montgomery Blair, later wrote with Olympian authority, "The case of Dred Scott does not furnish a judicial precedent or judicial decision."[29]

These dogmatic assertions gained credence all the more readily because there was a plausible aspect to the view that Taney had decided adversely on Scott's plea after first ruling that the plea need not be decided. If Scott was not a citizen, he could not bring a case, and if he could not bring a case, the Court apparently had no business deciding what effect the Act of 1820 had had upon him while he was living in Wisconsin Territory. Any observations which it made on this point, therefore, would be mere side comments or obiter dicta, with no force as law.

This view of the case readily gained acceptance, and for more than half a century the dismissal of the most important part of the decision as dicta was repeated by virtually every authority, either historical or legal. Yet the sequence of reasoning involved was extremely intricate, for the effect of the Act of 1820 upon Scott when he had lived in Wisconsin Territory had a bearing not only upon the substance of Scott's claim but also upon his right to sue. It involved not only the merit of the claim which he sought to bring as a litigant, but also his right to appear as a litigant. If the Act of 1820 was void, it not only worked against the freedom Scott sought to claim, but also against the citizenship by which he sought to claim it. Since the Court could not fully consider whether Scott was a citizen without determining whether the Act of 1820 had set him free, it was in the paradoxical position of having to decide whether the Missouri Compromise was constitutional before it could fully decide whether Dred Scott was legally qualified to bring a case involving its constitutionality. What the Court decided was that the act was unconstitutional, and that for this reason, among others, Scott was not free; and since not free, certainly not a citizen, which perhaps he would not have been even if he were free; and since not a citizen, not entitled to bring a case involving the question of constitutionality. In short, the Court had to answer the question of constitutionality

29. Benton, *Historical and Legal Examination of . . . the Dred Scott Case* (New York, 1857), pp. 7–8; Curtis, *Constitutional History of the United States* (2 vols.; New York, 1889–96), II, 270; Frederick S. Allis, Jr., "The Dred Scott Labyrinth," in H. Stuart Hughes (ed.), *Teachers of History: Essays in Honor of Lawrence Bradford Packard* (Ithaca, 1954), pp. 347–349.

for itself before it could be sure that it did not have to answer the question for Dred Scott.

These subtle relationships are confusing in any case, and it is not strange that condemnation of the decision as a dictum should have won wide acceptance, especially among those who wished to reject it anyway. But there is no more conclusive proof of the overwhelming success of the publicity campaign against the decision than the fact that for half a century no qualified authority ventured to affirm that the Court's statements concerning the Missouri Compromise had been a perfectly regular and legitimate judicial finding—whether correctly reasoned or not—and were not dicta at all. When Edward S. Corwin took this position in an important study in 1911, he found the basis for his reasoning in Taney's opinion of 1857. It had been there all the while.[30]

The real problem for the historians—widely overlooked—is not whether Taney's opinion was dictum, but why the question of dictum has been blown up to such vast proportions and has overshadowed the discussion of all other aspects of the case. Here was a situation in which the chief justice advanced some highly vulnerable arguments about the citizenship of Negroes and about the power of Congress over the territories. But for decades, historians passed over these flaws with a minimum of analysis, while devoting elaborate emphasis to the claim that Taney had no right to rule on a question on which they would clearly have been delighted to have him rule if only he had ruled differently. This intense preoccupation with the technical and legalistic question of dictum, at the expense of a concern for the substantial question of the force of the Act of 1820, is itself an anomaly which calls for explanation. This anomaly,

30. Corwin, "The Dred Scott Decision," pp. 56–59. "The matter of the validity of the Chief Justice's mode of proceeding then comes down to this question: Is it allowable for a court to base a decision [that Scott was not a citizen] upon more than one ground [that he was not a free Negro as well as that a free Negro was not a citizen] and if it does so, does the auxiliary part of the decision become *obiter dictum?*" To answer this question, Corwin quoted *The American and English Encyclopaedia of Law.* "Where the record presents two or more points, any one of which, if sustained, would determine the case, and the court decides them all, the decision upon any one of the points cannot be regarded as *obiter.*" Corwin concludes that Taney "had . . . an undeniable right to canvass the question of Scott's servitude in support of his decision that Scott was not a citizen of the United States and he had the same right to canvass the question of the constitutionality of the Missouri Compromise in support of his decision that Scott was a slave." Allis, "Dred Scott Labyrinth," is the best survey of what historians have written about what the Court "really" decided.

in turn, may perhaps be explained by examining the purpose of those who opposed the decision.

Antislavery leaders were in the position of wishing to defy the court, but not wishing overtly to defy the law. The New York *Independent,* a major Congregational periodical, declared in a headline, "The Decision of the Supreme Court is the Moral Assassination of a Race and Cannot be Obeyed." The New York *Tribune* said that no man who really desired the triumph of freedom over slavery in the territories would submit to the decision of a bench with "five slaveholders and two doughfaces" on it.[31] Thus the opposition was committed to defying the decision of the Court. Since, under the American system, the decision of the Supreme Court duly rendered is law, even if it is wrong, defiance of the Court's decision constitutes defiance of the law. Yet there are strong inhibitions against open defiance of the law, even when the law is believed to be wrong. Therefore, if people were to defy the decision in *Scott* v. *Sandford* without violating their inhibitions concerning the law, they had to find a way to regard the decision as lacking in ordinary judicial force. They could do this only by categorizing the decision as a dictum. Such a categorization was a psychological godsend to them; it got them out of an intolerable dilemma.

A curious side effect of the general obsession with the concept of dictum has been the assumption that since a broad decision was unwarranted, therefore anyone who contributed to it was blameworthy. Thus historians have engaged in a spirited dispute as to which justices were "responsible" (i.e., at fault) for creating the situation which led to a broad decision. Frank H. Hodder in 1933 indicted Justices McLean and Curtis as having forced the majority reluctantly to tackle an improper question by taking it up themselves in a way that necessitated a reply. Allan Nevins, in 1950, retorted that "the truth seems rather to be that the responsibility falls upon a number of the judges and that Wayne must be included among those whose share in bringing about a broad decision was greatest. . . . Responsibility must be widely distributed." Both writers suggested that the motive for supporting a broad decision must somehow be selfish or partisan.[32]

31. For these quotations and numerous others of the same tenor, see Warren, *Supreme Court,* II, 304–309.

32. Hodder, "Some Phases of the Dred Scott Case," pp. 10–16; Nevins, *Emergence,* II, 473–477.

This false focus on the question of dictum has led historians to criticize the Court for deciding a crucial question at all, when perhaps the real criticism ought to be for deciding it incorrectly. Such criticism of the breadth rather than the quality of the decision loses sight of the fact that for ten years the country had been convulsed by a controversy as to the constitutionality of legislation regulating slavery in the territories. Twice during this decade, Congress had adopted legislation designed to secure a judicial decision on this question, but no case lending itself to such a decision had come to the Supreme Court. Finally, in 1857, a case from an unexpected direction gave the Court occasion to decide whether the Act of 1820 had given a Negro plaintiff the freedom which was one prerequisite to the citizenship which would enable him to appeal as a litigant. The Court responded by deciding that the Act had not so operated. In choosing to confront this question, the Court had done what all parties said that it ought to do, and had shown a sense of courageous responsibility, in striking contrast with the protracted equivocation of Congress. If the purpose of the critics had really been to evaluate the judgment, therefore, they would have assailed the answer which the Court gave, rather than the Court's resolve that it must give an answer. But the purpose was more than this: it was to deny the binding character of the judgment, which could be done only by accusing the Court of usurping power by deciding a question it had no right to decide. This accusation, made for political purposes at the time, shifted the ground of discussion, and the shift was later accepted even by historians sympathetic with Taney, who tried to defend him by blaming McLean and Curtis for the broad decision, instead of upholding the validity of a broad decision as such.

Recognition is also needed that much of the criticism of the Dred Scott decision was formulated at a time when it was regarded as a judicial virtue for judges to show a due hesitance in overruling the acts of legislative bodies. In 1857 the justices were acutely conscious of this tradition, and their first impulse, accordingly, had been to decide Dred Scott's case on the narrow basis of *Strader* v. *Graham*. But even the advocates of judicial restraint do not deny the responsibility of the courts to decide questions of constitutionality when concrete circumstances bring these questions before them. If Corwin is correct, the question of constitutionality was inescapably before the Court in 1857, and as Carl B. Swisher has remarked, "most of the Supreme Court decisions which stand out as landmarks in the minds of students of American history and American constitutional law have been policy-making decisions."[33]

It is a long way from *Dred Scott* v. *Sandford* to *Brown* v. *Board of Education* (1954), and to suggest any similarity between these two landmark decisions may seem perverse, since they were polar opposites in their treatment of the American Negro and perhaps also in their degree of moral enlightenment. Yet in a purely judicial sense, there are certain interesting parallels between them, and also between the attacks upon them. Both were, in fact, broad decisions in which the Court deliberately sought to face a major public question and to exercise an influence on public affairs rather than to retreat into narrow legalism or reliance on limiting precedents (such as *Strader* v. *Graham*). Both issued from Courts which probably underestimated the opposition that would be aroused by their judgments, and which hoped that their action would ease the tension of an acute national problem. The two decisions met fiercer opposition than scarcely any other judgments of an American Court.

If these parallels seem paradoxical, the height of the paradox appears in the similarity between the attitudes of the slaveholders of 1857 and the civil rights advocates of 1954 in upholding the authority of the Supreme Court, and in the likeness of the abolitionists of 1857 and the White Citizens' Councils of 1954 in rejecting it. In 1857, proslavery men took a high moral tone in reminding abolitionists of the obligation of the citizen to accept the rulings of the Supreme Court as law, whether he liked them or not, while a century later, the proponents of Negro rights were pointing out this obligation to white southerners. It is, of course, not a coincidence that this righteous concern for the sanctity of the law was expressed in both cases by groups who were pleased with what the law said. Conversely, the pro-Negro groups of 1857 and the anti-Negro groups of 1954 were alike in their attitudes toward the judiciary, if in nothing else, for both found ways to deny the legality of the decisions which they rejected. White Citizens' Councils did this by asserting that the Brown decision was "unconstitutional" and the product of a Communist conspiracy. Horace Greeley and the Republican press did the same thing by asserting that the Scott decision was an obiter dictum and the product of a slaveholders'

33. Swisher, *Taney*, p. 505.

conspiracy. Their accusation of conspiracy was the second major theme in the attack on the Court, and it was a striking manifestation of the psychological tendency to interpret the behavior of the opposition in conspiratorial terms.[34]

The accusations of conspiracy were predicated upon a sequence of events on March 4 to 6, 1857. On March 4, at Buchanan's inauguration, the president-elect exchanged a few perfunctory words with the chief justice in the presence of a throng of spectators to whom the conversation was inaudible. Then, in his inaugural address, Buchanan stated that the issue of the status of slavery in the territories was "a judicial question which legitimately belongs to the Supreme Court of the United States, before whom it is now pending, and will, it is understood, be speedily and finally settled. To their decision, in common with all good citizens, I shall cheerfully submit, whatever this may be, though it has ever been my individual opinion that, under the Kansas-Nebraska Act, the appropriate period will be when the number of actual residents in the Territory shall justify the formation of a Constitution with a view to its admission as a state." Finally, two days later, Taney handed down the decision in which he declared the Missouri Compromise unconstitutional.[35]

Buchanan had indeed committed a breach of propriety in urging Justice Grier to support a broad decision rather than a narrow one, and there was an element of hypocrisy in his promising to "cheerfully submit" as if he did not know what the ruling would be, when in fact he distinctly understood from Grier what was impending.[36] But the critics of the decision pictured something far worse than this. To them, the whole case was a contrivance that had been trumped up by the slave power—a dummy case from the beginning, with proslavery forces controlling the plaintiff's counsel as well as

34. See Richard Hofstadter, *The Paranoid Style in American Politics and Other Essays* (New York, 1965), pp. 3–40; David Brion Davis, *The Slave Power Conspiracy and the Paranoid Style* (Baton Rouge, 1969).

35. James D. Richardson (ed.), *A Compilation of the Messages and Papers of the Presidents, 1789–1902* (11 vols., New York, 1907), V, 431.

36. It was not unusual at this time for justices to inform their close friends as to the tenor of impending decisions. In fact, as mentioned on page 273 above, Justice Curtis in 1856 had informed his uncle as to the consultations among the justices. Grier's letter to Buchanan, Feb. 23, 1857, had stated explicitly, "There will, therefore, be six, if not seven (perhaps Nelson will remain neutral) who will decide the Compromise law of 1820 to be of *non-effect.*"

the defendant's.[37] As William H. Seward expressed it in 1858, in a classic exposition of the conspiracy thesis, the Supreme Court took advantage of the fact that the territorial question had been raised, seized upon this "extraneous and idle forensic discussion," and decided it in a way "to please the incoming President." Then,

> the day of inauguration came—the first one among all the celebrations of that great national pageant that was to be desecrated by a coalition between the executive and the judicial departments to undermine the national legislature and the liberties of the people. The President arrived . . . and took his seat on the portico. The Supreme Court attended him there in robes which yet exacted public reverence. The people, unaware of the import of the whisperings carried on between the President and the Chief Justice, and imbued with veneration for both, filled the avenues and gardens far away as the eye could reach. The President addressed them in words as bland as those which the worst of all the Roman emperors pronounced when he assumed the purple. He announced (vaguely indeed, but with self-satisfaction) the forthcoming extra-judicial exposition of the Constitution, and pledged his submission to it as authoritative and final.

Later, continued Seward, the justices made their customary formal call on the president, and he "received them as graciously as Charles the First did the judges who had, at his instance, subverted the statutes of English liberty."[38]

The foremost Republican in the country, then, did not hesitate to accuse the president and the Supreme Court of conspiracy, tyranny, deception, and subversion—comparable to the worst villainies of recorded history. It is very interesting, by comparison with this melodramatic accusation, to note how Abraham Lincoln couched his criticism in a tone of ridicule and contemptuous amusement at detecting a rascality. Lincoln admitted that there was no evidence to prove absolutely that the policy of the Pierce administration in Kansas, the election of Buchanan, the decision of the Court, and the endorsement of the Court's decision by Douglas, Pierce, and Buchanan were parts of a concerted plan. However, he said, "When we

37. On the accusations and counteraccusations that the case was not a real contest, but had been deliberately set up by proslavery or antislavery groups, as the case might be, see Warren, *Supreme Court,* II, 301, 326–327; Beveridge, *Lincoln,* IV, 95, 131, 133, 135–136; Hopkins, *Dred Scott's Case,* pp. 24, 177, 179, 180–182; Swisher, *Taney,* pp. 486–487.

38. *Congressional Globe,* 35 Cong., 1 sess., pp. 939–945; discussion in Warren, *Supreme Court,* II, 324–329.

see a lot of framed timbers, different portions of which we know have been gotten out at different times and places and by different workmen—Stephen, Franklin, Roger and James, for instance—and when we see these timbers joined together . . . or, if a single piece be lacking, we can see the place in the frame exactly fitted and prepared to yet bring such a piece in—in such a case we find it impossible to not believe that Stephen and Franklin and Roger and James all understood one another from the beginning and all worked upon a common plan or draft drawn up before the first lick was struck."[39]

When Lincoln said that Stephen (Douglas), Franklin (Pierce), Roger (Taney), and James (Buchanan) understood one another, he spoke a truism, for so they did, just as Sumner and Chase, for instance, understood each other. But when he said that they all worked upon a common plan, that was quite a different assertion. Although Buchanan behaved improperly in writing to Grier, the very fact that he felt impelled to do so shows how far the case was from being a prearranged coup by the "Slave Power." Until the last moment, the justices were in doubt whether to rule on the congressional power to prohibit slavery in the territories, and though Wayne and Daniel may have wanted a broad decision, they were never able to bring the majority to agree with them until the last moment and until the antislavery justices had served notice of their intent to ventilate this question. Some of the Republican accusations were palpably untenable: for instance it was said that Buchanan added his comments on the impending decision to his inaugural address at the last moment after "whispering" with Taney on the inaugural platform. In fact the address, with this comment included, had been set up in type before the inauguration.[40] But the antislavery men came to believe their own propaganda. Charles Sumner, at the time of Taney's death seven years later, declared: "The name of Taney is to be hooted down the page of history . . . an emancipated country will fasten upon him the stigma which

39. "House Divided" speech at Springfield, June 16, 1858, in Roy P. Basler (ed.), *Collected Works of Abraham Lincoln* (8 vols., New Brunswick, N.J., 1953), II, 465–466.

40. Philip Shriver Klein, *President James Buchanan* (University Park, Pa., 1962), pp. 271–272. The prompt release of their dissenting opinions for newspaper publication by Justices Curtis and McLean led to a bitter exchange between Curtis and Taney, after which Curtis resigned. For a summary, see Charles Grove Haines and Foster H. Sherwood, *The Role of the Supreme Court in American Government and Politics, 1835–1864* (Berkeley, 1957), pp. 425–429.

he deserves. He administered justice, at last, wickedly, and degraded the judiciary of the country, and degraded the age."[41] So successfully had Sumner's views been propagated that for nine years Congress refused to vote to place a bust of Taney in the Supreme Court room along with the busts of other chief justices, and for half a century Taney's valuable contributions to American constitutional development remained unrecognized because of the Dred Scott decision.

During the storm of Republican anger immediately following the decision, Democratic newspapers exposed the fact that Dred Scott was probably still the property of the former Mrs. Emerson, now the wife of Calvin Chaffee, an antislavery congressman from Massachusetts. Soon after, John Sanford (Mrs. Chaffee's brother) died in an insane asylum, and the Chaffees hastened to end an association that was embarrassing not only for them but for the whole Republican party. They transferred ownership of Dred Scott and his family to Taylor Blow of St. Louis, son of Scott's original owner, and on May 26, 1857, Blow manumitted them. Dred Scott lived for only one more year, but he died a free man.[42] By then, his case had already become, and it has remained, one of the landmarks of American history.

Like a good many other measures during these years—for instance the Compromise of 1850, the Kansas-Nebraska Act, and the Ostend Manifesto—the Dred Scott decision conspicuously failed to accomplish what was expected of it, either by its advocates or by its opponents. Like these measures, also, it strangely combined theoretical significance with trivial consequences. Probably no other major judicial decision in history affected the daily lives of as few people as this one. It annulled a law which had in fact been repealed three years previously, and it denied freedom to the slaves in an area where there were no slaves. In some respects, it was about as abstract as a decision could be. Viewed in this aspect it seems less a divisive force in itself than a context in which the broader divisive forces found expression, a crossroads at which they chanced to meet, a sign of the troubled times.

Yet, in other respects, it was momentous in its meaning and its indirect results, and by all functional tests, it was a failure for those

41. *Congressional Globe,* 38 Cong., 2 sess., p. 1012; Swisher, *Taney,* pp. 581–582.
42. Hopkins, *Dred Scott's Case,* pp. 176–177. Chaffee's involvement lent false credence to Democratic charges that the case was contrived by abolitionists.

who supported it and a disaster for the American people. The extent of this failure and disaster can be measured in three ways.

First, it is legitimate to ask what effect the decision had in reducing sectional tensions. Clearly it had none, but instead it placed obstacles in the way of sectional adjustment. In the South, for instance, it encouraged southern rights advocates to believe that their utmost demands were legitimized by constitutional sanction and, therefore, to stiffen their insistence upon their "rights." In the North, on the other hand, it strengthened a conviction that an aggressive slavocracy was conspiring to impose slavery upon the nation, and that any effort to reach an accommodation with such aggressors was futile. While thus strengthening the extremists, it cut the ground from under the moderates. The sectional peacemakers had always sought to avoid the alternatives offered by Wilmot and by Calhoun, for either alternative meant a total sectional victory—slavery in none of the territories or slavery in all of them. The Missouri Compromise had sought to find ground between these alternatives and had relied on the power and moderation of Congress to occupy that ground. Taney's decision, of course, did not impair the Missouri Compromise, for it had already been destroyed, but it did impair the power of Congress—a power which had remained intact up to this time—to occupy middle ground. Concretely, it impaired Douglas's doctrine of popular sovereignty, for if Congress itself could not restrict slavery in the territories, a question arose at once as to whether it could authorize the territorial legislatures to do so. This implication concerning the power of the territorial legislatures was not so clear as it has appeared to some later writers in hindsight,[43] and it is an exaggeration to say, as such writers have said, that the Dred Scott decision destroyed Douglas[44] —a mistake because events in Kansas were developing in a way even

43. Although Taney specifically extended his opinion to deny the validity of popular sovereignty ("And if Congress itself cannot do this [exclude slavery]—if it is beyond the powers conferred on the Federal government—it will be admitted, we presume, that it could not authorize a territorial government to exercise them." 19 Howard 451), Don E. Fehrenbacher, *Prelude to Greatness: Lincoln in the 1850's* (Stanford, 1962), pp. 133–134, 190, shows that this was dictum, not part of the decision, and that even Justice Campbell disclaimed any ruling which would limit the power of territorial legislatures: "How much municipal power may be exercised by the people of the Territory, before their admission to the Union, the courts of justice cannot decide." 19 Howard 514. Other justices were silent on this question, and a majority certainly took no position on what the decision would do to the powers of a territorial legislature.

44. Hodder, "Some Phases of the Dred Scott Case," p. 21, makes this assertion.

more fatal for Douglas than the decision, and also because the blow which the Scott decision dealt to popular sovereignty was an indirect blow, and many contemporaries still went on believing that the territorial legislatures might exclude slavery even if Congress could not. But at best, the decision embarrassed Douglas's position. Thus it strengthened the forces working against sectional adjustment and weakened those working for it.

Second, in view of the last-minute nature of the shift from a narrow to a broad decision, a question arises as to the realism of the justices in supposing that they could settle the sectional fight by resolving a question which Congress had avoided. In concrete terms, those who believed that the decision would have a tranquilizing effect were proceeding on the theory that the northern public would accept a ruling favorable to the South by five southern justices accompanied by only one of their four northern associates.[45] This was a ruling which said that the Missouri Compromise—long venerated as a cornerstone of sectional adjustment—had not been valid, that the Wilmot Proviso was not valid, and that Popular Sovereignty was probably not valid. It was a ruling which invalidated a measure passed by Congress, and which sought to validate a position that Congress had repeatedly voted against. Apart from a general public veneration for the judiciary, there was nothing in the circumstances to warrant the belief that six justices could settle a question which a succession of Congresses had acknowledged their inability to settle. However admirable may have been the courage of the justices in facing the music, their tactical judgment was wretched.

Finally, the Dred Scott decision was a failure because the justices followed a narrow legalism which led them into the untenable position of pitting the Constitution against basic American values, although the Constitution in fact derives its strength from its embodiment of American values. Concretely, the American people wanted

45. The sectional character of the decision was accentuated, as many critics emphasized, by the fact that the South was heavily overrepresented on the Court. At a time when the justices rode circuit, Supreme Court appointments were, in effect, circuit appointments, and circuits were determined by the distances to be ridden more than by the populations involved. This disproportionately large area of the sparsely populated South as well, perhaps, as excessive southern influence had resulted in a situation where there were four free-state circuits, with a white population of 12,648,000, and five slave-state circuits with a white population of 6,026,000. Warren, *Supreme Court*, II, 289.

the United States to be a republic of free people and regarded the Constitution as essentially a charter for a free people. The utmost exception which they would make was to concede the right of local areas (states) to maintain slavery as a local—and, they hoped, temporary—institution. But always they regarded slavery as having only a local sanction, and freedom as having a national sanction. Yet by the 1850s the vicissitudes of sectional strife had brought about an anomalous situation. The South, while attempting defensively to ward off attacks on slavery, had adopted a position that went far beyond the defensive. Southern leaders had developed the doctrine that southern citizens with southern property (slaves) could not legally be kept out of the federal territory. The argument was not without legal plausibility, but it fatally reversed the place of slavery and freedom in the American system. It made freedom local—an attribute of those states which abolished slavery, but not of the United States; it made slavery national, in the sense that slavery would be legal in any part of the United States where a state government had not abolished it. Apart from the morality of it, this was a ruinous decision because, in the process of splitting logical hairs, it arrived at a result which converted the charter of freedom into a safeguard of slavery.[46]

The story of the Dred Scott decision has consistently overshadowed all other aspects of the judiciary's role in the sectional conflict. Yet this major case was, in one sense, not representative: by its circumstances, it did not lend itself to being defied; it could be denounced, but since no slaveholders now offered to take slaves into the territorial area, and since the Act of 1820 had been repealed three years previously in any case, the decision was really an abstraction, applying to proposed Republican legislation but not to any known persons held in slavery except Dred Scott and his family. Like most of the proslavery triumphs of 1846–1860, the Dred Scott decision was a hollow victory.

But there were other cases before the federal courts in these years which involved concrete enforcement of federal law—especially the Fugitive Slave Act—in cases involving slaves. In these cases, the antislavery forces launched a systematic assault both to prevent the

46. On this point and on proslavery doctrine generally, see Arthur Bestor, "State Sovereignty and Slavery: A Reinterpretation of Proslavery Constitutional Doctrine, 1846–1860," ISHS *Journal*, LIV (1961), 117–180.

enforcement of the law and to discredit the federal courts. The federal judiciary has come under severe attack, for very different reasons, on several occasions in American history—notably in 1801, 1896, 1934–1935, and after 1954—but one of the severest and most sustained of all these attacks began about 1850 and lasted into the Civil War. Foreshadowings of this attack began to appear as early as 1848 when Senator Clayton proposed leaving to the judiciary the question of the constitutionality of restrictions on slavery in the territories. Antislavery men recognized that such a decision was likely to go against them, and some of them mounted an anticipatory attack on the courts. As early as 1850, in the Senate, Salmon P. Chase denied that Congress must be bound by the courts' decisions, and John P. Hale made accusations which culminated a year later in his calling the Supreme Court "the very citadel of American slavery"—a phrase which he fondly repeated from time to time throughout the decade.[47]

These adverse criticisms gained in intensity. In *Strader* v. *Graham* (1851), the Court unanimously refused to reverse a Kentucky court decision denying freedom to a group of slaves that had been taken temporarily into Ohio, and it came under a barrage of antislavery criticism as a consequence.[48] There were other assaults the same year when several of the justices, acting separately in their respective circuits, upheld the constitutionality of the Fugitive Slave Act.[49] A few years later, the supreme court of Wisconsin sought to nullify the rulings of federal courts in a case involving the rescue of a fugitive slave. In 1854, Sherman M. Booth, an abolitionist editor, took part in inciting a mob to break down the door of the Milwaukee jail and rescue a fugitive slave who was being held in the custody of the federal marshal, Stephen V. R. Ableman. Booth was thereupon arrested and held for trial in federal court on the charge of violating the Fugitive Slave Act, but before his case could be heard, he appealed to the Wisconsin Supreme Court for a writ of habeas corpus, and this state court ordered his release on the ground that the Fugitive Slave Act was unconstitutional. Thereupon, Ableman sued out a writ of error, appealing the Wisconsin decision to the Supreme Court of the United States, and also secured the rearrest

47. Warren, *Supreme Court,* II, 207–224.
48. 10 Howard 305–310; Warren, *Supreme Court,* II, 224–226.
49. Warren, *Supreme Court,* II, 229–231.

of Booth by order of a federal judge, rather than by a commissioner, who had issued the first order of arrest. On this occasion, Booth was tried and convicted in federal district court. But again the supreme court of Wisconsin intervened, granted habeas corpus, heard the case, and freed Booth, on the ground of the unconstitutionality of the Fugitive Slave Act. What this amounted to was that a state court asserted its power to release a prisoner duly tried, convicted, and sentenced in a federal court for the violation of a federal law. In the history of resistance by the states to federal authority, few acts of defiance have approached this one, which involved nullification in a form that even John C. Calhoun had not advocated. Four years later, in 1859, Chief Justice Taney ended this case by issuing a strong opinion for a unanimous Supreme Court, declaring that the rulings of federal courts in cases involving federal law were not subject to review or interference by state courts.[50] By this time secession was less than two years away, and the Wisconsin court, perceiving that it was a tactical error for antislavery men to support doctrines of state sovereignty, acquiesced in the decision. But at the time of the Dred Scott decision, Ableman's appeal to the Supreme Court was still pending, and antislavery men were, in general, still supporting Wisconsin. The New York *Tribune*, for instance, had said, "The example which Wisconsin has set will be as rapidly followed as circumstances admit. By another year, we expect to see Ohio holding the same noble course. After that, we anticipate a race among the other free states, in the same direction, 'till all have reached the goal of state independence."[51]

Within another five years, events had shown that Wisconsin and Ohio could attain their goals better by controlling the federal machinery than by resisting it. Soon the South resumed its traditional role as the chief defender of states' rights and the North was again

50. On the Booth case, see Vroman Mason, "The Fugitive Slave Law in Wisconsin with Reference to Nullification Sentiment," State Historical Society of Wisconsin *Proceedings*, 1895, pp. 117–144; Bestor, "State Sovereignty and Slavery," pp. 136–142; Joseph Schafer, "Stormy Days in Court—the Booth Case," *Wisconsin Magazine of History*, XX (1936), 89–110; William Norwood Brigance, *Jeremiah Sullivan Black* (Philadelphia, 1934), pp. 57–60; James L. Sellers, "Republicanism and State Rights in Wisconsin," *MVHR*, XVII (1930), 213–229; Horace H. Hagan, "Ableman vs. Booth, Effect of Fugitive Slave Law on Opinions as to Rights of Federal Government and of States in the North and South," *American Bar Association Journal*, XVII (1931), 19–24; Warren, *Supreme Court*, II, 258–266, 332–334; Haines and Sherwood, *Role of Supreme Court*, pp. 224–244; Ableman *v.* Booth, 21 Howard 506.

51. Quoted in Warren, *Supreme Court*, II, 260–261.

identified with the acceptance of federal authority. But for a brief time when the law of the land was the law of the Taney Court, events had shown how little either northern or southern attitudes were governed by inherent devotion to the law of the land on one hand or to states' rights on the other. The events of the fifties offered a telling demonstration that the attitudes of various groups in a society toward upholding the law is in direct proportion to their approval or disapproval of the law which is to be upheld.

1. Anti-Whig cartoon, with Zachary Taylor on top of the heap. Lithograph by N. Currier.

(Library of Congress)

2. John Jordan Crittenden. Daguerreotype.

(Courtesy Chicago Historical Society)

3. Millard Fillmore. Daguerreotype taken by J. H. Whitehurst.

(Courtesy Chicago Historical Society)

4. The Polk family. Daguerreotype.

(International Museum of Photography at George Eastman House, Rochester, New York)

5. San Procopio (Retablo), by an unknown native New Mexican artist.

(Index of American Design, National Gallery of Art, Washington, D.C.)

6. "Justice's Court in the Backwoods." Oil on canvas, by Tompkins H. Matteson, 1850.

(New York State Historical Association, Cooperstown)

7. "Cider Mill," by William Tolman Carlton. Oil on Canvas.

(New York State Historical Association, Cooperstown)

8. Implements used in warfare, 1850s. From an illustration in Ballou's "Pictorial Drawing-Room Companion," 1856.

9. "Mexican News," by James Goodwyn Clonney. Oil on canvas, 1847. (Munson-Williams-Proctor Institute, Utica, New York)

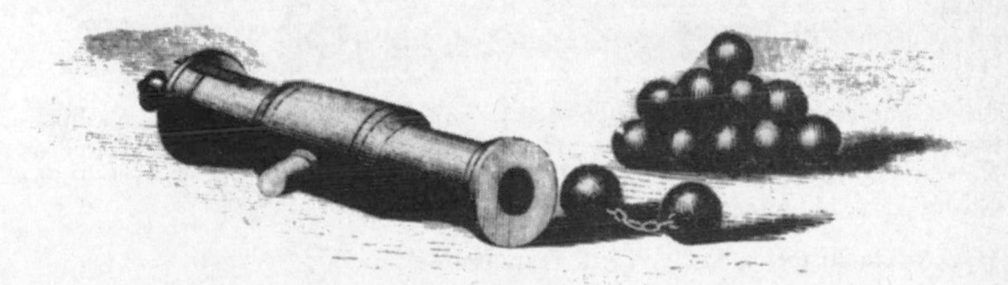

10. John Charles Frémont. Daguerreotype by Marcus A. Root.

(Courtesy Chicago Historical Society)

11. Characters of San Francisco, 1854. *Top Row:* Street Dentist; Washington Combs; The Dandy. *Second Row:* Fritz, Maguire's Fat Boy; Emperor Norton. *Third Row:* J. L. Martel; Unknown; Leon Chemis. *Fourth Row:* Gutter Snipe; Bummer and Lazarus; Drummer Boy.

(California Historical Society, San Francisco)

12. Cover for the sheet music of "Ho! for the Kansas Plains," dedicated to Henry Ward Beecher.

(Library of Congress)

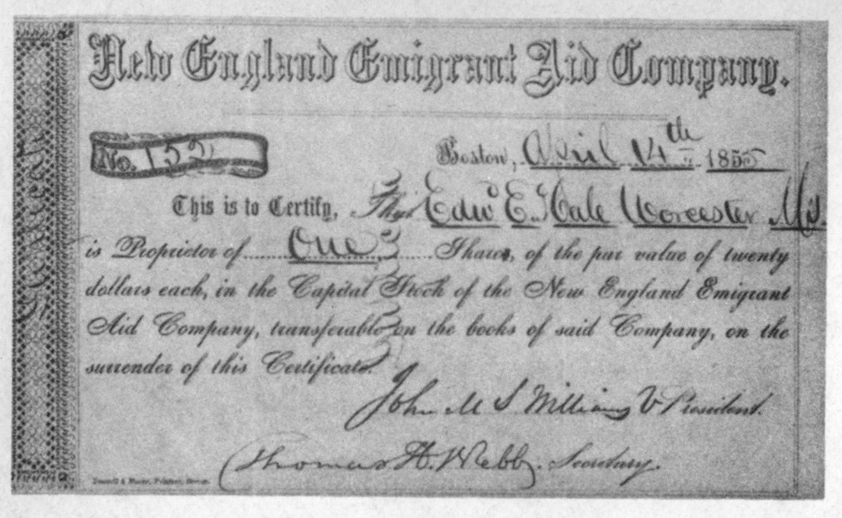

New England Emigrant Aid Company.

No. 152

Boston, April 14th 1855

This is to Certify, That Edwd E. Hale Worcester Ms

is Proprietor of One Share, of the par value of twenty dollars each, in the Capital Stock of the New England Emigrant Aid Company, transferable on the books of said Company, on the surrender of this Certificate.

John M S Williams V President.

Thomas H. Webb Secretary.

13. Land ownership certificate enabling free-stater Edward E. Hale to vote in Kansas.

(The Kansas State Historical Society, Topeka)

14. Slave-staters voting in Kansas.

(The Kansas State Historical Society, Topeka)

15. Free-staters in Kansas readying their cannon.
(The Kansas State Historical Society, Topeka)

16. Ruins of the Free State Hotel, Lawrence, Kansas. From a Daguerreotype.
(The Kansas State Historical Society, Topeka)

17. William Henry Seward. Daguerreotype taken by J. H. Whitehurst.

(Courtesy Chicago Historical Society)

18. The United States Senate, 1850. Engraving by R. Whitechurch after P. F. Rothermel, about 1855.

(Library of Congress)

19. Franklin Pierce. Daguerreotype.

(Courtesy Chicago Historical Society)

20. Daniel Webster. Daguerreotype.

(Courtesy, Chicago Historical Society)

21. Lewis Cass. Daguerreotype.

(Courtesy Chicago Historical Society)

22. "Slave Market at Richmond," by Eyre Crowe. Oil on canvas.

(Kennedy Galleries, New York)

23. "The Resurrection of Henry Box Brown at Philadelphia." Lithograph. Brown escaped from Richmond in a box three feet long, two and one-half feet deep and two feet wide.

(Library of Congress)

24. Scene from *Uncle Tom's Cabin*, by Harriet Beecher Stowe. Wood engraving by Cruikshank.

(Library of Congress)

25. Anti-slavery handbill.

(Library of Congress)

CAUTION!!

COLORED PEOPLE

OF BOSTON, ONE & ALL,

You are hereby respectfully CAUTIONED and advised, to avoid conversing with the

Watchmen and Police Officers of Boston,

For since the recent ORDER OF THE MAYOR & ALDERMEN, they are empowered to act as

KIDNAPPERS

AND

Slave Catchers,

And they have already been actually employed in KIDNAPPING, CATCHING, AND KEEPING SLAVES. Therefore, if you value your LIBERTY, and the *Welfare of the Fugitives* among you, *Shun* them in every possible manner, as so many *HOUNDS* on the track of the most unfortunate of your race.

Keep a Sharp Look Out for KIDNAPPERS, and have TOP EYE open.

APRIL 24, 1851.

26. Harriet Beecher Stowe. Daguerreotype portrait by Southworth and Hawes.

(The Metropolitan Museum of Art, Gift of I. N. Phelps Stokes, Edward S. Hawes, Alice Mary Hawes, Marion Augusta Hawes, 1937)

36 FOURTH OF JULY.

FOURTH OF JULY.

—— Men like household goods or servile beasts,
Are bought and sold, kidnapped and pirated;
Driven in droves e'en by the Capitol;
Then haul our striped and starry banner down;
Our cannon freight not; stop the noisy breath
Of heartless patriotism; be our praise unsung.
To-day we'll not discourse of British wrong,
Of valorous feats in arms by freemen bold,
Nor spit on kings, nor tauntingly call names;
But we will fall upon our bended knees,
And weep in bitterness of heart, and pray
Our God to save us from his gathering wrath;
We will no longer multiply our boasts
Of Liberty, till *All* are truly free.

W. L. Garrison.

27. A page from the *American Anti-Slavery Almanac*, published in 1844.

(Rare Book Department, University of Rochester Library)

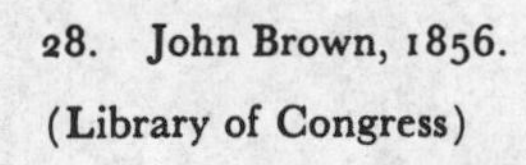

28. John Brown, 1856.

(Library of Congress)

29. Dred Scott.

(Missouri Historical Society)

30. "Southern Chivalry: Argument versus Club's." The "caning" of Senator Charles Sumner by Representative Preston S. Brooks of South Carolina, 1857. Lithograph by J. L. Ragee.

(Prints Division, The New York Public Library; Astor, Lenox, and Tilden Foundations)

31. Abraham Lincoln. Photograph taken by Alexander Hesler in Springfield, Illinois, June 3, 1860. Lincoln grew his beard after the election of 1860.

(Courtesy Chicago Historical Society)

32. Stephen A. Douglas. Daguerreotype.

(Courtesy Chicago Historical Society)

33. Cloth banner promoting the Republican ticket, 1860. By H. C. Howard. (Library of Congress)

34. Pro-Lincoln political cartoon. Originally published by J. Sage & Sons, Buffalo, New York.

(Library of Congress)

35. and 36. "Stephen A. Douglass, Patriot" and "A Secession Movement." Printed on envelopes, cartoons and portraits such as these made the mails a vehicle for political propaganda.

(Library of Congress)

37. "James Buchanan, Judas."

(Library of Congress)

38. "J. C. Breckinridge, Traitor."

(Library of Congress)

ANTI-SLAVERY

MASS MEETING!

Agreeably to a call, signed by about 50 persons, and published in the Lawrence Republican, a Mass Meeting of the friends of Freedom will be held at Miller's Hall, at 2 o'clock P. M., on Friday, Dec. 2d the day on which

CAPT. JOHN BROWN IS TO BE EXECUTED,

To testify against the iniquitous SLAVE POWER that rules this Nation, and take steps to

Organize the Anti-Slavery Sentiment

of the community. Arrangements have been made with prominent speakers to be present and address the meeting.

PER ORDER OF COMMITTEE OF ARRANGEMENTS.

Lawrence, Nov. 26, 1859.

39. Poster announcing an Anti-Slavery Meeting.

(The Kansas State Historical Society, Topeka)

40. "We Go for the Union." Unknown American Artist.

(National Gallery of Art, Washington, D.C. Gift of Edgar William and Bernice Chrysler Garbisch)

CHARLESTON MERCURY

EXTRA:

Passed unanimously at 1.15 o'clock, P. M. December 20th, 1860.

AN ORDINANCE

To dissolve the Union betwee[illegible] State of South Carolina and other States united with her [illegible]r the compact entitled "The Constitution of the United States of America."

We, the People of the State of South Carolina, in Convention assembled, do declare and ordain, and it is hereby declared and ordained,

That the Ordinance adopted by us in Convention, on the twenty-third day of May, in the year of our Lord one thousand seven hundred and eighty-eight, whereby the Constitution of the United States of America was ratified, and also, all Acts and parts of Acts of the General Assembly of this State, ratifying amendments of the said Constitution, are hereby repealed; and that the union now subsisting between South Carolina and other States, under the name of "The United States of America," is hereby dissolved.

THE UNION IS DISSOLVED!

41. Announcement of South Carolina's secession on the front page of the Charleston *Mercury,* December 20, 1860.

(Library of Congress)

42. The Sharps Rifle, known as a "sharpshooter," and the most common weapon used in the Civil War. (Index of American Design, National Gallery of Art, Washington, D.C.)

43. Fort Sumter after the first shelling, 1861.

(Library of Congress)

42. The Sharps Rifle, known as a "sharpshooter," and the most common weapon used in the Civil War. (Index of American Design, National Gallery of Art, Washington, D.C.)

43. Fort Sumter after the first shelling, 1861.

(Library of Congress)

44. "On to Liberty." Oil on Canvas, by Theodor Kaufmann.

(Hirschl & Adler Galleries)

CHAPTER 12

Lecompton: The Descent Grows Steeper

MORE than a century of historical writing has stamped the administration of James Buchanan as a failure. So familiar is this verdict that it is difficult to look at Buchanan as he seemed in 1857.[1] With forty years of experience in the House, the Senate, the foreign service, and the cabinet, he was past his prime but one of the best-trained men who has ever occupied the presidency. Thoroughly professional in his standards, and acutely conscious of his place in history, he believed he could avoid the mistakes of less experienced leaders like Taylor and Pierce, and he meant to be an illustrious president. Specifically, he intended to bring the "long agitation" of the slavery question, as a political issue, to "its end" and thus, also, to bring about the extinction of sectional strife and sectional political parties. He thought that this would not be as difficult as it might appear, for no one disputed the status of slavery in the states. Only in the territories had it been contested, and there needlessly, he said in his inaugural. For the principle of popular sovereignty had provided an answer: Leave the people of the territories "perfectly free to form and regulate their domestic institutions in their own way, subject only to the Constitution." To implement this principle, the government must "secure to every resident inhabitant the free and independent expression of his opinion by his vote."[2]

When Buchanan added that nothing could be "fairer" than such

1. Philip Shriver Klein, *President James Buchanan* (University Park, Pa., 1962), p. xiv.
2. Inaugural Address, March 4, 1857, in James D. Richardson (ed.), *A Compilation of the Messages and Papers of the Presidents, 1789–1902* (11 vols.; New York, 1907), V, 431–432.

a settlement,[3] he by-passed the fact that the Republicans believed in congressional exclusion and that the northern and southern Democrats were divided as to the time at which local exclusion might become effective. But he perceived that the issue was concrete, not theoretical. It was the issue of statehood in Kansas, and if he could create a situation that would give every resident of Kansas a free and independent voice in deciding whether the territory should be a slave state or a free state, he would be well on the way to a solution. But to accomplish this, he needed to produce something better than the existing uneasy truce in Kansas. He had to give the territory a governor strong enough to impose order and gain ascendancy over the warring elements, and impartial enough to win the confidence of antislavery men who felt such deep-seated distrust of the territorial government that they refused to participate in the elections.

But as Buchanan knew very well, this task required a man of stature, unlike ordinary territorial governors, who were usually political hacks. He found such a man in Robert J. Walker of Pennsylvania and Mississippi, his former colleague in the Polk cabinet and one of the nation's foremost Democrats. Walker had a weakness for grandiose and questionable speculative ventures, but he was hard-driving and skillful in political matters and was capable of taking a broad view of public questions. He did not want the onerous Kansas assignment, and it took much persuasion to win his acceptance.[4] Also, he was far too shrewd to take this post without publicly committing the administration to the support of the policies he proposed to follow. Knowing that Buchanan was susceptible to southern influence, he wrote a public letter in which he took pains to state his understanding that the president and cabinet "cordially concurred in the opinion . . . that the actual *bona fide* residents of the territory of Kansas, by a fair and regular vote, unaffected by fraud or violence, must be permitted, in adopting their State Constitution, to decide for themselves what shall be their social institutions."[5]

3. *Ibid.*, p. 431.

4. Testimony of Walker before the Covode Committee, April 18, 1860, *House Reports,* 36 Cong., 1 sess., No. 648 (Serial 1071), pp. 105–106, hereafter cited as *Covode Committee;* James P. Shenton, *Robert John Walker: A Politician from Jackson to Lincoln* (New York, 1961), pp. 147–149; Allan Nevins, *The Emergence of Lincoln* (2 vols.; New York, 1950), I, 144; Roy F. Nichols, *The Disruption of American Democracy* (New York, 1948), pp. 96–98.

5. Walker to Buchanan, March 26, 1857, in KSHS *Transactions,* V (1891–1896), 290.

When he wrote his inaugural address as governor before leaving Washington, Buchanan came to his house and went over it with him, presumably approving the passage which said: "In no contingency will Congress admit Kansas as a slave state or free state, unless a majority of the people of Kansas shall first have fairly and freely decided this question for themselves by a direct vote on the adoption of the Constitution, excluding all fraud or violence."[6] Armed with these understandings, the new governor set out for Kansas better prepared than any of his three predecessors to impose a firm control over all factions. Mending his political fences as he traveled west, Walker stopped in Chicago to confer with Douglas and to gain his approval also for the inaugural.[7] Soon after arriving in Kansas, he sent Buchanan an analysis of the political composition of the territory, clearly indicating the strategy he intended to follow: "Supposing the whole number of settlers to be 24,000," he said, "the relative numbers would probably be as follows: Free State Democrats, 9,000, Republicans, 8,000, Proslavery Democrats 6,500, Proslavery Know-Nothings, 500."[8] This meant that the antislavery elements outnumbered the proslavery group 17,000 to 7,000, but that the Democrats outnumbered the adherents of other parties 15,500 to 8,500. Walker assumed accordingly that the administration could easily bring Kansas into the Union as a Democratic state, which Buchanan was eager to do, if it did not make the mistake of trying to create a slave state, which the majority, including many Democrats, would oppose.

The principal obstacle to the victory of a free-state program lay in the division of the free-state men. Many of the Democrats among them participated in the established government of which Walker

6. Testimony of Walker, April 18, 1860, *Covode Committee*, p. 106; Walker's inaugural address, KSHS *Transactions*, V, 339, also 329. Klein, *Buchanan*, p. 292, accepts a letter of Alfred Iverson to Howell Cobb, Sept. 17, 1857, reporting a conversation with Buchanan, and states that "Buchanan . . . had never seen a draft of the inaugural." Walker's testimony asserted that when he conferred with Buchanan the "address was not then complete, except that portion of it that related to the constitution being submitted to the vote of the people, and what I said on the subject of slavery in Kansas. What I said on the subject of submitting the constitution to a vote of the people, Mr. Buchanan fully approved. As regards what I said on the subject of slavery in Kansas, he suggested a slight modification which, with some little variation from the words he suggested, but embodying substantially the same idea, was adopted by me. That modification . . . applied to only a single sentence."

7. See letter of Douglas to Walker, July 21, 1857, in Robert W. Johannsen (ed.), *The Letters of Stephen A. Douglas* (Urbana, Ill., 1961), pp. 386–387. For relations of Walker and Douglas, see Shenton, *Walker*, pp. 152, 251.

8. Walker to Buchanan, June 28, 1857, in *Covode Committee*, pp. 115–119.

now became the head, but some of the free-state Democrats and almost all of the Republicans refused to recognize this government on the ground that it was a "bogus" regime, based upon fraudulent elections and committed to fraudulent practices.[9] When Walker arrived, he found that the violence had subsided. Governor Geary had succeeded in restoring a measure of order, and the enticements of land speculation had seduced many of the former belligerents. The majority of settlers were land-hungry pioneers who valued peace and prosperity for themselves above either slavery or freedom for the Negro. But the deep division between the free-state "government" at Topeka and the recognized government at Lecompton kept men apart. Moreover, the old antagonisms showed signs of flaring up again because in February 1857, shortly before Buchanan became president, the Kansas territorial legislature authorized an election in June of a constitutional convention to meet in September and set the territory on the road to statehood.[10]

Walker did not reach Kansas until the end of May, too late to work out a basis on which the free-state faction would consent to vote in the election. They contended that they had been disfranchised, gerrymandered, and simply counted out by stuffed ballots, and that if they should vote, they would then be bound by the results of an election which they knew was going to be fraudulent, or, at best, unrepresentative. Only if Walker would set aside the procedure specified in the act calling the election would they take part. But he had no power to do this, and though he warned them that their abstention would give the proslavery party a victory by default, they refused to be moved.

In June, precisely what he had foreseen took place. In a quiet election, with many proslavery candidates unopposed and only 2,200 out of 9,000 registered voters going to the polls, a large majority of extreme proslavery men won election as delegates to the constitutional convention in September.[11]

This election, at the very beginning of Walker's governorship, placed him in difficulties from which he was never able to extricate himself. It saddled him with a constitutional convention which, as a representative body for the people of Kansas, was a farce, but

9. Above, pp. 204–206.

10. *House Reports,* 35 Cong., 1 sess., No. 377 (Serial 966), pp. 17–21.

11. Registry of qualified voters, in *ibid.*, pp. 22–23; election result stated in message of Acting Governor Frederick P. Stanton to Kansas legislature, in KSHS *Transactions,* V, 415; New York *Tribune,* July 11, 1857.

which had been elected in an entirely legal manner. He could hardly deny the validity of the election after having warned the free-staters that they ought to vote in it because they would be bound by it.[12] Yet this situation completely undercut his effort to unite all Democrats—both free-state and proslavery—in a combination which would create another Democratic state and would provide a popularly acceptable settlement for the eternal slavery question.

Worse still for Walker, in his effort to win the confidence of the free-staters he had shown them a partiality which now excited a reaction against him in the South. At the outset, southerners had thought of him as a southerner and had urged Buchanan to include him in the cabinet, for he had served as a senator from Mississippi and was a vigorous advocate of the annexation of Cuba. But he was a native of Pennsylvania, had lived in Mississippi only nine years, and like that other eminent slaveholder, Zachary Taylor, he was not a proslavery man politically. During Buchanan's election campaign he had publicly stated in a widely circulated pamphlet that he did not believe Kansas would become a slave state. He had repeated this in his inaugural address as governor of Kansas, with learned allusion to an "isothermal" line north of which slavery could not flourish, and with a suggestion to the proslavery men that they seek a compensatory slave state in the Indian territory to the south of Kansas. To men who had hoped Kansas itself would be compensatory for the imminent admission of Minnesota as a free state, talk of a slave state in the Indian Territory was pure pie-in-the-sky—"vile hypocrisy" and "flimsy twaddle," as one southerner protested.[13] Not only did Walker make these outspoken statements but he was hobnobbing in a most alarming way with the very men who denied the legitimacy of the government he headed. When he entered Kansas, he went to the free-state headquarters at Topeka and there had a love feast with Jim Lane and others who had engaged

12. Walker's inaugural address, May 27, 1857, in KSHS *Transactions,* V, 328.

13. On southern backing for Walker, Shenton, *Walker,* pp. 141–144; Walker's inaugural, in KSHS *Transactions,* V, 336–337. The southern reaction to the inaugural was expressed by Thomas W. Thomas of Georgia in a letter to Alexander H. Stephens, June 15, 1857: "I have just read Walker's inaugural in Kansas and if the document I have seen is genuine, it is clear Buchanan has turned traitor. . . . He, Walker, is reported as travelling to that country through the North, gathering up a free-soil suite . . . and attempting to mask his vile hypocrisy with the flimsy twaddle of a slave state in the Indian country south of Kansas." Ulrich Bonnell Phillips (ed.), *The Correspondence of Robert Toombs, Alexander H. Stephens, and Howell Cobb,* in AHA *Annual Report,* 1911, II, 400.

in armed hostilities against the recognized government. He bought them drinks and went to church with them. He spoke from the same platform with the free-state pretender to the governorship and, in short, behaved with a degree of bland tolerance that was viewed very sourly in the South.[14]

As late as June 1857, southerners could not have entertained very serious objections to Walker's believing that Kansas would become a free state, for most of them believed it themselves, as did Buchanan.[15] But still they objected when the man whom Buchanan had sent to Kansas to apply popular sovereignty with strict impartiality began to make public statements which prejudged the outcome. It was one thing to believe that Kansas would become a free state, but quite another for the man presiding over the dispute to assure one of the contending parties that their victory was a foregone conclusion. Moreover, victory in the June election confronted the proslavery party with an overwhelming temptation. At a time when most of them had regarded their cause as hopeless, they suddenly realized that the creation of a new slave state was within their grasp. The legally elected convention needed only to draft a constitution and send it to the Democratically controlled Congress for acceptance. The biggest obstacle was Walker's embarrassing promise to let the voters decide.

In these circumstances, many proslavery Democrats mounted a heavy assault upon Walker from one end of the South to the other. Newspapers in Richmond, New Orleans, Vicksburg, Jackson, and elsewhere assailed him. Party leaders such as Jefferson Davis, Albert G. Brown, and Robert Toombs joined in the attack. The Democratic state conventions of Georgia and Mississippi solemnly adopted resolutions of censure.[16]

Thus, within a few weeks of his arrival in Kansas, Walker found himself in serious trouble. Eager as he was to induce all Democrats

14. New York *Tribune*, June 1, 4, 1857; *New York Times*, June 1, 1857; Shenton, *Walker*, pp. 152–154; Nichols, *Disruption*, p. 107.

15. Klein, *Buchanan*, p. 290.

16. For the southern assault on Walker, see Nevins, *Emergence*, I, 163, 165–167, 169–170; George Fort Milton, *The Eve of Conflict: Stephen A. Douglas and the Needless War* (Boston, 1934), pp. 266–268; Shenton, *Walker*, pp. 165–168; Phillips, *Toombs, Stephens, Cobb Correspondence*, pp. 400–408; and Avery O. Craven, *The Growth of Southern Nationalism, 1848–1861* (Baton Rouge, 1953), pp. 284–285. Nichols, *Disruption*, pp. 113–114, reviews evidence of this southern protest, but concludes that by mid-July "the effort to make Walker an issue in the Southern . . . elections of that summer apparently had failed."

to agree on a free-state constitution for Kansas, he had failed to prevent the election of a proslavery convention. Wanting to uphold the principle of popular sovereignty, he had himself violated it by his open assurances of a free state. He had hoped to gain a broad basis of support by persuading the free-staters to vote and by uniting the proslavery and antislavery Democrats, but he had failed to win the free-staters, while at the same time antagonizing the proslavery group so deeply that their leaders had broken with him publicly.

In these circumstances, some tension naturally began to develop between Walker and the administration. Buchanan must have been disappointed in the results, and he had specific reason to object to two developments: first, Walker's abandonment of any pretense of impartiality as to whether Kansas should be free or slave; and second, Walker's increasing insistence that the voters of Kansas should have an opportunity not merely to choose between a slavery clause and a no-slavery clause in the proposed constitution, but that they should have the chance to accept or reject the constitution *in toto*. This distinction later became so controversial that historians have tended to regard it as an issue from the beginning, but the evidence suggests otherwise. For instance, on July 12, Buchanan assured Walker, "On the question of submitting the constitution to the *bona fide* resident settlers of Kansas, I am willing to stand or fall."[17] Walker, on his part, was apprehensive, as he had been from the outset, as to the support of the administration. He knew the fate of Geary and his other predecessors in the Kansas governorship; he knew Buchanan's prosouthern predilections; he was troubled by the rising criticism from southern Democrats; and when Buchanan made an appointment to a judgeship in Kansas without consulting him, he was alarmed by the implications.[18]

Henry S. Foote, who was working very actively in the South to uphold Walker, later asserted that Buchanan surrendered to the southern pressure and turned against the governor in July 1857. Many historians have agreed that this was indeed what happened.[19] But in fact Buchanan seemed to be sustaining Walker vigorously. In his "stand-or-fall" letter in July, the president commended Walker for trying "to build up a great democratic party in Kansas" regard-

17. Buchanan to Walker, July 12, 1857, in *Covode Committee*, p. 112.
18. Walker to Buchanan, June 28, 1857, *ibid.*, pp. 117–118.
19. Henry S. Foote, *Casket of Reminiscences* (Washington, 1874), pp. 116–118; Milton, *Eve of Conflict*, pp. 268–269; Nevins, *Emergence*, I, 172.

less of whether it went for or against slavery, assured him that "the strictures of the Georgia and Mississippi conventions" would pass away, and held out to him the prospect of returning "triumphantly" from his "arduous, important and responsible mission." In October, Buchanan's position had apparently not changed, for he then wrote, "I am rejoiced . . . that the convention of Kansas will submit the Constitution to the people. . . . I think we may now fairly anticipate a happy conclusion to all the difficulties. . . . I am persuaded that with every passing day the public are more and more disposed to do you justice." Walker himself testified three years later that he believed Buchanan had been innocent of any plans to undercut him.[20]

Buchanan actually supported Walker better than Pierce had supported Reeder or Shannon or Geary. But if the president himself stood by his appointee in Kansas, the cabinet became very dissatisfied, and, as developments showed, Buchanan could not always control his cabinet. From the outset, the four southerners—Howell Cobb of Georgia at the Treasury, John B. Floyd of Virginia in the War Department, Aaron V. Brown of Tennessee as postmaster general, and Jacob Thompson of Mississippi, secretary of the interior —all had reason to object to Walker's well-advertised alliance with the free-staters. But in fact Cobb supported Walker firmly, though with some misgivings,[21] and the northern members of the cabinet were as impatient with him as the southerners. This came out in July, when, with Buchanan away on a vacation, the cabinet had to deal with a request by Walker for two thousand troops. The people in the antislavery community of Lawrence had set up a municipal government without Walker's consent. He responded by treating this as a defiance of his authority, issuing a bellicose proclamation, marching dragoons to the town, and calling on the administration for soldiers.[22] Buchanan's advisers regarded Walker's reaction as

20. Note 17, above; Buchanan to Walker, Oct. 22, 1857, in John G. Nicolay and John Hay, *Abraham Lincoln, A History* (10 vols.; New York, 1890), II, 110–112; Walker's testimony, *Covode Committee*, pp. 111, 114. The view that Buchanan still continued to support Walker is maintained in Nichols, *Disruption*, pp. 114, 127, and Klein, *Buchanan*, pp. 293–295. Shenton, *Walker*, pp. 163–165, sees Buchanan as beginning to cool toward Walker, but continuing a qualified support at this time.

21. Cobb to Alexander H. Stephens, June 18, July 21, 1857, in Phillips (ed.), *Toombs, Stephens, Cobb Correspondence*, pp. 402–408.

22. See Walker's proclamation, July 15, 1857, to the people of Lawrence asserting, "A rebellion so iniquitous . . . has never before disgraced any age or country," and

excessive and flatly turned down his request. According to Floyd, they thought that Walker hoped to throw the blame for his failure upon others. Lewis Cass, who, as secretary of state, probably presided in Buchanan's absence, informed the president, "We do not like Governor Walker's letter. We all fear that Governor Walker is endeavoring to make a record for the future."[23]

Whatever the members of the cabinet may have thought, none of them did anything about it until early in October. By that time the Lecompton convention had already met on September 7–11, but had decided not to remain in session because another of Kansas's frequent elections was in the offing. Having elected the constitutional convention in June, the voters were to elect a new territorial legislature in October. Having failed to induce the free-staters to vote in the June election, Walker was busily engaged in trying to persuade them to participate in the October balloting. The convention therefore adjourned until October 19 to see what would happen.[24] This was the situation when, on October 1, Secretary Thompson of Interior sent a clerk in the Land Office, Henry L. Martin, to Lecompton. Officially, Martin was assigned to examine land records in the basement of the building where the Lecompton convention would be meeting, but no one ever denied that an important part of his mission was political.[25]

Why he was sent and whether Thompson acted for the administration in sending him will probably remain in dispute as long as anyone cares about these events. But it seems fairly certain that the mission was not friendly to Walker, for Thompson was an old personal enemy. In 1845, Walker had deliberately refrained from deliv-

---

his letters to Secretary Cass (through whose department he reported), July 20, 27, and Aug. 3, stating, "The territorial government is in imminent danger of overthrow if I am not sustained by at least 2000 troops." KSHS *Transactions*, V, 355–360, 362–364, 370–371.

23. Nevins, *Emergence*, I, 171, citing Floyd to Buchanan, July 31, 1857, and Cass to Buchanan, same date, in Buchanan papers. On July 23, Howell Cobb wrote to Alexander H. Stephens, "There is no doubt of the fact that Walker is playing a bold game for the succession [to the presidency] and is strongly backed up in N.Y." Phillips (ed.), *Toombs, Stephens, Cobb Correspondence*, p. 408.

24. KSHS *Transactions*, V, 293–295, 341–348; Nichols, *Disruption*, pp. 111–112, 117; Journal of the Lecompton convention in *House Reports*, 35 Cong., 1 sess., No. 377 (Serial 966), pp. 23–73. (The record of this journal terminated on Nov. 3. The convention did not adjourn until Nov. 8. Thus there is no official record of the last five crucial days of the convention.)

25. *Covode Committee*, pp. 110, 114, 157–174, 314–323.

ering to Thompson a secretly prepared commission by the governor of Mississippi which would have made Thompson a United States senator; Thompson had been a supporter of Walker's, but Walker had wanted someone else to be appointed to the Senate instead.[26] Now, twelve years later, Thompson's agent turned up in Lecompton, at the very time of the convention, looking conspicuously like a special envoy of the administration. Furthermore, his arrival came at the very time when the October election shattered all remaining ties between Walker and the proslavery party.

In this election, the free-staters at last decided to take part, and the contest was close. When the votes were returned, however, they showed a proslavery majority for the new legislature, but this majority resulted largely from the astonishing tallies of 1,628 and 1,200 votes at two places, Oxford in Johnson County, and three precincts in McGee County. Walker investigated these returns and discovered another of Kansas's "electoral absurdities." In McGee, there were about twenty voters, but no election had been held at all; in Oxford, where there were six houses, fewer than thirty votes had actually been cast, but 1,601 names, all in one hand and all on one immense roll of paper, had been copied onto the voting list in consecutive order from Williams's *Cincinnati Directory*. Secretary Cass had already warned Walker that he had no legal power to review election returns—such review was a matter for the courts. But this was too much. Ignoring the legal point, the governor peremptorily threw out the returns from these districts. The result was that the free-state forces were left in a majority. For the first time, the recognized legislature of Kansas would be in the hands of the antislavery faction.[27]

Walker's bold response to this remarkable fraud showed that he had not lost his capacity for decisive action. But the evidence suggests that he was, by this time, sick of the whole Kansas business. He had gone to the territory with high expectations of achieving a masterful settlement and returning as a triumphant proconsul, perhaps to stand for the presidency. He had found, instead, that the free-staters refused to cooperate and the proslavery faction harassed him in every conceivable way; he was nowhere near his objective of a constitution which would tranquilize Kansas and initi-

26. Shenton, *Walker*, pp. 64–66.
27. *Covode Committee*, p. 109; *National Intelligencer*, Nov. 5, 1857; KSHS *Transactions*, V, 375–378, 382–384, 403–408.

ate a Democratic Pax Romana; and he had become the target for storms of criticism from the South. He was dissatisfied with the support he was receiving from the administration. Further, the Panic of 1857 had struck with sudden fury in August, endangering his speculative investments in a way that demanded his presence in the East. Kansas had proved a dreary, crude, inhospitable place, and his health was suffering. It may be, in fact, that the cabinet members were right in thinking that Walker knew he had failed and was seeking a way out, for on October 10, nine days before the Lecompton convention was due to reassemble, he requested a thirty-day leave—this at the very time when his presence might be most crucial.[28] His relations with the proslavery party were by this time so bad that he may have supposed matters would go better if he were absent, for he left Lecompton and went to stay with a friend at Leavenworth forty miles away. By November, he had heard from Buchanan that he might take his requested leave after the convention was over. It ended on November 8, and nine days later, he left Kansas, never to return.[29]

While Walker was at Leavenworth, the Lecompton convention proceeded to write a constitution which can easily be described. It departed from the routine pattern of new state constitutions at a number of points, including a prohibition of any amendment for a period of seven years and requirement of twenty years' citizenship for eligibility for the governorship. There was also a rigid restriction on the chartering of banks and a clause excluding free Negroes from the state. (The free-state Topeka constitution contained a similar provision.) With some strong rhetoric, the constitution guaranteed slaveholders their property rights in the two hundred or so slaves already in Kansas. On the central question of whether new slaves could be brought into Kansas, it left the decision to the voters in a referendum in which they would vote for "the constitution with slavery" or "the constitution without slavery." But they were not given a chance to accept or reject the entire constitution.[30]

28. Walker to Cass, July 15, complaining of the rebelliousness of the free-staters, the administration's removal of troops, the criticisms from the South, and his ill health; Oct. 10, asking for a 30-day leave of absence, in KSHS *Transactions*, V, 341–348, 401.

29. Nichols, *Disruption*, pp. 118–122; KSHS *Transactions*, V, 402–403, 408–410; *Covode Committee*, pp. 109–111. Nevins, *Emergence*, I, 241, cites the Chicago *Tribune* correspondent as reporting that Walker took all his books, papers, and personal property, well boxed, as if he did not expect to return.

30. Text of Lecompton constitution in *House Reports*, 35 Cong., 1 sess., No. 377

Yet if the provisions themselves are clear, their meaning is endlessly disputed. How this constitution was adopted, what went on behind the scenes, who was in control, and even where the victory lay are all matters of intricate dispute. Essentially, two versions of the Lecompton story have emerged.

In one version, the extreme proslavery faction, with the secret backing of the administration, seized control, violated all the promises that had been made to Walker, adopted a slavery constitution, and betrayed the pledge to give the voters a choice between acceptance and rejection, but concealed this betrayal by offering a spurious choice, which really forced the voters to accept a proslavery constitution in either a more obnoxious or a less obnoxious form.

This version is a dramatic one. It argues that the proslavery forces had never given more than lip service to the idea of popular sovereignty, and that when they unexpectedly won the June election a movement began in Washington, within the councils of the administration, to undercut Walker and push through a proslavery constitution for Kansas. As part of this movement, Henry Martin went to Kansas to work with the proslavery leaders in controlling the convention. Martin carried with him the message that Secretary Thompson favored submission of the constitution to the voters but would not object "if a pro-slavery constitution should be made and sent directly to Congress by the convention." Ostensibly, this statement adhered to the official position of the administration, but cryptically, with a wink and a nod, it encouraged the delegates to do just the opposite of what Walker wanted them to do. When Martin reached Lecompton, he was received as the agent of the administration. He attended the caucuses of the proslavery party and occupied a seat of honor on the convention floor. He joined forces with John Calhoun, local leader of the proslavery party and president of the convention. Acting as political managers, these two executed the plan concocted in Washington, substituting a kind of pretended referendum for the one promised. Of course, they still had to reckon with Walker, and Calhoun in an interview sought his support for the scheme of what came to be known as "partial submission" —i.e., a vote for or against the slavery clause rather than for or

(Serial 966), pp. 73–92. On the personnel of the convention, showing that it was a body of "ordinary respectability," see Robert W. Johannsen, "The Lecompton Constitutional Convention: An Analysis of Its Membership," *Kansas Historical Quarterly*, XXIII (1957), 225–243.

against the constitution as a whole. Walker replied that this would not do—that it was contrary to the policy of the administration. He quoted Buchanan's "stand or fall" letter of July and heatedly denounced Calhoun's plan as "a vile fraud" and "a base counterfeit." But Calhoun replied that the administration had changed its policy. When Walker asked whether Calhoun had a letter from Buchanan, Calhoun said he had not, but that the assurance came to him "in such a manner as to be entirely reliable," presumably meaning that it came from Martin. Then Calhoun and Martin proceeded to secure the passage of their plan in the convention, thus completing the betrayal of Walker and of the principle of popular sovereignty.[31]

There are evidently several points of validity in this version. No doubt the southerners badly wanted another slave state and overreached themselves in grasping for it. No doubt Buchanan did have a predilection for the southern viewpoint, and no doubt he was sometimes by-passed by members of his cabinet. Martin was almost certainly sent to Kansas to work with the proslavery faction, and without question he had an important hand in the result. Assuredly, no love was lost between Walker and Calhoun. But there are also certain points at which the theory of what might be called the Lecompton Conspiracy breaks down. These features lend themselves to a second version.

First of all, Calhoun was not a sottish mediocrity or a minion of the slavocracy, as antislavery writers have often pictured him. He was a capable politician and follower of Stephen A. Douglas; he wrote to Douglas for guidance in the Kansas situation and tried to infer Douglas's views from the Chicago *Times* when the Little Giant did not reply. He visited Washington in March, and Buchanan outlined to him the plan of submitting the constitution to the voters and told him that he "was expected to carry it out in good faith." Calhoun attempted to do so. He both voted and spoke in support of "total submission"—i.e., submission of the entire constitution for acceptance or rejection. At the time of their election, nearly all of the delegates had pledged themselves to support a referendum of this kind, but after Walker threw out the fraudulent election returns, they were so angry with him that many of them turned to the idea

31. Nicolay and Hay, *Lincoln*, II, 101–118, offers a good statement of this first version. Also see George D. Harmon, "President James Buchanan's Betrayal of Governor Robert J. Walker of Kansas," *PMHB*, LIII (1929), 51–91.

of framing a constitution and sending it directly to Congress. This, after all, was the procedure which had been followed in the admission of many states.[32] Thus Calhoun, though later stereotyped as an ultraproslavery man, actually was struggling against the extreme proslavery group in the convention. To his consternation, he found that they held a majority in the convention, and on November 6 they voted to put a slavery clause in the constitution and send it to Washington without any kind of referendum at all. At that point, it took all of Calhoun's resourcefulness to escape total defeat, but he hastily arranged an adjournment. It was only then that he and Martin turned to the plan of "partial submission." They sponsored this plan, not as a ruse to conceal the abandonment of a real choice between acceptance and rejection of the constitution, but as a way to save the essential element in the principle of "submission"—the voters would still have a choice of opening Kansas to slavery or keeping it free, save for a limited number of slaves already resident.[33]

Ultimately, the controversy reduced itself to a dispute as to whether "partial submission" offered the voters of Kansas a meaningful or a spurious choice. To the antislavery men, the facts were simple: The voters had been promised a chance to accept or reject the proposed constitution, and this promise had not been kept; they had been promised a chance to vote against slavery, and now their only option was to vote either for limited slavery or for unlimited slavery. Opponents of the constitution were not impressed by Democratic arguments that the number of slaves was small and that there was good precedent for recognizing title to slaves already held within a jurisdiction before an emancipating or prohibitory act went into effect. (For instance, slaves had been held in New York, Pennsylvania, and New Jersey for many years after these states became

32. For a good summary on this point, see Klein, *Buchanan*, pp. 305–306.

33. Adherents of the conspiracy theory (e.g., Milton, *Eve of Conflict*, p. 270) picture Calhoun as merely offering nominal support to the program of full submission and abandoning it at the earliest opportunity. Those who reject the theory (e.g., Nichols, *Disruption*, pp. 123–126, citing Kansas newspapers and the Douglas correspondence extensively) picture him as fighting for full submission first and for a compromise when he could not get his initial objective. See testimony of Walker, Martin, and A. J. Isaacs, in *Covode Committee*, pp. 111, 162–163, 174–176, indicating that Calhoun thought the result would please Douglas. Even if there was a conspiracy, Calhoun could have been a victim of it, rather than a party to it, especially in view of his adherence to Douglas.

"free"; and Illinois, admitted as a free state in 1818, had specifically maintained the continued servitude of unfree labor already within the state.) Antislavery men pointed out that the states cited as precedents had carefully avoided the use of the term "slavery" and had specifically provided for the freedom of persons born after a stated date; but the Lecompton convention did neither, and it aggressively flaunted a clause that "the right of the owner of a slave to such slave and its increase is the same, and as inviolable as the right of the owner to any property whatever."[34]

Defenders of the Lecompton constitution argued with considerable cogency that state constitutions, by this time, had become somewhat standardized, and that the whole constitution was, in a sense, the package containing the slavery or no-slavery choice. If voters had to accept or reject a proslavery constitution, this would mean paying the penalty of losing statehood as the price of rejecting slavery, but if they voted only on a slavery clause in a constitution which was otherwise not at issue, they might reject slavery without sacrificing statehood. Viewed in this way, the promise of statehood would become a kind of bribe to voters to accept the constitution, and it was infinitely preferable to avoid making statehood contingent upon the action on slavery. Democrats believed that the antislavery faction rejected the choice offered by the convention because they wanted ammunition for propaganda and did not want an honest settlement. These views had some merit, but their ultimate weakness lay in the fact that the voters were not permitted to cast their ballots for a clear-cut no-slavery clause. The only option open to an antislavery voter was one which excluded the importation of slaves, but affirmed the principle of continuing slavery for all human chattels already in Kansas and for their descendants as well, and this was not option enough.

The supporters of Lecompton also contended that the administration had never intended to promise the voters a chance to vote for or against the Constitution as a whole. Apparently this was true, for there had been much ambiguity in what was thought and what was said. Most of the discussion of "submission" had not been

34. Article VII. See note 30 above. Nevins, *Emergence,* I, 235, discusses the antislavery view of the constitution, pointing out that the presence of a limited number of slaves "would facilitate the smuggling of new slaves across the border." Nevins apparently fails to note the phrase "and their increase," for he says a question was left: "Would their progeny also be held in servitude?"

explicit, but had simply alluded to "the right of the people to decide," without specifying how. In view of later antislavery claims that the original intent—later perverted—had been to offer a choice between full acceptance and full rejection, it is significant to note that a number of Democratic spokesmen (including the editors of the Washington *Star,* Senator William Bigler of Pennsylvania, and the territorial secretary, Frederick Stanton) had all suggested, beginning as early as May, that the Lecompton convention ought to prepare a separate article on slavery for submission to the voters.[35] These proposals did not at the time elicit any protests from the antislavery party, which leads to the inference that the issue of total or partial submission had not yet come into focus. President Buchanan himself later insisted that it had not. He argued that the Kansas-Nebraska Act, the very embodiment of popular sovereignty, had not required that the voters should be given the opportunity to accept or reject a constitution as a whole, but only that they should be (quoting the act) "perfectly free to form and regulate their domestic institutions [a euphemism for slavery] in their own way." His own statements, said Buchanan, had been "in general terms," and while he had meant that the convention was "bound to submit this all-important question of slavery to the people," he had never meant that "they would have been bound to submit any portion of the Constitution to a popular vote in order to give it validity."[36]

This may well have been true as a statement of what Buchanan had intended, but it was, in fact, seriously misleading as to what he had actually said. For his written instructions to Walker had stated categorically that the people of Kansas "must be protected in the exercise of their right of voting for or against that instrument." In December Buchanan quoted his own language publicly in a message to Congress. Yet ten days later, in a letter accepting Walker's resignation, he denied that he had ever "entertained or expressed the opinion that the convention were bound to submit any portion of the constitution to the people, except the question of slavery."

35. Nichols, *Disruption,* pp. 105, 115, 123, citing the following: statement of Stanton: "I think the convention ought to prepare a separate article on the subject of slavery, either for or against it . . . [and] submit this to the people"; New York *Herald,* May 6, 1857; Bigler to Buchanan, July 9, 1857, in Black Manuscripts; and Washington *Star,* Sept. 1, 1857.

36. Buchanan, messages to Congress, Dec. 8, 1857, Feb. 2, 1858, in Richardson (ed.), *Messages and Papers,* V, 450, 477.

There was a curious blindness in his failure to recognize that his later statement was clearly inconsistent with his earlier one—such blindness that he did not even try to conceal the inconsistency.[37]

Quite possibly, Buchanan took too favorable a view of the "partial submission" clause, as adopted, because, instead of comparing it with what the antislavery men wanted, he was comparing it with what he himself had feared—namely that the Lecompton convention would refuse to submit any question whatever to the voters. The danger that the convention would concede nothing whatever made the administration supporters unduly grateful for its conceding anything at all. Thus "partial submission" could be viewed as a partial victory over the proslavery extremists, and John Calhoun apparently thought that he deserved the gratitude of his chief, Buchanan, and his patron, Douglas. Even Buchanan himself seems to have felt somewhat euphoric about the result. He persuaded himself that his pledge had been kept and that the Kansas crisis was about to be terminated. Let Kansas be admitted, free or slave, he thought, and the excitement which had "for some years occupied too much of the public attention" would "speedily pass away."[38]

This was a dangerous frame of mind for a man who now faced a critical and terribly difficult decision. Buchanan was indeed in a dilemma. If he refused to support Lecompton, he would be in the untenable position of rejecting the work of a convention whose legality he had stoutly defended. By such an act, he would alienate almost the entire southern contingent in the party, and this was no mere wing of the party but almost the party itself. One hundred and twelve of his 174 electoral votes had come from the South. In

37. Buchanan's instructions, conveyed by Cass to Walker, March 30, 1857, in KSHS *Transactions*, V, 322–323. That they were Buchanan's own words is indicated by his repetition of them on Dec. 8. Henry S. Foote later declared that the instructions were in Buchanan's own handwriting (*Casket of Reminiscences*, p. 114). Acceptance of Walker's resignation, Cass to Walker, Dec. 18, 1857, KSHS *Transactions*, V, 431. Buchanan's defense rested upon two distinctions: First, that in advocating submission to the voters, he had not specified acceptance or rejection of the whole constitution; this defense seems untenable in view of the language quoted above. Second, that while he had been ready to *advocate* total submission, he had never believed that he had the right to *force* it upon the convention or to reject the constitution because the convention did not adopt it. But why he lacked the right to force total submission if he possessed the right to force acceptance of partial submission, he did not say.

38. Nichols, *Disruption*, p. 126; Buchanan, message to Congress, Dec. 8, 1857, in Richardson (ed.), *Messages and Papers*, V, 453.

Congress, 75 out of 128 Democratic representatives and 25 out of 37 Democratic senators were southerners. Moreover, he depended on his southern associates, with whom he had always been on cordial terms. William R. King of Alabama and John Slidell of Louisiana had been his closest friends. Three slave-state senators—Slidell, Bayard of Delaware, and Benjamin of Louisiana—and one emigrant slaveholder who lived on the north bank of the Ohio, Bright of Indiana, had engineered his nomination. Four members of the cabinet were southerners, and one of them, Howell Cobb, was always urged to stay at the White House as company for the lonely bachelor president during Mrs. Cobb's frequent absences in Georgia. Buchanan could not bring himself to break with the southerners. He would not be able to do so in 1861, even when they sundered the Union of which he was president. Certainly he could not do so over Lecompton.[39]

Taking them one by one, each of the steps leading to the Lecompton constitution had a certain plausibility, and legality might be claimed for each.[40] But the final result was untenable. Two thousand voters in a territory with 24,000 eligible for the franchise had elected a body of delegates whom no one seriously regarded as representative of majority opinion in Kansas. These delegates, acting in the name of popular sovereignty, had offered the voters a "choice" which affirmed the inviolability of slavery no matter which option was taken. Then, for good measure, they had placed the control of the balloting not in the hands of Governor Walker, but under the control of officials who had countenanced, if not perpetrated, repeated frauds.[41]

39. For Buchanan's southern susceptibilities, see Nevins, *Emergence,* I, 64–66; for southern influence in his nomination and for his friendship with Cobb, see Nichols, *Disruption,* pp. 2–18, 80.

40. For a defense of Buchanan's insistence upon the importance of the strictly legal aspects, see Klein, *Buchanan,* p. 304.

41. The Lecompton convention's provisions for the election may have been more crucial than the question of partial or total submission. Discussion tended to center on the latter, because it could be discussed in terms of principles, while election procedure could not. But the convention took control of the election entirely out of the hands of Walker, who had emphasized his commitment to honest elections more than anything else, and placed it in the hands of Calhoun and the "land office ring," who had been implicated in a succession of gross electoral frauds. If "partial" submission denied free-staters a chance to vote against the servitude of slaves already resident, the electoral provisions threatened them with the possibility that votes against any aspect of slavery would somehow be counted out. With these provisions, neither "total" nor "partial" submission of the slavery question could really have been acceptable to the free-staters.

Having permitted the dilemma to develop, Buchanan now had to choose one of its horns. Some historians have blamed him for making the wrong choice,[42] but in fact, either one would have led to disaster. His administration failed when he held back the support that Walker needed to control the forces at work in Kansas. There is no evidence, however, that Buchanan perceived this, for he showed no signs of the indecisiveness which sometimes characterized him when facing less difficult decisions. Instead he moved promptly and firmly to the support of the Lecompton formula.

Events marched swiftly in the seven weeks after the Lecompton convention adjourned on November 7. On November 18, only a day or two after the news reached the East, the Washington *Union* carried an editorial which Buchanan personally had approved, supporting the Lecompton arrangements.[43] On November 26, Walker, fresh from Kansas, had an interview with Buchanan and the cabinet, in which the clash of opinions was complete. Walker insisted that Lecompton did not fulfill the promise of popular sovereignty, that it cheated the people of Kansas, and that an attempt to ram it through would cause bloodshed. Cobb and Black retorted that it offered a chance to vote on the only real issue—namely, slavery—and that if lawless free-staters revolted against a perfectly legal procedure, they ought to be suppressed.[44] By December 2, Buchanan had completed a statement on Kansas to be included in his message to Congress on December 8. He had written it without consulting Douglas, and he sent off an advance copy to Frederick Stanton, in Walker's absence now the acting governor of Kansas, wanting it "as extensively published as possible throughout the territory, before the election of the 21st" of December. The message vigorously endorsed the action of the Lecompton convention; it affirmed that the question of slavery had been "fairly and explicitly referred to the people," that the exciting question could now be peacefully settled "in the very mode required by the organic law," and that if any Kansans should refuse this fair opportunity to vote, they alone would be "responsible for the consequences."[45]

42. E.g., Nevins, *Emergence,* I, 239–247.
43. Washington *Union,* Nov. 18, 1857; Buchanan to Black, Nov. 18, 1857, as quoted by Shenton, *Walker,* p. 174, with discussion of the original text and alterations in this editorial.
44. New York *Herald,* Nov. 28, 29, 1857; Nevins, *Emergence,* I, 242.
45. Cass to Stanton, Dec. 2, 1857, in KSHS *Transactions,* V, 413; Buchanan's message, Dec. 8, 1857, in Richardson (ed.), *Messages and Papers,* V, 449–454.

On December 3, Stephen A. Douglas appeared at the White House, and he and Buchanan had an angry conversation that ended, according to Douglas, in a classic riposte. Douglas urged Buchanan not to support Lecompton and threatened to oppose him if he did. This provoked Buchanan to warn, "Mr. Douglas, I desire you to remember that no Democrat ever yet differed from an administration of his own choice without being crushed. . . . Beware the fate of Tallmadge and Rives." Buchanan was alluding to two politicos who had supposedly made the fatal mistake of running afoul of Andrew Jackson. But Douglas retorted by giving the comparison an invidious turn. "Mr. President," he said, "I wish you to remember that General Jackson is dead."[46] By the time of this interview, Buchanan was fully prepared to go before Congress and stake the success of his administration on the Lecompton constitution.

Meanwhile, in Kansas, the acting governor, Frederick Stanton, had given the president what, in retrospect, may be regarded as his last chance to escape from the trap into which he was walking. Stanton, a proslavery Tennessean, had become disgusted with the frauds committed by the proslavery ring and feared another outburst of civil war. He knew that strife never lay far below the surface in Kansas. The free-staters constantly maintained their own "government" and their own armed forces. Walker's most urgent task had been to keep them at peace, and he had succeeded only by the vigor and persuasiveness with which he promised that all voters would have a chance to oppose the Lecompton constitution in an honest election. But now Walker had gone to Washington, and Kansans knew that governors who left did not come back. They were on the eve of an election conducted by parties whose previous elections had never been honest, and in which they were denied the promised opportunity to vote against the work of the "bogus" convention. Their excitement seemed about to boil over. Stanton regarded the situation as acutely dangerous; he learned of "designs of a most desperate character," and he feared "violent measures." Accordingly, on December 1, he called the legislature to convene on December 7. This was the body with a free-state majority which had been elected in October after the McGee and Oxford frauds had been thrown out. He did not inform Buchanan of his action until nine days later, with the result that the news first reached the president by way of the newspapers. On

46. Douglas, speech at Milwaukee, Oct. 14, 1860, in Chicago *Times and Herald,* Oct. 17, 1860.

December 8, Stanton sent the newly assembled legislators a message about the impending "evils and dangers" and recommended that they adopt a law submitting the entire constitution to the voters for outright acceptance or rejection. He insisted that such a law would not conflict with the action of the convention calling for an election on December 21. The latter was legal but not binding upon Congress, and it would also be legal, he said, for the Kansas legislature to provide for a separate vote which would show Congress whether the people of Kansas wanted the Lecompton constitution. The legislature adopted the recommended measure, and Stanton signed it on December 17.[47]

Thus the choice between acceptance and rejection of the Lecompton constitution was submitted to the voters after all. Walker's promise was kept. If Buchanan had supported this referendum, he might still have fulfilled his inaugural pledge. Such a decision by Buchanan would certainly have caused an outcry from the South, but there is much evidence that southerners were more concerned with maintaining their abstract rights in Kansas than with making it a slave state, and Buchanan might have rallied substantial strength in the South if he had upheld the newly scheduled election as an arrangement which gave all parties a fair chance.[48] Certainly he would have found it no worse as a basis for bisectional support than his narrow, legalistic defense of Lecompton. But his course was already set. He sent to Congress on December 8 his message supporting the Lecompton plan, and on the same day, which was also the day of Stanton's message to the Kansas legislature, he issued an order removing Stanton from the acting governorship. But because of the time required to communicate with Kansas, Stanton remained in office just long enough to sign the bill he had recommended.[49]

47. KSHS *Transactions,* V, 413–419, 459.

48. The position of the South at this time is one of the most controversial points in connection with the Lecompton story. There were threats of secession, and in this case, as in the case of the Compromise of 1850, the election of 1856, etc., historians disagree as to what would have happened if the fire-eaters had been defied. Thus, Hamilton thinks Zachary Taylor could have held the Union together in 1850 without any compromise (above, p. 122), and most historians think Buchanan could have done so without yielding to the South on Lecompton in 1858 (e.g., Nevins, *Emergence,* I, 302). South Carolina, Georgia, Alabama, and Mississippi were threatening secession; these threats were enough to impress Buchanan and, Nevins believes, to terrify him. But Craven, *Growth,* pp. 289–295, argues strongly that southern opinion was not deeply aroused and, by implication, that the danger of secession was small.

49. Buchanan issued the order for Stanton's removal on Dec. 10 and appointed

The rapid pace of events continued. The day after the president's message, Douglas rose in the Senate and assailed Buchanan's policy in a long, impassioned speech, which some observers considered the finest effort that he had ever made in Congress. On December 15, Walker sent the president a long letter defending his own course, reproaching Buchanan for abandoning his original ground, and offering a resignation which was curtly accepted.[50] On December 21, there was voting in Kansas, in the election called by the Lecompton convention. The free-staters abstained, and the official returns showed 6,226 votes for the Lecompton constitution with slavery and 569 for it without slavery. Charges of fraud were promptly made and later discovered to be justified in many cases.[51]

On January 4 there was more voting in Kansas, this time in the election called by the legislature. Now it was the turn of the proslavery men to abstain, and the new governor reported 10,226 votes against Lecompton, 138 for it with slavery, and 24 for it without slavery.[52] Walker had estimated 24,000 eligible voters, and the combined totals of these two polls seemed to confirm the relative accuracy of his estimate. The results indicated that the majority of the people of Kansas were opposed to Lecompton.

Buchanan nevertheless went grimly ahead. As he did so, he brought all the resources of forty years of political experience into his great effort. He had, of course, canvassed the Congress carefully, and he knew that he was sure of the Senate. There the Democrats had a majority of 14, and at most only three northern Democrats would follow Douglas into opposition. The House would be closer.

James W. Denver to succeed him. Stanton signed the bill on Dec. 17; on Dec. 19, Denver became acting governor, and he took the oath on Dec. 21. KSHS *Transactions*, V, 457, 459, 465.

50. *Congressional Globe*, 35 Cong., 1 sess., pp. 14–18; Gerald M. Capers, *Stephen A. Douglas, Defender of the Union* (Boston, 1959), p. 165; KSHS *Transactions*, V, 421–431. Most historians have treated Walker's resignation as a courageous act of conscience, but copious evidence shows that he was both an ambitious and a corrupt man. Nichols, *Disruption*, p. 154, says, "In spite of much historical writing to the contrary, Walker seems to have been getting out from under an impossible situation, taking refuge in the usually profitable role of martyr."

51. Letter of John Calhoun, certifying election returns, in *House Reports*, 35 Cong., 1 sess., No. 377 (Serial 966), pp. 93–94. Election frauds ran to the picturesque, as well as the hyperbolic, in territorial Kansas. In this election, a set of ballots which had disappeared when their validity was questioned were found in a candle box, buried under a woodpile, conveniently near the surveyor general's office. Leverett Wilson Spring, *Kansas, the Prelude to the War for the Union* (Boston, 1888), pp. 229–230.

52. *House Reports*, 35 Cong., 1 sess., No. 377 (Serial 966), p. 97.

There, 118 votes were needed, and he could be confident of only about 100. To get the additional 18 would require hard work, but the Democrats were in the majority, with 128 votes to the Republicans' 92 and the Americans' 14, and if many northern Democrats seemed reluctant, there were devices by which a few stray votes could be captured. The administration could crack the whip of party regularity, backed by patronage rewards for its supporters and by threats of dismissal for the friends of those who defected. There were government contracts, commissions of various kinds, and even cash. There were the social enticements of dinner parties, liquor, and feminine charm. None of these weapons would remain unused in the months to come. It would be an unpleasant business requiring steady nerves, but it would be worth a great deal if the everlasting Kansas question could be laid to rest.[53]

Buchanan's tactical reasoning was sound, and he was not unrealistic to think he might win. He conducted the fight shrewdly and ably. His basic mistake—part of the basic dilemma—was his failure to see how badly the northern wing of the Democracy would be damaged, even by a victory, and his failure to appreciate what a fearful handicap his northern followers would incur if they supported him on this issue.[54]

There had been running debate on the subject since the Congress convened, but the real struggle began on February 2 when Buchanan sent the Lecompton constitution to the two houses with a message urging its adoption, denouncing the free-staters in Kansas

53. One school of thought, of which Nevins is the most distinguished exponent, seems to doubt that Buchanan made any decisions himself and believes that he was controlled by a kind of "Directory" of southern advisers (*Emergence*, I, 239–240, 251–255). There can be no doubt that he did not hold a firm control over his cabinet, but the belief that others dominated him in the Lecompton situation seems to arise more from a general theory that he was a weak character and that he must have been swayed, since he was changing his position, than from specific evidence of his yielding to his advisers. Albert J. Beveridge, *Abraham Lincoln, 1809–1858* (4 vols.; Boston, 1928), IV, 169, says, "The President himself appears to have formed his policy; no evidence has been adduced to support the charge that he yielded to Southern influence."

54. Nevins, *Emergence*, I, 249 n., takes sharp exception to the "curiously myopic" view of Beveridge, *Lincoln*, IV, 172, that if only Douglas had not revolted, Lecompton "would have been adopted without much difficulty." The real question is not whether Douglas's support would have procured an easy adoption, but what such support would have done to his later career. In view of the close margin by which he won reelection to the Senate in 1858, even after opposing Lecompton, it appears that he must surely have been defeated if he had supported it.

for maintaining an illegal government, and foolishly asserting, "Kansas is . . . at this moment as much a slave state as Georgia or South Carolina." This message inaugurated a titanic contest, full of tension and drama. For weeks, the attention of the country was focused constantly upon this one issue. Both sides made heroic efforts, throwing all their resources of oratorical talent, parliamentary skill, and political acumen into the struggle. Both sides kept up a tremendous pressure on their adherents. Filibusters, late sessions, and fights on the floor marked the progress of the battle. The closeness and uncertainty of the divisions added a great deal to the excitement. For instance, on a critical vote to send the bill to conference, the Speaker broke a tie of 108 to 108. For weeks in the House, every vote was close enough to make the result uncertain until near the end of the roll call.[55]

In many respects, this was 1854 all over again. Once again a newly elected president, with all the influence a new president commands, had been induced, because of his southern sympathies, to support a bill that was highly objectionable to the northern members of his own party. Once again, a party revolt followed, leading once again to a pitched political battle, famous in the annals of party warfare. Once again, the administration prevailed first in the Senate, but faced a longer and harder fight in the House. There, Alexander H. Stephens of Georgia was again the floor manager, and Buchanan counted on him to gather, somehow, the votes that were lacking, just as he had brilliantly overcome a deficit of twenty-one votes in 1854.[56] Once more, the costly battle did great damage to the majority party and virtually ruined the administration that initiated it.

Along with these similarities, there were two important differences. First, Stephen A. Douglas, previously the Senate floor leader for the administration, was now the floor leader for the opposition. The same tireless energy and the same matchless readiness and resourcefulness in debate which had carried Kansas-Nebraska to victory were now devoted to the defeat of Lecompton. Whereas Buchanan could not face the revolt of southerners if he opposed

55. Richardson (ed.), *Messages and Papers*, V, 471–481; *Congressional Globe*, 35 Cong., 1 sess., *passim;* excellent narratives in Nevins, *Emergence*, I, 256–301, which is a little fuller on the national picture, and Nichols, *Disruption*, pp. 150–176, which follows the intricate parliamentary contest more fully.

56. *Ibid.*, p. 161.

Lecompton, Douglas could not face the hostile response of Illinois and of the North generally if he supported it. Hence Congress presented a new spectacle. Day after day, Douglas voted on the same side with Chase and Wade and the men who had treated him in 1854 as if he were the Antichrist. Stranger political bedfellows no one had ever seen, but for a season it was seriously believed that Douglas might become a Republican. Some of the eastern Republicans, especially, took up the idea of supporting him and bringing him into the party. Henry Wilson believed Douglas would join the Republicans, and praised him as being "of more weight to our cause than any other ten men in the country." Horace Greeley, for all his professions of idealism, now declared: "The Republican standard is too high; we want something practical." His idea of practicality was to throw Republican support behind Douglas in the pending Illinois election. He called on Douglas in Washington, and his *Tribune* began to praise Douglas extravagantly. To the end of his life, he believed that it would have been sound Republican strategy to support Douglas. In Massachusetts, Nathaniel P. Banks urged the Illinois Republicans to "sustain" Douglas. In Washington, as early as December 14, Douglas talked with Anson Burlingame and Schuyler Colfax about forming a great new party to oppose the southern disunionists.

Some of the more seasoned figures in the party like Seward and Lyman Trumbull recognized that their relation with Douglas was an alliance and not a union. He opposed Lecompton because it violated popular sovereignty; they opposed it because it permitted slavery. They were prepared to act in good faith as temporary allies —and nothing more. But the movement for Republican support of Douglas gained enough momentum to worry Abraham Lincoln, who needed solid Republican support if he were to challenge Douglas successfully in the Illinois senatorial contest that autumn. Lincoln wrote anxiously to Trumbull, asking: "What does the New York Tribune mean by its constant eulogizing and admiring and magnifying Douglas? Does it, in this, speak the sentiments of the Republicans at Washington? Have they concluded that the Republican cause, generally, can be best promoted by sacrificing us here in Illinois? If so, we would like to know it soon; it will save us a great deal of labor to surrender at once." Also, Lincoln's partner, William H. Herndon, made a trip east to see Greeley, Banks, and Douglas himself. In the long range, of course, nothing came of this temporary coalition, but it is symptomatic of the degree to which the northern Democrats had been alienated from their fellow party members by the disruptive force of the Lecompton contest.[57]

The second great difference between the two political crises is that in 1854 the administration victory was an inordinately costly one, but in 1858 the administration, despite its immense effort, did not win at all. During February and March, the northern public seemed to turn increasingly against Lecompton. Newspapers in every northern state denounced it; the legislatures of New Jersey, Rhode Island, and Michigan passed resolutions against it; the New York Assembly invited the recently dismissed Stanton to deliver an address; and the Ohio legislature instructed Senator George Pugh to vote against admission. The governor of Pennsylvania publicly expressed the opinion that the people of Kansas ought to have an opportunity to reject the constitution. The resolutions of mass meetings, the proceedings of party conventions, and the vote in local elections all indicated that revolt was sweeping the North.[58] Against this tide, and against the power of Douglas in combat, the administration held steady, maintaining an unremitting pressure in the Senate until Lecompton was adopted on March 23 by a vote of 33 to 25. But everyone knew that the crucial action would take place on the other side of the Capitol.

The House of Representatives now witnessed one of the fiercest struggles in its history. Here the northern Democrats were far more responsive to Douglas's leadership than they had been in the Senate, and a bloc of between nineteen and twenty-four anti-Lecompton Democrats banded together against the administration. In the early test votes on parliamentary questions, they won, but by such narrow margins that they feared ultimate defeat, and on March 29 they offered to vote for Lecompton if the administration would insert a provision that the people of Kansas might alter their constitution at any time instead of waiting, as the Lecompton instrument required, until 1864. At this point, the administration forces might have had the admission of Kansas under the Lecompton constitu-

57. Don E. Fehrenbacher, *Prelude to Greatness: Lincoln in the 1850's* (Stanford, 1962), pp. 59–61, 78; Beveridge, *Lincoln,* IV, 183–189; Milton, *Eve of Conflict,* pp. 280–285; Nevins, *Emergence,* I, 261–264; Roy P. Basler (ed.), *The Collected Works of Abraham Lincoln* (8 vols.; New Brunswick, N.J., 1953), II, 430; David Donald, *Lincoln's Herndon* (New York, 1948), pp. 112–116.

58. Nevins, *Emergence,* I, 270–275.

tion, slave clause and all, and it is difficult to understand why they rejected this dazzling opportunity, which gave them so nearly the substance of what they were fighting for.[59] But adding one more to a sequence of errors, they refused, and, as they did so, plied whip and spur to drive Lecompton "naked," as Buchanan expressed it, "through the House." But as indications of resentment against Lecompton in the northern states multiplied, they found their task more difficult. Moreover, a competing measure appeared. John J. Crittenden of Kentucky had offered in the Senate a bill to resubmit the Lecompton constitution to the people of Kansas in a carefully controlled vote. This effort had failed, but on April 1, with administration leaders hoping to squeeze the Lecompton bill through and dispose of the Kansas question forever, the House, by a margin of 120 to 112, voted to put the Crittenden-Montgomery resolution (as it was now called) in the place of Lecompton, and then carried the substitute measure to final passage by the same vote.[60]

Senate and House were now deadlocked, and the sole remaining hope of the party regulars was a conference committee where some sort of adjustment might be worked out to save the face of the administration. But now, the aroused opposition, stimulated by their victory in the House, refused even to agree upon a conference with the Senate, and when the issue was put to a roll call the administration won only by using the Speaker's vote to break a tie.[61]

The administration arranged to have William H. English of Indiana appointed as one of the three conferees for the House. English was the representative of an anti-Lecompton district but a friend of the administration. Therefore he sincerely wanted an amicable settlement, and members of the administration party, including Stephens, made some pregnant suggestions to him. Thus was born a scheme that would send the Lecompton constitution back to the voters of Kansas, but would avoid overt acceptance of the principle of resubmission. This arrangement turned upon the fact that the Lecompton constitution had been accompanied by an extraordinary request for more than 23 million acres of public land, about six times the normal grant to new states. The Crittenden substitute

59. New York *Tribune*, March 27, 29, 30, 1858; Nevins, *Emergence*, I, 292, has pointed out the significance of this long-overlooked episode. As drafted by the convention, the Lecompton constitution could not be amended for seven years.

60. *Congressional Globe*, 35 Cong., 1 sess., pp. 1435–1438.

61. *Ibid.*, pp. 1589–1590.

would have cut it to approximately 4 million acres. Why not then reduce the land grant and submit to the voters of Kansas the question of whether they would accept or reject the constitution with the reduction? To the South, this offered several advantages. It avoided the "principle" of straight-out resubmission which southerners resisted so bitterly, and it loaded the alternatives somewhat in favor of slavery by offering the Kansans statehood if they would accept slavery and denying it to them if they would not. Also, it assured the southerners that while they might possibly gain a slave state they ran no immediate risk of admitting a free state, for if Kansas rejected this proposal, the English measure stipulated that she could not apply again for statehood until the census showed a population of 90,000. To the North, the great inducement was simply that this would at last give the voters of Kansas a federally sanctioned chance to vote against the Lecompton constitution.[62]

The conference committee agreed on the English bill, as it was soon called, and apparently hoped that all parties would accept it. But some of the radical southerners were at first opposed, and the southern opposition might have been more general if Douglas had agreed to it. In fact, Douglas almost did so, but at the last moment some of his more militant supporters persuaded him to remain in opposition. By this time, many southerners had reached a point where they believed that anything which Douglas opposed must be all right, and they rallied to the new and somewhat less inspiring banner which the administration now raised. With their support, the English bill passed the Senate by a vote of 31 to 22 on April 30 and the House by 112 to 103 on the same day. Buchanan signed it and it became law.[63]

62. A minor historical fallacy which has long confused the history of the English bill is the assertion that the bill offered an unusually large land grant, and that this grant, in effect, constituted a bribe to the people of Kansas to accept the Lecompton constitution. Henry Wilson, *History of the Rise and Fall of the Slave Power in America* (3 vols.; Boston, 1872–77), II, 558–559, first incorporated this partisan accusation in a historical context, and it was later repeated by Hermann von Holst and even by James Ford Rhodes. In 1906, Frank H. Hodder, "Some Aspects of the English Bill for the Admission of Kansas," AHA *Annual Report,* 1906, I, 199–210, demonstrated clearly that the land grant offered under the English bill was, as stated above, only about one-sixth of the grant asked in the original Lecompton constitution, and that it was computed on exactly the same basis as other land grants for other states in this period. The old error is still repeated occasionally despite Hodder's disproof.

63. For the vacillation of Douglas, see Nichols, *Disruption,* pp. 173–174, with

On August 2 the voters of Kansas followed the not unfamiliar practice of going to the polls, and for the third time in less than eight months they voted on the Lecompton constitution, this time in the guise of a plebiscite on a land grant. They turned it down cold, by a vote of 11,300 to 1,788.[64] Kansas was to remain a territory until 1861.

This ended a political battle which had convulsed the country and virtually destroyed two administrations, but the full consequences of the prolonged struggle had yet to become evident. Not until the Civil War which the Kansas issue had done so much to precipitate, not until William Quantrill had led his savage raids along the border, not until the James boys—Jesse and Frank—had run their course of crime, did the nation know the final toll which it was to pay for Bleeding Kansas.

In July, Buchanan wrote Representative English a bland letter, thanking him for his measure as if it had been a victory for the administration.[65] Indeed, the opposition of Douglas and the Republicans lent a semblance of plausibility to this polite fiction. But in fact, the administration and the South had sustained a smashing defeat, and all that they had been spared was the admission of Kansas as a free state.

For ten years, the Union had witnessed a constant succession of crises; always these ended in some kind of "victory" for the South, each of which left the South with an empty prize and left the Union in a weaker condition than before. In 1850 the South had paid a dear price for the Fugitive Slave Act; in 1853 it squandered some of its influence to procure the Ostend Manifesto; in 1854 it sacrificed the bisectional ascendancy of the Democratic party for the sake of Kansas-Nebraska; in 1857 it prepared to pay whatever the cost might be for upholding the Dred Scott decision. In 1858 it sacrificed what was left of the northern Democracy in a vain attempt to force the adoption of the Lecompton constitution. Such were the trophies of victory. Not one of them added anything to the area, the strength, the influence, or even the security of the southern system. Yet each had cost the South a high price, both in alienating the public opinion of

---

extensive citations; for final debates and enactment, *Congressional Globe*, 35 Cong., 1 sess., pp. 1880–1906.

64. Governor J. W. Denver to Secretary Cass, Aug. 24, 1858, in KSHS *Transactions*, V, 540.

65. Buchanan to English, July 2, 1858, cited in Nevins, *Emergence*, I, 301.

the nation and in weakening that one great bulwark of bisectionalism, the Democratic party, which alone stood between the South and sectional domination by the Republicans. When Pierce came to office, there had been 92 free-state Democrats in the House and 67 slave-state Democrats. Kansas-Nebraska, along with the Know-Nothing sweep, had cost the Democrats 70 of those northern seats in 1854. They made some recovery in 1856, so that when Buchanan came to office, there were 53 free-state Democrats and 75 slave-state Democrats. But now the 53 were to face another election ordeal similar to the one after Kansas-Nebraska. When the survivors were counted, only 32 free-state seats were left, and 12 of these were held by men who had saved themselves by repudiating the administration's Lecompton policy. The Democratic party in the House, when Congress met in 1859, consisted of 69 southerners, 19 party regulars from the free states, and 12 anti-Lecompton Democrats.[66] Obviously, the sectional balance within the party had been demolished, and the concentration of strength in the South had led to the adoption of proslavery policies which further reinforced the sectional concentration of strength, in a vicious circle. In fact, the sectional maldistribution of strength in the Democratic party in Congress had reached such a point that only in the national party convention, where every state had representation, could the northern Democrats exercise any power. This convention met, of course, only once every four years, and the northern supporters of Douglas would not have a chance to assert themselves again until 1860. When this time came, the Democratic party proved too weak to stand the strain, and the final crisis of the Union followed.

But before the northern Democrats could challenge the southern leadership of their party in 1860, they first had to fight the battles for political survival in their own constituencies. By the time that Kansas voted under the English measure in August 1858, such battles were already in progress. Many of these contests were dramatic and significant. But in the one that overshadowed all the others, the great adversary of Lecompton, the Little Giant of Illinois, fought to save his Senate seat, his career, and his party. Douglas faced a truism of American politics, namely that a man cannot be a national leader unless he has demonstrated his ascendancy in

66. Compiled from lists in *Congressional Globe,* 35 Cong., 1 sess., pp. 1–2, and 36 Cong., 1 sess., pp. 1–2.

his own locality. Only if he retained his strength in Illinois could he continue to be a potent figure in the national Democracy. But Douglas's position in Illinois was now challenged both by the administration Democrats who hated him for daring to defy Buchanan and by the Republicans who wanted free soil and not popular sovereignty. The Republican contender against him, although a relatively unknown Springfield lawyer, was politically formidable enough to threaten the famous senator very seriously, and formidable enough intellectually to make the test of strength between them an occasion for the classic exposition of the ideas that lay behind all of this protracted sectional combat.

CHAPTER 13

Lincoln, Douglas, and the Implications of Slavery

BETWEEN 1850 and 1858, Stephen A. Douglas had occupied a crucial command post in three titanic conflicts in Congress. First, in 1850, he had led the forces of conciliation against proslavery and antislavery militants who resisted Henry Clay's proposals for compromise. Then, in 1854, he had led the southern Democrats in applying popular sovereignty in Kansas, despite the opposition of many northern Democrats, who wanted to keep the slavery exclusion adopted in 1820. Then, finally, in 1858, he had led the northern Democrats and many Republicans against the southern Democrats in a fight to show that popular sovereignty could mean slavery exclusion as well as slavery extension.[1]

Douglas's position in the summer of 1858, at the end of this third contest, was a singular one. The administration Democrats and the southern junta, who, in 1854, had regarded him as their champion, now looked upon him as a traitor trying to deny them the fruits of electoral victory in Kansas. Just as militant antislavery men in 1854 had seen popular sovereignty as a trick to promote the hidden purposes of the slavery extensionists, so the militant proslavery men in 1858 saw it as a trick to promote the hidden purposes of the slavery exclusionists. They hated Douglas worse than they hated the Black Republicans. Soon, they would seek to defeat him in his bid for reelection to the Senate, and when the next session of Congress met, they would strip him of his prized chairmanship of the Committee on Territories.

1. See above, pp. 108–113, 151–165, 316–326.

The antislavery men, on the other hand, saw him in a new light as a crusading opponent of an imposed proslavery constitution for Kansas—as a tireless fighter battling against heavy odds to block aggression by the doughfaces and the slavocrats of his own party. William H. Seward, the Republican leader in Congress, had been working with him closely, Horace Greeley had praised him; and Republicans in Congress had looked to him for leadership against the Lecompton forces.[2] Four years after receiving his unique excoriation in the Appeal of the Independent Democrats, Douglas seemed in strong position to assume command of the antislavery hosts.

In some respects, this potentiality appears realistic, even in historical perspective. There is evidence that Douglas was, in his private and personal thinking, an antislavery man. To say this is not to say that he throbbed with aspirations for the betterment of the Negro's lot. On the contrary, he seemed to go out of his way to express a certain callous scorn for the blacks, as people with whom he did not recognize any affinity. His thinly veiled disdain for slavery seemed primarily to reflect a sense that it was a rather shabby, unattractive institution, unworthy of a society as progressive as that of the United States.[3] Thus, without much concern for the slaves, and without believing that the slavery issue was worth a political crisis, Douglas did regard the restriction of slavery as desirable, and thought of popular sovereignty as an effective device by which to restrict it without precipitating a constant running battle in Congress.

For a man with sentiments such as these, the events of 1857 and 1858 might have been enough to cause a basic recasting of his views. Not only had the slaveholders overreached themselves and tried to pervert popular sovereignty by imposing a proslavery constitution upon the incipient state of Kansas, but Chief Justice Roger B. Taney, in the Dred Scott decision, had moved to destroy popular

2. See above, pp. 321–322.

3. Douglas purposely abstained from expressing a public opinion on slavery because, as he said, "I hold that under the Constitution of the United States, each state of this Union has a right to do as it pleases on the subject of slavery. In Illinois we have exercised that sovereign right by prohibiting slavery. . . . I approve of that line of policy. . . . We have gone as far as we have a right to go under the Constitution. . . . It is none of our business whether slavery exists in Missouri. . . . Hence I do not choose to occupy the time allotted to me in discussing a question that we have no right to act upon. . . ." Speech at Quincy, Oct. 13, 1858, in Roy P. Basler (ed.), *The Collected Works of Abraham Lincoln* (8 vols., New Brunswick, N.J., 1953), III, 266–267.

sovereignty by denying that a territorial legislature could exclude slavery at all. Thus, Douglas found himself involved in a disruptive fight within his own party and forced to find a way to nullify the effect of Taney's opinion without seeming to deny the authority of the Court. Many a man, at such a point, might have decided to scuttle the popular sovereignty doctrine and to look for a vehicle by which to move into the antislavery camp. Especially so, if he faced, as Douglas did, the immediate necessity of gaining reelection to the Senate in the preponderantly antislavery constituency of Illinois, with both the administration Democrats and the antislavery Republicans assailing him.

But Douglas, for all his tactical opportunism, all his consorting with spoilsmen, all his scorn for moralists in politics, was deeply committed to certain attitudes which had become, with him, matters of principle. He believed that the integrity of the Union was more important than the solution of the slavery question, but that the two were not incompatible in any case, for the slavery question could be taken out of national politics and could be resolved at the level of local self-government. Further, he believed that local self-government was the truest form of democracy.[4] Hence, both democracy and Union would be saved by localizing the slavery question, and if the Supreme Court placed obstacles in the way of localizing it, he would have to find some way of by-passing the obstacles. In short, Douglas elected to stand by the essential idea of popular sovereignty. He had defended it against the slaveholders during the Lecompton contest; now he would defend it against the Republicans in the senatorial contest.

Douglas's position carried important implications both for the administration Democrats and for the Republicans. His bid for reelection on a basis that both the Supreme Court and the southern Democrats had rejected was a bid to seize control of the Democratic party in the North and, from that base, to challenge the southern

4. On Douglas's philosophical commitment to the principle of popular sovereignty, see Robert W. Johannsen, "The Kansas-Nebraska Act and Territorial Government in the United States," *Territorial Kansas* (Lawrence 1954), pp. 17–32; Johannsen, *Frontier Politics and the Sectional Conflict: The Pacific Northwest on the Eve of the Civil War* (Seattle, 1955), pp. 132–134; Johannsen, "Stephen A. Douglas, 'Harpers Magazine,' and Popular Sovereignty," *MVHR*, XLV (1959), 606–631; Johannsen, "Stephen A. Douglas, Popular Sovereignty, and the Territories," *Historian*, XXII (1960), 378–395. Damon Wells, *Stephen Douglas, the Last Years, 1857–1861* (Austin, Tex., 1971), pp. 55–80.

domination of the party which had been growing steadily ever since the defeat of Van Buren for the nomination in 1844. The Democratic regulars could not fail to see that the division within the party, which had begun with the Lecompton struggle, either would be terminated by the defeat of Douglas in Illinois in 1858 or would be extended to a climactic showdown in the national convention in 1860. Buchanan and the custodians of the party machine therefore set about to defeat Douglas, and they did not stop short of measures that would lend aid to the Illinois Republicans during the campaign.[5]

The Republicans, on the other hand, had to decide whether to oppose the senator, who, although a Democrat, had been an indispensable leader in the Lecompton contest. A good number of eastern Republicans believed that their party ought to help the Little Giant in his grim struggle against Buchanan. If the Illinois Republicans were going to run a candidate against him, it urgently behooved them to convince the eastern party leaders that Douglas's antislavery was basically different from Republican antislavery, and that they were not opposing him in a spirit of mere partisanship.

The Republicans, of course, did decide to contest the election. They formally nominated, as their candidate for the Senate, Abraham Lincoln, a former Whig from Springfield, who had served one term in Congress in 1845–1847 but who had never been prominent in national politics. The people who knew him realized that Lincoln was a resourceful person with a capacity for spare and muscular logic that came into full play during the campaign.

The result was one of the most important intellectual discussions of the slavery question that occurred during three decades of almost uninterrupted controversy. Much of the discussion by the abolitionists seems unrewarding today because it resolved itself into the

5. For the efforts of the administration Democrats to defeat Douglas in 1858, see George Fort Milton, *The Eve of Conflict: Stephen A. Douglas and the Needless War* (Boston, 1934), pp. 271–275, 279, 282, 284, 286, 294–304, 309, 326–328, 345–348, 351–352; Roy F. Nichols, *The Disruption of American Democracy* (New York, 1948), pp. 210–215; Philip Shriver Klein, *President James Buchanan* (University Park, Pa., 1962), pp. 328–329; Philip G. Auchampaugh, "The Buchanan-Douglas Feud," ISHS *Journal*, XXV (1932), 5–48; O. M. Dickerson, "Stephen A. Douglas and the Split in the Democratic Party," MVHA *Proceedings*, VII (1913–14), 196–211; Louis Martin Sears, "Slidell and Buchanan," *AHR*, XXVII (July, 1922), 712–724; Reinhard H. Luthin, "The Democratic Split During Buchanan's Administration," *Pennsylvania History*, XI (1944), 13–35.

denunciation of sin, and much of the discussion by politicians seems even more sterile because it did not deal with slavery directly, but focused on such legalistic points as the powers of a territorial legislature. But Lincoln and Douglas debated what America ought to do about slavery, and that is what gives especial historic significance to the contest between them. Yet it must be added that even in their case, there were many trivialities and repetitions, and though the debates have been famous ever since they were held, they have sometimes been famous for the wrong reasons. It is necessary to cut through a layer of folklore before one can get to the essence of the debates.

Lincoln and Douglas both campaigned very actively in 1858. Lincoln made sixty-three speeches; Douglas claimed to have made one hundred and thirty.[6] Further, both men took the stump in Ohio the following year, and though not meeting on the same platform, they responded to each other's arguments as if in debate, and they continued to contest some of the points first developed during the 1858 campaign.[7] Thus the occasions on which the two met in personal confrontation were only a part of a far more extensive operation. But on July 24, Lincoln proposed to Douglas by letter that they "divide time and address the same audiences during the present canvass." Douglas, the greater drawing card of the two, did not welcome the idea of providing audiences for his opponent, but agreed to one joint debate in each of the nine congressional districts, except the Springfield and Chicago districts in which both men had already spoken. Thus seven joint debates took place between August 21 and October 15.[8]

6. Don E. Fehrenbacher, *Prelude to Greatness: Lincoln in the 1850's* (Stanford, 1962), pp. 100–101. Basler (ed.), *Works of Lincoln,* II, 461 to III, 335, contains reports on 26 of Lincoln's speeches in this campaign (in addition to the joint debates), and five sets of notes which Lincoln made for such speeches. Most reports are only a page or two in length.

7. Harry V. Jaffa and Robert W. Johannsen (eds.), *In the Name of the People: Speeches and Writings of Lincoln and Douglas in the Ohio Campaign of 1859* (Columbus, 1959).

8. The joint debates were reported with a high standard of completeness and accuracy in the Chicago *Press and Tribune* (Republican) and the Chicago *Times* (Democratic). Lincoln kept a complete file of both, and from this file Follett, Foster and Company of Columbus, Ohio, published the first edition of the debates in book form in 1860. See Jay Monaghan, " 'The Lincoln-Douglas Debates': The Follett, Foster Edition of a Great Political Document," *Lincoln Herald,* XLV (June 1948), 2–11. Of modern, scholarly editions, the first is Edwin Erle Sparks (ed.), *The Lincoln-Douglas Debates of 1858* (Springfield, Ill., 1908), containing extensive press commentary as well as text of the debates. There have been three subsequent editions: Basler (ed.),

These encounters made the Illinois senatorial election of 1858 perhaps the most famous local political contest in American history. Everyone knows, or is supposed to know, that Lincoln and Douglas campaigned vigorously across the Illinois prairies, that they struggled mightily in dramatic tests of strength with one another, and that Douglas won the immediate stakes, the Senate seat, but that Lincoln gained a position from which he was able to defeat Douglas for the greater prize—the presidency—only two years later. In many respects the popular image of these debates is a true one. They were, to begin with, almost a perfect exemplification of nineteenth-century American democratic practice at its best. Over the dusty roads of rural Illinois, farmers drove their teams into the country towns—to Ottawa, to Freeport, to Jonesboro, to Charleston, to Galesburg, to Quincy, to Alton—so that they might hear the candidates speak. Amid the heat of the late August and September days, these plain folk, men of limited education, listened for two and a half to three hours to the arguments and the rebuttals by the two candidates. A note of festivity prevailed, and the occasions have been compared to the day of the Big Game in any American college town.[9] Band music stirred the sultry air, and the candidates enlivened the occasion with jokes and with animated parry-and-thrust. In these face-to-face encounters, the rivals sometimes assailed each other with the blunt combativeness of men who believed in their cause and were not afraid of a fight, but always in the American fashion of being able to shake hands after they had traded blows. This was what laymen have called good sportsmanship and what scholars have called consensus, and what it meant at bottom was that the values which united them as Americans were more important than those which divided them as candidates, or if not that, at least that the right to fight for one's ideas involved an obligation to fight fair and to recognize a democratic bond with other fighters for other ideas. Lincoln and Douglas both spoke with force in a direct,

---

Works of Lincoln (1953), III, 1–325, with a limited number of textual notes and notes of identification; Paul M. Angle (ed.), *Created Equal? The Complete Lincoln-Douglas Debates of 1858* (Chicago, 1958), with a 25-page introduction, headnotes for each debate, press reports, etc; Robert W. Johannsen (ed.), *The Lincoln-Douglas Debates of 1858* (New York, 1965), with 13-page introduction and notes. Among the many secondary accounts, the most thorough is Richard Allen Heckman, *Lincoln vs. Douglas: The Great Debates Campaign* (Washington, 1967).

9. Harry V. Jaffa, *Crisis of the House Divided: An Interpretation of the Issues in the Lincoln-Douglas Debates* (New York, 1959), p. 432.

unpretentious style, but though they sometimes salted their discourse with homely anecdote and rustic wit, they did not condescend to their rural auditors. Indeed, the debates examined some profound questions of democracy with intellectual rigor.

All in all, it is no wonder that twentieth-century Americans, who sit in armchairs in air-conditioned rooms and watch television screens as candidates deal with life-or-death issues in a few bland minutes, should look back with nostalgia upon the Lincoln-Douglas debates. Nor is it a wonder that the civic competence of the relatively unschooled Illinoisans of the mid-nineteenth century should seem impressive to a later generation in which prolonged schooling, ignorance, and political apathy have often gone together. Nor that the personal tolerance with which Lincoln and Douglas could agree to disagree should prove attractive in a time when "tolerance" is often equated with indifference and when agreement on underlying values is hard to find. There are many reasons why the Lincoln-Douglas contest has become a symbol of democracy at the grass-roots level, and it is natural that the contest should have become a set piece in the American memory and part of the national folklore.

When folklore appropriates a scene, however, it begins at once, unfortunately, to improve upon history by adding certain characteristic fictitious touches. First of all, it dramatizes an ordinary contest into an epic struggle between a virtuous and apparently defenseless hero on one hand and an evil, seemingly invincible adversary on the other. In this struggle, virtue invariably overcomes wickedness by some simple but supernaturally effective device—a silver bullet, a magic phrase, a sling for David against Goliath.

When the contest in 1858 for the United States Senate seat from Illinois was thus dramatized, the record underwent some astonishing modifications. Although Lincoln had been for twenty years a prosperous Springfield lawyer and a recognized leader among the Whigs, the legend converted him into a simple rail-splitter, a pioneer fresh from the forest. Although Douglas was cornered and fighting for his political life against the bosses of his own party, and though he came to the campaign straight from a battle to give the citizens of Kansas a chance to vote against slavery, he was necessarily cast as the villain, the instrument of the slave power, armed with all the unfair advantage that fame, influence, and financial resources could give. More than this, a Senate seat is not a great enough prize for folklore to celebrate, and soon a new version of the contest told how Lincoln perceived a strategy by which he would lose in the short run but would win the presidency beyond. Against the advice, of course, of his worldly wise counselors, our innocent but preternaturally far-seeing hero decided to play for bigger game. Thus he lost the Senate seat, but by that native cleverness which always compensates folk heroes for their lack of sophistication, he gained the White House. Finally, he slew his Little Giant with one very small and harmless-looking weapon, whose magic only he had perceived. This was a simple but artfully contrived question which the Giant could not answer without destroying himself.

For many decades, the deposit of folklore lay heavy upon the Lincoln-Douglas debates, and when it was at last removed, skeptics overreacted by saying that there was little underneath—that the debates were of negligible significance. In fact, both the traditionalists and the skeptics were wrong, but to evaluate the story of the debates, one must begin by examining the version which finds the heart of the whole campaign in the question Lincoln asked at Freeport on August 27: "Can the people of a United States Territory, in any lawful way, against the wish of any citizen of the United States, exclude slavery from its limits prior to the formation of a State Constitution?"[10]

The Freeport question was, of course, posed against the background of Taney's Dred Scott opinion, which had asserted that the territories could not exclude slavery. It presented a dilemma: if Douglas answered with an unqualified affirmative, he would reaffirm the doctrine of popular sovereignty and would repudiate the Scott decision, which would cost him support among the southern Democrats and impair his chances for the presidency in 1860. But if he replied with an unqualified negative, it would mean accepting the Scott decision and abandoning his own doctrine of popular sovereignty; this would cost him support among northern Democrats and probably prevent his reelection to the Senate.

The dilemma was a real one, but instead of being content to emphasize the difficulty it made for Douglas, some historians also accepted a story, first recorded in 1860, that Lincoln's advisers had all counseled him not to ask the question, for fear Douglas would devise a response that would help him win the Senate race. But when they told him in chorus, "If you do [put the question], you can

10. Basler (ed.), *Works of Lincoln,* III, 43.

never be Senator," Lincoln responded, "Gentlemen, I am killing larger game. If Douglas answers, he can never be President, and the battle of 1860 is worth a hundred of this."[11]

The "larger game" anecdote is part of a tradition of Lincoln's readiness to sacrifice his own career for high principle. Thus, several months earlier, when he was similarly advised not to deliver his "House Divided" speech, he replied, according to William H. Herndon, "The time has come when these sentiments should be uttered, and if it is decreed that I should go down because of this speech, then let me go down linked to the truth—let me die in the advocacy of what is just and right."[12] There is, of course, some reason to doubt that any politician talking informally with his close advisers would indulge in such bathos, or that the man who later wrote the Gettysburg address would use such melodramatic rhetoric, or that Lincoln would have expected to improve his prospects for the presidency by losing the senatorial contest in his own state. But apart from plausibility, there is the more tangible question of evidence. Did any witness ever claim to have been present, or to have talked with anyone else who was present, when Lincoln announced his quest for "larger game"? The answer is that there was one belated "eyewitness," Joseph Medill, writing thirty-seven years later, and two others who claimed to have heard the story from eyewitnesses since deceased—Horace White (1892) from Charles H. Ray and William H. Herndon (1890) from Norman B. Judd. But it has been demonstrated that Ray could not have been present at the conference, and that both Judd and Medill, contrary to the later recollections, encouraged Lincoln to use the Freeport question.[13]

The legend of Lincoln proving his superhuman prescience by seeing ahead to 1860 has obscured the real problem of the importance of the Freeport interrogatory. In fact, if Freeport had been the first place at which it was asked or answered, the question might better deserve the prominence historians have given it. But the record is clear that the question had already been asked, and Douglas had already answered it. In fact, he may have answered it before

11. John Locke Scripps, *Life of Abraham Lincoln* (Chicago, 1860), p. 28. Later copied (as shown by Fehrenbacher, *Prelude*, p. 187, n. 4) by a long succession of Lincoln biographers, including Nicolay and Hay, who, however, spoke of it as "a tradition." Albert J. Beveridge, *Abraham Lincoln* (4 vols.; Boston, 1928), IV, 294, doubted the story.

12. Paul M. Angle (ed.), *Herndon's Life of Lincoln* (New York, 1930), p. 326.

13. Fehrenbacher, *Prelude*, pp. 123–124.

it was asked, for on June 12, 1857, in a speech at Springfield, Douglas had found a way to avoid the horns of the dilemma. The Dred Scott decision, he said, guaranteed the right of a master to take a slave into a territory. But this right remained "barren and worthless . . . unless sustained, protected and enforced by appropriate police regulations and local legislation. . . . These regulations . . . must necessarily depend entirely upon the will and wishes of the people of the territory, as they can only be prescribed by the local legislatures."[14] In short, slavery could not exist without positive legislation to support it, and the people of a territory could effectively exclude slavery by refraining from the adoption of positive legislation.

Probably Douglas would have had occasion to enlarge upon this view in the months that followed, but the rancor of the Lecompton contest tended to eclipse other questions. As soon as the Illinois campaign of 1858 opened, however, Lincoln took care to raise this point again, and at Chicago on July 10, he put a query which involved the essence of the Freeport question: "Can you get anybody to tell you now that the people of a territory have any authority to govern themselves in regard to this mooted question of Slavery, before they form a State Constitution?" Perhaps Douglas would have taken up the question even if Lincoln had not asked it, but in any case, he was quick to respond. Twice within the following week, first at Bloomington and then at Springfield, with Lincoln in the audience on both occasions, Douglas again offered his formula: "Slavery cannot exist a day in the midst of an unfriendly people with unfriendly laws."[15] Lincoln himself already realized that Douglas had worked out a response: "He will instantly take ground," predicted Lincoln, "that slavery cannot actually exist in the territories unless the people desire it and so give it protective territorial legislation." More important, Lincoln understood correctly that the Lecompton contest was what had destroyed Douglas's standing in the South and that he knew it: "[Douglas] cares nothing for the South; he knows he is already dead there."[16]

Lincoln had no major purpose in asking the Freeport question.

14. *New York Times*, June 23, 1857; Milton, *Eve of Conflict*, p. 260; Fehrenbacher, *Prelude*, p. 134.

15. Angle, *Created Equal?* pp. 28, 59; Fehrenbacher, *Prelude*, pp. 136–137; Milton, *Eve of Conflict*, p. 344.

16. Lincoln to Henry Asbury, July 31, 1858, in Basler (ed.), *Works of Lincoln*, II, 530–531.

He intended simply to keep a spotlight on the already established fact that Douglas could reconcile the Dred Scott decision with popular sovereignty only by the lame expedient of telling the South that it possessed constitutional rights which it could not enforce, and the North that it had constitutional obligations which it need not fulfill. But even after posing his question and receiving the expected answer, that "slavery cannot exist a day or an hour anywhere unless it is supported by local police regulations,"[17] Lincoln did not pursue this issue very vigorously in any of the five joint debates that followed.[18] It might be said, then, that the Freeport question was one of the great nonevents of American history, both in the literal sense that Lincoln was not the first to ask it, and Douglas had already answered it repeatedly, and also in the deeper sense that the question of popular sovereignty was not the major issue of the debates. In fact Lincoln wanted to move the focus away from the territorial question, because he knew that this was a point on which Douglas and the Republicans might arrive at the same answer for quite different reasons—Douglas supporting exclusion of slavery because he believed in the right of a local majority to decide the question, the Republicans supporting it because they considered slavery morally wrong. Lincoln was painfully aware that many Republicans, like Horace Greeley, felt willing to support Douglas on this expedient.

Hence Lincoln wanted to shift attention from the policy aspects of the question, where the positions of Douglas and the Republicans might converge, to the philosophical aspects, where he believed that their differences were conspicuous and fundamental. Lincoln had sought from the beginning of the campaign to hold the focus on these aspects. On the day of his nomination as Republican candidate for the Senate, he had sought, in his famous "House Divided" speech, to define the basic philosophical difference which he would

17. *Ibid.*, III, 51. In responding, Douglas said that when Lincoln asked the question, "he knew I had answered that question over and over again. He has heard me answer a hundred times from every stump in Illinois, that in my opinion the people of a territory can, by lawful means, exclude slavery from their limits prior to the formation of a State Constitution." Also, Douglas at Jonesboro, p. 143.

18. Lincoln's basic criticism of the so-called Freeport doctrine was that Douglas had taken refuge in the absurdity of saying that "a thing may be lawfully driven away from where it has a lawful right to be." Speech at Columbus, Ohio, Sept. 16, 1859, *ibid.*, p. 417. There were five joint debates with Douglas after Freeport. At Jonesboro, Lincoln commented at some length on the Freeport doctrine (*ibid.*, pp. 128–133); at Quincy, briefly (pp. 278–279); at Alton, briefly (pp. 316–318); and at Charleston and Galesburg, not at all.

seek to develop in the campaign speeches that followed. On the one hand were the opponents of slavery who wanted to arrest its further spread and "place it where the public mind shall rest in the belief that it is in course of ultimate extinction." On the other were the advocates of a "care not policy," who first, in 1854, had opened all the national territories to slavery and then, in 1857, had denied that Negroes could ever be citizens, had provided constitutional guarantees for slavery in the territories, and had paved the way, as he saw it, for a constitutional guarantee of slavery in the states. In the presence of such a division, said Lincoln, "Our cause must be intrusted to, and conducted by, its own undoubted friends"—meaning men who regarded slavery as wrong, and not merely men who opposed the Lecompton constitution only because it lacked ratification by a popular vote in Kansas.[19] As Lincoln later expressed it, the question was one of right and wrong: the framers of the Constitution, recognizing the wrong, had carefully avoided explicit, verbal recognition of slavery and had restricted it so that it might ultimately wither away. The Founding Fathers, by the exclusion of slavery from the Northwest and by their preliminary arrangements for the abolition of the African slave trade, had given a clear indication that they "intended and expected the ultimate extinction" of slavery.[20] Douglas and the Democrats, refusing to recognize the wrong, had provided constitutional sanctions for slavery and made possible its extension. In the last of the debates, Lincoln was still hammering this point: "The real issue in this controversy—the one pressing upon every mind—is the sentiment on the part of one class that looks upon the institution of slavery *as a wrong,* and of another class that *does not* look upon it as a wrong. . . . The Republican party . . . look upon it as being a moral, social and political wrong . . . and one of the methods of treating it as a wrong is to *make provision that it shall grow no larger.* . . . That is the real issue. That is the issue that will continue in this country when these poor tongues of Judge Douglas and myself shall be silent. It is the eternal struggle between these two principles—right and wrong—throughout the world."[21]

On the whole, Lincoln succeeded in making the debates an open

19. Text of "House Divided" speech and preliminary draft, *ibid.* II, 448–454, 461–469. For an analysis of its significance, see Fehrenbacher, *Prelude,* pp. 70–95. Also see Beveridge, *Lincoln,* IV, 181–225.

20. Lincoln at Chicago, July 10, 1858. Basler (ed.), *Works of Lincoln,* II, 492.

21. Lincoln at Alton, *ibid.,* III, 312–313, 315.

and direct examination of the place of slavery in American society. He and Douglas fell short in a number of ways, but they came closer than any two public men of their generation to confronting the need for a consideration of the slavery anomaly in its relation to American democratic thought.

Fundamentally, Douglas began with a conviction of the inferiority of the Negro, and he had a habit of stating it with brutal bluntness: "I do not doubt that he [Lincoln] . . . believes that the Almighty made the Negro equal to the white man. He thinks that the Negro is his brother. I do not think that the Negro is any kin of mine at all. . . . I believe that this government of ours was founded, and wisely founded, upon the white basis. It was made by white men, for the benefit of white men and their posterity, to be executed and managed by white men."[22] In the same speech: "I am utterly opposed to any political amalgamation or any other amalgamation on this continent."[23] And at another time: "The Negro is not a citizen, cannot be a citizen, and ought not to be a citizen."[24] This did not mean that Negroes should necessarily be slaves, for, "The Negro, as . . . an inferior race, ought to possess every right, every privilege, every immunity, which he can safely exercise, consistent with the safety of the society in which he lives. . . . Humanity requires and Christianity commands that you shall extend to every inferior being, and every dependent being all the privileges, immunities and advantages which can be granted to them consistent with the safety of society."[25]

But though Douglas spoke of rights, clearly he did not mean intrinsic rights, carrying their own claim to fulfillment. He thought, instead, of "rights" granted as a gift, at the discretion of the state, and he did not believe they ought to be very extensive. In Illinois, they included freedom, but not citizenship or the ballot. As for equality, that "they never should have, either political or social, or in any other respect whatever."[26]

Although Douglas became almost obsessively committed to the doctrine of popular sovereignty, the key to his thought lay not in his political theory but in his belief in the inferiority of Negroes and Indians. Since they were inferior, he thought, they must be subordi-

22. Douglas at Springfield, July 17, 1858, in Angle (ed.), *Created Equal?* pp. 62, 60.
23. *Ibid.*, p. 64; also 112, 156, 294–295.
24. *Ibid.*, p. 295; also 60, 112.
25. *Ibid.*, pp. 295 and 112; also 23, 60, 201.
26. *Ibid.*, pp. 22, 23.

nate. Slavery seemed an excessively severe form of subordination, and privately he wished that slaveowners would abandon the institution. Also, he sincerely believed that popular sovereignty would prevent the extension of slavery to the territories. But he did not think the choice between slavery and some other form of subordination for an inferior people was important enough to make an issue of it at the risk of disrupting the Union. As he translated his values, objectives, and priorities into political formulas, popular sovereignty served his purposes to perfection. It promised to remove the dangerous slavery issue from the national arena. It appeared likely to keep slavery out of the territories.[27] And it offered a flexible and democratic mode of decision. Regarding slavery as a purely optional adjustment to a distinctive set of physical or economic circumstances, he argued that the United States presented too much variety to admit of a uniform policy concerning the institution: "It is neither desirable nor possible," he said, "that there should be uniformity in the local institutions and domestic regulations of the different states of this Union. The framers of our government never contemplated uniformity in its internal concerns. . . . They well understood that the great varieties of soil, of production and of interests, in a republic as large as this, required different local and domestic regulations in each locality. . . . Diversity, dissimilarity, variety in all our local and domestic institutions is the great safeguard of our liberties."[28]

As an additional argument against uniformity, Douglas pointed out that under a system which left each state free to decide for itself, six former slave states (Massachusetts, Connecticut, Rhode Island, New York, New Jersey, and Pennsylvania) had adopted emancipation, but that no free states had adopted slavery. A doctrine of uniformity in 1787 would have resulted in making all the states slave states. State autonomy had been conducive to freedom.[29]

It has been suggested that the democracy of Stephen A. Douglas was a majoritarian democracy, in which the dominant forces impose a coercive will upon the minority, rather than a democracy of freedom, in which the liberties of individuals are cherished.[30] As far as

27. See above, p. 171.
28. At Chicago, July 9, 1858, in Angle (ed.), *Created Equal?* pp. 18–20. Also see pp. 54–55, 110, 112–114, 200, 364, for Douglas's frequent recurrence to this theme.
29. *Ibid.*, pp. 110, 296–297, 364.
30. Jaffa, *Crisis of the House Divided*, pp. 304–305, 332–335; Jaffa, *Equality and Liberty: Theory and Practice in American Politics* (New York, 1965), pp. 82, 88–90, 95–96.

it goes, this is true, and yet, fundamentally, it was not that majoritarianism made him ready to subordinate the blacks, but that a readiness to subordinate the blacks made him responsive to majoritarianism. Further, his majoritarianism was qualified by his intent to apply it at the local (or territorial) rather than the national level. Small local majorities would be less prone to arbitrary action, executed without regard for local interests, than great monolithic majorities. To a man who, as Lincoln observed, had "no very vivid impression that the Negro is a human,"[31] slavery did not appear either as a great moral issue or as an agonizing dilemma. The most important thing about it was to avoid a violent national quarrel about it, and that could be done best by treating it as a local question.

For Lincoln, in contrast, slavery did present both a moral issue and a dilemma, and since this experience was common in the North, Lincoln's difficulties with the question reflect a great deal of the perplexity and ambiguity with which a considerable segment of the northern public approached it.

If the slavery question had been exclusively a matter of ethics, and if it had not impinged upon other primary values, Lincoln would have found it clear-cut and simple, for his ethical views were unqualified: "I . . . contemplate slavery as a moral, social, and political evil"; "If slavery is not wrong, then nothing is wrong. I cannot remember when I did not so think, and feel."[32]

Lincoln hated slavery because he regarded Negroes as humans and because he believed, philosophically at least, in the equality of all men. Naturally, he appealed to the Declaration of Independence as a criterion, and in a speech at Chicago, even before the joint debates began, he gave a ringing affirmation of the creed of equality. If one should take the Declaration, with its assertion that all men are equal, and start making exceptions to it, "where," he asked, "will it stop? If one man says it does not mean a Negro [which Douglas did say] why may not another say it does not mean some other man? . . . Let us discard all this quibbling about this man and the other man—this race and that race and the other race being inferior, and therefore they must be placed in an inferior position. . . . Let us discard all these

31. Speech at Bloomington, Oct. 16, 1854, in Basler (ed.), *Works of Lincoln,* II, 281.
32. Speech at Galesburg, Oct. 7, 1858, Lincoln to Albert G. Hodges, April 4, 1864, *ibid.*, III, 226; VII, 281.

things and unite as one people throughout this land until we shall once more stand up declaring that all men are created equal."[33]

From a somewhat abstract belief in equality Lincoln moved to a conviction that slavery, as a violation of equality, must not be permitted to spread, and as he stated in the "House Divided" speech, that it must be placed "where the public mind shall rest in the belief that it is in the course of ultimate extinction." Apparently, he set much store by this phrase, "ultimate extinction," for it echoed through all the joint debates and all Lincoln's speeches of the senatorial campaign.[34]

As soon, however, as he passed beyond the ethical absolute and the somewhat vague ultimate purpose, Lincoln began to encounter complications. For one thing, he was trapped in a conflict of values. He valued freedom, which impelled him toward emancipation, but he also valued the Union, which repelled him from emancipation because any attempt to achieve it was likely to produce in the South a reaction against the Union. A single-minded abolitionist like Garrison could say, "So much the worse for the Union," but Lincoln shrank from "doing anything to bring about a war between the free and slave states."[35] Living in an age of romantic nationalism nowhere more intense than in the United States, Lincoln had become a devotee of the cult of Union as preached by Webster and Clay. Regarding this Union as the chief bulwark of freedom in the world, he could not knowingly take a position which would weaken the harmony of its sections. With his legal orientation, he directed these nationalistic impulses into constitutional channels. The guarantees of the Constitution were almost like the wedding vows which North and South had taken in agreeing to their union. Burdensome as some of these guarantees might be, they were promises given and must be kept. Hence Lincoln accepted the obligation to leave slavery undisturbed in those states which chose to retain it, and even the obligation to enforce a law for the return of fugitive slaves.[36]

33. *Ibid.*, II, 500, 501. Other invocations of the Declaration of Independence are at II, 519–520; III, 16, 220, 249, 280, 300–304, but after the first joint debate, these were decidedly subdued.

34. *Ibid.*, II, 491 (twice), 492 (3 times), 493, 494, 514, 515; III, 18 (4 times), 181, 276, 305, 306 (3 times), 307, 308 (twice), 316.

35. *Ibid.*, III, 19.

36. *Ibid.*, pp. 41, 131 (on fugitive slaves), 16, 116, 255, 277, 311 (on noninterference with slavery in the states). At Alton (*ibid.*, p. 300) Lincoln quoted his own previous statement: "We had slaves among us [in 1787], we could not get our

The goal of ultimate extinction seems more or less incompatible with Lincoln's acquiescence in southern retention of slavery. Douglas exploited this conflict by suggesting that Lincoln wanted to amend the Constitution, removing the guarantees upon which the slave states relied, or even to take more drastic action against slavery.[37] Lincoln denied it, saying, "I propose nothing but what has a most peaceful tendency."[38] He even acknowledged that the desired end might still be far off: "I do not suppose that . . . ultimate extinction would occur in less than a hundred years at the least."[39] This would have meant emancipation completed in about 1958 instead of 1865. Such a relaxed approach to the monstrous injustice of slavery was in sharp contrast to the abolitionists' demand for "immediate" emancipation.

The second major complication for Lincoln was pointed out by Douglas in the Ottawa debate. "Slavery," he said to the audience, "is not the only question which comes up in this controversy. There is a far more important one to you, and that is, what shall be done with the free negro?"[40] Lincoln, in fact, had no satisfactory answer. He had said as much at Peoria in 1854, and he repeated it at Ottawa: "If all earthly power were given me, I should not know what to do."[41] Three times during the debates Lincoln declared his belief that there was "a physical difference between the white and black races" which would "forever forbid the two races living together on terms of social and political equality." That being the case: "While they do remain together, there must be the position of superior and inferior, and I as much as any other man am in favor of having the superior position assigned to the white race."[42]

This left Lincoln, ordinarily a man of rigorous logic, in the difficult position of reconciling subordination with equality. He set about it doggedly, by specifying a number of rights which he would not accord to the blacks: he would not allow them legally to intermarry with whites; he would not permit them to serve as jurors or to hold office; he would not accord them citizenship in the state of

---

Constitution unless we permitted them to remain in slavery, we could not secure the good we did secure if we grasped for more."

37. Angle (ed.), *Created Equal?* pp. 51–52.
38. Basler (ed.). *Works of Lincoln,* III, 309.
39. *Ibid.,* p. 181; also p. 18.
40. *Ibid.,* p. 11.
41. *Ibid.,* II, 255; III, 14–15.
42. *Ibid.,* p. 146; also pp. 16, 249.

Illinois; and he would not accord them the right to vote.[43] Here he lagged conspicuously behind his fellow Republican, William H. Seward, who had long been an advocate of Negro citizenship and suffrage in New York.[44]

Now, if political and social equality were denied, and if black men were relegated to a position of inferiority, how did this accord with discarding all quibbles about race and standing up to reaffirm that all men are created equal? Lincoln made the best he could of this question by declaring that the black was "entitled to all the natural rights enumerated in the Declaration of Independence, the right to life, liberty, and the pursuit of happiness. . . . In the right to eat the bread, without leave of anybody else, which his own hand earns, *he is my equal and the equal of Judge Douglas, and the equal of every living man.*"[45]

No matter how often and how vigorously Lincoln intoned this assertion, what it meant was that equality amounted to little more than a right not to be a chattel, and not to have one's labor owned by somebody else. Without the vote, without citizenship, without social parity with his fellow man, the Negro's "equality" would be a strangely ambiguous status, a no man's land somewhere between freedom and slavery. Lincoln had recognized for at least four years that it would not be very satisfactory to "free them [the blacks] and keep them among us as underlings"[46] and that it was questionable whether this would "really better their condition." At Springfield in 1858, in a burst of candor, he said, "What I would most desire would be the separation of the white and black races."[47] Because of this desire, he entertained for more than ten years the idea of colonizing the blacks outside the United States. His first impulse, he said in 1854, "would be to free all the slaves and send them to

43. On all these disabilities except citizenship, *ibid.*, p. 145; on citizenship, p. 179.

44. Under the constitution of New York, Negroes were permitted to vote if they met special property qualifications which were not imposed upon whites. In 1838, Seward had opposed any change in the suffrage, but in 1846 he declared in favor of giving the ballot to "every man, learned or unlearned, bond or free." Glyndon G. Van Deusen, *William Henry Seward* (New York, 1967), pp. 51, 94; Frederic Bancroft, *The Life of William H. Seward* (2 vols.; New York, 1900), I, 70, 162; Leon F. Litwack, *North of Slavery: The Free Negro in the Free States, 1790–1860* (Chicago, 1961), pp. 87–88; Dixon Ryan Fox, "The Negro Vote in Old New York," *Political Science Quarterly*, XXXII (1917), 253–256.

45. Basler (ed.), *Works of Lincoln*, III, 16. Also, II, 520.

46. Speech at Peoria, Oct. 16, 1854, *ibid.*, pp. 255–256.

47. *Ibid.*, p. 521.

Liberia—to their own native land."[48] Apparently he ignored the fact that Liberia was not, in fact, the birthplace of these native-born Americans, and not even the land of their ancestors, but he did recognize that even "a moment's reflection" would expose the idea of colonization as a fantasy—there was not enough money and not enough shipping.[49] Since he did not know what should be done with the free Negro, perhaps it would be just as well for emancipation to be decidedly gradual.

But perhaps this is entirely the wrong way to go about evaluating Lincoln's position. The central point, some would argue, is that he was engaged in a contest for votes, in a constituency where strong anti-Negro sentiment prevailed. He could best serve the antislavery cause by winning the election, and tactically the best way to win the election was to take a minimum antislavery position—one which would make him preferable to Douglas in the eyes of all antislavery men, but which would antagonize as few as possible of those who cared little about slavery. According to this view, which emphasizes that politics is the art of the possible, it was enough for him to assert the principle of equality, and the qualifications and ambiguities which surrounded the assertion should be discounted as necessary political opportunism.

Opportunism can have either a selfish or a disinterested aspect. Viewed as selfish, it means that the candidate's sole objective is to get elected and that he will say and do whatever serves that objective. In fact, Douglas made this accusation against Lincoln throughout the joint debates—that his attitude depended upon his latitude, that he talked about equality one way in Chicago and quite another way in Charleston, far downstate. Lincoln denied these charges quite as strenuously as Douglas asserted them. Reading the debates more than a century later, one can hardly doubt that Lincoln's equality, as defined at Chicago, had been badly eroded by the time he got to Charleston, but instead of explaining the change in terms of geographical opportunism, one might argue that Lincoln became more cautious in his equalitarianism as the campaign progressed—more anxious to stress the abstract side of his antislavery position and to deemphasize the practical problems associated with it.[50]

48. *Ibid.*, p. 255.
49. *Ibid.*, pp. 255–256.
50. Douglas pressed relentlessly the accusation that Lincoln shifted his position as he moved from northern to southern Illinois, raising the subject in at least five of

Viewed as disinterested, opportunism can mean that a man recognizes the limitations of the situation in which he is working, and that he chooses to accept them realistically. Certainly Lincoln understood that most of his fellow citizens, both in Illinois and in the North generally, might support the abstract idea of emancipation, but not the idea of racial equality. As he said in 1854 and again in 1858, "What next? Free them, and make them politically and socially our equals? My own feelings will not admit of this; and if mine would, we well know that those of the great mass of white people will not." Then followed a most significant comment: "Whether this feeling accords with justice and sound judgment is not the sole question, if indeed, it is any part of it. A universal feeling, whether well or ill-founded, can not be safely disregarded. We can not, then, make them equals."[51] This statement was not unlike another grimly discouraging statement which Lincoln was later to make to a committee of five blacks, on August 14, 1862, after he had already told his cabinet of his intent to issue the Emancipation Proclamation. To the committee he said, "Even when you cease to be slaves, you are yet far removed from being placed on an equality with the white race. . . . on this broad continent not a single man of your race is made the equal of a single man of ours. . . . I cannot alter it if I would. It is a fact."[52]

To Lincoln, public attitudes were part of the complex of deterministic forces which set the limits of possible action—just as real a part as constitutional guarantees, economic arrangements, and the physical dissimilarities of blacks and whites. These attitudes were "a fact," something no realist could safely disregard and no idealist could alter. This was the disinterested opportunism which says that politics is the art of the possible.[53]

---

the joint debates, sometimes at length. *Ibid.*, III, 5, 105, 174–176, 213–216, 237–239, 323. Response by Lincoln, 247–251.

51. Speech at Peoria, Oct. 16, 1854, *ibid.*, II, 256.

52. *Ibid.*, V, 372.

53. Litwack, *North of Slavery*, p. 278, says, "Lincoln and the Republican Party had correctly gauged public opinion. Protect the Negro's life and property, but deny him the vote, jury service, the right to testify in cases involving whites, and social equality, and—if possible—colonize him outside the United States." Eric Foner, *Free Soil, Free Labor, Free Men: The Ideology of the Republican Party before the Civil War* (New York, 1970), pp. 261–267, shows most effectively that the Republicans were under a constant barrage of attacks from the Democrats for being "amalgamationists." The Republicans often responded, as he shows, by expressions of racism (such as one finds in some places in Lincoln's speeches in the joint debates) which were "political replies

In the first flush of campaign enthusiasm in 1858, Lincoln had made his appeal for an unqualified equalitarianism—for once more standing up and declaring that all men are created equal. But as the realities of public response and of tactical necessity began to reassert themselves during the campaign, Lincoln, in effect, separated what he would do for the slave from what he would do for the Negro. For the slave, he would offer ultimate emancipation, in some unspecified way at some unspecified time, but not soon enough, presumably, to alarm anyone. For the Negro, he would offer no rights of the franchise, jury box, or citizenship, no promise of political or social equality. As a limited step toward his far-distant goals, Lincoln would exclude slavery from the territories by federal action. But even this token took on something of an ambiguous meaning when Lincoln talked about placing the territories legally "in such a condition that white men may find a home. . . . I am in favor of this not merely for our own people who are born amongst us, but as an outlet for *free white people everywhere,* the world over."[54] Thus, while neatly repudiating Know-Nothingism by indirection, he also endorsed a racism which would give a priority to foreign-born whites over native-born blacks, and placed himself in a situation which made it possible for historians later to say that Lincoln was skillfully combining the antislavery vote with the anti-Negro vote—the votes of those who would free the blacks with the votes of those who would segregate the territories for whites.[55]

The more closely one examines Lincoln's approach to Negro subordination, the more attenuated his proposals appear. Beyond his advocacy of the "ultimate extinction" of slavery, devoid of any concrete plans for bringing it about, the distinction between his position and that of Douglas seems to have been slight. Lincoln probably was aware of and embarrassed by this close parallelism,

---

to Democratic accusations rather than gratuitous insults to the black race." Foner sees Lincoln as the architect of a "shaky consensus within the party," between the westerners who thought he had gone too far in his abstract affirmations of equality and easterners who thought he should have gone further in extending specific rights of citizenship, the franchise, etc. Fehrenbacher, *Prelude,* p. 111, declares, "Lincoln's first [i.e., primary] principle of racial relations—that the Declaration of Independence belongs to all Americans—was actually subversive of the existing order [of racial inequality] which he endorsed."

54. Basler (ed.), *Works of Lincoln,* III, 312; also see II, 498.

55. Richard Hofstadter, *The American Political Tradition and the Men Who Made It* (New York, 1948), pp. 110–113.

and, perhaps in order to make a sharply defined issue, he discovered a covert purpose of Douglas and the Democrats to nationalize slavery with another Supreme Court decision, one that would deny to the states any constitutional power to prohibit the institution within their boundaries. Here he fitted Douglas neatly into his picture by maintaining that Douglas's attitude of not caring whether slavery was voted up or down would blunt the moral opposition to slavery, and thus in its final effect would help just as much "to nationalize slavery as the doctrine of Jeff Davis himself."[56]

Lincoln postulated this danger in the "House Divided" speech, which served as a kind of blueprint for his entire campaign. "We may, ere long," he said, "see . . . another Supreme Court decision, declaring that the Constitution of the United States does not permit a *state* to exclude slavery from its limits."[57] In the subsequent speeches of the debates, he developed this warning. Thus, at Ottawa he asked: "What is necessary for the nationalization of slavery? It is simply the next Dred Scott decision. It is merely for the Supreme Court to decide that no *State* under the Constitution can exclude it, just as they have already decided that under the Constitution neither Congress nor the territorial legislature can do it."[58]

This suggestion seemingly infuriated Douglas, partly no doubt because it accused him of entering into a conspiracy, and partly because he regarded it as preposterous. No point in the joint debates aroused him to quite such strong language as he used in declaring that he "did not suppose there was a man in America with a heart so corrupt as to believe such a charge could be true," and that Lincoln had accused the Supreme Court of "an act of moral treason that no man on the bench could ever descend to." When Lincoln, at Freeport, asked Douglas whether he would acquiesce in a Supreme Court decision declaring that "states cannot exclude slavery from their limits," he replied that he was "amazed that Lincoln should ask such a question. . . . Mr. Lincoln . . . knows that there never was but one man in America, claiming any degree of intelligence or decency [namely the editor of the Washington *Union*] who ever for a moment pretended such a thing. . . . a schoolboy knows better."[59]

56. Fehrenbacher, *Prelude*, pp. 79–82; Basler (ed.), *Works of Lincoln*, IV, 21.
57. Basler (ed.), *Works of Lincoln*, II, 467.
58. *Ibid.*, p. 518; III, 27, 29–30, 233, 316, 369.
59. *Ibid.*, pp. 24, 43, 53.

For a long period, historians tended to agree with Douglas that Lincoln was raising a fictitious issue. One eminent scholar has described the conspiracy which Lincoln suggested as "quite fanciful and non-existent," and has characterized the argument that the judicial decision to protect slaves in the territories would lead to a decision to protect it also in the states as "something of a non-sequitur." Another has said that it would have been more realistic to discuss the dangers from future annexation of potential slave territory, or from the categorical southern demands for positive protection for slavery in the territories, but neither of these issues was examined, and the "absurd bogey" of legalization of slavery in all the states was conjured up instead.[60]

One can now see that no plans existed to nationalize slavery by a second Dred Scott decision which would legalize it in the states, and one may even regard the fears of such a plan as another instance of the paranoid factor in American politics. But Lincoln, of course, did not have the advantage of hindsight, and several recent scholars have shown that the circumstances of 1858 gave some plausibility to his fears. For example, the Washington *Union* had been contending that state legislation forbidding slavery was a violation of property rights and, in fact, unconstitutional. The *Union,* though only one paper, was not just any paper, but the organ of the Buchanan administration. It was also true that Chief Justice Taney had declared in the Dred Scott decision, "The right of property in a slave is distinctly and expressly affirmed in the Constitution." Lincoln, in the debate at Galesburg, pointed out this declaration and added his own rebuttal: "I believe that the right of property in a slave *is not* distinctly and expressly affirmed in the Constitution." Justice Nelson, in his Dred Scott opinion, had included the cryptic remark that, "except in cases where the power is restrained by the Constitution . . . , the law of the state is supreme over slavery." But what did he mean, Lincoln wanted to know, by the words "except in cases where the power is restrained by the Constitution"? The Fourteenth Amendment did not at that time exist with its restriction on the powers of the states to deprive persons of property, but the Fifth Amendment, with its property-protecting clause, might have been

60. J. G. Randall, *Lincoln the President: Springfield to Gettysburg* (2 vols.; New York, 1945), I, 108, 116; Allan Nevins, *The Emergence of Lincoln* (2 vols.; New York, 1950), I, 361–363.

construed to extend to the states, and Article IV, Section 2, of the Constitution ("The citizens of each state shall be entitled to all the privileges and immunities of citizens of the several states") also might have been applied.[61]

In short, the juridical ingredients for a decision to legalize slavery nationally were by no means wholly lacking, but it seems incredible that nine sane justices could have contemplated such a decision. Yet if the Dred Scott decision itself had not been rendered, it might have seemed incredible that the Court could deny the power of Congress to regulate slavery in the territories despite the fact that it had been doing so since 1789 under Article IV, Section 3, of the Constitution, which specified that "the Congress shall have power to . . . make all needful rules and regulations respecting the Territory or other property belonging to the United States." Also, one must remember the general fear of the slave power and the rather ominous specific context that Lincoln so skillfully exploited—a context in which the moral objections to slavery would first be eroded away by Douglas's "don't care" policy, and then, when the way had been thus prepared, the legal obstacles to the nationalization of slavery would be removed by the Court.[62]

"Within its context," the fear of the nationalization of slavery was "far from absurd,"[63] and perhaps the main thing to be said against it was that it treated a potentiality as if it were an actuality. As one writer has said, perhaps Lincoln "ought to have been content to denounce the [Dred Scott] decision for what it was, rather than to predict an imaginary new decision."[64]

Lincoln wanted to assail the slave power in a way that would sharply differentiate his position from that of Douglas. He did so more by attributing to Douglas a sinister design for future expansion of slavery than by criticizing Douglas's concrete proposals. One can recognize the fact that some of Lincoln's fears for the future were by no means preposterous, and at the same time realize that he was strongly motivated by the fact that, in his lack of specific

61. Basler (ed.), *Works of Lincoln*, III, 230 (quoting Taney), 231, 251 (on Justice Nelson). See Jaffa, *Crisis of the House Divided*, pp. 275–293; Arthur Bestor, "State Sovereignty and Slavery: A Reinterpretation of Proslavery Constitutional Doctrine, 1846–1860," ISHS *Journal*, LIV (1961), 162–172.

62. Foner, *Free Soil*, pp. 97–112 (on the slave power); Fehrenbacher, *Prelude*, pp. 80–81 (on the context of Douglas's arguments and the Court decision).

63. Fehrenbacher, *Prelude*, p. 81.

64. Randall, *Lincoln the President*, I, 116.

policies for freeing the slaves or for removing racial discriminations against the blacks, his position was embarrassingly close to that of Douglas.[65]

The major objection to concluding simply that Lincoln, like Douglas, was a "white supremacist" is not that the conclusion is literally false, but that a categorization so loose that it fits Douglas as well as it fits Lincoln does not say very much.[66] There were really several significant differences between the two men but perhaps none more profound than the fact that Lincoln constantly appealed to his hearers to recognize that they shared a common humanity with the blacks, while Douglas was tickling the racist susceptibilities of the same audiences with charges that Lincoln regarded the Negro as "his brother."

This concern for humanity runs through a large part of Lincoln's writing and speeches, but it is mixed up, as we have seen, with his acceptance of the practices of an American culture which treated Negroes as inferior. Hence his attitudes frequently seem ambiguous and, to a hostile critic, hypocritical. But occasionally, one glimpses clear evidence that when Lincoln thought most intensively about the slavery question, he did not think about blacks specifically as blacks; he thought more broadly in terms of the ownership of men by other men. As he wrote, but did not say publicly,

> If A. can prove, however conclusively, that he may, of right, enslave B. —why may not B. snatch the same argument, and prove equally, that he may enslave A?—
>
> You say A. is white, and B. is black. It is *color*, then; the lighter having the right to enslave the darker? Take care. By this rule, you are to be slave to the first man you meet, with a fairer skin than your own.
>
> You do not mean *color* exactly?—You mean the whites are *intellectually* the superiors of the blacks, and, therefore have the right to enslave them? Take care again. By this rule, you are to be slave to the first man you meet, with an intellect superior to your own.
>
> But, say you, it is a question of interest; and, if you can make it your *interest*, you have the right to enslave another. Very well. And if he can make it his interest, he has the right to enslave you.[67]

Here, clearly, Lincoln saw blacks and whites together, caught indiscriminately in the web of injustice which society often weaves. His

65. Compare *ibid.*, pp. 123–126, with Fehrenbacher, *Prelude*, pp. 109–112.

66. See Lerone Bennett, Jr., "Was Abe Lincoln a White Supremacist?" *Ebony*, Feb. 1968, pp. 35ff.

67. Basler (ed.), *Works of Lincoln*, II, 222–223.

own personal situation and the situation of the slave were potentially interchangeable; it was only the random chance which had made him free and had made "Sambo" (Lincoln's term) a slave.[68]

The same concern with basic humanity was also reflected in Lincoln's acute insight that even slaveholders, although they wanted to regard slaves as property rather than as humanity, nevertheless could not repress their recognition that slaves were their fellow men. As he expressed it,

> While you require me to deny the humanity of the negro, I wish to ask whether you of the south yourselves, have ever been willing to do as much? The great majority, south as well as north, have human sympathies, of which they can no more divest themselves than they can of their sensibility to physical pain. These sympathies in the bosoms of the southern people, manifest in many ways, their sense of the wrong of slavery, and their consciousness that after all, there is humanity in the negro. If they deny this, let me address them a few plain questions. In 1820 you joined the north, almost unanimously, in declaring the African slave trade piracy, and in annexing to it the punishment of death. Why did you do this? If you did not feel that it was wrong, why did you join in providing that men should be hung for it? The practice was no more than bringing wild negroes from Africa, to sell to such as would buy them. But you never thought of hanging men for catching and selling wild horses, wild buffaloes or wild bears.[69]

In this same connection, Lincoln argued that the southern tendency to avoid social contact with slave traders reflected a sense that they were engaged in an inhuman sort of business. He also observed that in the slave states there were more than 500,000 free blacks, potentially worth more than $200 million. All of them either had been slaves themselves or were the descendants of slaves. Why were they not in slavery? It was because of "*something* which has operated on

68. *Ibid.*, III, 204.

69. Speech at Peoria, Oct. 16, 1854, *ibid.*, II, 264. Also see a fragment by Lincoln, Oct. 1, 1858, *ibid.*, III, 204–205: "For instance, we will suppose the Rev. Dr. Ross has a slave named Sambo, and the question is, 'Is it the Will of God that Sambo shall remain a slave, or be set free?' The Almighty gives no audible answer to the question, and his revelation—the Bible—gives none—or, at most, none but such as admits of a squabble, as to its meaning. No one thinks of asking Sambo's opinion on it. So, at last, it comes to this, that *Dr. Ross* is to decide the question. And while he considers it, he sits in the shade, with gloves on his hands, and subsists on the bread that Sambo is earning in the burning sun. If he decides that God wills Sambo to continue a slave, he thereby retains his own comfortable position; but if he decides that God wills Sambo to be free, he thereby has to walk out of the shade, throw off his gloves, and delve for his own bread. Will Dr. Ross be actuated by that perfect impartiality, which has ever been considered most favorable to correct decisions?"

their white owners, inducing them, at vast pecuniary sacrifices, to liberate them. What is that *something*? Is there any mistaking it? In all these cases it is your sense of justice, and human sympathy, continually telling you, that the poor negro has some natural right to himself."[70]

The difference between Douglas and Lincoln—and in a large sense between proslavery and antislavery thought—was not that Douglas believed in chattel servitude (for he did not), or that Lincoln believed in an unqualified, full equality of blacks and whites (for he did not). The difference was that Douglas did not believe that slavery really mattered very much, because he did not believe that Negroes had enough human affinity with him to make it necessary for him to concern himself with them. Lincoln, on the contrary, believed that slavery mattered, because he recognized a human affinity with blacks which made their plight a necessary matter of concern to him. This does not mean that his position was logically consistent or that he was free of prejudice. In fact he was a classic illustration of Gunnar Myrdal's American dilemma: philosophically and abstractly he believed in the humanity of blacks and the equality of humans; concretely and culturally he accepted the prevailing practices of Negro subordination. In a very real sense his position was ambiguous. But even an ambiguous position was vastly different from that of Douglas. And, one must add, an ambiguous position is by definition one in which opposing values conflict with one another. It is hard to believe that, in Lincoln's case, the conflicting values were really of equal force. In the long-run conflict between deeply held convictions on one hand and habits of conformity to the cultural practices of a biracial society on the other, the gravitational forces were all in the direction of equality. By a static analysis, Lincoln was a mild opponent of slavery and a moderate defender of racial discrimination. By a dynamic analysis, he held a concept of humanity which impelled him inexorably in the direction of freedom and equality.

On November 2, 1858, the voters of Illinois cast about 125,000 votes for the Republicans, 121,000 for the Douglas Democrats, and 5,000 for the Buchanan Democrats. When broken down by legislative districts, this balloting resulted in the election of forty-six Democratic legislators and forty-one Republicans. This result assured the

70. *Ibid.*, II, 265.

reelection of Douglas by the legislature. Since the Republicans did not gain legislative seats in proportion to their popular vote, some historians have assumed that Lincoln lost because of the lack of a truly proportional representation. But this is not true. There were thirteen state senators who held over from a previous election, and eight of these were Democrats. If the Republicans had won legislative seats in exact proportion to the popular vote (forty-four Republicans to forty-three Democrats), these holdovers would still have given Douglas a victory.[71]

The defeat for Lincoln was also a defeat for James Buchanan and the Democratic regulars. By winning another term in the Senate, Douglas had solidified his leadership of northern Democrats in a way that would enable him to make his supreme bid in 1860 for control of the party.

Lincoln had gained a kind of success by preventing Douglas from bringing the antislavery forces to an opportunistic support of popular sovereignty, which had operated against slavery in Kansas, but which was not intrinsically antislavery at all. Lincoln had demonstrated his own stature as an antislavery leader, and he had provided part of the American public with an overarching discussion of the real problems of slavery in American society—a discussion such as all the moralists in the abolition crusade and all the constitutional lawyers in politics had not supplied. But this was perhaps no great consolation to him, for he remained a defeated candidate who had not held public office for ten long years.

71. Fehrenbacher, *Prelude*, pp. 118–120, explains the electoral circumstances of Douglas's victory. For the election post-mortems, see Heckman, *Lincoln vs. Douglas*, pp. 137–142.

CHAPTER 14

Harpers Ferry: A Revolution That Failed

IF Lincoln and Douglas in 1858 caused a considerable part of the American public to think about the philosophical aspects of slavery, John Brown in 1859 focused attention dramatically upon its emotional aspects. The emotional aspects proved to be much the more powerful of the two.

Hardly anything can be said with certainty about John Brown, but it appears that he was taught to hate slavery by his father and gradually became increasingly committed to the fight against it, although until his Kansas adventures at the age of fifty-six, he spent most of his life in pursuits such as farming, running a tannery, raising sheep, speculating in land, driving cattle, and acting as agent for a wool company.[1] It also appears that he was a man of very high abstract standards—rigorously moral, condemning wrong, despising weakness. But he did not live up to his rigid standards, and his life was checkered with episodes that must have been very hard on the self-respect of a man of such exacting righteousness. He gave a bill of exchange on the Bank of Wooster for money which he did not have. He secretly mortgaged a piece of land which he had already pledged as security to a man who had sustained a loss of $6,000 by signing a note for him. He was subsequently sent to jail for refusing to relinquish the land to the legal owner. He induced a woolen company to make him its agent and to advance him $2,800 with which to buy wool, then used the money for his own purposes.

1. For general works on John Brown, see Chapter 9, note 27.

He escaped criminal prosecution for this act by promising to make restitution, which was never made. He was sued no less than twenty-one times, usually for defaulting on financial obligations.[2]

Throughout the years when these episodes occurred, he constantly expressed the most pious ideals and high-minded convictions. In some men such a discrepancy between words and acts would indicate deliberate deception and knavery, and indeed this has been attributed to Brown. From the full record of his life, however, it appears that he did have very high standards but was unable to live up to them. Until his fifty-sixth year, "Old Brown" had failed at every enterprise to which he set his hand and had repeatedly violated his own principles. It further seems likely that in order to avoid facing this reality, he began to create an image of himself as a man who was immune to ordinary human frailities—a man of iron in physical endurance; a deeply purposeful man, with dedication unrelieved by any element of levity, or self-indulgence, or even casualness; a man of deeds and not of words. Perhaps most people have attributed such superhuman qualities to themselves in occasional compensatory fantasy, but the remarkable fact about John Brown is that he began to act as if he really had these characteristics, so that after a while, the fantasy became, in a sense, the reality, except that the outward qualities of strength—of heroic endurance, inspiring leadership, and relentless purpose—concealed inner qualities of weakness—of flawed judgment, homicidal impulse, and simple incompetence.[3] John Brown never did develop the basic

2. Stephen B. Oates, *To Purge This Land With Blood: A Biography of John Brown* (New York, 1970), pp. 35–39, 44–45, 48–49, 76; Oswald Garrison Villard, *John Brown, 1800–1859: A Biography Fifty Years After* (Boston, 1910), pp. 26–41; Hill Peebles Wilson, *John Brown, Soldier of Fortune: A Critique* (Lawrence, Kan., 1913), pp. 28–34.

3. Note the accounts of Brown shooting a dog which he could not discipline; his severe corporal punishment of his son and his requirement that his son should, in turn, whip him until he bled; his thrashing of his son Jason, age four, for "lying" by insisting that a dream he had had was "real"; his rejection of anything humorous; his generally grim and "relentless sternness"; the statement of George Gill that "I had it from Owen [Brown's son] in a quiet way, and from other sources in quite a loud way, that in his family his methods were of the most arbitrary kind." John Brown, Jr., in F. B. Sanborn (ed.), *The Life and Letters of John Brown* (Boston, 1891), pp. 91–93; memoir in 1859 by James Foreman, an employee of Brown between 1820 and 1825, in Louis Ruchames (ed.), *A John Brown Reader* (London, 1959), pp. 163–168; George Gill, an associate of Brown, July 7, 1893, *ibid.*, pp. 231–234; Salmon Brown, "My Father, John Brown," *Outlook*, CIII (Jan. 25, 1913), 212–217; Villard, *Brown*, pp. 8–9, 19, 20, 24, 36; Oates, *To Purge This Land*, pp. 14–24; Jules Abels, *Man on Fire: John Brown and the Cause of Liberty* (New York, 1971), pp. 6–7.

human capacity of making his means serve his ends, and his ultimate triumphant failure was built upon the accident of his survival to face trial after Harpers Ferry.

The more Brown yearned to dedicate himself, the more he turned to antislavery as the overriding purpose for his life.[4] To say this is not to suggest a doubt that the antislavery cause could take possession of a man by its own inherent moral strength. But in any case, John Brown really began to make a career of antislavery after the Pottawatomie massacre. For three years, from 1856 to 1859, he gave up all other pursuits and devoted himself exclusively to developing his plans for military operations against slavery, either in Kansas or elsewhere.[5]

Brown had no money of his own, and since a military company cannot function without equipment and supplies, he soon discovered the irony that his dedication to a life of military action had in fact committed him to an occupation which was one part fighting and several parts fund raising. For some thirty months, between January 1857 and July 1859, he spent approximately half his time traveling about soliciting money. He made seven trips to Boston, five trips to Peterboro, New York, to see Gerrit Smith, and numerous visits to other places, so that he became a kind of circuit rider, frequently forced, as he himself felt, to beg in a humiliating way for the support that would enable him to operate. With this aid, he was able to keep together a little band of about a dozen devoted young men, to hire at inadequate wages an English adventurer named

4. James Foreman (see note 3) described Brown as strongly antislavery in the 1820s; in 1832 he was said to have had on his farm a hiding place for fugitives; in 1834 he wrote to his brother telling of his and his wife's plans "to get at least one negro boy or youth and bring him up as we do our own"; in 1849 he moved to North Elba, New York, to live in a colony of Negroes whom Gerrit Smith was trying to establish there—this residence was broken off in 1851 because of Brown's wool business, but resumed in 1855. Villard, *Brown*, pp. 25–26, 43, 71–74; Oates, *To Purge This Land*, pp. 30–33, 41–44, 65–67.

5. On Brown's career in Kansas, see above, pp. 211–213. He arrived in Kansas on Oct. 6, 1855. Pottawatomie was in May 1856. During the following months, Brown operated as captain of a guerrilla band. His son Frederick was shot and killed by proslavery guerrillas on Aug. 30, 1856. In Oct., Brown left Kansas and spent the first half of 1857 in the East, the second half in Kansas and Iowa (where Tabor was his headquarters). By early 1858, he was back in New England, revealing his plans and seeking support for his Virginia adventure. But in June he returned to Kansas for the third time, participated in a raid on Fort Scott (Dec. 16), and led a raid into Missouri (Dec. 20–21) that put his name in headlines again. In Jan. 1859, Brown left Kansas for the last time, taking eleven captured Missouri slaves for release in Canada.

Hugh Forbes as an instructor in military drill, and to order one thousand pikes for a purpose which seemed obscure when he bought them. Between himself and his financial backers there was fairly constant tension, for he kept waiting for them to give enough to enable him to act, and they kept waiting for him to do something with what they had already given before they gave more.[6]

When Brown embarked upon this career early in 1857, he was fresh from nearly four months of "military service" of the bushwhacking variety in Kansas, and his purpose was not at all unusual. Kansas was full of free-lance fighting men, operating with bands which they had raised themselves. Brown was one of them, and what he wanted initially was to equip and lead a crack military company of about fifty men to continue fighting the battles which were then being waged in the territory. His own experiences in the Kansas strife and the killing of his son Frederick by a proslavery man may have fortified his purpose, or he may by this time have developed an *idée fixe* unrelated to ordinary emotions. In any case, antislavery men in the East had given him very limited financial help when he first migrated to Kansas in 1855, and he now conceived the idea of appealing to these same sources for support in his project. He procured two letters from Charles Robinson, the free-state "governor" of Kansas, expressing thanks for "your prompt, efficient and timely action against the invaders of our rights" and urging all "settlers of Kansas" to "please render Captain John Brown all the assistance he may require in defending Kansas from invaders and outlaws."[7] Armed with these, he set out for the East in October 1856. At Chicago, he met with members of the National Kansas Committee; in Ohio, Salmon P. Chase provided him with a letter of general commendation; and in Springfield, Massachusetts, he obtained a letter of introduction to Franklin B. Sanborn, a well-connected young schoolteacher and antislavery worker in Boston. He arrived in Boston on January 4, 1857.[8]

Brown's reception was an immense personal success and a great financial disappointment. The elite of Boston had a deep ideological commitment to the cause of freedom in Kansas, and they proba-

6. Villard, *Brown,* pp. 291–292; Oates, *To Purge This Land,* pp. 199–201; Tilden G. Edelstein, *Strange Enthusiasm: A Life of Thomas Wentworth Higginson* (New Haven, 1968), pp. 207–220.

7. Robinson to Brown, Sept. 13, 15, 1856, in Villard, *Brown,* pp. 262–263.

8. *Ibid.*, pp. 269, 271; Oates, *To Purge This Land,* pp. 177, 181.

bly felt some guilt that most of their support had been merely rhetorical. They were prepared, therefore, to lionize a genuine Kansas fighting man, and John Brown filled the role to perfection with his grim silences, his expressions of contempt for words rather than deeds, and his picturesque frontier dress, including a bowie knife in his boot which he had taken from a notorious proslavery bushwhacker. Here was a man hunted by his enemies, who always went armed and who barricaded himself in his room at night, even in Boston. Sanborn, the young schoolteacher, was completely captivated, and became a disciple; he took Brown to see Dr. Samuel Gridley Howe, famous throughout the country for his work with the blind and for other philanthropies, and Theodore Parker, probably the foremost clergyman in the United States. Very soon, Brown had met many of the eminent figures of Boston: Amos A. Lawrence, the textile magnate; George L. Stearns, another man of property; Thomas Wentworth Higginson, a young Unitarian parson of Brahmin family; Dr. Samuel Cabot, Wendell Phillips, William Lloyd Garrison (whose doctrine of nonresistance prevented a close relation with Brown), and a little later Henry David Thoreau and Ralph Waldo Emerson (in both of whose homes Brown stayed as a guest), as well as Bronson Alcott.

John Brown's stiff angularity of posture, of manners, and of speech reminded the highly literate Bostonians of certain familiar literary, historical, and biblical images. Brown was a Highland chief, a Cromwellian Covenanter, an Old Testament prophet. They saw him as, by nature and instinct, a man of action, utterly devoid of artistry and rhetoric, and they never sensed at all that he was, in some ways, more of an artist and a man of words than any of them. He had romanticized himself quite as much as others romanticized him, and though not widely educated, he was aware of the relevance of Highland chiefs and prophets as models for his own image, and as alternative personae for the John Brown whose earlier persona had been a shabby and unsatisfactory one. John Brown's nature, holding the mirror up to art, captivated the literati by his consummate "naturalness." Thus, Thoreau saw him as a man of "rare common sense and directness of speech," and Bronson Alcott wrote, transcendentally, "I am accustomed to divine men's tempers by their voices—his was vaulting and metallic, suggesting repressed force and indomitable will." Emerson made him virtually a noble savage: "A shepherd and herdsman, he learned the manners of the animals and knew the secret signals by which animals communicate."[9]

Personally, Brown in Boston was a *succès fou.* The Boston intellectuals suspended their ordinary critical faculties where he was concerned, and ultimately this suspension was to have grave consequences. But though they idealized him and welcomed him in their homes, they did not raise very much money for him. Once the effort fell through to get $100,000 for him by act of the Massachusetts legislature, he was reduced to small gifts—a little better than handouts—and to a contingent promise from George Stearns of $7,000 to subsist one hundred volunteer-regulars *if* it became necessary to call that number into service in Kansas.[10] As limited gifts came in, he found himself under increasing pressure to go back to the territory and engage in some of the direct action which was supposed to be his forte. Therefore, by June he was on his way west to Iowa, and in November he crossed over into Kansas again.

Kansas in November 1857 was a very different place from the territory he had left in October 1856. Robert J. Walker had replaced John W. Geary as governor; fighting had died down; and free-staters had won a majority in the new legislature, thanks to Walker's decisive action in throwing out fraudulent returns. The antislavery party had nothing whatever to gain at this point from a resumption of the border wars. They remembered unpleasantly what Brown had done at Pottawatomie (something the Bostonians did not know); they regarded him as a troublemaker; and they conspicuously failed to welcome his return. Brown saw that Kansas was no place for him, that his career as a Kansas guerrilla was played out, and he left the territory after less than two weeks, going back again to his base at Tabor, Iowa.[11]

At this point, Brown faced a difficult and crucial decision. He had

9. Villard, *Brown,* pp. 271–274, 398–400; Oates, *To Purge This Land,* pp. 181–192.

10. Oates, *To Purge This Land,* pp. 194–195, 203. Brown had collected about $1,000 in cash and received pledges for about $2,000 more. But he had also been promised some $13,000 worth of guns and supplies by the Massachusetts Kansas Committee, and George L. Stearns had undertaken to pay for 200 pistols. In April 1857, as he prepared to head west again, he expressed his bitter disappointment in a kind of open letter to New England, titled: "Old Browns Farewell to the Plymouth Rocks, Bunker Hill monuments, Charter Oaks, and Uncle Thoms Cabbins." It was in response to this document that Stearns pledged his $7,000. Text in Ruchames (ed.), *Brown Reader,* p. 106.

11. Villard, *Brown,* pp. 305–308, quoting Brown to Stearns, Nov. 16, 1857.

either to abandon his role as an antislavery warrior, admitting another failure, or to redefine his mission. He gave his answer at Tabor in late November or early December to the nine men who had accompanied him there. His ultimate destination, he told them, was the state of Virginia.[12] This must have come as a shock to them, and several of the men were disposed to argue, but Brown's hypnotic eloquence won them over.

At first glance, it would appear that Brown had seized upon the Virginia scheme as a desperate alternative when the adventure in Kansas drew toward an unavoidable close. But upon closer scrutiny, one perceives that the Allegheny Mountains had long held great fascination for this strange, disguised romanticist. Kansas was only a detour on the path of his destiny. Apparently, the possibility of basing himself in the mountains and operating from there to emancipate the slaves in Virginia had been the main topic of discussion when he first visited with Frederick Douglass, the foremost Negro in America, in 1848. Also, Brown's daughter, half a century later, asserted that the plan of an invasion from the mountains had been freely discussed in their home as early as 1854.[13] Brown was gather-

12. Confession of John E. Cook (Charlestown, Va., 1859), printed in Richard J. Hinton, *John Brown and His Men* (rev. ed.; New York, 1894), p. 702; Villard, *Brown*, p. 308; testimony of Richard Realf, Jan. 21, 1860, in *Senate Reports*, 36 Cong., 1 sess., No. 278 (Serial 1040), cited hereafter as *Senate Report on Harpers Ferry*, p. 92: "During our passage across Iowa, Brown's plan in regard to an incursion into Virginia gradually manifested itself." Also, Sanborn, *Brown*, p. 425, quoting Edward Coppoc, and p. 541, quoting Owen Brown.

13. Douglass wrote in the *North Star*, Dec. 8, 1848, of his recent interview with Mr. John Brown, but he did not indicate what they discussed. Years later, in the *Life and Times of Frederick Douglass Written by Himself* (1881; rev. 1892; reprint, 1962), pp. 271–275, Douglass told of his meeting with Brown at Springfield, Massachusetts, in 1847 (his memory was wrong by one year), and how Brown had unfolded to him a plan to operate in the Allegheny Mountains to emancipate the slaves of the South. "These mountains are the basis of my plan. God has given the strength of the hills to freedom; they were placed here for the emancipation of the Negro race; they are full of natural forts where one man for defense will be equal to a hundred for attack; they are full also of good hiding places." Douglass seems to have been half convinced. His story is accepted by Villard, *Brown*, pp. 47–48; Oates, *To Purge This Land*, pp. 62–63, 372; Benjamin Quarles, *Frederick Douglass* (Washington, D.C., 1948), pp. 170–171; Arna Bontemps, *Free at Last: The Life of Frederick Douglass* (New York, 1971), pp. 176–180. However, Abels, *Man on Fire*, pp. 26–27, disagrees: "Sanborn is apparently on firm ground in stating that several decades later when he wrote his autobiography Douglass was confused as to the time and this disclosure actually came eleven years later." See Sanborn, *Brown*, p. 421 n. Villard, *Brown*, p. 54, cites statement by Brown's daughter Annie, made in 1908, that she first heard of the plan to raid Harpers Ferry in 1854.

ing information on slave insurrections as early as 1855. But there is no evidence of any explicit plans or commitments until August 1857, just before his return to Kansas. At this time he told his associate, the English soldier of fortune Hugh Forbes, of a plan to invade Virginia and free the slaves, and Forbes challenged the practicability of the plan.[14] But Brown continued with his project nevertheless, and after November it emerged as a grandiose and revolutionary scheme, wholly unlike his participation in the homemade wars of Kansas. Again he would need money, and this time it was a project which could not be advocated before a legislature. Many of the people he had appealed to previously were too mild to be approached on this matter, and Brown despised the timidity of most of the abolitionists, in any case. But there were some men in Boston whom he felt that he could trust. He started east again in January 1858.

Early in February, he revealed his scheme to Frederick Douglass, who had been both a slave and a fugitive and who had a realistic understanding of what was involved. Douglass warned him against the plan, but Brown did with this as he did with all advice—he ignored it.[15] Later in the same month, at Gerrit Smith's home in Peterboro, New York, he unfolded to Smith and to Franklin Sanborn a plan for a campaign in slave territory somewhere east of the Alleghenies to set up a government that would overthrow slavery. Sanborn accurately described it as "an amazing proposition, desperate in its character, wholly inadequate in its provision of means," and he might have added, profoundly illegal in its purposes. Smith and Sanborn tried to induce him to give it up, but when he proved unyielding, they rallied to his support, and as he soon wrote to his family, "Mr. Smith & family go *all* lengths with me."[16]

14. Forbes to Samuel Gridley Howe, April 19, 1858, in New York *Herald,* Oct. 27, 1859; Franklin B. Sanborn to Forbes, Jan. 15, 1858, in Sanborn, *Brown,* pp. 429–430.

15. Douglass, *Life and Times,* pp. 315–320.

16. Brown to his wife and children, Feb. 24, 1858, cited in Villard, *Brown,* p. 320. Of the "six" who supported Brown, the subsequent conduct of Smith was perhaps least admirable. Before Brown's raid, Smith was publicly predicting insurrections, but immediately after the raid, he destroyed all evidence in his possession bearing on Brown's plan, and sent to Boston and to Ohio to have evidence there destroyed also. Five days after Brown was sentenced to death, Smith, who had been accused by the New York *Herald,* Oct. 21, 1859, of being an accessory before the fact, and who expressed acute fear of indictment, was taken to the New York State Asylum for the Insane. Subsequently, Smith showed an almost obsessive impulse to deny any real connection with Brown's enterprise. Ralph Volney Harlow, *Gerrit Smith, Philan-*

From Peterboro, Brown went on to Boston, where he met five of his staunchest supporters, George L. Stearns, Franklin B. Sanborn, Thomas Wentworth Higginson, Theodore Parker, and Samuel Gridley Howe. To them also he unfolded his plan, and all of them agreed to raise money for his support. These five, together with Gerrit Smith, became known as the "Secret Six," and it was their rather limited aid which finally enabled Brown to strike his blow at Harpers Ferry.

The five are remembered chiefly as genteel intellectuals and philanthropists: Howe, a pioneer in care for the blind and the mentally retarded; Parker, a Unitarian clergyman of astonishing erudition and scholarly eminence; Higginson, another Unitarian, living at the very hub of the Brahmin society to which he had been born, and later the "dear preceptor" of Emily Dickinson; Stearns, the richest man in Medford, the husband of Lydia Maria Child's niece, close friend of Sumner, and the patron of all good causes; Sanborn, a younger man, a hard worker who became secretary of every group he joined, and who ultimately made a career of exploiting his relation with great men whom he had hero-worshiped—Brown, Howe, Emerson, Thoreau, and Bronson Alcott. But at the time, all were notable as unusually militant antislavery men. Howe, Higginson, and Sanborn had all been to Kansas. Stearns had been one of the foremost raisers of funds for the purchase of Sharps rifles. Parker had been head of the Boston Vigilance Committee which was committed to resisting the Fugitive Slave Law by violence if nonviolent methods failed. The other four were also members. Stearns and Parker had concealed fugitives in their homes. Higginson, the most extreme, personally led an attack to rescue Anthony Burns from the Boston Court House in 1854, and three years later, he sponsored a "Disunion Convention" at Worcester.[17]

---

thropist and Reformer (New York, 1939), pp. 407–422, 450–454, gives details and evidence on the controversies and litigation between Smith and (1) Watts Sherman and others, and (2) the Chicago *Tribune,* because of their statements that he had been a party to Brown's activities.

17. On Higginson, see his own *Cheerful Yesterdays* (Boston, 1898); Edelstein, *Strange Enthusiasm;* Howard N. Meyer, *Colonel of the Black Regiment: The Life of Thomas Wentworth Higginson* (New York, 1967). On Parker: John Weiss, *Life and Correspondence of Theodore Parker* (2 vols.; New York, 1864); Henry Steele Commager, *Theodore Parker, Yankee Crusader* (Boston, 1936), an admirable scholarly study, but brief on John Brown. On the other three Bostonians: Frank Preston Stearns, *The Life and Public Services of George Luther Stearns* (Philadelphia, 1907); F. B. Sanborn, *Recollections of Seventy Years* (2 vols.; Boston, 1909); Harold Schwartz, *Samuel Gridley Howe, Social Reformer* (Cambridge,

In the end, news of the Harpers Ferry raid threw four of the Secret Six into panic at the thought of being implicated. Parker was dying in Europe, and only Higginson stood firm, neither disclaiming his association with Brown nor taking flight, nor destroying his correspondence. Yet even Higginson in later years tended to minimize the revolutionary character of Brown's plans, and indeed it was never certain whether he had any carefully formulated plans, or, if he did, to what extent he actually adhered to them after he went to Maryland. Further, he was so secretive—and so distrustful of some of his supporters—that one cannot assume that he revealed his plans—especially to men who frankly stated that they did not want to know too much in detail. Thus, controversy turns on three questions: (1) whether Brown had fixed plans, (2) whether he revealed them, and (3) whether the revelations were understood by those to whom they were disclosed. These questions will always leave some aura of uncertainty, but the fact is that there was never as much uncertainty about *what* Brown proposed to do as about *how to interpret it.* He proposed to take an armed force into Virginia, rally the slaves, place weapons in their hands, and resist by force any effort to prevent their being freed. Such action could hardly have failed to result in a bloody slave insurrection, and indeed Howe, Smith, and Parker all talked about it in those terms. It would, on the other hand, be possible to argue that the slaves would not resort to violence unless the whites made efforts to subjugate them, in which case the slave masters and not the slaves would be responsible for any violence that ensued. Also, one could cling to the idea that Brown intended to recruit a large number of slaves and hurry them north to freedom, rather than to precipitate a large-scale insurrection in the slave states. Brown's own statements illustrate the ambiguity in the project, for soon after his capture he asserted that freeing the slaves was "absolutely our only object," but he admitted in the next breath that he had taken a prisoner's money and watch, and that "we intended freely to appropriate the property of slaveholders to carry out our object."[18] Again, in the famous speech on the occasion of being sentenced to death, he admitted a "design on

---

Mass., 1956). For a brisk, irreverent, and, in my opinion, very acute analysis of the role of the Secret Six in the Harpers Ferry affair, see J. C. Furnas, *The Road to Harpers Ferry* (New York, 1959), pp. 327–382.

18. Questioning of Brown by Senator Mason, Governor Wise, and others, Oct. 19, 1859, in New York *Herald,* Oct. 21, 1859, reprinted in Sanborn, *Brown,* pp. 562–569.

my part to free the slaves," suggesting that this might be accomplished simply by spiriting the slaves away. "I never did intend murder or treason," he said, "or the destruction of property, or to excite or incite slaves to rebellion, or to make insurrection."[19] Later still, he amended this second statement by saying, "I intended to convey this idea, that it was my object to place the slaves in a condition to defend their liberties, if they would, without any bloodshed, but not that I intended to run them out of the slave states."[20] No doubt Brown meant to make the point that his primary purpose was to free slaves and not to kill slaveholders. Still, it was a tenuous distinction to say that slaves would be encouraged to defend their freedom but not incited to insurrection; or that a governmental arsenal would be seized, its defenders overpowered, and its arms taken but that no treason was intended; or that bloodshed would be avoided but that the property of slaveowners would be seized. Brown's disclaimers amounted to the assurance that no persons would be killed unless they interfered with what Brown was engaged in doing. In this sense, any one of the Six could have asserted that he had not meant to support an insurrection. But all of them knew that Brown intended to strike with armed men, to take slaves from their masters by force if necessary, to take hostages, and to prevent the masters from regaining control over the slaves.[21] They should have known, and probably did know, that this amounted to starting a servile insurrection, whatever it might be called. Parker and Higginson—and, for a while, Howe and Smith—seem to have been willing to recognize this reality frankly. They regarded slavery itself as a kind of war which gave philosophical justification to resistance by the slave.

It is pertinent that the word "treason" was first applied to them not by their accusers but by themselves. Not only was Higginson, in his own words, "always ready to invest in treason,"[22] but Sanborn, at almost the same time, said, "The Union is evidently on its last legs and Buchanan is laboring to tear it in pieces. Treason will

19. See below, pp. 377–378.

20. Brown to Andrew Hunter, Nov. 22, 1859, in *Senate Report on Harpers Ferry*, "Testimony," pp. 67–68.

21. This is the substance of John Brown, Jr.'s later deposition about his father's plans, made July 19, 1867, cited in Harlow, *Gerrit Smith*, p. 398. See Oates, *To Purge This Land*, pp. 233–238.

22. Higginson to Brown, Feb. 8, 1858, quoted in Edelstein, *Strange Enthusiasm*, p. 208.

not be treason much longer, but patriotism."[23] The fact that the Six did not reveal the plan to other antislavery men suggests their awareness that it was strong medicine; their skittish insistence that Brown should refrain from informing them about details of his plan testified to their recognition of the illegality of his intended measures.

Perhaps the clearest indication of how much they knew, however, is an indirect one. Early in 1858, two bombshells, in the form of letters from Hugh Forbes, hit Howe and Sanborn.[24] Brown had never told them about Forbes, but clearly he had told Forbes about *them,* for Forbes related that Brown had employed him to drill troops and had spoken of his financial support in Boston, but had not paid him what was promised. Forbes held Brown's backers responsible for this default. He also disparaged Brown's judgment, demanded to be paid or put in charge of the whole operation, and threatened to sell his secrets to the New York *Herald* if he were not recompensed. The Six did not yield to this blackmail, but the important point was that they learned, almost inadvertently, not what Forbes knew about Brown, but what he knew about them—which only Brown could have told him. Sanborn, explaining the whole matter to Higginson, wrote that Forbes knew "what very few do [know]—that the Dr. [Howe], Mr. Stearns and myself are informed of it—How he got this knowledge is a mystery."[25] In short, Sanborn and Howe knew enough about Brown's plans to be deeply concerned that anyone else should be aware of their knowledge.

In spite of many subsequent efforts to make it appear that Brown was engaged simply in a "raid," in which he intended only to snatch a few slaves and quickly slip away to some hideout in the Virginia mountains, it is clear that his enterprise was meant to be of vast magnitude and to produce a revolutionary slave uprising throughout the South. The first proof of this lies in a "Provisional Constitution" which Brown imprudently presented to a group of some thirty-five Negroes and a few white men at Chatham, Ontario, in April 1858. This document was so strange that a question must arise as to the sanity of its framer. But with its provisions for confiscating all the personal and real property of slaveowners, and for imposing

23. Sanborn to Higginson, Feb. 11, 1858, quoted in *ibid.*, p. 209.
24. Schwartz, *Howe,* pp. 227–230.
25. Sanborn to Higginson, May 5, 1858, quoted in Sanborn, *Brown,* p. 458. On Hugh Forbes and his threats see Villard, *Brown,* pp. 285–318.

martial law, and for maintaining an elaborate government over a large area, it clearly contemplated a sustained military occupation of an extensive region in which slavery would be overthrown. Since Brown never expected to have more than fifty or a hundred men in his striking force, and since he later gave military commissions to thirteen out of seventeen of his white followers (though to none of the five Negroes), it is evident that the large army necessary to this operation would have to be composed of slaves who had thrown off their bondage.[26] The second proof lies in his decision to seize the armory at Harpers Ferry, Virginia. Harpers Ferry was in a difficult location, and it was clearly more risky to attack federal property than private property. The only thing to be gained by seizing the armory was weapons, and since Brown's own little band already had far more guns than they needed, one can only conclude that he intended to place arms in the hands of large numbers of slaves.

If all had gone according to plan, Brown would have struck in the summer of 1858, but at the last moment Hugh Forbes threatened to sink the project by revealing all of the secret plans. Forbes had joined the enterprise in the belief that Brown could make it highly lucrative by tapping large wealth in New England (Brown may once have believed this himself). Later, when Brown could give him only a few hundred dollars for many months' service, and when he became disillusioned by the mistakes in Brown's planning, he defected and went to Senators Henry Wilson (in person) and William H. Seward (by letter) with information about the plot. Wilson reacted by sending Howe a very sharp letter, with pointed inquiries about why the Kansas Committee was mixed up in an affair like this, and with warnings that it would seriously injure the antislavery cause.[27] The Six promptly held hurried meetings at which they transferred property from the Kansas Committee to Stearns so they could deny that the Committee was involved, and then, over the protests of Howe and Higginson, they instructed Brown that he must suspend his plan and go west.[28]

26. Text of this constitution in *Senate Report on Harpers Ferry,* pp. 48–59.

27. Testimony of Wilson and Seward, *ibid.*, "Testimony," pp. 140–145, 253–255; Also Wilson to Howe, May 9, 1858, in Stearns, *Stearns,* p. 168; Howe to Wilson, May 12, May 15, in Sanborn, *Brown,* p. 462; Stearns informed John Brown, May 14, May 15, and Brown replied in an undated letter, all in Stearns, *Stearns,* pp. 169–170.

28. Sanborn, *Brown,* p. 463; telegram May 24, 1858, Sanborn to Smith, in Harlow, *Smith,* p. 402; Edelstein, *Strange Enthusiasm,* pp. 210–212.

Higginson thought the enterprise would never be revived, and with anyone but Brown as leader, it probably would not have been. Certainly the postponement was dangerous, not only because of the difficulty of keeping the little band together during an indefinite delay, but also because it seemed unlikely that the precarious secrecy of the plot could be preserved much longer. A number of Brown's young followers had talked and written indiscreetly; many Ontario Negroes must have known about the "convention" at Chatham; Forbes had already let his tongue wag; and the security precautions of the Secret Six might have seemed amateurish even to a small boy. Moreover, one of Brown's followers, John H. Cook, was already at Harpers Ferry, where he soon found a job and a wife. Brown was acutely fearful that Cook would talk too much.

Perhaps only a mad project could have survived, but, in any case, this one did. Brown went back to Kansas for the third and last time in June 1858, and in December he led a raid into Missouri in which his followers killed one slaveholder, took a certain amount of livestock and property, and liberated eleven slaves whom they then carried east in midwinter, across the northern prairies, all the way to Ontario.[29] This was perhaps the most successful operation Brown ever engaged in. After another three and a half months of fund raising and delay, he then went to Maryland and rented a farm five miles from Harpers Ferry. There he settled down and waited three and a half months longer for additional men and money, which, for the most part, never arrived. By mid-October he had twenty-two followers and probably recognized that his little force never would be any stronger. On the evening of October 16, he set out with all but three of these men, marched down toward the Potomac with a wagonload of arms, cut the telegraph wires, crossed the bridge, captured the watchman guarding the bridge, and moved into Harpers Ferry. With no difficulty whatever, he seized the armory and rifle works. He then sent out a detail to capture two slaveholders of the neighborhood along with their slaves. One of these was Colonel Lewis Washington, a great-grandnephew of George Washington, and Brown told his men to be sure to bring in one of the family heirlooms, the sword which Frederick the Great had presented to George Washington. This mission was accom-

29. Villard, *Brown*, pp. 346–390; Oates, *To Purge This Land*, pp. 260–264; Allan Nevins, *The Emergence of Lincoln* (2 vols.; New York, 1950), II, 23–26.

plished, and the detail, with its prisoners, was back at the armory by daybreak. Meanwhile, at about 1:00 A.M., Brown's men had stopped a Baltimore and Ohio train and inadvertently killed the Negro baggage master, but had later allowed the train to go on its way.

At morning, as the employees at the armory straggled in for their day's work, Brown took a number of them prisoner, and he tried to send a detail back to his farm to move some of the military equipment from there to a schoolhouse nearer the Ferry. But otherwise he sat down and waited. In his own mind, he was waiting for the slaves to rise, but in reality, he was waiting for the slow-moving forces of organized society to get into motion and to overwhelm him. By midmorning, the local militia of nearby towns in Maryland and Virginia were on their way to the Ferry, and the president of the Baltimore and Ohio Railroad had decided to risk being made a laughingstock by reporting to Washington the incredible information that an insurrection was in progress at Harpers Ferry. Also the local inhabitants began to seize the initiative. At first they had lain very low, assuming quite logically that no one would dare seize a government arsenal without a large force at his back. But now they began a desultory firing in the direction of the armory.

By midafternoon of October 17, the militia companies had arrived and gained control of both bridges. Outpost details that Brown had placed were killed or driven in or had escaped, and Brown himself was forced to hole up in the engine works. By ten o'clock that night, Lieutenant Colonel Robert E. Lee, United States Cavalry, with his aide Lieutenant J. E. B. Stuart, had come to command all federal forces in the area.

The engine house could have been captured that night, within twenty-four hours of the beginning of the raid, but Lee, very much the professional soldier, was in no hurry. He preferred to observe protocol, giving the Virginia troops a chance to lead the assault if they wanted to (which they did not), giving the insurrectionists a chance to surrender, and taking precautions to avoid shooting any of Brown's prisoners. The next morning, he sent Stuart to parley with the leader of the insurrectionists, and as they talked through a crack in the engine house door, Stuart, who had served in Kansas, was astonished to recognize John Brown of Osawatomie. Up to this time, no one on the outside had known who was attacking. A few moments later, when Brown refused to surrender, Stuart stepped aside and waved in a detachment of twelve marines who charged with fixed bayonets, without firing a shot. In a few moments it was all over. One marine and two of Brown's men were killed in the assault. Brown himself would have been killed if his assailant, Lieutenant Israel Green, in command of the detachment, had not been armed only with a decorative dress sword which inflicted some painful but not very serious wounds. Altogether, Brown's men had killed four people and wounded nine. Of his own small force, ten were dead or dying; five had escaped the previous day, and seven were captured.[30]

Technically, Brown's operations had been almost incredibly bad. Leading an army of twenty-two men against a federal arsenal and the entire state of Virginia, he had cut himself off from any chance of escape by moving into a position where two rivers walled him in, as if in a trap. Leading what purported to be an utterly secret operation, he had left behind him on the Maryland farm a large accumulation of letters which revealed all his plans and exposed all his confederates; as Hugh Forbes wrote, "the most terrible engine of destruction which he [Brown] would carry with him in his campaign would be a carpet-bag loaded with 400 letters, to be turned against his friends, of whom the journals assert that more than forty-seven are already compromised."[31] After three and a half months of preparation, he marched at last without taking with him food for his soldiers' next meal, so that, the following morning, the commander in chief of the Provisional Army of the North, in default of commissary, was obliged to order forty-five breakfasts sent over from the Wagner House. For the remaining twenty-four hours, the suffering of Brown's besieged men was accentuated by acute and needless hunger. His liaison with allies in the North was so faulty that they did not know when he would strike, and John Brown, Jr., assigned to forward additional recruits, later stated that the raid took him completely by surprise. If, as is sometimes suggested, this indicated the disordered condition of Brown, Jr.'s mind rather than lack of information from his father, it still leaves the question of why such a crucial role should have been entrusted to one whose mental instability had been conspicuous ever since Pottawatomie. Finally, the most bizarre feature of all is that Brown tried to lead a slave insurrection without letting the slaves know about it. It is as clear

30. Villard, *Brown*, pp. 402–455; Oates, *To Purge This Land*, pp. 288–301.
31. Forbes, quoted in Villard, *Brown*, p. 467.

as it is incredible that his idea of a slave insurrection was to kidnap a few slaves, thrust pikes into their hands while holding them under duress, and inform them that they were free. He then expected them to place their necks in a noose without asking for further particulars.[32] As Abraham Lincoln later said, with his disconcerting accuracy, "It was not a slave insurrection. It was an attempt by white men to get up a revolt among slaves, in which the slaves refused to participate. In fact, it was so absurd that the slaves, with all their ignorance, saw plainly enough it could not succeed."[33]

Lincoln also said, "John Brown's effort was peculiar." With all that has been written about whether John Brown was "insane," this is perhaps as exact as it is possible to be. But let it be said briefly, first, that insanity is a clear-cut legal concept concerning a mental condition which is seldom clear-cut; and second, that the insanity explanation has been invoked too much by people with ulterior purposes—first by those who hoped to save Brown's life, then by Republicans who wanted to disclaim his act without condemning him morally, and finally by adverse critics who hoped to discredit his deeds by calling them the acts of a madman. The evidence shows that Brown was very intense and aloof, that he became exclusively preoccupied with his one grand design, that he sometimes behaved in a very confused way, that he alternated between brief periods of decisive action and long intervals when it is hard to tell what he was doing, that mental instability occurred with significant frequency in his family, and that some believed he had a vindictive or even a homicidal streak with fantasies of superhuman greatness. Also, Pottawatomie should be borne in mind. From all this, one may clearly infer that Brown was not, as we now say, a well-adjusted man.[34] But the strongest element in the case for his madness is the seeming irrationality of the whole Harpers Ferry operation. In lay terms, a man who tried to conquer the state of Virginia with twenty-two men might be regarded as crazy. Was Brown crazy in these terms?

This question presents a difficulty, for if a belief in the possibility

32. David M. Potter, "John Brown and the Paradox of Leadership among American Negroes," in his *The South and the Sectional Conflict* (Baton Rouge, 1968), pp. 201–218.

33. Lincoln, Cooper Union Address, Feb. 27, 1860, in Roy P. Basler (ed.), *Collected Works of Abraham Lincoln* (8 vols.; New Brunswick, N.J., 1953), III, 541.

34. On Brown's psychological condition, see very able discussions in Nevins, *Emergence*, II, 5–11; C. Vann Woodward, "John Brown's Private War," in his *The Burden of Southern History* (Baton Rouge, 1960), pp. 45–49.

of a vast, self-starting slave insurrection was a delusion, it was one Brown shared with Theodore Parker, Samuel Gridley Howe, Thomas Wentworth Higginson, and a great many others whose sanity has never been questioned at all. It was an article of faith among the abolitionists that the slaves of the South were seething with discontent and awaiting only a signal to throw off their chains. Gerrit Smith believed it, and two months before Brown's attempted coup he wrote, "The feeling among the blacks that they must deliver themselves gains strength with fearful rapidity."[35] Samuel Gridley Howe believed it, and even after Brown's failure and when war came, he wrote that twenty to forty thousand volunteers could "plough through the South & be followed by a blaze of servile war that would utterly and forever root out slaveholding and slavery."[36] Theodore Parker believed it, and wrote after Harpers Ferry, "The Fire of Vengeance may be waked up even in an African's heart, especially when it is fanned by the wickedness of a white man; then it runs from man to man, from town to town. What shall put it out? The white man's blood."[37] Thomas Wentworth Higginson believed it and suggested that white men were foolish to be shaved by Negro barbers. "Behind all these years of shrinking and these long years of cheerful submission," he added, "there may lie a dagger and a power to use it when the time comes."[38] As J. C. Furnas has expressed it, there was a widespread "spartacus complex" among the abolitionists, a fascinated belief that the South stood on the brink of a vast slave uprising and a wholesale slaughter of the whites. "It is not easy, though necessary," says Furnas, "to grasp that Abolitionism could, in the same breath warn the South of arson, rape, and murder and sentimentally admire the implied Negro mob leaders brandishing axes, torches, and human heads."[39] If Brown believed that the South was a waiting pyre, and that twenty-two men without rations were enough to put a match to it, the belief was one of the least original notions in his whole stock of ideas. Thus the Boston

35. Smith to the chairman of the Jerry Rescue Committee, Aug. 27, 1859, in Octavius Brooks Frothingham, *Gerrit Smith,* (New York, 1879), p. 240. "For many years," said Smith, "I have feared and have published my fears that slavery must go out in blood. . . . These fears have grown into belief."

36. Schwartz, *Howe,* p. 250, citing Howe to Martin F. Conway, Dec. 10, 1860.

37. Henry Steele Commager (ed.), *Theodore Parker: An Anthology* (Boston, 1960), p. 267.

38. Quoted in Edelstein, *Strange Enthusiasm,* p. 211.

39. Furnas, *Road to Harper's Ferry,* p. 232.

Post spoke much to the point when it said, "John Brown may be a lunatic [but if so] then one-fourth of the people of Massachusetts are madmen."[40]

The *Post* certainly did not intend to shift the question from one concerning Brown's personal sanity to one concerning the mass pathology of the abolitionists. A historian may, however, regard the latter as a legitimate focus of inquiry, especially now that it is recognized that rationality is by no means a constant in human society. But any question about whether the abolitionists were in touch with reality must carry with it a recognition that the Spartacus complex was by no means confined to the abolitionists. Southerners shared it in the sense that they were ever fearful of slave insurrection and were immensely relieved to learn that the slaves had not flocked to Brown's support. Clearly they had felt that it might be otherwise.[41]

A year and a half later, when the Civil War came, experience proved that the slaves were not as resentful or as bloodthirsty as the abolitionists thought, and though they decamped in droves from their plantation homes, the path which they chose to freedom was not the path of insurrection, rapine, and butchery. In the light of the Civil War experience, it seems justifiable to say that Brown had been wrong in supposing that the slaves were ripe for revolt.[42] Yet even this conclusion has to be qualified by the fact that Brown did not submit his own hypothesis to a fair test. He did not give the slaves a chance to show how they would react to an insurrection. In spite of all Brown's pretense of having made a deep study of Spartacus, Toussaint, and other practitioners of the art of slave revolt, he managed his plans in a way which Toussaint or Gabriel Prosser, not to mention Denmark Vesey, would have scorned. More than a year before he struck, Hugh Forbes warned him that even slaves ripe for revolt would not come in on a plan like his. "No preparatory notice

40. Quoted in Woodward, "John Brown's Private War," p. 48.

41. Henry A. Wise, in a speech at Richmond, said, "And this is the only consolation I have to offer you in this disgrace, that the faithful slaves refused to take up arms against their masters. . . . Not a slave around was found faithless." Richmond *Enquirer*, Oct. 25, 1859.

42. On the larger question of the extent to which American slaves were predisposed to revolt, see Eugene D. Genovese, *In Red and Black: Marxian Explorations in Southern and Afro-American History* (New York, 1972), pp. 73–101, 129–157. Genovese remarks that "the staggering truth is that not one full-scale slave revolt broke out during a war in which local white police power had been drastically reduced" (p. 139).

having been given to the slaves," he said, "the invitation to rise might, unless they were already in a state of agitation, meet with no response, or a feeble one."[43] But Brown brushed this aside: he was sure of a response, and calculated that on the first night of the revolt, between two hundred and five hundred slaves would rally to him.[44] This expectation explains a great deal—why Brown dared to start a war with an army of twenty-two men, why he wanted the weapons at Harpers Ferry, why seventeen of his men held officers' commissions, why he carried no rations with him, why he had taken the trouble to frame a provisional constitution and get it adopted, and most of all, why he did nothing but wait at the arsenal on October 16 while his enemies gathered to beset him.

To Brown and the abolitionists, the plan seemed perfectly reasonable, and the literati of Boston admired him extravagantly as a man of action for attempting it. But to Frederick Douglass and the Negroes of Chatham, Ontario, nearly every one of whom had learned something from personal experience about how to gain freedom, Brown was a man of words trying to be a man of deeds, and they would not follow him. They understood him, as Thoreau and Emerson and Parker never did.

Two of Brown's sons were killed at Harpers Ferry. If he had been killed also, as he certainly would have been but for the inadequacy of Israel Green's dress sword, the impact of his coup would probably have been very much diminished, for the general public did not sympathize with promoters of slave insurrections, and it might quickly have dismissed Brown as a mere desperado. But he was not killed, and he surpassed himself as few men have ever done, in the six weeks that followed. The most striking testimony to his superb behavior was the fact that he extorted the complete admiration of the Virginians. They had regarded all abolitionists as poltroons, but Brown showed a courage which captivated southern devotees of the cult of courage in spite of themselves. Governor Henry A. Wise, a Virginian far gone in chivalry, was perhaps worse smitten than any of them. "He is a bundle of the best nerves I ever saw, cut and thrust and bleeding and in bonds," said Wise. "He is a man of clear head,

43. Forbes to S. G. Howe, May 14, 1858, in New York *Herald,* Oct. 27, 1859.

44. Brown expressed to Frederick Douglass his conviction that when he invaded Harpers Ferry the slaves would flock to his support, and he implored Douglass to join the expedition: "When I strike, the bees will begin to swarm, and I shall want you to help hive them." Douglass, *Life and Times,* pp. 319–320.

of courage, fortitude, and simple ingeniousness. He is cool, collected, and indomitable, and it is but just to him to say that he was humane to his prisoners."[45] Later, refusing to have Brown examined for insanity, he said, "I know that he was sane, and remarkably sane, if quick and clear perception, if assumed rational premises and consecutive reasoning from them, if cautious tact in avoiding disclosures and in covering conclusions and inferences, if memory and conception and practical common sense, and if composure and self-possession are evidence of a sound state of mind."[46]

The admiration of the Virginians for Brown's gameness, of course, would not prevent them from trying him and hanging him for his offense, and he recognized this fact calmly without waiting for sentence to be pronounced. As he did so, he had composure and unselfishness enough to recognize that the manner of his death might be a great service to antislavery, and he prepared to die in a way which would glorify his cause. Harpers Ferry had been another failure after a lifetime of failures, but he still faced one more test—the wait for the gallows—and, while this might seem a harsher one than all the others, he knew that this was a test he would not fail. "I have been *whiped* as the saying *is,*" he wrote to his wife, "but am sure I can recover all the lost capital occasioned by that disaster, by only hanging a few moments by the neck; & I feel quite determined to make the utmost possible out of a defeat."[47]

Description can hardly do justice to his conduct. He was arraigned with excessive promptness, while still suffering from his wounds, and was indicted and brought to trial on the day of the arraignment, one week after his capture. The trial lasted one week, after which he was sentenced to be hanged one month from the date of sentence. This haste was shocking by any standards and appalling by modern standards of infinite prolongation, but it was generally

45. Governor Wise, speech of Oct. 21, 1859 (see note 41 above).

46. Message of Wise to Virginia Legislature, Dec. 5, 1859; quoted in Villard, *Brown,* p. 509. See also Nevins, *Emergence,* II, 92–93.

47. Brown to Mrs. Brown, Nov. 10, 1859, in Villard, *Brown,* p. 540. Brown's supporters, too, were quick to recognize the tactical usefulness of his death. Thomas Wentworth Higginson declared, "I don't feel sure that his acquittal or rescue would do half as much good as his being executed," Mary Thacher Higginson (ed.), *Letters and Journals of Thomas Wentworth Higginson* (Boston, 1921), p. 85; Thoreau wrote on Oct. 22, 1859: "I almost fear to hear of his deliverance, doubting if a prolonged life, if any life, can do as much good as his death." Bradford Torrey and Francis H. Allen (eds.), *The Journal of Henry D. Thoreau* (14 vols.; Boston, 1906), XII, 429.

agreed by Brown and others that the trial was conducted fairly and with a rough justice.[48] During the trial, where Brown lay wounded on a pallet, and later, while awaiting execution, he handled himself with an unfailing dignity and composure. Apparently he never flinched from the hour of his capture until the moment of his death. His conduct deeply affected his jailer, won the hearts of his guards, and made a profound impression on millions of people who stood the death watch vicariously with him as his execution approached. On the occasion of his sentence, he responded with one of the classic statements in American prose:

> . . . it is unjust that I should suffer such a penalty. Had I interfered in the manner in which I admit, and which I admit has been fairly proved—for I admire the truthfulness and candor of the greater portion of the witnesses who have testified in this case—had I so interfered in behalf of the rich, the powerful, the intelligent, the so-called great, or in behalf of any of their friends, either father, mother, brother, sister, wife or children, or any of that class, and suffered and sacrificed what I have in this interference, it would have been all right. Every man in this Court would have deemed it an act worthy of reward rather than punishment.
>
> This Court acknowledges, too, as I suppose, the validity of the law of God. I see a book kissed, which I suppose to be the Bible, or at least the New Testament, which teaches me that all things whatsoever I would that men should do to me, I should do even so to them. It teaches me, further, to remember them that are in bonds as bound with them. I endeavored to act up to that instruction. I say I am yet too young to understand that God is any respecter of persons. I believe that to have interfered as I have done, as I have always freely admitted I have done, in behalf of His despised poor, is no wrong, but right. Now, if it is deemed necessary that I should forfeit my life for the furtherance of the ends of justice, and mingle my blood further with the blood of my children and with the blood of millions in this slave country whose rights are disregarded by wicked, cruel, and unjust enactments, I say, let it be done.
>
> Let me say one word further. I feel entirely satisfied with the treatment I have received on my trial. Considering all the circumstances, it has been more generous than I expected. But I feel no consciousness of guilt. I have stated from the first what was my intention, and what was not. I never had any design against the liberty of any person, nor any disposition to commit treason or incite slaves to rebel or make any general insurrection. I never

48. On the trial, the fullest reports were in the daily papers such as the New York *Herald*, the *National Intelligencer*, etc. A good collection of such reportage is *The Life, Trial, and Execution of Captain John Brown* (New York: Robert M. De Witt, Publisher; reprinted 1969), pp. 55–95.

encouraged any man to do so, but always discouraged any idea of that kind.[49]

In its broad historical effects, John Brown's death was significant primarily because it aroused immense emotional sympathy for him in the North, and this sympathy, in turn, caused a deep sense of alienation on the part of the South, which felt that the North was canonizing a fiend who sought to plunge the South into a blood bath.

When John Brown was hanged at Charlestown, Virginia, on December 2, 1859, the organized expressions of sympathy in the North reached startling proportions. Church bells tolled, black bunting was hung out, minute guns were fired, prayer meetings assembled, and memorial resolutions were adopted. In the weeks following, the emotional outpouring continued: lithographs of Brown circulated in vast numbers, subscriptions were organized for the support of his family, immense memorial meetings took place in New York, Boston, and Philadelphia, a memorial volume was rushed through the press, and a stream of pilgrims began to visit his grave at North Elba, New York. The death of a national hero could not have called forth a greater outpouring of grief.

If this outburst of national mourning—for it was nothing less—had been confined merely to expressions of admiration for Brown's courage and sorrow for his death, perhaps the ultimate significance might have been less. Society allows everyone a considerable measure of eulogy in lamenting a death, and probably no one would have objected very seriously when young Louisa May Alcott wrote:

> No breath of shame can touch his shield
> Nor ages dim its shine.
> Living, he made life beautiful,
> Dying, made death divine.

But it quickly appeared that the celebration of the memory of John Brown was not so much a matter of mourning for the deceased as it was of justifying his purposes and damning the slaveholders. Two days after he was sentenced, the *Liberator* exhorted its readers to "let the day of his execution . . . be the occasion of such a public moral demonstration against the bloody and merciless slave system as the

49. Villard, *Brown*, pp. 498–499, adopts the text as it appeared in New York *Herald*, Nov. 3, 1859. Other texts show very minor variations.

land has never witnessed,"[50] and this is, in fact, what it became. Wendell Phillips struck the note of castigation, which was sounded almost endlessly, when he declaimed, before Brown's death, "Virginia is a pirate ship, and John Brown sails the sea, a Lord High Admiral of the Almighty, with his commission to sink every pirate he meets on God's ocean of the nineteenth century. . . . John Brown has twice as much right to hang Governor Wise as Governor Wise has to hang him."[51]

The moral effect of condemning the slave system was achieved partly in an indirect way by extravagant veneration of Brown. In phrases which are well remembered, Emerson declared that Brown would "make the gallows as glorious as the cross." Thoreau compared him to Christ and called him "an angel of light."[52] Garrison said that the huge assembly at Tremont Temple in Boston was gathered to witness John Brown's resurrection. But in many cases, abolitionist speakers and writers went beyond the mere glorification of Brown to an explicit approval of the idea of slave insurrection. Garrison announced, "I am prepared to say 'success to every slave insurrection at the South and in every slave country.' And I do not see how I compromise or stain my peace profession in making that declaration."[53] Wendell Phillips, speaking on "The Lesson of the House," said, "The lesson of the hour is insurrection." The Reverend George B. Cheever thought it "were infinitely better that three hundred thousand slaveholders were abolished, struck out of existence," than that slavery should continue to exist; the Reverend Edwin M. Wheelock believed that Brown's mission was to "inaugurate slave insurrection as the divine weapon of the antislavery

50. *Liberator,* Nov. 4, 1859. For accounts of the demonstrations of mourning, see Villard, *Brown,* pp. 558–564; Nevins, *Emergence,* II, 98–101; James Redpath, *Echoes of Harper's Ferry* (Boston, 1860).

51. Speech at Brooklyn, Nov. 1, 1859, *ibid.,* pp. 51–52.

52. Emerson delivered two memorial addresses for Brown at Boston, Nov. 18, 1859, and at Salem, Jan. 6, 1860, Ralph Waldo Emerson, *Miscellanies* (Boston, 1904), pp. 267–281, but his famous remark, quoted above, was made in a lecture on "Courage" on Nov. 8, and was omitted from the published version. See Ralph L. Rusk, *The Life of Ralph Waldo Emerson* (New York, 1949), p. 402. Thoreau's "Plea for Captain John Brown" and his "The Last Days of John Brown" are in *A Yankee in Canada* (Boston, 1866), pp. 152–181, 278–286, esp. 179: "Some eighteen hundred years ago, Christ was crucified; this morning, perchance, Captain Brown was hung. These are the two ends of a chain which is not without its links. He is not Old Brown any longer; he is an angel of light." Also, Torrey and Allen (eds.), *Journal of Thoreau,* XII, 406, 429, 432, 437, 447; XIII, 6, 7.

53. *Liberator,* Dec. 9, 1859.

cause" and that people should not "shrink from the bloodshed that would follow." To him, Brown's activities had been a "sacred and radiant treason," and to the Reverend Fales H. Newhall the word treason had been "made holy in the American language."[54]

The Albany *Argus* tried to assure the public in general and the South in particular that these pronouncements were not at all representative. "It is the fashion," it said, "to impute to the Clergy as a body, sympathy with the sectional intolerance of the day. Nothing can be more false or more unjust. . . . The divines who preach 'killing no murder' are few indeed. In the city of New York, Cheever (who is a pensioner upon the British Anti-Slavery Societies); in Brooklyn, Beecher; in Boston, one or two of the same kidney, and in the interior some scattered imitators, are all of the clergy engaged in this crusade."[55] To support the *Argus*, a great deal of evidence could be adduced to show that responsible opinion in the North did not support the devotees of insurrection. Two leading Republicans, Abraham Lincoln and William H. Seward, both repudiated Brown's act—Lincoln saying that although Brown "agreed with us in thinking slavery wrong, that cannot excuse violence, bloodshed and treason," and Seward that Brown's execution was "necessary and just," although pitiable. Within a year, the Republican platform of 1860 would characterize Brown's coup as "among the gravest of crimes."[56] Also, many men not Republicans organized Union meetings, at which such eminent figures as John A. Dix and Edward Everett tried to offset the impression that all northerners sympathized with Brown.[57]

But to the South, these reassurances were not convincing. The Republican disclaimers smacked of tactical maneuvers to avoid losing the votes of moderates; it was hard to see in Republican ranks any real regret about Brown except regret that he had failed. As for the Union meetings, they helped very little. They were too obviously inspired by northern merchants, partly motivated by fear of losing southern trade; and they took too much of a proslavery

54. Statements by Phillips and Cheever in Redpath, *Echoes of Harper's Ferry,* pp. 43–66, 141–175; by Wheelock and Newhall, quoted in Woodward, "John Brown's Private War," p. 122.

55. Albany *Argus,* reprinted in *National Intelligencer,* Dec. 7, 1859.

56. Lincoln, speech at Leavenworth, Kansas, Dec. 3, 1859, and at Cooper Union, Feb. 27, 1860, in Basler (ed.), *Works of Lincoln,* III, 502, 538–542; George E. Baker (ed.), *The Works of William H. Seward* (5 vols.; Boston, 1887–90), IV, 637; see below, p. 422.

57. Nevins, *Emergence,* II, 105–106.

tone.[58] By defending slavery, they made it appear that the North was divided between advocates of slavery and advocates of slave insurrection, with no middle group which opposed slavery but also opposed insurrection and throat-cutting as remedies for slavery.

Despite all efforts to explain Brown away, the South knew that what had happened at Harpers Ferry represented far more than the fanatical scheme of one man and a handful of followers. It knew that vast throngs of people had turned out to honor Brown's memory; knew that the Massachusetts legislature had very nearly adjourned on the day of his execution; and knew that Joshua Giddings could count on thousands of votes in Ohio whenever he ran for office, in spite of the fact—or perhaps because of the fact—that he had said he looked forward to the hour "when the torch of the incendiary shall light up the towns and cities of the South, and blot out the last vestiges of slavery."[59] The discovery of much of Brown's correspondence at the farmhouse in Maryland quickly brought to light the fact that he had enjoyed support in high quarters in the North. Indeed, with the names of Howe, Parker, Emerson, and Thoreau among his supporters, it was clear that he had had the backing of the cultural aristocracy of New England. It was also clear from the behavior of the Secret Six that a significant part of this elite was committed to attitudes which went far beyond mere opposition to slavery and made the union of the states seem a questionable relationship indeed. These attitudes included hostility to the Union and hatred of the white South. The southerners of 1859, of course, did not know all that was later revealed. They were not aware that Franklin Sanborn had praised Brown by calling him "the best Disunion champion you can find,"[60] or that if Thomas Wentworth Higginson had had his way, Brown would have followed through on his original plan to seize Harpers Ferry a year earlier.[61] But it *was* publicly known that Higginson had collaborated with Garrison in his disunion convention in 1857, that Gerrit Smith had advised antislavery men in Kansas to fight the federal troops, and that Wendell Phillips was denouncing the American eagle as an American vulture.[62] They

58. *Ibid.*, pp. 106–107; Philip S. Foner, *Business and Slavery: The New York Merchants and the Irrepressible Conflict* (Chapel Hill, 1941), pp. 156–164; William Dusinberre, *Civil War Issues in Philadelphia* (Philadelphia, 1965), pp. 83–94.

59. Quoted in Nevins, *Emergence*, II, 104.

60. Sanborn to Higginson, Sept. 11, 1857, in Villard, *Brown*, p. 303.

61. Edelstein, *Strange Enthusiasm*, pp. 210–211.

62. Harlow, *Gerrit Smith*, pp. 348, 350, 351–352, 355, 357–361; *Proceedings of State Disunion Convention held at Worcester, Massachusetts, January 15, 1857* (Boston, 1857);

knew Theodore Parker as one of Brown's supporters, but they did not know that Parker, after Harpers Ferry, had written: "It is a good antislavery picture on the Virginia Shield: a man standing on a tyrant and chopping his head off with a sword; only I would paint the sword-holder black and the tyrant white, to show the immediate application of the principle."[63] They did not know how far the Six had shared Brown's guilt, but they did know that Gerrit Smith had had a mental breakdown because of his fears that someone would find out, and that Franklin Sanborn and Samuel Gridley Howe had fled to Canada to escape interrogation, while Higginson refused to go before a congressional committee. They did not know that Higginson had seriously discussed an expedition to rescue Brown, although they did know that rescue rumors had been rife in the North.[64]

Plainly, some of the leading intellectuals in the North had subsidized Brown to lead a slave insurrection, and when he paid the penalty for this act, he had been mourned more than any American since Washington. The South, realizing this fact, questioned whether the American Union was a reality or merely the shell of what had once been real. As for Brown's courage, said the Baltimore *American*, this proved nothing. "Pirates have died as resolutely as martyrs." As for Brown's high principles, said Jefferson Davis, his actual mission was "to incite slaves to murder helpless women and children."[65]

If we may believe John W. Burgess, there was a revolution of opinion in the South within six weeks after Harpers Ferry. Unionist sentiment, which had remained robust up to that time, suddenly began sinking as the South saw itself isolated and beset in a union with fellow citizens who would turn loose upon it the horror which it dreaded too much to name. For many southerners, this hazard

---

Louis Filler remarks, "The fact that disunion sentiments were not a Garrisonian vagary but a popular Northern view has been obscured for decades," *The Crusade Against Slavery, 1830–1860* (New York, 1960), p. 303; Phillips, speech at Brooklyn, Nov. 1, 1859, in Redpath, *Echoes of Harper's Ferry*, pp. 43–66.

63. Letter from Parker at Rome to Francis Jackson, Charles W. Wendte (ed.), *Saint Bernard and Other Papers* (Vol. XIV of Centenary ed. of the works of Theodore Parker; Boston, 1911), p. 425.

64. For the numerous plans and rumors of plans for the rescue of Brown, and the elaborate precautions in Virginia to prevent a rescue, see Villard, *Brown*, pp. 511–517; Higginson, *Cheerful Yesterdays*, pp. 223–234.

65. Baltimore *American*, Dec. 3 and 7, 1859, quoted in Villard, *Brown*, p. 569. Davis in *Congressional Globe*, 36 Cong., 1 sess., p. 62.

meant only one thing: those who are not for us are against us. "We regard every man in our midst an enemy to the institutions of the South," said the Atlanta *Confederacy*, "who does not boldly declare that he believes African slavery to be a social, moral, and political blessing."[66]

By this definition, nearly every man in the North was an enemy. James M. Mason of Virginia said in the Senate that "John Brown's invasion was condemned [in the North] only because it failed." Jefferson Davis declared that the Republican party was "organized on the basis of making war" against the South. A Mississippi legislator warned his constituents, "Mr. Seward and his followers . . . have declared war on us." The governor of South Carolina informed the legislature that the entire North was "arrayed against the slaveholding states." These were well-worn phrases, but John Brown gave them a new meaning. It was hard for a southern Unionist to answer the statement of the Richmond *Enquirer* that "the Northern people have aided and abetted this treasonable invasion of a Southern state," hard to refute C. C. Memminger of South Carolina when he said, "Every village bell which tolled its solemn note at the execution of Brown proclaims to the South the approbation of that village of insurrection and servile war."[67]

But if the South was friendless externally, at least it had solidarity internally. "Never before, since the Declaration of Independence," proclaimed the Sumter, South Carolina, *Watchman*, "has the South been more united in sentiment."[68] This unity must now be used to protect the South: Governor William H. Gist felt that if the South did not "now unite for her defense," southern leaders would "deserve the execration of posterity."[69] Robert Toombs was more specific: "Never permit this Federal government to pass into the traitorous hands of the black Republican party."[70] The Mississippi legislature passed resolutions declaring that the election of a president by a party unprepared to protect slave property would be a cause for the southern states to meet in conference and that Missis-

66. John W. Burgess, *The Civil War and the Constitution* (2 vols.; New York, 1901), I, 36; Atlanta *Confederacy*, quoted in Nevins, *Emergence*, II, 108 n.

67. Villard, *Brown*, pp. 565–567. See also Harold S. Schultz, *Nationalism and Sectionalism in South Carolina, 1852–1860* (Durham, N.C., 1950), pp. 190–199.

68. Dec. 24 1859, quoted in Nevins, *Emergence*, II, 110.

69. Quoted in Henry D. Capers, *The Life and Times of C. G. Memminger* (Richmond, 1893), p. 239.

70. *Congressional Globe*, 36 Cong., 1 sess., appendix, pp. 88–93.

sippi stood ready to help Virginia or other states to repel such assailants as Brown.[71] With a presidential election only nine months away, this injunction was neither vague nor abstract. But the Baltimore *Sun* did not even need to wait for an election. It announced that the South could not afford to "live under a government, the majority of whose subjects or citizens regard John Brown as a martyr and a Christian hero, rather than a murderer and robber."[72] The governor of Florida also thought that he had seen enough: he favored an "eternal separation from those whose wickedness and fanaticism forbid us longer to live with them in peace and safety."[73]

Two Richmond newspapers effectively summarized what had happened in Virginia and the South. On October 25 the *Enquirer* observed, "The Harpers Ferry invasion has advanced the cause of disunion more than any other event that has happened since the formation of its [*sic*] government." A month later, the *Whig* declared, "Recent events have wrought almost a complete revolution in the sentiments, the thoughts, the hopes, of the oldest and steadiest conservatives in all the Southern states. In Virginia, particularly, this revolution has been really wonderful. There are thousands upon . . . thousands of men in our midst who, a month ago, scoffed at the idea of a dissolution of the Union as a madman's dream, but who now hold the opinion that its days are numbered, its glory perished."[74]

Certainly the psychological ties of union were much attenuated at the end of 1859. Harpers Ferry had revealed a division between North and South so much deeper than generally suspected that a newspaper in Mobile questioned whether the American republic continued to be a single nation or whether it had become two nations appearing to be one.[75]

71. Percy Lee Rainwater, *Mississippi: Storm Center of Secession* (Baton Rouge, 1938), p. 105.

72. Nov. 28, 1859, quoted in Villard, *Brown*, p. 568.

73. *Ibid.*, p. 584, quoting *Liberator*, Dec. 23, 1859.

74. Henry T. Shanks, *The Secession Movement in Virginia, 1847–1861* (Richmond, 1934), p. 90, quoting *Enquirer*, Oct. 25, and *Whig*, Nov. 22.

75. Mobile *Register*, Oct. 25, 1859, quoted in Avery O. Craven, *The Growth of Southern Nationalism, 1848–1861* (Baton Rouge, 1953), p. 309. Pages 305–311 give copious evidence of the psychological impact of Harpers Ferry in the South: "A wave of indignation, hatred, and fear swept across the whole South to give it a unity it had never known before."

CHAPTER 15

Southern Maneuvers on the Eve of Conflict

AT the time of John Brown's raid, the Buchanan administration still had sixteen months to run, during which the Thirty-sixth Congress would hold both its long and its short sessions. Measured by the congressional sequence, the administration was only at half-way mark, owing to the curious time lag between the election and the meeting of a Congress. Unless called into special session, Congress did not meet for its first session until thirteen months after its election, nor for its second session until after its successor had been elected. In a sense, nearly half of any Congress was out of phase. This anomaly always showed up even more conspicuously in the second Congress of any administration, for the first of its sessions took place during the early part of the presidential campaign, with Congress usually in session during the party conventions and frequently subordinating legislative business to campaign activity, both on and off the floor. The second session did not meet until after the new president had been elected.

The only session likely to be fully functional was the first session of the first Congress of any administration. President Polk had secured his Walker Tariff, his Oregon settlement, and his war with Mexico at the first session of the Twenty-ninth Congress, and then, running onto the reef of the Wilmot Proviso, nothing thereafter. Fillmore had won adoption for the compromise measures of 1850 at the first session of the Thirty-first, and virtually nothing thereafter. Pierce had spent a handsome majority to purchase the enactment of Kansas-Nebraska at the first session of the Thirty-third, and had gained nothing for the rest of his term. Buchanan, who understood the political system as well as anyone, had nevertheless also used up his leverage at the first session of the Thirty-fifth Congress, in a vain effort to force the acceptance of the Lecompton constitution. The legislative history of the second session, as has been shown, was a shambles. By the end of 1859, the process of choosing Buchanan's successor was already in full swing, but half of the congressional activity of his presidency was still ahead of him.

The first session of the Thirty-sixth Congress was important not for what it did, but for what it symptomized. It met on December 5, exactly three days after John Brown was hanged. The atmosphere was still tense, and the circumstances of the new session contributed nothing to relax it. The Democrats controlled the Senate, but no one knew who controlled the House. One hundred and nineteen votes were needed to elect a Speaker, but the Republicans had only 109, and the Democrats claimed 101, but of these, 13 were anti-Lecompton men, unlikely to support a proslavery Democrat. Twenty-seven Whigs or Americans, mostly from the South, would probably give the bulk of their support to a proslavery man, but their own electoral successes in 1859 and the disarray of the Democratic party made them reluctant to support a Democrat. Since the Speaker at that time appointed all the committee chairmen, the contest promised to be as fierce as those of 1849–1850 and 1855–1856.[1]

The Republicans quickly concentrated their support on John Sherman of Ohio. Sherman, entering his third term in the House, was a thoughtful, moderate man, primarily interested in finance, and he was not a militant on the slavery question. As he himself remarked, he had repeatedly stated that he was "opposed to any interference whatever by the people of the free states with the relations of master and slave in the slave states."[2] But Sherman had laid himself open to vigorous southern attack. Ten months previously he had agreed in a routine way to lend his support to a digest of a book which was arousing violent antagonism in the South, *The Impending Crisis,* by Hinton R. Helper, published in 1857. Helper, a

1. Ollinger Crenshaw, "The Speakership Contest of 1859–1860," *MVHR,* XXIX (1942), 323–338; Roy F. Nichols, *The Disruption of American Democracy* (New York, 1948), pp. 273–276.

2. *Congressional Globe,* 36 Cong., 1 sess., p. 21; Allan Nevins, *The Emergence of Lincoln* (2 vols.; New York, 1950), II, 123.

rather obscure nonslaveholding white from North Carolina, had taken a firm grasp of the idea that the North was rapidly outstripping the South in the race for economic progress, and that the South was, in fact, falling into a state of economic decline. The southerners who suffered most, as he saw it, were the nonslaveholding whites, more and more of whom were lapsing into the wretched status of "poor whites." Slavery, with its wastefulness and inefficiency and its monopolistic aspects, was the curse of the South and especially of the nonslaveholders. Helper wasted no sympathy on the slaves; in fact, he called stridently for their deportation, and later became one of the country's most violently anti-Negro writers. But his attack on slavery was particularly alarming to the South because he appealed to class divisions between the slaveholding and the nonslaveholding whites. No dogma of the southern creed was held more sacrosanct than the tenet that race transcended class and, indeed, extinguished it—that all whites were on the same footing, simply by virtue of their status as whites. And no form of attack—not even the appeal for a slave insurrection—found the South more vulnerable than did an appeal to the nonslaveholders to reject the slave system. Southerners had denounced Helper as "incendiary and insurrectionary," as a traitor, a renegade, an apostate, a "dishonest, degraded, and disgraced man."[3] Now the Republican party was getting ready to flood the North with 100,000 copies of a handy abridgment of Helper's book, and to make matters worse, they added some offensive captions, such as: "The Stupid Masses of the South" and "Revolution—Peacefully if we can, Violently if we must."[4] John Sherman was one of about sixty Republican congressmen who had signed a letter endorsing the plan for a compendium of Helper's work.[5]

Immediately after the first, inconclusive ballot for Speaker, John

3. Quotation from Hugh T. Lefler, "Hinton Rowan Helper: Advocate of a White America," in Joseph D. Eggleston, *Southern Sketches*, No. 1 (Charlottesville, Va., 1935). On Helper, see introduction to George M. Fredrickson (ed.), *The Impending Crisis of the South: How to Meet It*, by Hinton R. Helper (Cambridge, Mass., 1968); Hugh C. Bailey, *Hinton Rowan Helper, Abolitionist-Racist* (University, Ala., 1965). On the southern reaction to Helper, see Avery O. Craven, *The Growth of Southern Nationalism, 1848–1861* (Baton Rouge, 1953), pp. 249–252, which gives additional citations; Edward Channing, *A History of the United States* (6 vols.; New York, 1905–25), VI, 203–210.

4. Hinton R. Helper, *Compendium of the Impending Crisis of the South* (New York, 1860).

5. *Congressional Globe*, 36 Cong., 1 sess., p. 16.

B. Clark of Missouri introduced a resolution declaring that "no member of this House who has endorsed . . . [*The Impending Crisis*] or the compend from it, is fit to be Speaker of this House."

Clark's resolutions were never adopted, and Sherman told the House that he had never seen either Helper's book or the compendium, but the issue turned enough border state men and South Americans against him to prevent his election, though the Republicans continued steadily to support him for eight weeks, and he came within three votes of victory.[6]

While the Republicans were standing by Sherman, the Democrats were trying a number of candidates—beginning with Thomas S. Bocock of Virginia and including John A. McClernand of Illinois, the foremost Douglas Democrat in the House. As it turned out, McClernand could have been elected if a small group of Democrats from the lower South had not refused to support him.[7] His defeat was a kind of prologue to the intensification of the quarrel between the Buchanan Democrats and the Douglas Democrats, which would soon have disastrous consequences for the party. This conflict had already cost the Democrats control of the House.

The speakership contest lasted for two months before the Democrats learned that they could not unite, and the Republicans that they could not elect Sherman. At that point, Sherman withdrew, and two days later the Republicans were able to elect William Pennington of New Jersey with exactly the number of votes required to win. Pennington was incompetent to be Speaker, but the Republicans found him acceptable because he had steadily supported exclusion of slavery from the territories, while he picked up crucial votes among South Americans because he was a conservative who had supported the Fugitive Slave Law and a long-time Whig who had only recently turned Republican.[8]

In terms of results, the contest was not very decisive, but it revealed a deeper estrangement on the part of the South than any previous crisis had shown. To begin with, it was ominous that many

6. *Ibid.*, pp. 3, 21, 430; Crenshaw, "Speakership Contest," pp. 323–328.

7. *Congressional Globe,* 36 Cong., 1 sess., pp. 649–650; Victor Hicken, "John A. McClernand and the House Speakership Struggle of 1859," ISHS *Journal,* LIII (1960), 163–178.

8. *Congressional Globe,* 36 Cong., 1 sess., pp. 651–652; Crenshaw, "Speakership Contest," p. 328; James Ford Rhodes, *History of the United States from the Compromise of 1850* (7 vols.; New York, 1892–1906), II, 421–426.

of the southern members did not really want to organize the House, which means that they were quite willing to paralyze the federal government. They engaged in protracted and disorderly debate, frequently resorted to dilatory tactics, allowed only forty-four ballots to be taken in forty days of session (compared with 130 ballots in a comparable period in 1855–1856), and resisted to the end a rule such as had been adopted in 1850 and 1856, permitting election by a plurality of the votes cast. Ultimately they deadlocked the House from December 5 to February 1, the second longest such paralysis in its history.[9]

During this time, members displayed such hostility as to make the House simply an arena, and scarcely a deliberative body at all. Speeches reached an unprecedented level of acrimony, and apparently many members carried weapons. During one bitter debate, a pistol fell from the pocket of a New York congressman, and other members, thinking that he had drawn it intending to shoot, almost went wild. Senator Hammond said, "The only persons who do not have a revolver and a knife are those who have two revolvers," and Senator Grimes wrote, "The members on both sides are mostly armed with deadly weapons, and it is said that the friends of each are armed in the galleries." The widespread expectation of a shoot-out on the floor of Congress seemed not unrealistic.[10]

In such an atmosphere as this, it is not surprising that men talked of disunion in plainer terms than ever before. Though few southerners seemed prepared to secede on the speakership issue, many were now ready to state that the South ought to leave the Union if the Republicans should win the presidency. Thus, a Georgia congressman said his constituents were ready for "independence now and forever"; a member from Alabama predicted that his state and indeed "most if not all of the Southern states, with Old Virginia in the lead" would go out, peaceably by preference, but fighting if need be. Lawrence Keitt of South Carolina was prepared to "shatter this Republic from turret to foundation stone." Thaddeus Stevens grimly responded that he did not blame southern members for threatening to secede: "They have tried it fifty times, and fifty times they have found weak and recreant tremblers in the north . . . who

9. Nevins, *Emergence*, II, 120; Rhodes, *History*, II, 427.

10. Crenshaw, "Speakership Contest," pp. 332–334; Rhodes, *History*, II, 424; Nevins, *Emergence*, II, 121–122; William Salter, *The Life of James W. Grimes* (New York, 1876), p. 121.

have acted from these intimidations." Stevens implied that all this was an empty bluff, but even at the time, Governor Gist of South Carolina was writing to Congressman Miles of that state: "I am prepared to wade in blood rather than submit to inequality and degradation; yet if a bloodless revolution can be effected, of course it would be preferable. If, however, you upon consultation decide to make the issue of force in Washington, write or telegraph me, and I will have a regiment in or near Washington in the shortest possible time."[11] Nothing came of Gist's remarkable proposal, and many Republicans continued to believe, along with Stevens, that the talk of secession was all wind. But one year to the day after Gist's letter, South Carolina would adopt an ordinance of secession.

If the speakership contest offered a portent of disunion, it also dramatized the irreparable split within the Democratic party. After the withdrawal of Bocock as a candidate, McClernand became the Democratic choice. Although a Douglas man, McClernand worked hard for conciliation within the party, and he was approved even by Jefferson Davis, who came over from the Senate to rally support for him. McClernand received 91 votes on the forty-third ballot, and stood within 26 votes of election. This was the closest that any Democrat came to gaining the speakership, but Senator James Green of Missouri, an inveterate foe of the Douglas wing of the party, appeared in the House to stop the McClernand bandwagon. Nine Democrats from Alabama and South Carolina voted against McClernand and thus prevented his election. Evidently they preferred to lose the speakership altogether rather than have a supporter of Douglas win it.[12]

The struggle for the speakership had, by its bitterness, illustrated the depth of the sectional division. The legislative session that followed illustrated the same division in another way. Northern members were primarily concerned with enacting a new economic program appropriate to an emerging industrial society, while southern members were preoccupied with vindicating the slave system symbolically by forcing their territorial doctrine on the northern wing

11. *Congressional Globe,* 36 Cong., 1 sess., pp. 23, 24, 25, 71, 72, 164, 165; Henry Wilson, *History of the Rise and Fall of the Slave Power in America* (3 vols.; Boston, 1872–77), II, 643–654; Crenshaw, "Speakership Contest," pp. 334–335, for Gist quotation; Rhodes, *History,* II, 422.

12. *Congressional Globe,* 36 Cong., 1 sess., p. 641; Hicken, "McClernand and the House Speakership Struggle," pp. 174–175.

of their party—though they might destroy the party in the process. In short, North and South were simply moving in opposite directions, and the South was almost obsessively defining its position in terms that isolated it from the North and identified it with policies that, because of the tendencies of the modern world, were foreordained to defeat.

The steady growth in strength of the Republican party was demonstrated in this session by action on the protective tariff and on a homestead bill. In the previous session, a homestead measure, allowing a person to acquire 160 acres of public land simply by settling on the property, had passed the House but had been blocked in the Senate when Vice-President Breckinridge cast a tie-breaking vote against it. Now, however, a homestead bill passed both houses, only to be vetoed by Buchanan.[13] The sharpness of sectional alignment had been exhibited in the vote in the House, when 114 of the 115 affirmative votes were cast by free-state members; 64 of the 65 negative votes, by slave-state members. In the previous session, the Republicans had struggled in vain to pass a bill for a protective tariff. Now they carried such a measure in the House, 105 to 64, but the Senate killed it with a vote to postpone. The Republicans also struggled for a Pacific railroad bill and for a bill to improve navigation on the Great Lakes, but without success in either case.[14]

Southerners had logical reasons for opposing all of these measures. They recognized that no one could establish a plantation on 160 acres, but that the lure of free land might attract immigrants who would add to the already great preponderance of the free-state population. They regarded a protective tariff as a form of subsidy which would enable Yankee manufacturers to increase their exploitation of all agricultural producers, and especially cotton producers who sold in an open world market and had nothing to gain by buying in a protected domestic one. They foresaw that a Pacific railroad would, in effect, link the Pacific coast with the North. And they regarded large federal appropriations for internal improvements as measures to aggrandize a central government, which they had no desire to strengthen, and to foster a highly articulated domestic commerce, which they had no desire to build.

13. Nevins, *Emergence*, I, 444–445, 453–455; II, 188–191.

14. *Ibid.*, I, 455–457; II, 193–196. For analysis of votes in the House in the first session of the 36th Congress, see Thomas B. Alexander, *Sectional Stress and Party Strength* (Nashville, 1967), pp. 253, 257, 260, 262.

But the southern opposition was almost too logical, for it placed the South in a posture not only of defending slavery but also of resisting progress. In effect, by blocking the dynamic economic forces which were at work in the North and West, the South impelled the proponents of those forces to join in a coalition, which might not otherwise have materialized, with the antislavery forces. The logical vehicle for such a coalition was the Republican party, and in fact the Republican platform of 1860 laid the foundations for the coalition even before Buchanan had vetoed the Homestead bill or the Senate had blocked the protective tariff.

During this session of Congress, the Republicans also gathered some effective campaign material by holding one of the first major investigations ever conducted by a congressional committee. The faction-rent Democratic party was vulnerable on several counts: The party had voted large appropriations to the public printer, Cornelius Wendell, and then had expected him to make large "contributions" when the party needed funds. The secretary of war, John B. Floyd, had favored friends with government contracts which were not properly scrutinized, and when congressional appropriations were slow in coming, he had encouraged banks to advance funds to contractors on bills of theirs which he had endorsed. The president had denied ever approving Governor Walker's pledge that there should be a plebiscite on the Kansas constitution, but Walker possessed a letter from Buchanan stating his approval and was willing to appear before a committee.

The House appointed such a committee, headed by John Covode of Pennsylvania, which investigated extensively, calling numerous witnesses and looking into every sordid transaction of which it could get wind. Ultimately, the committee discovered enough to indicate a pervasive taint of financial laxity and scandal in the administration. Its report appeared in June 1860, five months before the election, or just in time to make the issue of corruption a significant factor in the campaign.[15]

While the Republicans were busy broadening the basis of their popular appeal and exposing the dirty linen of the Democrats, the latter seemed to be spending most of their energies narrowing the

15. Report of Covode Committee in *House Reports,* 36 Cong., 1 sess., No. 648 (Serial 1071). See David E. Meerse, "Buchanan, Corruption, and the Election of 1860," *CWH,* XII (1966), 116–131; Nichols, *Disruption,* pp. 190, 284–287, 328–331.

basis of their appeal and discrediting one another. During the winter and spring of 1859–1860, the prolonged process by which the Democratic party ceased to be a single national party reached its culmination.

As late as 1852, the party had possessed enough strength in both North and South to maintain a bisectional equilibrium. But the northern wing had first been decimated by the Kansas-Nebraska Act, and then by the refusal of the southern wing, at the time of the Lecompton contest, to give popular sovereignty a fair trial in Kansas.

The weakening of the northern wing had shown up most conspicuously in what James McGregor Burns has called the "Congressional Party"—that is, the apparatus of party caucus, committee structure, and so on, in the Senate and the House. These passed under southern domination, and indeed the northern congressional Democrats were so weak that when Douglas waged the Lecompton contest, he had to rely on Republican votes to compensate for the lack of strength in northern Democratic ranks.

Another consequence of the decreasing strength of the Democratic party in the North was that in states where it no longer stood much chance of winning elections, it tended to become, as parties do in such circumstances, primarily a patronage organization, perpetuated for the purpose of distributing postmasterships and other political largess, rather than an organization to contest elections. At its worst, a patronage organization even discourages new supporters, keeping its numbers small so that the controlling insiders can monopolize the plums for themselves. This was the pattern that Republican state organizations in the South were later to follow for more than half a century after Reconstruction.[16] Such organizations are, of course, especially susceptible to the influence of the administration, and this was true in 1859, in the sense that nearly every northern state had a "regular" Democratic organization which acted as a pliant tool of the Buchanan administration.

This meant that insofar as there was a popularly based northern Democracy, with Stephen A. Douglas leading it, it operated under the twin handicaps of opposition within the northern states by the

16. The prospect of having a Republican patronage party established in the South was one of the principal reasons that antebellum southerners feared the election of a Republican president.

mercenaries of the administration and of domination in Congress by a southern wing which imposed proslavery policies that weakened the northern wing even further.

In a sense, then, there were two Democratic parties: one northern, one southern (but with patronage allies in the North); one having its center of power in the northern electorate and in the quadrennial party convention (where all states had full representation, whether they ever actually voted Democratic or not), the other with its center of power in Congress; one intent on broadening the basis of support to attract moderate Republicans, the other more concerned to preserve a doctrinal defense of slavery even if it meant driving heretics out of the party.

The basic structure of the Democratic party was, in itself, enough to assure intraparty strife, but such antagonism developed even more intensely because of the aftermath of bitterness from the Lecompton struggle and because of the personal incompatibility of the opposing leaders, Buchanan and Douglas. Both had strength in a way—Buchanan the strength of stubborn defensiveness and shrewd inertia; Douglas the strength of great energy, imagination, and impetuosity. Both believed in loyalty—Buchanan, blind loyalty to a party hierarchy; Douglas, sacrificial loyalty to his party teammates. Both cared for power—Buchanan cherishing it as something to be hoarded and transmitted through an ordained succession; Douglas regarding it as something to be won in combat. Buchanan was a custodian who loved safety; Douglas was an innovator who loved risks.

To Douglas, Buchanan seemed a cold, selfish, conventional-minded party hack, domineering, yet at the same time timid and obsequious to the aristocratic southern leaders, and obsessed with party regularity in its most stultifying form. To Buchanan, Douglas seemed a hard-drinking brawler, a political freebooter, an ambitious upstart, a disturber of the peace, and worst of all, a disloyal Democrat who had allied himself with the Republicans against the Lecompton policy of his party's administration.[17]

17. The best general treatment of the party strife is Nichols, *Disruption*. Important aspects of party conflict and pinpointed in Philip G. Auchampaugh, "The Buchanan-Douglas Feud," ISHS *Journal*, XXV (1932), 5–48; Richard R. Stenberg, "An Unnoticed Factor in the Buchanan-Douglas Feud," *ibid.*, XXV (1933), 271–284 (Buchanan's private hope of being renominated); O. M. Dickerson, "Stephen A. Douglas and the Split in the Democratic Party," MVHA *Proceedings*, VII (1913–14), 196–211; Reinhard H. Luthin, "The Democratic Split During Buchanan's Administration," *Pennsylvania History*, XI (1944), 13–35; William O. Lynch, "Indiana in the Douglas-

Thus, the strife within the Democratic party, after reaching a new level of intensity during the Lecompton contest, continued to rage as an intraparty feud during the second session of the Thirty-fifth Congress and the first session of the Thirty-sixth, and reached its climax in the convention, or rather the conventions, of 1860. For months, while the Buchanan-Douglas feud was at its height, Buchanan used the patronage as a weapon to break down the Douglas organization, and Douglas made powerful appeals urging the public to repudiate policies which were, as he saw it, destroying the Democratic party in the North.

In the first stage of this contest, the main arena of combat was the Congress. There the party regulars and the southern Democrats held the ascendancy. The southerners had long manifested an excessive interest in symbolic triumphs, and as sectional antagonisms grew progressively more heated, the voters of the lower South showed an increasing propensity to reward those candidates who could display the greatest degree of ardor in the proslavery cause. Abstractions might be futile at the national level, but they paid off handsomely at the state level. Southern political candidates responded accordingly, and giving popularity at home priority over the maintenance of a broad, national basis of party strength, they became ever more ready to make issues on any and all aspects of slavery, and to devise doctrinal tests by which to measure the orthodoxy of northern Democrats.

In 1858–1859, some of the defenders of slavery, in search of a demand so extreme that no one else could top it, found their issue in a proposal for the reopening of the slave trade with Africa—a trade which had been prohibited in 1808, as soon as prohibition was possible under the Constitution. To evaluate the significance of this demand, it should be understood at the outset that it never commanded any important body of support, but it reveals in a striking way certain important aspects of the situation on the eve of the Civil War.[18]

Buchanan Contest of 1856," *IMH*, XXX (1934), 119–132. The fullest treatment of Douglas has long been George Fort Milton, *The Eve of Conflict: Stephen A. Douglas and the Needless War* (Boston, 1934), now superseded by Robert W. Johannsen, *Stephen A. Douglas* (New York, 1973). On Buchanan, the best study is Philip Shriver Klein, *President James Buchanan* (University Park, Pa., 1962).

18. On the question of reopening the slave trade, see Ronald T. Takaki, *A Pro-Slavery Crusade: The Agitation to Reopen the African Slave Trade* (New York, 1971); Harvey Wish, "The Revival of the African Slave Trade in the United States, 1856–1860," *MVHR*, XXVII (1941), 569–588; Barton J. Bernstein, "Southern Politics and At-

The reopening of the African trade had been suggested in 1839 by the New Orleans *Courier*. In 1853 Leonidas W. Spratt, editor of the Charleston *Standard*, had begun systematic advocacy of the repeal of the prohibition on the trade. Robert Barnwell Rhett's Charleston *Mercury* took up the cry in 1854. Two years later, Governor James H. Adams of South Carolina declared, "The South at large does need a reopening of the African slave-trade." But the state legislature in 1857–1859 rejected a series of attempts to bring the issue to a vote, as did the Texas legislature in 1857. Perhaps the high tide of the effort to secure legislation came in March 1858, when the Louisiana house of representatives voted 46 to 21 to authorize the importation into Louisiana of "twenty-five hundred free Africans" as apprentices. The use of "apprentices" was already well known in the West Indies, where Hindu and African labor had been brought in under an apprentice system, after slavery had been abolished. Theoretically, "apprentices" might be imported without violating the prohibition on the African slave trade, but practically, they would become the equivalent of slaves. This bill would have passed the Louisiana senate if opposition senators had not prevented it by absenting themselves and thus breaking a quorum.[19]

As it became clear that no representative body was going to endorse the reopening of the trade, supporters of the idea turned increasingly to agitation in the Southern Commercial Convention, an organization designed to promote southern economic development. Originally, such conventions had been irregular, but beginning in 1852, large gatherings with a certain measure of continuity were held annually through 1859. As the meetings continued, their

---

tempts to Reopen the African Slave Trade," *JNH*, LI (1966), 16–35; W. J. Carnathan, "The Proposal to Reopen the African Slave Trade in the South, 1854–1860," *SAQ*, XXV (1926), 410–429. The first modern critical treatment—brief but comprehensive—of this topic was Robert R. Russel, *Economic Aspects of Southern Sectionalism, 1840–1861* (Urbana, Ill., 1924; reissued New York, 1960), pp. 212–224, and see notes immediately following.

19. For South Carolina, see Ronald T. Takaki, "The Movement to Reopen the African Slave Trade in South Carolina," *South Carolina Historical Magazine*, LXVI (1965), 38–54; Takaki, *Pro-Slavery Crusade*, pp. 184–199; Laura A. White, *Robert Barnwell Rhett, Father of Secession* (New York, 1931), pp. 139–158; Harold S. Schultz, *Nationalism and Sectionalism in South Carolina, 1852–1860* (Durham, N.C., 1950), pp. 130–133, 142–144, 157–164, 183–185. For Texas, see W. J. Carnathan, "The Attempt to Reopen the African Slave Trade in Texas, 1857–1858," Southwestern Political and Social Science Association *Proceedings*, 1925, pp. 134–144; Earl Wesley Fornell, *The Galveston Era: The Texas Crescent on the Eve of Secession* (Austin, 1961), pp. 215–230. For Louisiana, see James Paisley Hendrix, Jr., "The Efforts to Reopen the African Slave Trade in Louisiana," *Louisiana History*, X (1969), 97–123.

economic and commercial constituency diminished, and they passed increasingly under the control of extreme southern rights editors and politicians. Thus, in 1855, at New Orleans, a delegate introduced a resolution urging southern congressmen to work for the repeal of all laws suppressing the slave trade, but the convention refused to act on this proposal. Again in 1856, 1857, and 1858 it refrained from action, though the pressure of demands was constantly rising, and in 1858 Spratt, William L. Yancey (both in favor), and Roger A. Pryor (opposed) engaged in a lengthy and spirited debate. Finally, in May 1859, at Vicksburg, a majority of 40 to 19 approved the statement: "In the opinion of this Convention, all laws, State or Federal, prohibiting the African Slave Trade, ought to be repealed."[20]

Nothing more was ever actually accomplished. The movement had the support of a few fire-eating politicians and a few newspapers, including the New Orleans *Delta,* the Charleston *Standard,* the Houston *Telegraph,* and for a time, two Galveston papers and the Charleston *Mercury,* with some editorial encouragement from others. But its greatest legislative triumph was to carry one vote, in one house, of one state legislature. It never reached Congress at all except in the form of resolutions repudiating it. Such slaves as came in from Africa were smuggled illegally, and it appears that their number has been greatly exaggerated.[21] Altogether, historians may

20. John G. Van Deusen, *The Ante-Bellum Southern Commercial Conventions* (Durham, N.C., 1926), pp. 56–69, 75–79; Herbert Wender, *Southern Commercial Conventions, 1837–1859* (Baltimore, 1930), pp. 177–181, 197–204, 211–235. A basic source is *De Bow's Review,* Vols. XXII–XXVII (1857–59).

21. Apparently everyone in the South in the late 1850s knew someone who knew someone else who had seen a coffle of slaves direct from Africa. But no one who had seen them has left any testimony. One ship, the *Wanderer,* did bring a cargo of slaves from Africa in 1858, and this bizarre event was apparently reenacted many times in imagination. W. E. Burghardt DuBois, *The Suppression of the African Slave Trade to the United States of America, 1638–1870* (Cambridge, Mass., 1896), pp. 168–193, believed that there was a major increase in the trade, both to Brazil and to the United States. The New York *Post* estimated that 30,000 to 60,000 Africans were imported in 1859, and Stephen A. Douglas thought there were 15,000—a number Wish considered "credible in the light of contemporary evidence." Wish, "Revival of the African Slave Trade," p. 582. Warren S. Howard, *American Slavers and the Federal Law, 1837–1862* (Berkeley, 1963), pp. 142–154, deals judiciously with both the evidence and the rumors, and he shows that while there may have been appreciable activity in the outfitting of slavers from American ports, they probably traded to Cuba or Brazil rather than to the United States. He makes a strong case that importations to the South were negligible, and that the phenomenon was a striking illustration of the nature of rumor. See also Takaki, *Pro-Slavery Crusade,* pp. 200–226; Tom Henderson Wells, *The Slave Ship Wanderer* (Athens, Ga., 1967).

have given this matter more attention than it is worth.

But the demand for the reopening of the trade and also the refusal of the South to support that demand both tell much about the problems which were preoccupying the region during this final phase of the sectional struggle. The possibility of reopening the trade promised, at first glance, to solve certain problems of the South, but upon further scrutiny, it also presented serious difficulties in connection with most of these problems.

To begin with, it seemed to meet a psychological need by providing a way to dramatize the intellectual defense of slavery. If slavery was, as Calhoun had asserted, a positive good, why was the bringing of slaves from Africa a positive evil, to be punished as piracy? "If it was right," as William L. Yancey asked, "to buy slaves in Virginia and carry them to New Orleans, why is it not right to buy them in Africa and carry them there?"[22] Yancey might have noted that buying them in Virginia did not add to the number of slaves and did not reduce anyone from freedom to slavery; he might even have reversed his question and asked if it was wrong to buy slaves in Africa, why was it not wrong to buy them in Virginia. But there was a certain logic in the contention that one could not condemn the ethics of the slave trade and still uphold the ethics of slavery.[23]

Another consideration of quite another kind related to the anxiety which Hinton Helper had touched so skillfully—the question why nonslaveholding whites should support a system in which they had no personal stake, and, indeed, whether they would continue to support it. With the price of slaves steadily rising, as it had since the beginning of the century, a prime field hand who could have been purchased for less than $400 in 1800 was worth $1,500 in 1857. Only men of property could pay so much; poor men had been priced out of the market. If the ownership of slaves should become too concentrated, too much a prerogative of the rich, the nonslaveholding whites might withdraw their support, which was vital in defending the planter regime against its northern enemies. But slaves from Africa would be cheap slaves, and their low price might enable the South to broaden the basis of slaveholding—to "democ-

22. Speech of Yancey in *De Bow's Review*, XXIV (1858), 473–491, 597–605.

23. For a critique of the arguments, see Bernstein, "Southern Politics and Attempts to Reopen the African Slave Trade," which quotes J. D. B. De Bow's assertion, "If slavery benefits the slave then our position drives us to . . . the reopening of the slave trade."

ratize" the practice of slaveholding. Thus, the New Orleans *Delta* asserted, "We would re-open the African slave trade that every white man might have a chance to make himself owner of one or more Negroes." Governor Adams declared, "Our true purpose is to diffuse the slave population as much as possible, and thus secure in the whole community the motives of self-interest for its support."[24]

Entirely apart from the motivation of the nonslaveholders, the high price of slaves meant a high price of labor. While the growing volume of immigration from Europe was supplying northern industry with plenty of workers to hire, rising slave prices in the South reflected a scarcity of labor, and increasing production costs. The reopening of the African trade would help to provide adequate labor at reasonable cost.

Finally, advocates of reopening the trade hoped that this issue might help to consolidate southern opinion against the North. Instead of continuing what promised to be a losing battle for the control of Kansas, why could not the South take a position of "active aggression"? Why not give "a sort of spite to the North and defiance of their opinions"? This would hearten southerners who had stood too long on the defensive.[25]

As the discussion of the issue developed, however, it became evident that every one of these positive propositions had a negative corollary. Instead of strengthening southern solidarity, the mere question of reopening the trade proved to have a divisive influence in a number of ways. It constituted a threat to the upper South, which found a market for its surplus slaves in the cotton states. As W. E. B. Du Bois cogently stated, "the whole movement represented the economic revolt of the slave-consuming cotton-belt against their base of labor supply."[26] The Richmond *Enquirer* made the same point more delicately, but with even sharper warning to the cotton states: "If a dissolution of the Union is to be followed by the revival of the slave trade, Virginia had better consider whether the South of a Northern Confederacy would not be far more prefer-

24. Quotations from Wish, "Revival of the African Slave Trade," pp. 571–572.

25. *Ibid.*, p. 571. J. J. Pettigrew, in a report to the South Carolina legislature, said with reference to this question, "A great many worthy persons are honestly disposed to make issue with the North from a spirit of pure combativeness." *De Bow's Review*, XXV (1858), 306.

26. DuBois, *Suppression of the African Slave Trade*, p. 173.

able for her than the North of a Southern Confederacy."[27]

Not only would a reopening of the trade have antagonized the entire upper South, but it also presented serious economic dangers, both for slaveholders, despite claims that they needed more labor, and for nonslaveholders, despite claims that lower prices would enable them to own slaves. For the slaveholders, a renewal of the trade meant a reduction in the price of slaves, which, in turn, meant a staggering loss in the value of the slave property which they already held. For nonslaveholders, slaves from Africa would mean the competition of cheap labor in their own labor market, and instead of an opportunity to own slaves, it might mean impoverishment. A nonslaveholder, writing to the Edgefield *Advertiser* in South Carolina, asked, "If we are to have negro labor in abundance, where will my support come from? If my labor is to be supplanted by that of negroes, how can I live?"[28]

But most fundamentally, perhaps, it became apparent that a considerable part of the southern public had real moral objections to the trade. It may now seem hard to believe that men in the South could have condemned the trade as morally wrong and at the same time have regarded slavery itself as morally right; but this is perhaps no more anomalous than the fact that men in the North condemned slavery as morally wrong and regarded racial discrimination as morally right. The fact was that the South thought of slavery and the slave trade not logically but in sets of images. Its images of slavery were somewhat idealized, and it was prepared to defend the ideal. Its image of the slave trade was odious, and it remained unmoved by the logic of Leonidas Spratt. Southern thought about slavery had its darker side, but ordinarily, it could be summed up in these words of Benjamin F. Perry: "At present we have in South Carolina two hundred and fifty thousand peaceable and civilized slaves, happy and contented in their slavery."[29] This was too appealing and idyllic

27. Quoted in Takaki, *Pro-Slavery Crusade*, p. 234.

28. Edgefield *Advertiser*, Feb. 2, 1859, quoted by Takaki, "Movement to Reopen Slave Trade," pp. 48–49. Benjamin F. Perry said, "It is nonsense to talk about a poor man's being able to purchase slaves if they were cheaper, when his labor is cheapened, too, by the same operation, and it is only by his labor that he can purchase." As for the effect upon slaveholders, Perry said, "It would immediately diminish the value of all the slaves in the Southern States from one-half to two-thirds of their present prices." Lillian Adele Kibler, *Benjamin F. Perry: South Carolina Unionist* (Durham, N.C., 1946), pp. 282–283, quoting in full from Greenville *Southern Patriot*, Oct. 12, 1854.

29. Kibler, *Perry*, pp. 282–283.

a picture to mar it by introducing what Roger A. Pryor called "cannibals," "kidnapped from Africa." The idea of bringing such savages from the Dark Continent, said Pryor, was "repugnant to the instincts of Southern chivalry."[30] So much for Leonidas Spratt and his labored reasoning. Apparently Pryor was correct in his reading of southern opinion, for James H. Hammond estimated that nine-tenths of the southern people opposed the reopening of the trade. Alexander H. Stephens, who personally favored a resumption of the trade, nevertheless advised against making an issue of it because, he said, "The people here [in Georgia] at present, I believe, are as much opposed to it as they are at the North."[31]

After a time, the southern rights men perceived that they had got hold of the wrong issue. Robert Barnwell Rhett, who had earlier thrown the support of the Charleston *Mercury* behind Spratt, saw by 1858 that the slave trade question was seriously dividing the Democrats in South Carolina and that it was a handicap, even within the South, for the cause of southern rights. He ended by becoming very hostile to what he had recently advocated. William L. Yancey, who as late as 1858 had spoken eloquently in support of reopening the trade, was by 1860 ready to admit that southern public opinion had never been in favor of doing so.[32]

When the southern Democrats, in their congressional stronghold, sought an issue on which to base their defense of southern rights, they turned again to vindication of an abstract principle rather than the adoption of a program. Now, finally, they felt, with the Dred Scott decision to sustain them, they could purge the Democratic party of heresy by extirpating once and for all the ambiguity which had for so long surrounded the question of the status of slavery in the territories. There had long been general agreement in the party that when a territory became a state it should determine the question of slavery for itself (this agreement had not been chal-

30. Speech of Roger A. Pryor at Montgomery, Alabama, Commercial Convention, May, 1858, in *De Bow's Review*, XXIV (1858), 579–583.

31. Hammond in Edgefield *Advertiser*, March 2, 1859, cited in Takaki, "Movement to Reopen Slave Trade," p. 52; Stephens to J. Henly Smith, April 14, 1860, in Ulrich Bonnell Phillips, (ed.), *The Correspondence of Robert Toombs, Alexander H. Stephens, and Howell Cobb*, in AHA *Annual Report*, 1911, II, 467. Takaki offers a neat summary: "The Unionists like Perry saw that it [the slave trade question] would divide the Union, the extremists like Rhett saw that it would divide the South, and the moderates like Hammond saw that it would do both."

32. White, *Rhett*, 139–144, 152–154; John Witherspoon Du Bose, *The Life and Times of William Lowndes Yancey* (2 vols.; Birmingham, Ala., 1892), II, 570.

lenged in the Lecompton contest; there the issue was whether the question should be determined by a popular vote or by an elected convention), but there had never been agreement about the power of a territorial government during the territorial phase. Lewis Cass himself had never unequivocally asserted that a territorial legislature could exclude slavery from the territory. For more than a decade, however, other northern Democrats had been making such an assertion, while southerners denied it. Then, when Taney in his Dred Scott opinion declared that Congress could neither exclude slavery from a territory nor empower a territorial legislature to do so, Douglas, acting as spokesman for the northern Democrats, had sought to salvage what he could by contending, in the Freeport Doctrine, that the territories could effectively exclude slavery simply by withholding laws that slavery needed to exist. Later, Douglas had reaffirmed his belief that popular sovereignty could still be applied to the territories, in spite of Dred Scott. In June 1859 he stated that he would not accept a presidential nomination on a platform maintaining "the doctrine that the Constitution . . . either establishes or prohibits slavery in the territories beyond the power of the people legally to control it as other property."[33] In September, in an ambitious but poorly argued article in *Harper's Magazine,* he sought to save popular sovereignty by proving that the relation of the territories to the Union was parallel to the relation of the American colonies to the British Crown; also he attempted to reconcile his doctrine with the Dred Scott decision by making elaborate constitutional distinctions between the powers which Congress could exercise but could not confer and those which it could confer but could not exercise. The flimsiness of these contentions showed what an untenable position the Supreme Court had put the northern Democrats into, but it also showed that they still were clinging desperately to the idea that the people of a territory might exclude slavery.[34] The administration and the majority of the southern Democrats were determined to force them to abjure this doctrine.

33. Robert W. Johannsen (ed.), *The Letters of Stephen A. Douglas* (Urbana, Ill., 1961), pp. 446–447.

34. Robert W. Johannsen, "Stephen A. Douglas, 'Harpers Magazine,' and Popular Sovereignty," *MVHR,* XLV (1959), 606–631; also Harry V. Jaffa and Robert W. Johannsen (eds.), *In the Name of the People: Speeches and Writings of Lincoln and Douglas in the Ohio Campaign of 1859* (Columbus, 1959), pp. 58–125, 173–199, containing the *Harper's* essay and a reply to it by Jeremiah S. Black, U.S. attorney general.

At the opening of Congress in 1859, when President Buchanan sent his annual message to the Senate (the House not yet being organized), he commented on the Dred Scott decision as "the final settlement . . . of the question of slavery in the Territories," establishing "the right . . . of every citizen" not only "to take his property of any kind, including slaves into the common territories," but also "to have it protected there under the Federal Constitution."[35] What did Buchanan mean when he said "to have it protected there"? Jefferson Davis gave the South's answer in a set of resolutions introduced in the Senate on February 2, the day after the House elected its Speaker: "It is the duty of the Federal Government there to afford . . . the needful protection, and if experience should at any time prove that the judiciary does not possess power to insure adequate protection, it will then become the duty of Congress to supply such deficiency."[36] What Davis was demanding was a federal slave code for the territories.

It is significant that Davis moved vigorously to have his resolutions approved by the Senate Democratic caucus, but that he did not press for prompt action by the Senate itself.[37] What he really wanted was a doctrinal test to impose upon the Douglas Democrats in the national convention which was less than three months away. This strategy had already manifested itself. In naming delegates to the national convention, the Alabama Democracy had instructed them to insist upon a declaration of the federal government's obligation to keep the territories open "to all the citizens of the United States, together with their property of every description [i.e., slaves], and that the same should remain protected by the United States while the territories are under its authority." If the convention refused to adopt such a declaration, the Alabama delegates were "positively instructed" to withdraw.[38]

The Senate Democratic caucus adopted the Davis resolutions in a slightly modified form. But Democrats who wanted to preserve the remaining strength of their party in the North, and who even hoped

35. Richardson, *Messages and Papers*, V, 554.

36. *Congressional Globe*, 36 Cong., 1 sess., p. 658; Nevins, *Emergence*, II, 179; Nichols, *Disruption*, pp. 281–284.

37. Milton, *Eve of Conflict*, pp. 409–411. Nichols, *Disruption*, p. 284.

38. Clarence Phillips Denman, *The Secession Movement in Alabama* (Montgomery, Ala., 1933), pp. 80–81; text of resolutions in *Official Proceedings of the Democratic National Convention Held in 1860 at Charleston and Baltimore* (Cleveland, 1860), pp. 56–57.

for a possible victory in the 1860 election, deplored such resolutions. There was no possibility that Congress would enact slave codes for the territories, and the resolutions could have no effect except to injure the party in the North. Douglas complained bitterly that "the integrity of the Democratic party [was] to be threatened by abstract resolutions." Wigfall of Texas protested against crippling the Democracy on the eve of its great contest with the Republicans. And Toombs of Georgia wrote, "Hostility to Douglas is the sole motive of movers of this mischief. I wish Douglas defeated at Charleston, but I do not want him and his friends crippled or driven off. Where are we to get as many or as good men in the North to supply their places?"[39]

The Davis resolutions had proposed to turn the southern rights position into binding party doctrine. The southern rights leaders had made the most of their ascendancy in the congressional Democratic party. But in the peculiar dualism of the Democratic organization, southern rights dominated only in the congressional party. In the nationwide complex of state organizations, the Douglas Democrats still retained immense power, and in the quadrennial national convention, they would meet the southern Democrats on equal terms. The party had become so schizoid by this time that southern members tended to deny the claims of the Douglas men to a voice, on the ground that they came from states which were certain to vote Republican. But the Douglas supporters replied that southern intransigence was what had weakened them in states which the party had so recently dominated.

Less than ten weeks after the Senate Democratic caucus adopted the Davis resolutions, and while the Congress was still in session, the other Democratic party—the party of state organizations—met at Charleston on April 23, 1860. For the fourth time since James K. Polk had carried a bisectional Democratic party to victory on a program of territorial expansion—with no questions asked about slavery—the country faced a presidential election.

39. Douglas, in *Congressional Globe,* 36 Cong., 1 sess., p. 2156; Wigfall, *ibid.*, p. 1490; Toombs to Alexander H. Stephens, Feb. 10, 1860, in Phillips (ed.), *Toombs, Stephens, Cobb Correspondence,* p. 461.

CHAPTER 16

The Election of 1860

BY 1860, the United States had completed the development of a series of arrangements, both formal and informal, by which a president is chosen every four years. Some of these arrangements, though commonly taken for granted, are singular in the extreme, and make American presidential elections unique as a way of choosing a head of state.

Under the Constitution, the president was chosen by electors, rather than by voters, with each state having a number equal to the total of its senators and representatives. The mode of choosing electors was left by the Constitution entirely to the state legislatures, which might have proceeded in any of several ways: They might have chosen the electors themselves, which all of the thirteen original states except Virginia did at one time or another, and which eight states were still doing as late as 1820. Or they might have provided for choice by popular election, which is what all states except South Carolina were doing by 1832. In the process of popular election, a state might have given to each candidate electoral votes in proportion to the popular votes received in the state, but no states have ever done this. They might have chosen electors by district, and in fact, ten states, at various times between 1788 and 1832, used this method. But in general, the states were jealous of the political power they could wield by casting their vote as a block, and by 1836 every state (again, except South Carolina) was holding a popular, "general ticket" election, by which it cast its total electoral vote for whoever won a majority or even a plurality in the state election. Although conducted simultaneously throughout the nation, the November election was not a national election, but a multiplicity of statewide elections, in which popular votes had no value toward an electoral total unless the candidate receiving them carried the state in which they were cast. In short, election depended not upon winning popular votes but upon winning a combination of states which held a majority of electoral votes. In 1860, as it turned out, 39 percent of the vote was enough to provide such a combination.

These electoral arrangements are well known, but they have had profound effects not always recognized. Since, at the state level, a vote for a candidate was "wasted" unless he had some realistic chance of winning more votes than any other candidate, minor candidates tended to be squeezed out, and elections tended to resolve themselves into contests between two leading candidates. This was true both within a state and among the states, for it did a candidate no good to win popular votes unless they might be converted into electoral votes, and it did a state no good to give a candidate electoral votes unless he stood a good chance of winning enough electoral votes in other states to constitute a majority.

The iron logic of these circumstances tended, from a very early time, to make the American political system a bipartisan one and the party structure a federated one. In a situation in which minority votes were "wasted," third parties had short lives and supporters who often wanted merely to express a protest or to help defeat one of the major candidates by drawing away some of his votes. Thus, even when there were three or more parties in the race, the election in any given state tended to become a two-way contest, as in 1856, when the effective rivalry was between Buchanan and Frémont in the northern states and between Buchanan and Fillmore in the southern states. At the same time, every state political organization, while jealous of its own autonomy, was anxious that its party counterparts in other states should be strong enough to provide favorable prospects for winning the "national" election. In the 1830s, the national conventions, first of the Democrats and then of the Whigs, had been instituted, thus giving each state party a chance to express its voice in national party councils, and even more to see visible proof of the vigor and compatibility of the party organizations in other states.

Once the system of conventions and elections had been established, devices had to be developed for instructing and arousing the electorate. Party candidates were not expected to participate in this process, for the office was supposed to seek the man and not the man the office. But there were swarms of editors, officeholders, and party leaders available to publicize the issues, to organize supporters, and to galvanize the electorate with glee clubs, marching clubs, and other such activities for voters who responded to excitement more than to reason. By 1860, the colors were set. The quadrennial choosing of a president was accomplished in the context of a ritualized "campaign," which began in the summer with the national conventions and ended in November with the election.

It was part of the ritual that the Democrats should make their nominations first, and in 1860 they prepared to do so at Charleston, South Carolina. Just when it needed bisectional harmony more than ever before, the party met in the city least likely to support the cause of bisectional harmony. The atmosphere of Charleston—physically a miserable place for such a large convention—heightened the tensions within the party. Less than a year later, military warfare between North and South would begin in this same city where, in April 1860, some of the party leaders were seeking to avert political warfare.

The Democratic convention of 1860 remained in session for ten days in Charleston and then adjourned for six weeks, to convene again at Baltimore on June 18 for another six-day session. No American party convention has exceeded it in length except the Democratic convention of 1924, and none has been the scene of such a bitter and complicated contest. Altogether, the convention took fifty-nine ballots on the nomination of a presidential candidate, in addition to many votes on parliamentary issues, and it witnessed two major scenes of disruption by the withdrawal of delegates from the South. It ended with a schism which not only destroyed the last remaining party with a nationwide constituency, but also foreshadowed with remarkable accuracy the schism that appeared in the Union itself less than a year later.[1]

1. Probably the best narrative account of the Democratic conventions of 1860 is in Roy F. Nichols, *The Disruption of American Democracy* (New York, 1948), pp. 288–322. But Allan Nevins, *The Emergence of Lincoln* (2 vols.; New York, 1950), II, 203–228, 266–272, is also excellent. Also see Robert W. Johannsen, "Douglas at Charleston," in Norman A. Graebner (ed.), *Politics and the Crisis of 1860* (Urbana, Ill., 1961), pp. 61–90; Avery O. Craven, *The Growth of Southern Nationalism, 1848–1861* (Baton Rouge, 1953), pp. 323–334; George Fort Milton, *The Eve of Conflict: Stephen A. Douglas and the Needless War* (Boston, 1934), pp. 409–449, 458–479; Dwight L. Dumond, *The*

Despite all the elaborate maneuvering for advantage in the convention, and all the hairbreadth votes which seemed so crucial at the time, the basic situation was fairly simple: Douglas had the support of a bare majority of the delegate votes. With this majority, he could prevent the adoption of a platform calling for a congressional slave code such as the Davis resolutions had demanded; but, because of the two-thirds rule (which had blocked the nomination of another northern candidate, Martin Van Buren, in 1844), he could not gain the nomination. Further, neither side was prepared for the kind of concessions which so commonly resolve the deadlock in party conventions. As the southern rights supporters saw it, the Supreme Court, in the Dred Scott decision, had validated their claims, and they were not going to barter them away in an equivocating platform. But Douglas could reply that he was not insisting on a divisive doctrinal test—it was the South which took a rigid attitude. And as for the two-thirds rule, he felt that his majority placed the opposition under a moral obligation to acquiesce in his nomination; he had twice stepped aside, first for Pierce in 1852 and then for Buchanan in 1856, though he could definitely have blocked Buchanan; now, he would not let a minority deny him the nomination by creating a deadlock.

The convention assembled in an atmosphere of acute tension and excitement, for the participants all sensed that a disruption was imminent. Delegates from the Northwest were determined to resist southern demands for a platform with a plank calling for a slave code, and Henry B. Payne of Ohio had written to Douglas in the month preceding the convention that if such a platform were adopted the Ohio delegation would "be prepared to retire from the convention. I have no reason to doubt that this will be the course of seven Northwestern States."[2] But the South was equally resolute. The Alabama Democratic convention had explicitly instructed its delegates to withdraw if a slave code platform were not adopted,

---

Secession Movement, 1860–1861 (New York, 1931), pp. 35–91; Emerson David Fite, *The Presidential Campaign of 1860* (New York, 1911), pp. 106–116. The major sources for the convention are *Official Proceedings of the Democratic National Convention Held in 1860, at Charleston and Baltimore* (Cleveland, 1860); Murat Halstead, *Caucuses of 1860* (Columbus, 1860; reprinted with minor abridgments in an edition ed. by William B. Hesseltine, under the title *Three Against Lincoln: Murat Halstead Reports the Caucuses of 1860* [Baton Rouge, 1960]).

2. Payne to Douglas, March 17, 1860, quoted in Percy Lee Rainwater, *Mississippi: Storm Center of Secession* (Baton Rouge, 1938), p. 121.

and other state conventions in the lower South had instructed their delegations to insist upon such a platform. Three days before the formal opening of the convention, the delegates of Georgia, Arkansas, and the five Gulf Coast states met in caucus and agreed to withdraw from the convention if Douglas should be nominated.[3]

With a break so imminent, the moderate elements in the convention made desperate efforts to find a formula on which the opposing forces could agree, and they also resorted to dilatory tactics in an effort to avoid the dreaded showdown. But the delays gave more opportunity for impassioned speeches, delivered in theatrical circumstances, to packed galleries of ardent southerners. Some of this oratory even spilled over outside the convention and was delivered from hotel balconies or in public parks. As delays continued, tension rose, and an overwhelming sense of drama pervaded the atmosphere.

The oratorical climax came on the evening of the fifth day, as William L. Yancey of Alabama took the spotlight. Already well known in the South as the most silver-tongued of a race of uninhibited orators, and the most fervent exponent of southern rights, Yancey was greeted with a prolonged ovation, after which he launched into a vigorous, unqualified defense of the extreme southern position. Brushing aside all the peripheral questions about rights in the territories, he told the northern delegates that their initial error had been to accept the view that slavery was evil and then to acquiesce in its containment. Instead, they should have defended slavery on its merits. The South, he asserted, would now at last insist upon its rights, including the plank for a territorial slave code.[4]

The response of the North came from Senator George E. Pugh of Ohio. Blunter than Yancey, but no less forceful, Pugh reproached the South for first causing the ruin of the northern Democrats and then taunting them with their weakness. Now, he said, they were told that they must put their hands on their mouths and their mouths in the dust. "Gentlemen of the South," he said, "you

3. See above, p. 404; Halstead, *Caucuses of 1860*, p. 11; Austin L. Venable, "The Conflict Between the Douglas and Yancey Forces in the Charleston Convention," *JSH*, VIII (1942), 237.

4. John Witherspoon Du Bose, *The Life and Times of William Lowndes Yancey* (2 vols.; Birmingham, Ala., 1892), II, 457–460; Halstead, *Caucuses of 1860*, pp. 42–43, 52–54; Nevins, *Emergence*, II, 216–217.

mistake us—you mistake us. We will not do it."[5]

The lines of battle were drawn, and on the next day, the convention took up the question of a platform. After much preliminary skirmishing and one referral back to the platform committee, the issue finally resolved itself into a choice between a majority report by the southern group (which controlled a majority of the states' delegations and therefore a majority of the platform committee, in which each state had one vote) and a minority report by the Douglas forces. The southern wing proposed a platform which affirmed the "duty of the Federal government, in all its departments [meaning Congress also] to protect, when necessary, the rights of persons and property [meaning slaves] . . . in the territories." The Northern wing proposed to leave to the Supreme Court the question "as to the nature and extent of the powers of a territorial legislature, and as to the powers and duties of Congress . . . over the institution of slavery within the territories." To the Douglas forces, this meant that they were willing to leave the question open, instead of forcing a categorical doctrine; to the South, it was just one more refusal to recognize clearly established southern rights.[6]

When the question came to a vote, the minority report won adoption, 165 to 138 (free states, 154 to 30; slave states, 11 to 108). As this result was announced, the process of disruption began. Alabama formally withdrew from the convention, followed by Mississippi, Louisiana, South Carolina, Florida, Texas, one-third of the Delaware delegation, part of the Arkansas delegation, and the next day by Georgia and most of the remaining delegates from Arkansas.[7] Twelve years previously, William L. Yancey, with but one supporter, had walked out of the Democratic convention on this same issue. But now he carried the entire lower South with him.

The split had been long expected and was staged with strong dramatic effect. But once it took place, a kind of anticlimax ensued. If the remaining delegates had proceeded to nominate Douglas, the bolting delegates would no doubt have moved at once to set up a rival ticket of their own. But instead, the convention remained deadlocked. Meanwhile, the bolters organized a rival convention and adopted the rejected majority platform, but they stopped short of

5. Halstead, *Caucuses of 1860*, pp. 54–55; Milton, *Eve of Conflict*, pp. 435–436.
6. *Official Proceedings*, pp. 47, 48; Nichols, *Disruption*, pp. 298–302.
7. *Official Proceedings*, pp. 55–66; Halstead, *Caucuses of 1860*, pp. 74–88.

making nominations, and simply hung about, listening to speeches very much as if they secretly yearned to go back to the regular convention.

The Douglas Democrats had not been sorry to see a limited withdrawal by the delegates from the lower South, for they now appeared to have a better chance of getting the two-thirds majority necessary for a nomination.[8] The withdrawal had taken 50 delegate votes out of the convention, leaving 253. The Douglas supporters expected that this would reduce the required two-thirds from 202 (out of 303) to 169 (out of 253). But they received a rude shock when the chairman of the convention, Caleb Cushing, ruled that two-thirds of the original number of delegate votes was still required to nominate, and an even ruder shock when the New York delegation, which was otherwise supporting Douglas, cast 35 decisive votes to help sustain the decision of the chair, 144 to 108. Laboring under this handicap, the Douglas forces finally brought the convention to the stage of making nominations. On the eighth and ninth days of the convention, there were fifty-seven ballots, in the course of which Douglas's strength never fell below 145½ and never rose above 152½ (which was one vote more than a majority of the whole convention). The opposition vote was badly scattered among R. M. T. Hunter of Virginia, James Guthrie of Kentucky, and others, none of whom ever succeeded in concentrating more than 66½ votes.[9] Historians have often attributed Douglas's defeat to the refusal to permit nomination by two-thirds of the sitting delegates, but the fact is that Douglas never got closer than 16½ votes to attaining two-thirds of the sitting delegates. Of course, it might have

8. Murat Halstead reported on April 25 that the Douglas supporters "want about forty Southern delegates to go out for that would insure the nomination of Douglas and help him in the North. Their fear is that the secession will be uncomfortably large. A slight secession of merely the 'shred of Gulf States' would be a help." On April 30, Halstead described a strong antisouthern speech by a Douglas supporter, Stuart of Michigan, delivered just before the southern withdrawal: "If his object was to produce irritation, he succeeded admirably. But there was more powder in the explosion than Stuart calculated upon. Instead of merely blowing off a fragment or two, and producing the long-coveted reaction in the North, one half of the South—the very citadel and heart of Democracy—was blown away." *Caucuses of 1860*, pp. 40, 74. Also, Venable, "Conflict Between Douglas and Yancey Forces," p. 239.

9. *Official Proceedings*, pp. 73–89. The New York delegation was obviously following a somewhat devious course, voting for Douglas but voting against a construction of the two-thirds rule which might have facilitated his nomination. Apparently the New Yorkers hoped that a deadlock might result in a compromise, which they wanted to promote, between the Douglas forces and the South.

been impossible to stop him once he got that close, but in this case, the opposition was especially resolute, and it is hard to see where he would have got additional votes. Although eight states had withdrawn, enough of his opponents had remained to defeat him under any application of the two-thirds rule. His supporters recognized this fact, and on the tenth day of the convention they adjourned to meet again at Baltimore on June 18. As they did so, the bolters, somewhat disconcerted by their failure to force the nomination of a compromise candidate, adjourned their convention to Richmond, to convene on June 11.

During the interim, the Douglas forces made strenuous efforts in the lower South to develop new state organizations which would send new pro-Douglas delegations to Baltimore. These efforts were not wholly successful anywhere—in no case did the Douglas forces capture the regular state conventions which were reconvened. But in Louisiana, Alabama, and Georgia they were able to organize impressive meetings which appointed new delegations to replace the delegations that had bolted at Charleston.[10] When the Richmond convention assembled, it did nothing, and when the Baltimore convention met, the Yancey forces from Alabama and the fire-eaters of Louisiana and Georgia were there clamoring to come back in, while the Douglas forces supported the rival delegations. The divisive question at Baltimore, therefore, was one of contested delegations. The pro-Douglas majority brought in a report to award all of the seats from Alabama and Louisiana to the new Douglas delegations, as well as half of the Georgia vote and two votes from Arkansas.[11]

The anti-Douglas forces perceived that if this report were adopted, it would result in the nomination of Douglas, who had also picked up most of the few northern votes that had previously gone against him. The southern delegations struggled grimly, but they were outnumbered.[12]

The majority report was adopted, and this started the second disruption of the convention. Virginia took the lead in announcing her withdrawal, followed by North Carolina, Tennessee, more than

10. For these activities between the Charleston and the Baltimore conventions, see Dumond, *The Secession Movement*, pp. 62–75; Milton, *Eve of Conflict*, pp. 464–468; Nichols, *Disruption*, pp. 306–314.

11. *Official Proceedings*, pp. 113–116.

12. *Ibid.*, pp. 116–144.

half of Maryland, California, and Oregon, with most of Kentucky, Missouri, and Arkansas following.[13] This time the party was really split in two. But the Douglas forces proceeded to nominate. On the first ballot, Douglas received 173½ votes to 17, with 113 votes not cast. He still did not have two-thirds of the full vote as required under the rule adopted at Charleston, but the convention, after a second ballot, adopted a resolution declaring Douglas unanimously nominated. For the vice-presidency, Benjamin Fitzpatrick of Alabama was nominated, but he subsequently declined to run, and Herschel V. Johnson of Georgia replaced him on the ticket.[14]

The day after the disruption at Baltimore, a gathering of delegates of claimants to delegate status, with 23½ votes from the free states and 81½ from the slave states, met at another hall in Baltimore, adopted the majority platform as reported at Charleston, and nominated Vice-President John C. Breckinridge for the presidency and Senator Joseph Lane of Oregon for the vice-presidency, both on first ballots and by overwhelming majorities.[15]

Thus, sectional dissension had at last shattered the one remaining national political party, and historians have speculated ever since as to what motives impelled the southerners to adopt a course that seemed to guarantee a Republican victory, and why the Douglas supporters did not make concessions to hold the party together. The southerners probably had various motives. Some no doubt hoped to snatch victory from defeat by creating an electoral deadlock, thus throwing the election into the House of Representatives and perhaps even into the Senate.[16] Probably most southerners did not count on any such intricate process, but continued to hope that their tactics would unnerve the Douglas supporters and compel them to make concessions. By the time that the futility of such strategy had been demonstrated, the southerners were left with no tenable alternative. An indeterminate number of ultraradical south-

13. *Ibid.*, pp. 144–160.

14. *Ibid.*, pp. 156–174; Percy S. Flippen, *Herschel V. Johnson of Georgia: State-Rights Unionist* (Richmond, 1931), pp. 121–160.

15. Halstead, *Caucuses of 1860*, pp. 265–278.

16. *Ibid.*, p. 36, April 25 (before the disruption): "Southern Secession here would give Douglas strength in some of the Northern states. There would be no possibility of his election, however, for he would certainly lose several Southern States. He might, and the chances are that he would, carry Northern States enough to defeat the election of Seward. Thus the election would be thrown into Congress—and eventually into the Senate. This is, beyond question, the game of the Southern men."

erners had a different reason. They wanted to split the Democratic party in order to ensure a Republican victory, which, they believed, would precipitate the South into secession. The importance of this factor of "conspiracy" to bring about disunion is difficult to evaluate, partly because it was almost certainly exaggerated by the political opponents of the fire-eaters at the time and by pronorthern historians later. But it was unquestionably an element in the situation. At Charleston, southern radicals did not hesitate to declaim their willingness to quit either a party or a federal union which denied them their rights under the Constitution. On the day of the disruption at Charleston, a delegate from Mississippi made an "impassioned and thrilling" speech in which he declared, "with piercing emphasis, that in less than sixty days there would be a United South; and at this declaration there was the most enthusiastic shouting yet heard in the Convention." That evening, the southerners held an immense mass meeting at which there was "a Fourth of July feeling—a jubilee." No such occasion would have been complete without some eloquence from William L. Yancey, and the Alabama orator told the crowd that "perhaps even now the pen of the historian was nibbed to write the story of a new revolution."[17]

Arguments about whether the bolters wanted to throw the election into Congress, or to wring concessions from the Douglas forces, or to break up the Union all suffer from one common defect: They are too rational. The delegates at Charleston and at Baltimore were operating in an atmosphere of extreme excitement, in which gusts of emotion constantly swept the floor as well as the galleries. In the midst of this turmoil, men took positions which led on to consequences that they did not visualize. Men obsessed with the idea of stopping Douglas at all costs quite readily walked out of the convention with a hope that, in some undefined way, they could walk back in again in a stronger position. Other men, equally single-minded in a determination to nominate Douglas, were glad to see some of his opponents go, if it would make the nomination easier. Many on both sides clung to an opportunistic notion that later on, some persons unknown, in some fashion unknown, would somehow patch up the split.

But whether or not the disruption of the Democratic party was deliberately intended to lead to the disruption of the American

17. Halstead, *Caucuses of 1860*, pp. 84, 87, 86.

Union, it did foreshadow the disruption with remarkable precision. Seven of the eight states whose delegates walked out of the convention at Charleston were the same seven that constituted the original Southern Confederacy when Jefferson Davis was inaugurated as president. In the second disruption at Baltimore, the Virginia, North Carolina, and Tennessee delegates walked out, while Kentucky and Missouri were divided. Precisely the same pattern recurred after Fort Sumter, when the first three of these states went over wholly, and the other two partially, to the Confederacy. Only Arkansas, which had been in the first wave of party disruption, shifted to the second wave of secession a year later.

The Democrats had, of course, been drifting toward this debacle ever since the Lecompton contest. As they did so, their troubles had begun to give new hope to the apparently moribund body of old Whigs. In 1852 the Whigs had virtually ceased to be a bisectional, national party. The abrupt decline of party strength in the lower South had seemed to mark the destruction of southern Whiggery, and this had hastened the abandonment of the party by discouraged northern Whigs who despaired of regaining enough strength to maintain their status as a major party. But the defection of the northern wing had enabled the southern Whigs to regain control of what was left of the national organization, and the rise of the nativist American party had given them important allies, though at a considerable cost. Thus the combined Whigs and Americans had nominated in 1856 the man whom they had been prevented from nominating in 1852—that is, Millard Fillmore. With most northern Whigs by this time absorbed into the Republican party, Fillmore had developed political strength primarily in the South. But there, although he won only Maryland, he had run stronger than Scott in the lower South, carrying more than 40 percent of the vote in ten southern states. Though his strength was sectional, his political position was national, for he was the Unionist candidate in the South, opposing the Southern Rights Democrats. Thus, the Whig-American contingent survived in the South as the "Union-saving" group in that part of the country, as opposed to the fire-eating Democrats, while in the North it was the Democratic party that claimed a Union-saving role as opposed to the Republicans. In proportion as extremists gained control of the southern Democratic organization, southern moderates were driven to the support of the Whigs. Meanwhile, the rapid gains of the Republicans and the increasing sectionalization of politics in the North led many old Whigs after 1856 to feel that there was no longer a party of the Union. The Republicans, who had carried all but five of the free states, were utterly without any following in the South, and the Democratic party was increasingly dominated by militant southerners who dealt freely in threats of disunion. The party of Henry Clay, who stood first among Union-savers, still had a mission. It could no longer be called the Whig party, for that might keep away Democrats sympathetically inclined, and it could not be called the American party, for that smacked too much of Know-Nothingism, which had fallen into disrepute. But they were going to try to rehabilitate it, and they could easily call back many of the old Whigs who in 1856 had voted reluctantly for Buchanan in order to save the Union.

The bitter division within the Democratic party over the Lecompton constitution furthered this impulse. Senator John J. Crittenden of Kentucky, Henry Clay's successor, had opposed Lecompton and rallied a considerable amount of support among southern conservatives. Meanwhile, northern conservatives were organizing, and in December 1858 representatives from thirteen states had convened in Washington, where they discussed plans for a ticket in 1860 that would draw conservatives away from both the Republican and the Democratic parties. During 1859 this movement gained momentum, as American party candidates made a very strong race for the governorship of Virginia and gained several congressional seats in North Carolina, Tennessee, and Kentucky. In December 1859, Senator Crittenden called a conference of about fifty "opposition" members of Congress. This meeting established a liaison with the central committees of both the Whig and the American parties, and by January plans had been perfected for a new "Constitutional Union" party. On Washington's birthday, the organizers issued an "Address," denouncing both existing parties, appealing for the cause of Union, and calling for delegates to meet at Baltimore on May 9 to nominate a presidential ticket.[18]

18. The fullest and best accounts of the Constitutional Union party are in Dumond, *Secession Movement*, pp. 92–112, and in biographies of men prominent in its organization: Joseph Howard Parks, *John Bell of Tennessee* (Baton Rouge, 1950), pp. 339–360; Albert D. Kirwan, *John J. Crittenden: The Struggle for the Union* (Lexington, 1962), pp. 336–365. Also, see Arthur Charles Cole, *The Whig Party in the South* (Washington, 1913), pp. 328–338; Halstead, *Caucuses of 1860*, pp. 118–140; *National Intelligencer*, May 10, 11, 12, 1860.

The Constitutional Union convention proved to be the most "harmonious" in a year of strife-ridden party conclaves. Delegates appeared from twenty-three states and readily agreed not to adopt a platform but to stand on the Constitution (however it might be construed) and the Union. When they turned to the nominations, their first choice would unquestionably have been Crittenden, but he was seventy-four years old and had declined to be a candidate. The more willing possibilities were not much younger. They included Winfield Scott, also seventy-four, who had proven a very inept candidate at age sixty-six; Sam Houston, who at sixty-seven was handicapped partly by certain fantastic plans for establishing a protectorate over Mexico, and even more by being too much of an old Democrat for a party of old Whigs; Edward Bates of Missouri, also sixty-seven; and John Bell of Tennessee, aged sixty-four. Bates, as an old Whig and a border-state man with mild antislavery sentiments, had some appeal both for the Constitutional Unionists and for very moderate or very opportunistic Republicans who saw the expediency of muting their antislavery outcries. For many months, Bates had skillfully (or indecisively) ridden both horses, but late in March he had at last yielded to demands that he clarify his position. His assertion that Congress controlled slavery in the territories, and that slavery could not go there without congressional assent, made him a Republican and not a Constitutional Union contender.[19]

John Bell was not a man of much stature. His personality was cold, his manner formal, his speech calculated and uninspiring. But he had the right credentials. A lifelong Whig from a border state, he had voted against both the Kansas-Nebraska Act and the Lecompton bill. He was a large slaveholder, but not a vigorous exponent of the political rights of slavery. The convention nominated him on the second ballot, and then chose a man who overshadowed him, Edward Everett of Massachusetts, aged sixty-seven, as the vice-presidential nominee. Thus, Bell and Everett were the first candidates nominated in the 1860 campaign.[20]

19. Marvin R. Cain, *Lincoln's Attorney General: Edward Bates of Missouri* (Columbia, Mo., 1965), pp. 90–105, shows Bates's care to avoid breaking with his supporters in the American party, but scarcely indicates that he was a potential candidate for the Constitutional Unionists; Kirwan, *Crittenden*, p. 354, shows evidence that Bates, at one time, "seemed eager for the nomination"; on Houston's aspirations, see Llerena B. Friend, *Sam Houston: The Great Designer* (Austin, 1954), pp. 311–320.

20. Parks, *John Bell*, pp. 353–355.

Six days after the adjournment of the Constitutional Union convention, the Republican convention assembled at Chicago in the Wigwam, a new hall built especially for this meeting.[21] The party had come a long way since its nomination of Frémont four years earlier. At that time it was a new and untried organization, fencing with the Americans for a position as the second major party and engaged in the extremely ticklish business of lining up nativist support without overtly adopting nativist attitudes. It was so emphatically identified with antislavery that it suffered the disadvantage of being regarded as a one-idea party. In 1856 the Republicans had not really expected to win the election, and Thurlow Weed, a master of realistic politics, had not even wanted his associate, William H. Seward, to get the nomination.[22]

Even after the lapse of four years, the colors of the party had not quite set, and the boundaries which separated antislavery Republicans from anti-Lecompton Democrats, or which differentiated moderate Whigs, who mildly opposed slavery, from conservative Republicans, who opposed slavery only mildly, were far from clear. Thus, Horace Greeley had seemed ready in 1858 for a political marriage with the Douglas Democrats, and in 1860 he was eager for the Republicans to join the border state Whigs in the support of Edward Bates, a lifelong Whig who had voted for Fillmore in 1856. Other Republicans, notably Abraham Lincoln, opposed such political combinations and sought to define and distinguish the Republican position.

But despite continued blurring at the margins, the party had gained immensely both in establishing a broad policy base and in strengthening its organizational structure. The Republicans in Congress had supported bills for a protective tariff, for internal improvements, and for free homesteads of 160 acres. Such home-

21. On the Republican nomination and the events leading up to it, the best general accounts are William Baringer, *Lincoln's Rise to Power* (Boston, 1937); Reinhard H. Luthin, *The First Lincoln Campaign* (Cambridge, Mass., 1944), pp. 3–119, 136–167; J. G. Randall, *Lincoln the President* (2 vols.; New York, 1945), I, 129–177; Nevins, *Emergence,* II, 229–260; Don E. Fehrenbacher, "The Republican Decision at Chicago," in Graebner (ed.), *Politics and the Crisis of 1860* (Urbana, Ill., 1961), pp. 32–60; Fehrenbacher, *Prelude to Greatness: Lincoln in the 1850's* (Stanford, 1962), pp. 143–161. Basic sources include *Proceedings of the First Three Republican National Conventions,* C. W. Johnson, compiler (Minneapolis, 1893); Halstead, *Caucuses of 1860,* pp. 141–177.

22. Glyndon G. Van Deusen, *William Henry Seward* (New York, 1967), pp. 175–178. Seward to Mrs. Seward, June 6, 1856, in Frederick W. Seward, *Seward at Washington as Senator and Secretary of State, 1846–1861* (New York, 1891), p. 276.

steads were to be available to immigrants even though they were not citizens, and this went far to clear the Republicans of the stigma of nativism. The opposition of the Buchanan administration and the Democratic party to these measures had tended to make Republicans the friends of the extensive interests supporting them. No longer were the Republicans a one-idea party. Further, in the election of 1856, Frémont had carried eleven free states with 114 electoral votes, which was only 35 short of a majority. Since then, Minnesota and Oregon had entered the Union, and the Republicans were reasonably certain of Minnesota, but in order to obtain a majority they still had to win 34 additional votes from the five free states which they had lost to Buchanan in 1856. They had little hope in California, and this meant that they had to carry Pennsylvania with its 27 votes, and either Illinois (11), Indiana (13), or New Jersey (7). In short, all they needed in order to win the presidency was to turn the tables in Pennsylvania and any one of three other strategic states.

All four were border states, in the sense that they adjoined slave states. As such, they were more moderate on the slavery question than the states of the "upper North" (New England, New York, Michigan, Wisconsin, and Minnesota). To carry them, therefore, the Republican party had to adjust itself accordingly. This had been evident ever since 1856, but it was increasingly evident after the events at Charleston showed that the Douglas Democrats intended to stay in the race, and after the nomination of John Bell gave all "moderates" a candidate whom they could vote for if they regarded the Republican nominee as too radical.

In various ways, therefore, Republican aspirants and organizers had begun moving toward a moderate position or seeking moderate candidates. The most prominent figure to do so was William H. Seward. After a career of four years as governor of New York and twelve years as senator, Seward was undoubtedly the foremost Republican in the nation. He had dealt with a broader range of public questions than most of the Republican leaders, many of whom, like Salmon P. Chase and Charles Sumner, were identified somewhat restrictively with antislavery alone. Yet Seward had claimed the leadership in antislavery also, proclaiming in 1850 that there was "a higher law than the Constitution" and in 1858 that there was an "irrepressible conflict between freedom and slavery." By 1860, it began to appear that these phrases had succeeded too well, and on February 29, Seward delivered a major speech in the Senate appealing for "mutual toleration" and "fraternal spirit." Even the dualism of "free states" and "slave states" disappeared, and "labor states" and "capital states" replaced them. But Seward's conciliatory gestures were too patently opportunistic to win the confidence of the moderates. For more than a decade he had been building an image of himself as the antislavery leader. Southern fire-eaters, who had accepted the image literally, were not going to let him escape from it now. Almost the only people influenced by the speech were some of the radical antislavery men, who were antagonized.[23]

While Seward was converting himself into a moderate, the Republicans' foremost journalistic leader had been seeking an established moderate as a nominee. Horace Greeley of the New York *Tribune,* erratic, impulsive, and at most times vigorously antislavery, had made up his mind. "I want to succeed this time," he wrote privately, "yet I know the country is not Anti-Slavery. It will only swallow a little Anti-Slavery in a great deal of sweetening. An Anti-Slavery man *per se* cannot be elected; but a Tariff, River-and-Harbor, Pacific Railroad, Free-Homestead man *may* succeed *although* he is Anti-Slavery. . . . I mean to have as good a candidate as the majority will elect."[24] By the time Greeley stated this formula, he had long since settled upon Edward Bates of Missouri as a man who would meet the specifications. During 1859, partly at Greeley's urging, Schuyler Colfax of Indiana and the two Francis P. Blairs, Jr. and Sr., of Missouri and Maryland had begun to groom Bates as a man who could carry the border states. At the same time, Greeley's *Tribune* had begun to publicize Bates as a "practical emancipationist" who had freed his own slaves. "The Tariff men," said Greeley, "cannot object to him, for he is fully with them. The River and Harbor men will be glad to hail as a candidate the President of the Chicago River and Harbor Convention. As to the Pacific Railroad, the word St. Louis [Bates's home] tells all that need be said on that subject."[25] In fact, however, Bates had serious liabilities. He was sixty-seven

23. *Congressional Globe,* 36 Cong., 1 sess., pp. 910–915; Van Deusen, *Seward,* pp. 217–220; Seward, *Seward at Washington,* pp. 443–444.

24. Greeley to Mrs. R. M. Whipple, April 1860, quoted in Jeter Allen Isely, *Horace Greeley and the Republican Party, 1853–1861: A Study of the New York Tribune* (Princeton, 1947), p. 266.

25. New York *Tribune,* Feb. 20, 1860, quoted in *ibid.,* p. 273.

years old, had been openly a nativist, and had remained a Whig as late as 1856. He was also a colorless personality, and his views on slavery seemed equivocal. His conviction that Congress controlled slavery in the territories was not announced in clear-cut terms until two months before the Chicago convention. Prior to that, he had labeled the slavery issue a "pestilent question, the agitation of which has never done good to any party, section, or class, and never can do good." Greeley and the Blairs must have been well aware of Bates's deficiencies, but they kept their misgivings to themselves and went to Chicago supporting him.[26]

A third person who had been adjusting his position in a conservative direction was Abraham Lincoln. Ever since the debates with Douglas, knowledgeable Republicans had recognized Lincoln as a resourceful figure of some stature, and a small group of Illinosians had been quietly working to advance his candidacy for the Republican nomination. In October 1859 he received an invitation to deliver a lecture in New York City which he accepted eagerly. Thus, on February 27, 1860, two days before Seward's speech on "capital" states and "labor" states, Lincoln spoke at the Cooper Union to a good-sized audience heavily sprinkled with influential Republicans.

In the Cooper Union address, Lincoln was, in effect, still replying to Douglas—this time to his argument in *Harper's* that popular sovereignty had been a principle of the American Revolution. Lincoln had worked out, by sounder historical research, a case for the view that the founders of the Republic had regarded slavery as an evil and had "marked [it] as an evil, not to be extended, but to be tolerated and protected only because of and so far as its actual presence among us makes that toleration and protection a necessity." The Republicans would continue to leave southern slavery unmolested, "due to the necessity arising from its actual presence in the nation," but they would not give up their conviction that slavery was wrong, or their effort to exclude it from the territories. The Republicans,

26. On Bates's candidacy and Greeley's and Blair's sponsorship of it, see Cain, *Bates*, pp. 90–116; Marvin R. Cain, "Edward Bates and the Decision of 1860," *Mid-America*, XLIV (1962), 109–124; Reinhard H. Luthin, "Organizing the Republican Party in the 'Border Slave' Regions: Edward Bates's Presidential Candidacy in 1860," *MHR*, XXXVIII (1944), 138–161; Luthin, *First Lincoln Campaign*, pp. 51–68; Howard K. Beale (ed.), *The Diary of Edward Bates, 1859–1866*, AHA *Annual Report*, 1930, IV, 127–131; William Ernest Smith, *The Francis Preston Blair Family in Politics* (2 vols.; New York, 1933), I, 464–469; Willard H. Smith, *Schuyler Colfax* (Indianapolis, 1952), pp. 116–117, 133–134; Isely, *Greeley and the Republican Party*, pp. 255–286.

in Lincoln's view, were not denying to slavery any rights which had not been denied by the founders.[27]

Thus a number of significant realignments in a moderate direction had occurred in the months preceding the Chicago convention. As the clans gathered, it was clear that Seward stood far in the lead over all the other candidates. Thurlow Weed had arrived triumphantly in Chicago on a train of thirteen cars jammed with Seward supporters. He also brought large supplies of champagne and, it was reported, copious funds—all to be used to gain the nomination for the New York senator. Seward's prospects seemed brighter because none of the other candidates except Bates held the clear support of more than one state (Bates had Missouri, Maryland, and Delaware). The basic question, therefore, was whether Seward would gain the nomination by virtue of his great initial strength before the opposition could unite. It appeared to be Seward against the field.[28]

The tone of the convention itself indicated a change in the character of the Republican party since 1856. In 1856, the personnel of the convention had included a conspicuously high proportion of somewhat evangelical antislavery men. By 1860, the spirit of dedication had not disappeared. One reporter wrote, "The favorite word in the convention is 'solemn.' In Charleston, the favorite was 'crisis.' Here there is something every ten minutes found to be solemn." But if the decisions were solemn, the atmosphere certainly was not. Weed's thirteen carloads of Seward supporters were but a small part of the swarm of people that thronged Chicago, packed the Wigwam

27. Text of Cooper Union address in Roy P. Basler (ed.), *The Collected Works of Abraham Lincoln* (8 vols.; New Brunswick, N.J., 1953), III, 522–550. Luthin, *First Lincoln Campaign,* p. 81, described the Cooper Union address as "conservative," without explaining why he so regarded it. This view is in line with a tendency among some historians to emphasize Lincoln's moderation and to minimize the differences between his position and that of Douglas (e.g., Randall, *Lincoln the President,* I, 107–109, 117, 123–128). Fehrenbacher, *Prelude,* pp. 146–148, challenges Luthin's interpretation and makes an effective argument that "Lincoln's position in the Republican party remained the same" as in 1858 and "was neither on the left wing nor the right, but very close to dead center." It is, I believe, possible to accept Fehrenbacher's general interpretation and at the same time to point out that in the Cooper Union address Lincoln was putting less emphasis on the restriction of slavery in the territories, and more on the view that "we can afford to let it alone where it is." He had never denied this previously, but to the extent that he now emphasized it, he was adjusting his position in a moderate direction.

28. Luthin, *First Lincoln Campaign,* pp. 23–35; Van Deusen, *Seward,* pp. 213–227; Frederic Bancroft, *The Life of William H. Seward* (2 vols.; New York, 1900), I, 507–545.

to the rafters, overflowed in crowds of 20,000 outside the building, and made this convention the largest political gathering—perhaps the largest gathering of any kind—that the United States had ever seen up to that time. Brass bands and groups of delegates swinging hats and canes generated the exuberant spirit. Virtuous delegates who remembered the crusading spirit of 1856 were shocked at the free use of liquor, and Murat Halstead wrote, "I do not feel competent to state the precise proportions of those who are drunk and those who are sober. There are a large number of both classes."[29]

The change of tone was evident in the new platform. In 1856, the platform had devoted more than half of its nine brief resolutions to the slavery issue and had not addressed itself to any other public question except that of government aid for a Pacific railroad. But in 1860, the antislavery position was moderated while being reaffirmed. The platform denounced disunionism, efforts to reopen the African slave trade, and the extension of slavery into the territories; but it contained no language comparable to the earlier castigation of slavery as a "relic of barbarism." It denounced John Brown's raid as "among the gravest of crimes"; it promised the "maintenance inviolate of . . . the right of each state to order and control its own domestic institutions"; and in its original form it incorporated a mere general reference to the Declaration of Independence, whereas the 1856 platform had contained a specific quotation from the Declaration. Joshua Giddings, one of the party's antislavery patriarchs, secured on the floor a restoration of the quoted passage, but only after threatening to leave the convention.

After these adjustments of the antislavery position, the platform moved on to endorse a tariff which would promote "the development of the industrial interests of the whole country," to "demand" the passage of a homestead act, to denounce state or federal legislation which would impair the "rights of citizenship hitherto accorded to immigrants from foreign lands," and to advocate "immediate and efficient aid" in the construction of a Pacific railroad. It seemed significant that while the whole platform was received with uproarious enthusiasm, no part of it was greeted with louder cheers than the tariff plank, which sent Pennsylvania, especially, into "spasms of

29. Halstead, *Caucuses of 1860*, pp. 148, 142; Nevins, *Emergence*, II, 229–233, 247–251; William E. Baringer, *Lincoln's Rise to Power* (Boston, 1937), pp. 209–218, 246–247.

joy . . . her whole delegation rising and swinging hats and canes."[30]

While spectators were entranced by the noise and excitement—"a herd of buffaloes . . . could not have made a more tremendous roaring"—the political managers were engaged in a desperate struggle for the control of delegates. Fundamentally, the alignment, as it finally developed, was between the states of the upper North (New England, New York, Michigan, Wisconsin, Minnesota) and the states farther south, which bordered on slave territory (New Jersey, Pennsylvania, Ohio, Indiana, Illinois, Iowa). From the South, only the border slave states of Maryland, Delaware, Virginia, Kentucky, and Missouri (and a somewhat synthetic delegation from Texas) were represented in the convention. Otherwise, the only other delegates were those from California, Oregon, the two territories of Kansas and Nebraska, and the District of Columbia. Fundamentally, the southern, far western, and territorial delegations were marginal, and the real focus of the convention fell upon the two groups of states from the upper North and the lower North. The upper northern states all seemed safely Republican, no matter who was nominated. With this latitude of choice, they were heavily for Seward, and when the balloting began, they gave Seward 132 votes on the first ballot to 49 for all his opponents. Leaving out New England, which was influenced against Seward by doubts as to his capacity to win in the crucial states, the tally was 100 to 0. Among the states of the lower North, on the other hand, only two (Ohio and Iowa) had voted Republican in 1856, and all except Iowa were regarded as doubtful states in which the result might depend upon which nominee was chosen. In the initial balloting, these states gave only 3½ votes to Seward and 166½ to his rivals.[31] The number required to nominate was 233.

Seward started with some impressive assets. He enjoyed a national reputation which, among his opponents, only Salmon P.

30. Elting Morison, "Election of 1860," in Arthur M. Schlesinger, Jr., *et al.* (eds.), *History of American Presidential Elections* (4 vols.; New York, 1971), II, 1124–1127; Halstead, *Caucuses of 1860*, pp. 152–158. On the floor skirmish over the Declaration of Independence, *ibid.*, pp. 153–156. Fehrenbacher, *Prelude*, p. 156, shows that the contrast between the platforms of 1856 and of 1860 has been exaggerated, but it remains true that the 1860 platform was both more moderate in its language and more generalized in its content.

31. This analysis of the upper and lower North is from Fehrenbacher, *Prelude*, p. 158, except for the tally on the upper North without New England, which is computed independently.

Chase could begin to match. With the solid support of 70 votes from New York, and the undivided delegations of Michigan, Wisconsin, Minnesota, and California, he held a very long lead over all other aspirants. He was backed by Thurlow Weed, an astute political manager and a machine politician who reputedly commanded "oceans of money" with which he could tempt those in need of campaign funds.[32]

The only major obstacle to Seward's nomination was the persisting doubt as to whether he could win in the "battleground" states —a question of vital concern to party strategists not only in those states but in all others. Thus, Hannibal Hamlin of Maine, who did not attend the convention, had advised the states's delegates: "Appoint one of your members to canvass the delegates from the three doubtful states of Pennsylvania, Indiana, and Illinois. Have him obtain from them in writing the names of three men who can carry these states." In Massachusetts, Governor John A. Andrew made it clear that the Bay state delegates would be guided by any consensus which the crucial states might arrive at. Seward did have some supporters in the lower North, but these individuals may have been placing their hopes of personal advantage ahead of maximizing the chances of party victory. The Republican candidates for governor in Pennsylvania (Andrew Curtin) and Indiana (Henry S. Lane) were at the convention, both broadcasting their conviction that Seward would lose in their states—Lane was said to have repeated the assertion "hundreds of times."[33] Horace Greeley also used this argument, in a treacherous way, for he pretended to prefer Seward and to abandon him only under the compelling pressure of availability, when in fact he disliked Seward and relished his defeat.[34] But

32. Glyndon G. Van Deusen, *Thurlow Weed, Wizard of the Lobby* (Boston, 1947), pp. 235–254; Baringer, *Lincoln's Rise to Power,* pp. 213, 219–222, 234–238, 264–265, 270–273.

33. Charles Eugene Hamlin, *The Life and Times of Hannibal Hamlin* (Cambridge, Mass., 1899), pp. 335, 339–344; Thomas H. Dudley, "The Inside Facts of Lincoln's Nomination," *Century Magazine,* XL (1890), 477–479; Nevins, *Emergence,* II, 258 n.; Alexander H. McClure, *Abraham Lincoln and Men of War Times* (Philadelphia, 1892), pp. 24, 138–139; Charles Roll, "Indiana's Part in the Nomination of Abraham Lincoln for President in 1860," *IMH,* XXV (1929), 1–13; Reinhard H. Luthin, "Indiana and Lincoln's Rise to the Presidency," *IMH,* XXXVIII (1942), 385–405; Luthin, "Pennsylvania and Lincoln's Rise to the Presidency," *PMHB,* LXVII (1943), 61–82; Luthin, *First Lincoln Campaign,* pp. 141, 143, 145.

34. Greeley's role in defeating Seward was exposed to fierce publicity soon after the nomination when Henry J. Raymond, in the *New York Times,* May 24, 1860,

Greeley may not have influenced many delegates, while the negative views of Lane and Curtin undoubtedly influenced a large number.

The basic logic of the situation was working against Seward, and ultimately it defeated him, but his initial advantage was so great and the cloud of confusion was so dense that it appeared likely he would be nominated, especially since the opposition was divided, and some of its favorites were little known and were no more "available" than Seward would have been.

Salmon P. Chase, formerly governor of and now senator from Ohio, was one of the best-known, but Chase was even more radical on slavery than Seward, and therefore even more unacceptable to men like Lane and Curtin. Although a prominent candidate, Chase never really became a contender.[35] In terms of control of a large bloc of votes, Simon Cameron of Pennsylvania, looked formidable, and with suitable personal qualities, he might have been an irresistible claimant, for he was senator from the most crucial of the four decisive states; but Cameron had acquired—perhaps had earned—a national reputation as a spoilsman and machine politican *par excellence,* and this reputation made it difficult for him to gain any votes other than those he simply controlled.[36] There were also the "favorite sons"—Jacob Collamer of Vermont and William L. Dayton of New Jersey—but their candidacies were not entirely serious. If Seward were not nominated, the man in the strongest position to win was either Bates or Lincoln. Greeley and the Blairs were still pushing Bates, but he lacked strong appeal, for it seemed doubtful that

accused Greeley of treacherously causing Seward's defeat by pretending to give him up reluctantly because he could not be elected, while secretly nursing a private hostility toward him. The controversy that followed is well described in Harlan Hoyt Horner, *Lincoln and Greeley* (Urbana, Ill., 1953), pp. 178–181; Glyndon G. Van Deusen, *Horace Greeley, Nineteenth Century Crusader* (Philadelphia, 1953), pp. 246–251. Isely, *Greeley and the Republican Party,* pp. 276–281, generally favorable to Greeley, analyzes the evidence, then asks, "Did Greeley utilize his supposed friendship with Seward to help defeat the latter's nomination? The answer is, yes."

35. Reinhard H. Luthin, "Salmon P. Chase's Political Career before the Civil War," *MVHR,* XXIX (1943), 527–532; Luthin, *First Lincoln Campaign,* pp. 36–50. Chase perhaps suffered most from his own inordinate egocentrism, but he also suffered from the fact that he had two rivals from his own state—Benjamin F. Wade and John McLean. See H. L. Trefousse, *Benjamin Franklin Wade, Radical Republican from Ohio* (New York, 1963), pp. 121–128; Francis P. Weisenburger, *The Life of John McLean: A Politician on the United States Supreme Court* (Columbus, Ohio, 1937), pp. 211–214.

36. Lee F. Crippen, *Simon Cameron: Ante-Bellum Years* (Oxford, Ohio, 1942), pp. 204–221; Erwin Stanley Bradley, *Simon Cameron, Lincoln's Secretary of War* (Philadelphia, 1966), pp. 136–157; Luthin, *First Lincoln Campaign,* pp. 92–105.

he could carry Missouri; he had virtually repudiated the use of the slavery question as a campaign issue; and he was not in any clear sense a member of the Republican party—the Missouri slaveholders were the only people who regarded him as one.

Lincoln, on the contrary, was a Republican; he was more likely than anyone else to carry Illinois; and he combined moderation and antislavery in the most attractive combination possible by making full concession of the constitutional right of the southern states to maintain slavery, and by confining his attack to the view that slavery was morally wrong and that it ought not to go into the territories. Further, he was a native of the slave state of Kentucky and had been for many years a Henry Clay Whig. Lincoln's candidacy was being managed by a little-known but very shrewd and politically expert group of Illinosians—David Davis, Leonard Swett, Norman Judd, Stephen T. Logan, Jesse Fell, and others, who were engaged in a masterful campaign of gathering "second-choice" support for Lincoln. Thus holding their candidate back, antagonizing no one, they were in position to show startling gains as the first-choice candidacies evaporated.[37] So thoroughly had they avoided premature publicity for their candidate that Lincoln was not even listed in a contemporary booklet describing twenty-one possible choices for the presidency.[38] He was not endorsed as the candidate of Illinois until the state convention at Decatur, one week before the opening of the national convention at the Wigwam; and he did not acquire his mythic identity as a rail-splitter until the Decatur convention.[39]

In the last frantic days and hours before the balloting, seasoned political observers, including Greeley and Murat Halstead, issued predictions that Seward would be nominated, while the Lincoln forces and the leaders of the Pennsylvania and Indiana delegations made desperate efforts to unite the opposition. In this final, last-

37. There has been no adequate overall treatment of the remarkable work of the team of Illinoisans who managed Lincoln's nomination campaign. See Willard L. King, *Lincoln's Manager, David Davis* (Cambridge, Mass., 1960), pp. 133–142; Frances M. I. Morehouse, *The Life of Jesse W. Fell* (Urbana, Ill., 1916), pp. 58–62; Maurice Baxter, *Orville H. Browning: Lincoln's Friend and Critic* (Bloomington, Ind., 1957), pp. 95–102; David Donald, *Lincoln's Herndon* (New York, 1948), pp. 131–137; Mark M. Krug, *Lyman Trumbull, Conservative Radical* (New York, 1965), pp. 158–162.

38. Nevins, *Emergence*, II, 277.

39. Jesse W. Weik, *The Real Lincoln* (Boston, 1922), pp. 276–277; Baringer, *Lincoln's Rise to Power*, pp. 181–187; Benjamin P. Thomas, *Abraham Lincoln* (New York, 1952), pp. 206–207.

ditch phase, Lincoln's managers resorted to almost every stratagem known to politics. They packed the galleries with Lincoln supporters by using counterfeit tickets of admission; they planted men of notorious lung-power to shout upon a preconcerted signal; and more seriously, they made offers of cabinet posts to men from Indiana, Pennsylvania, and perhaps Maryland. Lincoln had twice instructed his managers to "make no contracts that will bind me," but David Davis is said to have disregarded these admonitions with the comment, "Lincoln ain't here, and don't know what we have to meet, so we will go ahead as if we hadn't heard from him, and he must ratify it."[40] These manipulative activities later gave rise to a kind of mythology that Lincoln gained the nomination by the sharp practices and unscrupulous maneuvers of his friends. This myth has been curiously juxtaposed with a countermyth that the unknown Lincoln was nominated by the direct intervention of Providence. These two legends of the mysterious workings of Providence and the no less secret but somewhat less mysterious workings of David Davis are congruent with the overall dualism in the folklore which has presented the alternative images of a godlike man of sorrows and an earthy frontier trickster. But, in fact, though the contest proved close enough to justify a belief that many small items may have been crucial to the result, there was really nothing very mysterious about the outcome. As for the promises of cabinet posts, certainly they were made, but it does not follow that they were instrumental in securing delegates. Weed no doubt made similar offers, but they would not have seemed worth much to men who did not believe that Seward could be elected. Further, shrewd politicians routinely try to get as much advantage as possible from agreeing to do what they have already decided that they are going to do in any case. The fact that promises were demanded and given does not prove that votes were changed.

On the first ballot, Seward received 173½ votes, of which 134 came from New England and the upper North. Lincoln was second —indicating that the scattered opposition was already concentrating in his support—with 102, including only two states, Illinois and Indiana, solidly for him. Cameron had 50½, Chase 49, Bates 48, and no one else more than 14. On the second ballot Seward gained only 11 votes, while Lincoln, receiving most of Cameron's support

40. Henry C. Whitney, *Lincoln the Citizen* (New York, 1907), p. 289.

in Pennsylvania, Collamer's in Vermont, and Bates's in Delaware, gained 79. On the third ballot, Lincoln gained most of Chase's support in Ohio, Bates's in Maryland, Dayton's in New Jersey, and many scattering. He was, at this point, only 1½ votes short of the nomination, and Ohio proved quicker than other states in switching four votes to make Lincoln the nominee.[41] Later on the same day Senator Hannibal Hamlin, a former Democrat from Maine, was nominated for vice-president to balance the ticket both politically and geographically.[42]

Historians have consistently emphasized the fact that Lincoln gained the nomination because of the argument of availability in the strategic states—an argument used against Seward by some men, like Horace Greeley, because they wanted to defeat him in any case and by others, like Curtin of Pennsylvania and Lane of Indiana, because they genuinely believed that he could not be elected and that they could not be elected on the ticket with him. But few historians have examined the question of whether this belief was realistic. It is, of course, impossible to determine how the votes that might have been cast for Seward would have varied from those actually cast for Lincoln, but it may be worthwhile to try to draw inferences from some of the election results. In Pennsylvania and Indiana, elections for the governorship were held in October, foreshadowing the presidential contest a month later, and tending to increase the advantage in November of the party that won. In Indiana, Henry Lane was elected, 135,000 to 125,000, and Andrew Curtin carried Pennsylvania, 262,000 to 230,000.[43] Both elections were close enough to inspire the belief that having Lincoln rather than Seward at the head of the ticket made a crucial difference. In November, Lincoln enjoyed the added advantage of a divided opposition. He carried Pennsylvania by 268,000 to 179,000 for the nearest rival ticket and 209,000 for the combined opposition. With this margin, it seems likely that Seward also could have carried Pennsylvania, contrary to Curtin's predictions. But Lincoln won three states and part of a fourth by relatively narrow margins: Illinois by 171,000 to 158,000 for his nearest opponent and 165,000 for the combined opposition; Indiana by 139,000 to 116,000 for his nearest opponent

41. Halstead, *Caucuses of 1860,* pp. 167–170.
42. H. Draper Hunt, *Hannibal Hamlin of Maine: Lincoln's First Vice-President* (Syracuse, 1969), pp. 116–118; Halstead, *Caucuses of 1860,* pp. 174–176.
43. Luthin, *First Lincoln Campaign,* pp. 200, 208.

and 133,000 for the combined opposition; California by 39,000 to 38,000 for his nearest opponent and 81,000 for the combined opposition; and four of the seven electoral votes in New Jersey with 58,000 votes against 63,000 for an opposition partially combined in a fusion ticket. By these narrow victories, Lincoln gained 32 of his ultimate total of 180 electoral votes. Without them, he would have had 148; his combined opposition would have had 155; and the election would have been thrown into the House of Representatives. Since Lincoln apparently gained a good many moderate votes that Seward might have lost, and lost very few that Seward might have gained, there seems good reason to believe that the Chicago strategists were realistic in thinking that Lincoln was the only genuine Republican who could be elected.

The nominations of Bell and of Lincoln in May 1860 were followed by those of Douglas and Breckinridge in June. Thereafter, the country moved into the strange quadrennial combination of uproar and organized effort which constitutes an American presidential campaign.

As elections go in the United States, the candidacies of 1860 presented the voters with choices which were more clear-cut than usual. On the primary issue, the distinctions, though limited to policy for the territories, were palpable. Breckinridge stood for congressional protection of slavery in the territories, Douglas was still committed to finding a way around the Dred Scott decision so that the inhabitants of a territory could determine the status of slavery locally, and Lincoln was pledged to exclude slavery from the territories altogether. On secondary issues, also, there was sharper definition than usual, with the Republicans supporting a protective tariff and a homestead act. Little doubt exists that, at a certain level, these genuine issues were crucial in determining the outcome of the election. The tariff, for instance, probably made an important difference in Pennsylvania.[44]

44. John R. Commons, "Horace Greeley and the Working Class Origins of the Republican Party," *Political Science Quarterly,* XXIV (1909), 468–488, argued that homestead policy was the primary motive force of the Republicans. Paul W. Gates, "The Homestead Law in Iowa," *Agricultural History,* XXXVIII (1964), 67–78, also stresses this issue. On the importance attached to the protective tariff, especially in Pennsylvania, see Arthur M. Lee, "The Development of an Economic Policy in the Early Republican Party" (Ph.D. dissertation, Syracuse University, 1953); Malcolm Rogers Eiselen, *The Rise of Pennsylvania Protectionism* (Philadelphia, 1932); Elwyn B. Robinson, "The 'North American': Advocate of Protection," *PMHB,* LXIV (1940), 345–355. The tariff issue could be presented with nativist overtones, as Eric Foner,

But apart from "issues" on which parties can formally choose up sides, presidential campaigns sometimes involve problems which the parties tend to avoid, simply because there is no way to take advantage of them. In 1860 there was an immense problem looming just beyond the election—the possible dissolution of the Union. Thousands of people in all parts of the country recognized the problem clearly, and in fact, it was the urgency of this matter which chiefly stimulated both the Bell and the Douglas candidacies. Douglas made titanic efforts to focus the campaign upon the danger to the Union. But the Breckinridge forces had nothing to gain by calling attention to the fact that, under certain circumstances, they would become disunionists; therefore, they insisted upon their devotion to the Union—meaning Union on their own terms—and failed to make voters aware that a crisis was at hand. The Lincoln forces, likewise, had nothing to gain by pointing out that the election of their candidate might produce the grimmest emergency the republic had ever seen, and so they consistently made light of the warnings that the crisis of the Union was at hand. They viewed the threats from the South as bluff and dismissed them with ridicule. Instead of recognizing that they might have to either permit the dissolution of the Union or wage war to prevent it, they laughed at "the old game of scaring and bullying the North into submission to Southern demands and Southern tyranny." James Russell Lowell called the threat of secession an "old Mumbo-Jumbo." Carl Schurz said that the South had already seceded twice, once when southern students left the Philadelphia Medical School and once when southern congressmen walked out of the House of Representatives after Pennington was elected Speaker. At that time, said Schurz, they took a drink and came back; after Lincoln's election, they would take two drinks and would again come back. The New York *Tribune*

Free Soil, Free Labor, Free Men: The Ideology of the Republican Party before the Civil War (New York, 1970), p. 203, suggests. According to William Dusinberre, *Civil War Issues in Philadelphia, 1856–1865* (Philadelphia, 1965), p. 78, Republicans emphasized the tariff in Philadelphia in order to play down the slavery issue and to conciliate the nativists without accepting their frank intolerance. Michael Fitzgibbon Holt, *Forging a Majority. The Formation of the Republican Party in Pittsburgh, 1848–1860* (New Haven, 1969), pp. 275–280, finds the Douglas Democrats just as pro-tariff as the Republicans in Pennsylvania and expresses doubt about the importance of the issue in Pittsburgh. Thomas M. Pitkin, "Western Republicans and the Tariff in 1860," *MVHR*, XXVII (1940), 401–420, finds a somewhat negative attitude in the West toward protectionism. Reinhard H. Luthin, "Abraham Lincoln and the Tariff," *AHR*, XLIX (1944), 609–629, shows how carefully Lincoln tried to meet the expectations of Pennsylvania without taking an unqualified high protectionist position.

scoffed that "the South could no more unite upon a scheme of secession than a company of lunatics could conspire to break out of bedlam." Seward, the foremost Republican campaigner, declared that the slave power, "with a feeble and muttering voice," was threatening to tear the Union to pieces. "Who's afraid?" he asked. "Nobody's afraid. Nobody can be bought."

As for Lincoln, he said nothing publicly on this or any other subject, but privately he expressed an insouciance that alarmed a well-known Ohio journalist, Donn Piatt, who talked with him at least twice during the campaign and later wrote:

He considered the movement South as a sort of political game of bluff, gotten up by politicians, and meant solely to frighten the North. He believed that when the leaders saw their efforts in that direction unavailing, the tumult would subside. "They won't give up the offices," I remember he said, and added, "Were it believed that vacant places could be had at the North Pole, the road there would be lined with dead Virginians."

Mr. Lincoln did not believe, could not be made to believe, that the South meant secession and war. When I told him, subsequently to this conversation, . . . that the Southern people were in dead earnest, meant war, and I doubted whether he would be inaugurated at Washington, he laughed and said the fall [in the price] of pork at Cincinnati had affected me.

Four years earlier, during the Frémont campaign, Lincoln had asserted bluntly, "All this talk about dissolution of the Union is humbug, nothing but folly. We do not want to dissolve the Union; you shall not." During the 1860 campaign he wrote to a correspondent that he had received "many assurances . . . from the South that in no probable event will there be any very formidable effort to break up the Union. The people of the South have too much of good sense, and good temper, to attempt the ruin of the government, rather than see it administered as it was administered by the men who made it. At least, so I hope and believe."[45]

45. John Wentworth, in New York *Herald,* Aug. 1, 1860; James Russell Lowell, *Political Essays* (New York, 1904), p. 50; New York *Tribune,* July 28, Sept. 22, 1860; Schurz quoted in Mary Scrugham, *The Peaceable Americans of 1860–1861* (New York, 1921), p. 46; Seward, quoted in Fite, *Presidential Campaign of 1860,* p. 189; Thaddeus Stevens, in *Congressional Globe,* 36 Cong., 1 sess., p. 24. Edwin D. Morgan, a Republican congressman, wrote on Dec. 22, 1855, "It is one of the most singular facts that some of our new members, when they hear the old Southern croakers talk about the dissolution of the Union, really believe them in earnest. We have adopted a rule that when one of them talks about dissolution in the House, we make our side of the House ring with laughter, sing out, 'Goodbye John,' & other things of the kind, which always turns their remarks & threats into ridicule." Temple R. Hollcroft (ed.), "A Congressman's Letters on the Speaker Election in the Thirty-fourth Congress,"

This total failure to perceive that the Union stood on the brink of dissolution was, in the words of Allan Nevins, "the cardinal error" of the Republicans. Tactically, it was perhaps shrewd, if not wise, to pretend that there was no serious danger. Yet tactics did not require them to deceive themselves with their own pretense.

Theoretically, the purpose of a political campaign is discussion of issues and the education of voters, just as, theoretically, capitalism exists for the maintenance of a competitive market economy. But, in both cases, the goal of the participants is quite different from the purpose of the institution. Campaigners seek to win votes, even by concealing the issues, as capitalists seek to make profits, even by eliminating their competition. In seeking votes, they know that whenever they clarify an issue they will probably lose some support and gain some. But when they substitute enthusiasm for issues, the effect upon voters may be almost pure gain. As politicians have understood better than historians, most voters are swayed less by reason than by emotion, by their group affiliations, by artificially generated excitement in which they can participate, and by the desire to be identified with power as personified in a man who projects a strong and appealing personality. Therefore, a successful campaign may totally omit attention to the most serious question of the day, but it must not omit group activities, excitements, a stereotype of victory, and an attractive image for the candidate.

Such were, therefore, the staple ingredients of the 1860 campaign. The parties formed marching clubs which paraded in uniform at rallies. The members of the Republican clubs were "Wide Awakes" who carried torches or oil lamps and wore glazed cloth to protect them from the dripping oil. The Constitutional Unionists carried not only torches but also bells, in subtle allusion to the name of their candidate. Douglas's followers were "Little Giants" or "Little Dougs," while the Breckinridge organizations took the less colorful name of "National Democratic Volunteers."[46]

MVHR, XLIII (1956), 444–458. On Lincoln, see Donn Piatt, *Memories of the Men Who Saved the Union* (New York, 1887), pp. 28–30; Basler (ed.), *Works of Lincoln,* II, 355; IV, 95. See also Fite, *Presidential Campaign of 1860,* pp. 187–189; David M. Potter, *Lincoln and His Party in the Secession Crisis* (New Haven, 1942), pp. 9–19; Nevins, *Emergence,* II, pp. 305–306.

46. On the campaign, including organization, ballyhoo, and oratory, see Nevins, *Emergence,* II, 272–317; Nichols, *Disruption,* pp. 334–350; Randall, *Lincoln the President,* I, 178–206; Luthin, *First Lincoln Campaign,* pp. 168–177; Baringer, *Lincoln's Rise to Power,* pp. 296–329; Baringer, "Campaign Techniques in Illinois—1860," ISHS *Transactions,* 1932, pp. 202–281; H. Preston James, "Political Pageantry in the Cam-

All parties resorted heavily to this kind of political showmanship, but historically the party that needed and relied upon it most had been the Whigs. They had perfected a "hurrah" type of campaign, characterized by mass celebrations, by picturesque symbols—such as the log cabin and the barrel of cider—which would emphasize their nominee's humble origins and democratic tastes, by attractive stereotypes of the candidate; and by keeping the candidate himself under wraps, lest he display his incompetence or make some tactless revelation of truth.

No one now thinks of Lincoln as a "hurrah" candidate, and hence the campaign of 1860 is seldom recognized as a "hurrah" campaign. But in fact, the Republicans were the natural legatees of the Whigs. They had used "hurrah" tactics to cover up for an inept candidate in 1856, and in 1860 they adopted "hurrah" devices again. They relied upon the "Wide Awakes" to provide noise, spectacle, and opportunity for participation. For jubilation, they incessantly sang, "Ain't I glad I joined the Republicans." They stereotyped Lincoln as "Honest Old Abe," son of the frontier. For symbolism they used rails or replicas of rails which he had split, carried in processions to remind everyone that although he had been a Whig, he bore no aristocratic taint. As for permitting the candidate to raise his voice, they did not wait to see whether he might be an effective and resourceful speaker, but told him at once what they had told Harrison and Taylor. William Cullen Bryant informed Lincoln firmly that his friends wanted him to "make no speeches, write no letters as a candidate." Lincoln complied, at least as far as public visibility was concerned, for he made no statements and stayed very close to Springfield.[47] But though inconspicuous, he was quite active, conferring with party chiefs, talking with newspapermen, directing campaign operations by letter, and smoothing frictions within the party organization. In fact, those who observed him began to see that he was a man of remarkable judgment and capacity.[48] But this fact was

paign of 1860 in Illinois," *Abraham Lincoln Quarterly*, IV (1947), 313–347; Holt, *Forging a Majority*, pp. 264–303—an especially valuable treatment; Fite, *Presidential Campaign of 1860*, pp. 132–235; Ollinger Crenshaw, *The Slave States in the Presidential Election of 1860* (Baltimore, 1945), pp. 74–298.

47. Bryant to Lincoln, June 16, 1860, quoted in Nevins, *Emergence*, II, 278; Lincoln to Samuel Galloway, June 19, 1860, in Basler (ed.), *Works of Lincoln*, IV, 80, said, "By the lessons of the past, and the united voice of all discreet friends, I am neither [to] write or speak a word for the public."

48. Nevins, *Emergence*, II, 273–279.

seldom suggested to the voters, most of whom knew him only as "Honest Old Abe." A Democratic newspaper complained that if his nomination had not been made by a major party, it would have been regarded as a farce: " 'He is honest!' Yes, we concede that. Who is not? 'He is old!' So are thousands. 'He has mauled rails!' What backwoods farmer has not? But what has he ever done for his country? Is he a statesman?"[49] It was not evident that proof of his statesmanship would have appreciably increased his appeal to the voters, and no special effort was devoted by speakers and editorial writers to demonstrating Lincoln's fitness for the presidency.

Underneath all the fun and excitement which were employed to generate voter enthusiasm, all parties depended primarily upon two means of communication—campaign speakers and violently partisan newspapers—to wage the actual battles with their rivals. In these aspects, the Republicans displayed a verve, initiative, and confidence in marked contrast to their adversaries. Republicans raised money easily, organized readily and effectively, and flooded the North with speakers and campaign literature. They also made strenuous efforts to improve the party's standing with immigrant voters. First, they put into the Chicago platform a plank condemning changes in the naturalization laws, or state legislation "by which the rights of citizenship hitherto accorded to immigrants . . . shall be abridged or impaired." Second, in advocating a homestead law, they proposed that noncitizen immigrants should be eligible for homesteads. Third, by rejecting Bates, who had been a Know-Nothing, and naming Lincoln as their candidate, they repudiated their nativist affiliations. Fourth, they appointed a special bureau within the campaign organization to convert immigrant voters to Republicanism, and they made Carl Schurz, a German Forty-eighter, head of this division. Schurz, whose energy was exceeded only by his self-esteem, worked hard at this operation, and there can be no doubt that he won many immigrants, especially Protestant immigrants, to the Republican cause. Later, after the apotheosis of Lincoln, when people of immigrant stock wanted to remember that they had contributed in a vital way to his election, and when Republicans wanted to forget how close they had been to the Know-Nothings, a legend arose that Lincoln had won the immigrant vote

49. Belleville, Illinois, *Democrat,* June 2, 1860, quoted in Baringer, *Lincoln's Rise to Power,* p. 310.

and that it had been crucial to his election. This legend crept into history. But as long ago as 1941, Joseph Schafer demonstrated that in fact the bulk of the immigrants, especially the German Catholics who substantially outnumbered German Protestants, voted against Lincoln. Since the Irish remained unswervingly Democratic, what this means is that nativist prejudice and immigrant reaction to that prejudice both cut deepest where religion, as well as "foreign" origin, was involved. More recent and more rigorous research has further confirmed that religious rather than ethnic prejudice was primary in nativism, and that while Lincoln may have gained the support of a larger proportion of the Protestant immigrant minority, he made very little headway with the Catholics, both German and Irish, who formed the bulk of the immigrant population. Only the heavy support by voters who were natives offset the substantial immigrant majorities against him.[50]

In several ways, the election of 1860 produced a "campaign like none other" in American history. For one thing, the fact that there were four major candidates in the race gave a new twist to a political system which had evolved in a context of two-party contests. Theoretically, the four candidates presented unusually clear alternatives to the voters on the issue of slavery, and to some extent even on the tariff and free land and Pacific railroad issues, but often a man's choice turned on identifying which candidate stood the best chance of defeating the candidate whom he opposed, rather than on

50. The legend that the votes of the foreign-born were crucial to Lincoln's election is set forth in William E. Dodd, "The Fight for the Northwest, 1860," *AHR*, XVI (1911), 774–788; Arthur Charles Cole, *The Era of the Civil War, 1848–1870* (Chicago, 1922), pp. 341–342; Donnal V. Smith, "The Influence of the Foreign Born of the Northwest in the Election of 1860," *MVHR*, XIX (1932), 192–204; Charles Wilson Emery, "The Iowa Germans in the Election of 1860," *Annals of Iowa*, 3rd series, XXII (1940), 421–453; Andreas Dorpalen, "The German Element and the Issues of the Civil War," *MVHR*, XXIX (1942), 55–76. The first important challenge to this view came from Joseph Schafer, *Four Wisconsin Counties* (Madison, 1927), pp. 140–158, and "Who Elected Lincoln," *AHR*, XLVII (1941), 51–63, followed by Hildegard Binder Johnson, "The Election of 1860 and the Germans in Minnesota," *Minnesota History*, XXVIII (1947), 20–36. Schafer was ineffectually challenged by Jay Monaghan, "Did Abraham Lincoln Receive the Illinois German Vote?" ISHS *Journal*, XXXV (1942), 133–139. Significant recent studies include: Robert P. Swierenga, "The Ethnic Voter and the First Lincoln Election," *CWH*, XI (1965), 27–43; George H. Daniels, "Immigrant Vote in the 1860 Election: The Case of Iowa," *Mid-America*, XLIV (1962), 146–162; Paul J. Kleppner, "Lincoln and the Immigrant Vote: A Case of Religious Polarization," *Mid-America*, XLVIII (1966), 176–195; Donald E. Simon, "Brooklyn in the Election of 1860," New York Historical Society *Quarterly*, LI (1967), 249–262; Holt, *Forging a Majority*, pp. 215–219, 299–303. Many of these essays have been assembled in Frederick C. Luebke (ed.), *Ethnic Voters and the Election of Lincoln* (Lincoln, Neb., 1971).

deciding which candidate he favored. It has already been explained that Lincoln, in order to win, needed only to hold the states carried by Frémont and gain 35 additional electoral votes from Pennsylvania (27), Indiana (13), Illinois (11), and New Jersey (7). And if his opposition was divided three ways, he would almost certainly win in all of these states. Thus, the electoral logic virtually compelled the state organizations of the three opposition parties to attempt what their national organizations had failed to accomplish, namely the forging of some sort of coalition. Yet, if the necessity was great, the obstacles to "fusion," as it was called, were immense. All the bitterness of the old Buchanan-Douglas feud stood in the way, enhanced by the fact that Douglas was denouncing the Breckinridge Democrats as disunionists; the numerous immigrant supporters of Douglas hated the Know-Nothingism of John Bell's followers; and ordinary voters wanted to vote for a candidate and not for a combination. In fact fusion may have lost more by the votes which it alienated than it gained by the votes which it merged. But the need to avoid dispersal of the opposition was overwhelming, and often fusion seemed important locally to Democrats who saw a chance to win state elections even if they could not win the national one. Thus, it finally developed that "fusion" tickets of all three opposition candidates were arranged in New York, New Jersey, and Rhode Island, and of Breckinridge and Douglas supporters in Pennsylvania. But splinter groups of irreconcilable Douglas Democrats persisted in running separate tickets in Pennsylvania and New Jersey, so that, in effect, Douglas was on two tickets in those two states. In Texas, a fusion ticket was arranged between the Douglas and the Bell supporters. In the crucial states of Indiana and Illinois, without fusion, the contest nevertheless tended to become a two-party affair between Lincoln and Douglas, with the combined strength of Breckinridge and Bell amounting to less than 7 percent in Indiana and 2 percent in Illinois. Yet even these amounts of dispersal were enough to make hopeless an opposition which would have been desperate even if concentrated.[51]

None of the opposition candidates had any real chance of winning

51. On fusion, Nichols, *Disruption*, pp. 341–350; Parks, *John Bell*, pp. 361–388; Kirwan, *Crittenden*, pp. 357–360; Louis Martin Sears, "New York and the Fusion Movement of 1860," ISHS *Journal*, XVI (1923), 58–62; Milledge L. Bonham, Jr., "New York and the Election of 1860," *NYH*, XXXII (1934), 124–143; Erwin Stanley Bradley, *The Triumph of Militant Republicanism: A Study of Pennsylvania and Presidential Politics 1860–1872* (Philadelphia, 1964), pp. 77–81; Charles Merriam Knapp, *New Jersey Politics During the Period of the Civil War and Reconstruction* (Geneva, N.Y., 1924), pp. 30–33; Friend, *Sam Houston*, pp. 319–320.

in the electoral college; they could hope, at most, to prevent an electoral majority and thus put the election into the House of Representatives. If this were to occur, only the three topmost candidates would be eligible, and they were reasonably certain to include Lincoln and Breckinridge, with either Bell or Douglas as the third. In an election in the House, each state delegation casts a single vote. The Republicans controlled fifteen such delegations; the Breckinridge Democrats thirteen (eleven slave states plus Oregon and California), the Douglas Democrats one (Illinois), and the Bell supporters one (Tennessee), while three (Kentucky, Maryland, and North Carolina) had delegations equally divided between Democrats and Americans.[52] In such a contingency, it seemed unlikely that Lincoln could gain the two states necessary for a majority, and the situation would have been auspicious for the southern Democrats. They had little reason to fear a combination of Bell and Lincoln supporters, for this would have required either slave-state support for Lincoln or Republican support for a large slaveholder from a slave state. If three congressmen from Tennessee and one each from the three equally divided slave states would swing over to Breckinridge, he would then have enough states to be elected. If the House became deadlocked, however, the vice-president elected by the Senate would become the acting president on March 4. The composition of the Senate was such that Breckinridge's running mate, Joseph Lane, stood to be elected. But the complexity of all these problems of fusion and alternative contingencies made tactics seem more important than questions of substance, and this partially neutralized the clarity of choice which the sharply defined positions of the candidates seemed to offer.[53]

In the end, a potential four-way contest was converted into two two-way contests, one between Lincoln and Douglas in the free states and the other between Bell and Breckinridge in the slave

52. Nichols, *Disruption*, p. 341. For a different calculation of the possible results if the election had gone into the House, see Crenshaw, *Slave States in Election of 1860*, pp. 68–69.

53. Crenshaw, *ibid.*, pp. 59–63, 69–73, presents extensive evidence that the chance of the election going into the House was recognized, but expresses doubt that the southerners made "a concerted drive" to accomplish this result. Also see Fite, *Presidential Campaign of 1860*, pp. 221–222; Nevins, *Emergence*, II, 211; Dumond, *Secession Movement*, p. 108; Craven, *Growth*, p. 339; Frank H. Heck, "John C. Breckinridge in the Crisis of 1860–1861," *JSH*, XXI (1955), 329; Parks, *John Bell*, p. 377; Alexander H. Stephens, *A Constitutional View of the Late War Between the States* (2 vols.; Philadelphia, 1868–70), II, 275–276; Milton, *Eve of Conflict*, p. 482.

states.[54] This situation represented a further extension of the tendency begun in 1856, for at that time, Buchanan had run against Frémont in the North and against Fillmore in the South. In 1860, Breckinridge received more than one-third of the vote cast in Oregon, more than a quarter of the vote in California, and more than one-fifth in Connecticut, but, except for these, he did not get as much as 6 percent in any free state. Bell received 13 percent of the vote in Massachusetts, but he received less than 5 percent of the vote in the free states as a whole. In the slave states, the concentration was equally heavy. Lincoln received 23 percent of the vote in Delaware and 10 percent in Missouri, but otherwise not as much as 3 percent in any slave state. South of Virginia, Kentucky, and Missouri he was not even on the ballot. As for Douglas, he carried the slave state of Missouri with 35.5 percent of the vote, received 17 percent of the vote in Kentucky, 15 percent in Alabama, 15 percent in Louisiana, and 11 percent in Georgia, but did not gain as much as 10 percent in any of the remaining ten slave states.[55]

It is not a very serious exaggeration to say that the United States was holding two elections simultaneously on November 6, 1860. This meant that each section remained somewhat insulated from what the other was doing. If the Republicans had been campaigning in the South, they would necessarily have stressed Lincoln's recognition of the right of the southern states to determine the question of slavery for themselves; they would have presented an image of him as an old-fashioned Henry Clay Whig, a native of Kentucky. Insofar as they had done this, it might have served to prevent the creation of a totally negative and fictitious image of Lincoln which was being developed in the South—the image of a "black Republican," a rabid John Brown abolitionist, an inveterate enemy of the South. Yet this picture prevailed during all the months of the campaign, and psychologically, it was not strange that southerners felt hostile to a candidate who was not even on the ticket in their part of the country. When Lincoln was elected, the result came to the South as a much greater shock than it would have if Republican speakers, or even Lincoln himself, had been ranging up and down and back and forth throughout the South, asking the voters to trust him. The Republicans would have had nothing to gain from such a campaign, and southerners would never have permitted it, but the

54. W. Dean Burnham, *Presidential Ballots, 1836–1892* (Baltimore, 1955), p. 77.
55. Computed from election returns, *ibid.*, pp. 246–256.

point is that the voters of the South were naturally prepared to believe the worst of a candidate when most of them had never seen even one of his supporters, much less the man himself, and when his party did not even seek their support. In fact, the American party system had ceased to operate in a nationwide context.

While the South failed to form a realistic impression of Lincoln, the North failed to understand the mood of the South. Preoccupied as they were with the exciting contest between Lincoln and Douglas, northern voters paid insufficient heed to the steady drumfire of disunionist editorials and speeches from the South. Perhaps such voters followed the Republican practice of dismissing all such statements as a bluff designed to prevent timid citizens from voting their principles. Perhaps they were too readily reassured by the one speech which Breckinridge made during the campaign. At Ashland, Kentucky, on September 5, he spent three hours affirming his unionism, without indicating that he meant Union on his own terms.[56] Perhaps they were too easily lulled by Unionists in the border states and by men trying to encourage moderation in the North who were in fact often no more ready to resist disunion than to advocate it. These misapprehensions of the Republicans were fostered by the dualism of the campaign, which raised barriers to communication between the North and the South.

The one person in public life who made a strenuous effort to break down these barriers was Stephen A. Douglas. Old at the age of forty-seven, weakened by drink, ill health, political reverses, and the reckless impulsiveness with which he threw his energies into political combat, Douglas was within a year of his death. His voice was hoarse, but his immense drive was undiminished, and he alone among the candidates was determined to carry to the American people the message that this election was a crisis and not just another hurrah campaign—to the northern people, that the Union was on the verge of dissolution; and to the southern people, that when they talked about secession they were flirting with both treason and disaster. It was, of course, in Douglas's interest to emphasize these realities, but in his manner of doing it, he exceeded himself and showed a sense of public responsibility unmatched by any of the other candidates. He decided early that, regardless of precedent, he would campaign vigorously, and in fact his campaign was not only

56. Heck, "Breckinridge in the Crisis," pp. 326–328; Crenshaw, *Slave States in Election of 1860*, pp. 160–161.

the first but also one of the greatest of campaigns by a presidential candidate. In July he ranged through upper New York and New England. In August he went to Virginia and North Carolina. At Norfolk he told his audience that the election of Lincoln would not justify a southern secession and that if secession occurred, he would do all in his power to maintain the supremacy of the laws. At Raleigh he stated that he would favor hanging anyone who attempted forcibly to resist the Constitution. In September he began at Baltimore, spoke in New York City, and then campaigned through Pennsylvania, to Cincinnati, Indianapolis, Chicago, and farther west. It was at Cedar Rapids, early in October, that he received dispatches from Pennsylvania and Indiana telling him of the Republican gubernatorial victories in those states. His response was to change at once his plans for the rest of the campaign: "Mr. Lincoln is the next President," he said. "We must try to save the Union. I will go South." He still had to fulfill speaking engagements in Milwaukee and in the old, familiar towns of Illinois—Bloomington, Springfield, Alton—but by October 19 he was in St. Louis, "not to ask for your votes for the Presidency . . . but to make an appeal to you on behalf of the Union." From there, he traveled on, at appreciable personal risk, into hostile territory. In Tennessee he spoke at Memphis, Nashville, Jackson, and Chattanooga; in Georgia, at Atlanta and Macon; in Alabama, at Selma and Montgomery. Election day overtook him and ended his odyssey at Mobile.

In all latitudes his message had been the same: The Union is in peril. Allan Nevins, by no means one of Douglas's warmest admirers, has well said, "Never did Douglas's claims to statesmanship stand higher than when he thus pointed to a danger which most Republicans were denying or minimizing, and defied the Southerners and border men who were attacking him on the ground that he was a brutal coercionist."[57]

57. Nevins, *Emergence*, II, 290–298; Milton, *Eve of Conflict*, pp. 480–500; Crenshaw, *Slave States in Election of 1860*, pp. 74–88; Robert W. Johannsen, "Stephen A. Douglas' New England Campaign, 1860," *NEQ*, XXXV (1962), 162–186; Johannsen, "The Douglas Democracy and the Crisis of Disunion," *CWH*, IX (1963), 229–247; Johannsen, "Douglas and the South," *JSH*, XXXIII (1967), 26–50; Lionel Crocker, "The Campaign of Stephen A. Douglas in the South, 1860," in J. Jeffery Auer (ed.), *Antislavery and Disunion, 1858–1861: Studies in the Rhetoric of Compromise and Conflict* (New York, 1963), pp. 262–278; David R. Barbee and Milledge L. Bonham, Jr. (eds.), "The Montgomery Address of Stephen A. Douglas," *JSH*, V (1939), 527–552; Rita McK. Cary, *The First Campaigner: Stephen A. Douglas* (New York, 1964); Quincy Wright, "Stephen A. Douglas and the Campaign of 1860," *Vermont History*, XXVIII (1960),

On November 6 the voters registered the result. Lincoln received approximately 1,865,000 votes and carried all of the eighteen free states except New Jersey, where he won four of the seven electoral votes, losing three to Douglas. This gave him a total of 180 electoral votes—27 more than he needed for victory. He received only 39 percent of the popular vote, which has led some writers to the mistaken belief that he won because his opposition was divided. But this was not the case; he won because his vote was strategically distributed. It was all located where it would count toward electoral votes, and virtually none of it was "wasted" in the states which he lost. In fact he won by clear majorities in every state he carried except Oregon, California, and New Jersey, and he could have lost those without losing the election. Douglas was second with about 1,000,000 votes plus a large but indeterminate share of nearly 600,000 fusion votes, nearly all of which were concentrated in the free states, where he was consistently beaten by Lincoln. He carried only one state (Missouri, by a whisker over Bell) and three electoral votes in New Jersey. Breckinridge ran third, and Bell fourth, both with totals which are indeterminate because their appropriate share of the fusion vote is incalculable. The strength of both was concentrated in the South, where Breckinridge carried eleven states, losing Missouri to Douglas, and Virginia, Kentucky, and Tennessee to Bell. But unlike Lincoln, he won statewide majorities in only a few states—Florida, Alabama, Mississippi, and Arkansas. South Carolina, if it had held popular elections for the presidency, would have been added to this list. But as a test of Unionism versus disunionism, the dominant fact was that the combined opposition to Breckinridge had won over 55 percent of the vote of the slave states and had gained majorities in ten of them. This fact probably helped to perpetuate Republican misconceptions about the strength and nature of Unionism in the South.[58]

A striking feature of the distribution of the vote was the strong

---

250–255; Damon Wells, *Stephen Douglas, the Last Years, 1857–1861* (Austin, 1971), pp. 241–258.

58. Popular vote figures for the election of 1860 do not indicate with complete accuracy the relative strengths of the candidates opposing Lincoln; for in a number of states, two or all three of them were combined on fusion tickets. In addition, Breckinridge's strength is somewhat understated in any total because South Carolina, which he carried, did not cast any popular votes. The following figures, compiled with minor corrections from Burnham, *Presidential Ballots*, pp. 246–256, show the dimensions of the fusion movement, which standard tables for the election of 1860 simply ignore entirely:

tendency of the cities to vote for "moderate" candidates. This tendency is especially revealing because disciples of Charles A. Beard at one time gained wide acceptance for the idea that the sectional conflict was essentially a struggle of northern business and industry against southern agriculture. If this had been so, one might expect to find northern cities to have been strongholds of Republicanism. But cities like Boston and New York were commercial centers that had strong ties with the South and much to lose if those ties were broken.[59] Furthermore, the urban population of the North had a high proportion of immigrants, a majority of whom clung to their Democratic allegiance. For these reasons and perhaps for others, Lincoln received much less support in the urban North than he did in the rural North. Whereas the North as a whole gave him 55 percent of its votes, in seven of the eleven cities with populations of 50,000 or more, he failed to get a majority.

Similarly, the urban South, perhaps because of commercial ties and a lesser interest in slavery, displayed a wariness of disunion. Of eighteen cities with populations of at least 10,000, Breckinridge carried only two by clear majorities and would have done the same in Charleston if there had been a popular vote in South Carolina. In cities like Richmond, Norfolk, Mobile, New Orleans, and Memphis, he received less than 30 percent of the total vote.[60]

In the southern states, the strongest centers of the Unionist vote

---

	Free States	*Slave States*	*Totals*
Lincoln	1,838,347	26,388	1,864,735
Opposition to Lincoln	1,572,637	1,248,520	2,821,157
Fusion	580,426	15,420	595,846
Douglas	815,857	163,568	979,425
Breckinridge	99,381	570,091	669,472
Bell	76,973	499,441	576,414

59. On the prosouthern attitudes of the northern mercantile community during the election, Philip S. Foner, *Business and Slavery: The New York Merchants and the Irrepressible Conflict* (Chapel Hill, 1941), pp. 169–207; Dusinberre, *Civil War Issues in Philadelphia,* pp. 87–94.

60. Ollinger Crenshaw, "Urban and Rural Voting in the Election of 1860," in Eric F. Goldman (ed.), *Historiography and Urbanization: Essays in American History in Honor of W. Stull Holt* (Baltimore, 1941), pp. 43–66; for county breakdown, Burnham, *Presidential Ballots,* pp. 235–243. Foner, *Free Soil,* pp. 226–227, 306 n., notes the statement of Carl Schurz that the Republican party was composed "chiefly of . . . the native American farmers" and that "the strength of our opponents lies mainly in the populous cities and consists largely of the Irish and uneducated mass of German immigrants." Also the lament of New York's Democratic congressman, Horace Clark, that "we have been, as it were, driven to take refuge within the walls of our Northern cities."

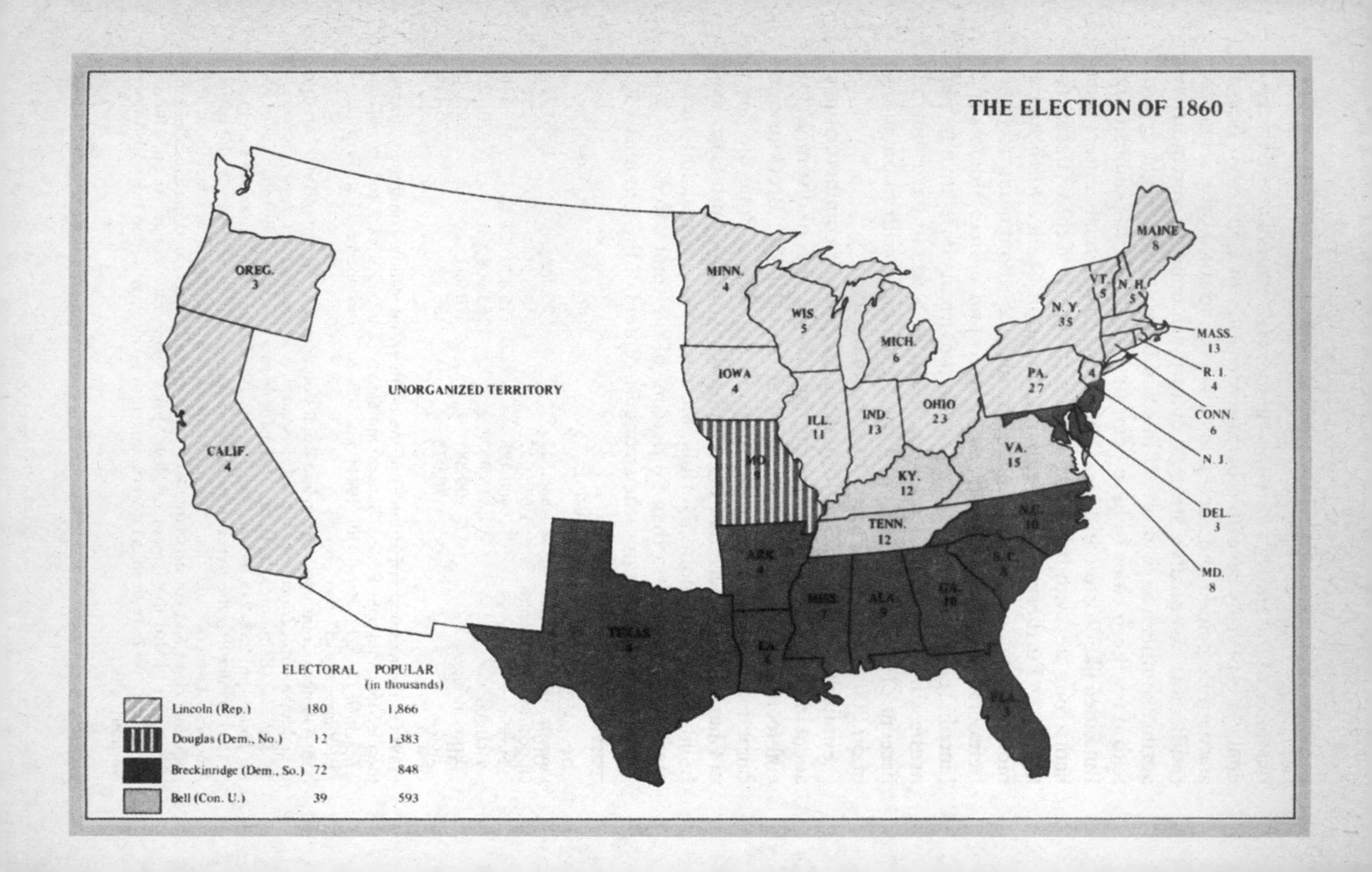
THE ELECTION OF 1860
UNORGANIZED TERRITORY
OREG.
3
CALIF.
4
MINN.
4
WIS.
5
MICH.
6
IOWA
4
ILL.
11
IND.
13
OHIO
23
KY.
12
TENN.
12
VA.
15
PA.
27
N. Y.
35
4
VT.
5
N. H.
5
MAINE
8
MASS.
13
R. I.
4
CONN.
6
N. J.
DEL.
3
MD.
8
ELECTORAL POPULAR (in thousands)
Lincoln (Rep.) 180 1,866
Douglas (Dem., No.) 12 1,383
Breckinridge (Dem., So.) 72 848
Bell (Con. U.) 39 593

besides the cities were the districts of heaviest slaveholding. An analysis of the election returns from 537 counties in Virginia, North Carolina, Tennessee, Georgia, Alabama, Mississippi, and Louisiana indicates that if these counties are separated into three groups according to the proportion of slaves in the population, those counties with the highest proportion of slaves cast 52 percent of their votes for Breckinridge; the counties with a medium proportion, 56 percent; and the counties with the lowest proportion, 64 percent.[61] Paradoxically, it appeared that areas with few slaves to defend were more zealous about defending them than areas which had many to defend. The result should have been reassuring to southerners who feared that the nonslaveholders did not have a sufficient stake in preservation of the "peculiar institution." To be sure, the vote on secession would be different, but at the presidential election the plain people apparently still supported the Democratic party which they had first supported as the party of Jackson, though it had now become the party of southern rights, while the large planters still voted for the successors to the Whigs, whom they had first supported as the party of property, though it had now become the party of compromise. One can see strong elements of traditionalism and political inertia in these political continuities.

Yet it was nothing less than a revolution that the country had committed itself electorally to a party which opposed slavery, at least to the extent of agreeing with Lincoln that the institution must "be placed in the course of ultimate extinction." How long this would take, and by what means it could be accomplished, Lincoln did not say. He hoped, evidently, for a gradual process, furthered by the use of persuasion rather than force. His policy seemed glacially slow to the abolitionists. But his election meant the triumph of a new attitude. During the seventy-two years from 1789 to 1861, slaveholders had held the presidency for fifty years. No major party had ever expressed clear opposition to slavery until 1856. But in 1860, the party which won the election was explicitly committed to the position that "the normal condition of all the territory of the United States is that of freedom." Even though the term "territory"

61. Seymour Martin Lipset, "The Emergence of the One-Party South—the Election of 1860," in *Political Man: The Social Bases of Politics* (New York, 1960), pp. 372–384. David Y. Thomas, "Southern Non-Slaveholders in the Election of 1860," *Political Science Quarterly,* XXVI (1911), 222–237, arrives at similar conclusions, by a somewhat different analysis for Mississippi, Louisiana, and Georgia.

was used here in a restricted sense, this statement was still symptomatic of a vast transformation.

In political structure, also, the election of 1860 marked the end of an era. For more than thirty years, bisectional parties had reinforced the cohesiveness of the Union. Within each party, strife between the sectional wings had sometimes been acute, but the wings had nevertheless remained dependent upon one another, and in working communication with one another. The elections of Jackson, Van Buren, Harrison, Polk, Taylor, and Pierce had all been bisectional victories, with a majority of the free states and a majority of the slave states voting for the winner. Though sectional strife raged, the parties served as buffers to contain it. But in 1854 a new party had emerged, with all its strength concentrated in the free states. It made no attempt to win southern support or even to make itself understood in the South. The South stood entirely apart from it, and insofar as this party became the government, the South would stand apart from the government. In 1856 the new party began to become the government; it won the speakership for Nathaniel P. Banks. Also, in 1856, it won control of one of the sections, for it carried all but five of the free states in the presidential election. But the totally sectionalized[62] party did not win the election, and Buchanan's victory was still the victory of a bisectional party, even though somewhat lopsidedly bisectional, without the old magic of majorities in both sections.

In 1860 the revolution was completed. In February the sectionalized party again won the speakership of the House. In April the bisectional party was torn apart at Charleston, and efforts to revive it at Baltimore in June proved unsuccessful. Between June and November two sectionalized parties conducted election campaigns in which they did not directly confront each other but merely worked to consolidate their sectional positions, while two Unionist parties made futile efforts to arrest the centrifugal tendency. In November the Unionist parties were overwhelmed, carrying only four states and part of a fifth. True, these two parties between them

62. The term "sectionalized" rather than "sectional" is used here because the term "sectional" was used controversially by Democrats who believed the Republicans were deliberately "sectional," while the latter argued that the South was "sectional" in refusing to give any support to the Republican position. There can be no controversy on the point that the parties, regardless of their own purposes, had become sectionalized by 1860.

gained a majority of the vote in the slave states as a whole, but while this may have shown a willingness to make concessions in the hope of avoiding the sectionalization of politics, it did not necessarily signify a readiness to acquiesce if sectionalization should triumph anyway. Sectionalization did triumph. Lincoln carried seventeen free states and no slave states; Breckinridge, eleven slave states and no free state; Bell, three slave states and no free states. The pitiful remnant of bisectionalism was Douglas's 10 electoral votes in Missouri and 3 in New Jersey.

The election marked the crystallization of two fully sectionalized parties. But it was the party of the northern section that won, and by winning the presidency, it became the government for ten states in which it had not even run a ticket. The process of sectional polarization was almost complete, and it remained to see what response would come from the section which was at the losing end of the axis.

CHAPTER 17

The Nature of Southern Separatism

TEN days after the election of Lincoln, the Augusta, Georgia, *Daily Constitutionalist* published an editorial reflecting on what had happened to American nationalism:

> The most inveterate and sanguine Unionist in Georgia, if he is an observant man, must read, in the signs of the times, the hopelessness of the Union cause, and the feebleness of the Union sentiment in this State. The differences between North and South have been growing more marked for years, and the mutual repulsion more radical, until not a single sympathy is left between the dominant influences in each section. Not even the banner of the stars and stripes excites the same thrill of patriotic emotion, alike in the heart of the northern Republican and the southern Secessionist. The former looks upon that flag as blurred by the stain of African slavery, for which he feels responsible as long as that flag waves over it, and that it is his duty to humanity and religion to obliterate the stigma. The latter looks upon it as the emblem of a gigantic power, soon to pass into the hands of that sworn enemy, and knows that African slavery, though panoplied by the Federal Constitution, is doomed to a war of extermination. All the powers of a Government which has so long sheltered it will be turned to its destruction. The only hope for its preservation, therefore, is out of the Union. A few more years of unquiet peace may be spared to it, because Black Republicans cannot yet get full possession of every department of the Government. But this affords to the South no reason for a moment's delay in seeking new guards for its future safety.[1]

1. Dwight Lowell Dumond (ed.), *Southern Editorials on Secession* (New York, 1931), p. 242.

When the *Constitutionalist* declared that not a single sympathy was left between the two sections, it exaggerated the degree to which Unionism had been eroded. The tenacity with which Maryland, Virginia, North Carolina, Kentucky, Tennessee, Missouri, and Arkansas clung to the Union during the next five months proved that Unionism retained much vigor. The great body of Americans, in both the North and the South, still cherished their images of a republic to which they could respond with patriotic devotion, and in this sense American nationalism remained very much alive—so much alive, in fact, that it was able to revitalize itself speedily after four years of devastating war. But though they cherished the image, the sectional conflict had neutralized their many affinities, causing antislavery men to depreciate the value of a Union which was flawed by slavery, and causing men in the slaveholding states to give the defense of the slave system such a high priority that they could no longer offer loyalty to a Union which seemed to threaten that system. As these forces of repulsion between North and South came into play, the southern states were, at the same time, drawn closer together by their common commitment to the slave system and their sense of need for mutual defense against a hostile antislavery majority. Southern separatism had been developing for several decades, and now it was about to end in the formation of the Confederate States of America. Historians have spoken of this separatism as "southern nationalism," and of the Confederacy as a "nation." Yet it is clear that much of the old devotion to the Union still survived among many citizens throughout the South and even dominated the action of some southern states until they found themselves forced to fight on one side or the other. Therefore, one must ask: What was the nature of southern separatism? What was the degree of cohesion within the South on the eve of the Civil War? Had the cultural homogeneity of the southern people, their awareness of shared values, and their regional loyalty reached the point of resembling the characteristics of nationalism? Were they drawn together by a sense of separate destiny which required separate nationhood, or were they rather impelled to united action by their common fears of forces that seemed to threaten the foundations of their society?[2]

2. "The only question is . . . can the Union and slavery exist together." William Henry Trescot to William Porcher Miles, Feb. 8, 1859, quoted in Steven A. Channing, *Crisis of Fear: Secession in South Carolina* (New York, 1970), p. 69.

The understanding of any so-called nationalism—indeed of any development involving sustained cohesive behavior on the part of a large group of people—is complicated by two kinds of dualism. One of these is the dualism of objective and subjective factors, or, one might say, of cultural realities and states of mind. Cohesion can scarcely exist among an aggregate of people unless they share some objective characteristics. Classic criteria are common descent (or ethnic affinities), common language, common religion, and most important and most intangible of all, common customs and beliefs. But these features alone will not produce cohesion unless those who share them also share a self-consciousness of what they have in common, unless they attach distinctive value to what is shared, and unless they feel identified with one another by the sharing. A second dualism lies in the interplay between forces of attraction and forces of repulsion. Wherever and whenever nationalism has developed in notably vigorous form, it has been in circumstances of conflict between the nationalizing group and some other group. In such a situation, the rejection of the out-group not only strengthens the cohesion of the in-group, but imparts to the members of the in-group a greater awareness of what they share. Indeed, it gives them new things to share—common danger, common efforts against the adversary, common sacrifice, and perhaps a common triumph. Sometimes it even impels them to invent fictitious affinities. Thus, conflict and war have been the great catalysts of nationalism, and forces of repulsion between antagonistic groups have probably done more than the forces of affinity within compatible groups to forge the kind of unity that translates into nationalism.

The problem of the South in 1860 was not a simple one of southern nationalism versus American nationalism, but rather one of two loyalties coexisting at the same time—loyalty to the South and loyalty to the Union. Because these loyalties were soon to be brought into conflict, they have often been categorized as "conflicting loyalties," with the implication that if a person has two political loyalties they are bound to conflict, that one of the two must be illegitimate, and that a right-minded person would no more maintain two loyalties then he would commit bigamy. But in fact, strong regional loyalties exist within many nations, and they existed in the United States in other areas besides the South. There was nothing inherently incompatible between regional loyalties and national loyalties as long as they could both be aligned in a pattern in which they remained congruent with one another instead of being at cross-purposes with one another. Any region which had enough power in the federal government could always prevent federal policy and regional policy from coming into any sort of major collision. But the South, by 1860, no longer had such power, or at least no longer had confidence of maintaining such power. Thus the loyalties of southerners became "conflicting loyalties," not necessarily because they loved the Union less but because they had lost the crucial power to keep them from conflicting.[3]

But what were the factors of affinity making for cohesion within the South in 1860, and what were the factors of repulsion between the South and the rest of the Union which gave negative reinforcement to southern unity?

The vast and varied region extending from the Mason-Dixon line to the Rio Grande and from the Ozarks to the Florida Keys certainly did not constitute a unity, either physiographic or cultural. But the whole area lay within what may be called the gravitational field of an agricultural economy specializing in staple crops for which plantations had proved to be effective units of production and for which Negro slaves had become the most important source of labor. This, of course, did not mean that all white southerners engaged in plantation agriculture and owned slaves—indeed only a small but very influential minority did so. It did not even mean that all of the states were heavy producers of staple crops, for the cotton states were only in the lower South. But it did mean that the economy of all of these states was tied, sometimes in secondary or tertiary ways, to a system of plantation agriculture.

Agricultural societies tend to be conservative and orthodox, with strong emphasis on kinship ties and on the observance of established customs. If land is held in great estates, such societies tend to be hierarchical and deferential. Thus, even without slavery, the southern states would have shared certain attributes to a high degree. But the presence of slavery had dictated conditions of its own, and these too were shared very widely throughout the South. Indeed they became the criteria for determining what constituted the South.

A slave system, since it means the involuntary subordination of a

3. David M. Potter, "The Historian's Use of Nationalism and Vice Versa," in his *The South and the Sectional Crisis* (Baton Rouge, 1968), pp. 34–83.

significant part of the population, requires a social apparatus distinctively adapted in all its parts to imposing and to maintaining such subordination. In the South, this subordination was also racial, involving not only the control of slaves by their masters but also the control of a population of 4 million blacks by 8 million whites. Such a system cannot be maintained simply by putting laws on the statute books and making formal records that one individual has acquired legal ownership of another. It is axiomatic that the enslaved will tend to resist their servitude and that the slaveowners must devise effective, practical means of control. The first requisite is that the system shall be able to deal with the contingency of insurrection. This alters the priorities, for though the system of subordination may have originated as a means to an end—to assure a permanent labor supply for the cultivation of the staple crops—the immediacy of the hazard of insurrection soon makes the subordination of the slaves an end in itself. This was what Thomas Jefferson meant when he said, "We have a wolf by the ears."

The question of the extent to which the South stood in real danger of slave insurrection is a most difficult one, complicated by the fact that the white South could never for a moment rid itself of the fear of insurrection, yet at the same time could never admit even to itself, much less to others, that its "civilized," "contented," and "loyal" slaves might some day massacre their masters.[4] The fear was probably out of proportion to the actual danger. But the point is that white southerners shared, subjectively, a fear of what the slaves might do, and, objectively, a social system designed to prevent them from doing it.

From the time of Spartacus, all slaveholding societies had lived with the danger of slave revolt. But for the South, no reminders from antiquity were needed. On the island of Santo Domingo, between 1791 and 1804, black insurrectionists under a series of leaders including Toussaint L'Ouverture and Jean Jacques Dessalines

4. On southern fears of slave insurrection, see especially Clement Eaton, *Freedom of Thought in the Old South* (Durham, N.C., 1940), pp. 89–117; Eaton, *The Growth of Southern Civilization, 1790–1860* (New York, 1961), pp. 72–97; John S. Kendall, "Shadow over the City" [New Orleans], *LHQ*, XXII (1939), 142–165; Harvey Wish, "The Slave Insurrection Panic of 1856," *JSH*, V (1939), 206–222; Ollinger Crenshaw, *The Slave States in the Presidential Election of 1860* (Baltimore, 1945), pp. 89–111; Channing, *Crisis of Fear*, pp. 17–62, 92–93, 264–273; Kenneth M. Stampp, *The Peculiar Institution: Slavery in the Ante-Bellum South* (New York, 1956), pp. 132–140; also, citations in note 6, below.

had risen in revolt, virtually exterminating the entire white population of the island and committing frightful atrocities, such as burying people alive and sawing them in two. Survivors had fled to New Orleans, Norfolk, and other places in the United States, and southerners could hear from their own lips the stories of their ordeal. Santo Domingo lived as a nightmare in the mind of the South.[5] Within the South itself, of course, there were also revolts or attempted revolts.[6] Gabriel Prosser led one at Richmond in 1800. Some sort of conspiracy under the leadership of Denmark Vesey apparently came near to hatching at Charleston in 1822. Nat Turner led his famous insurrection in Southampton County, Virginia, in 1831. All of these were negligible compared with Santo Domingo or even with revolts in Brazil,[7] but, each one hit an exposed nerve in the southern psyche. Also there were local disturbances. Altogether, one historian has collected more than two hundred instances of "revolts," and while there is reason to believe that some of these were wholly imaginary and that many others did not amount to much, still every one is a proof of the reality of southern apprehensions if not of the actual prevalence of the danger.[8] On

5. Winthrop D. Jordan, *White Over Black: American Attitudes toward the Negro, 1550–1812* (Chapel Hill, 1968), pp. 375–386.

6. Joseph Cephas Carroll, *Slave Insurrections in the United States, 1800–1865* (Boston, 1938); Herbert Aptheker, *American Negro Slave Revolts* (New York, 1943); Marion D. deB. Kilson, "Towards Freedom: An Analysis of Slave Revolts in the United States," *Phylon*, XXV (1964), 175–187; Harvey Wish, "American Slave Insurrections before 1861," *JNH*, XXII (1937), 299–320; Nicholas Halasz, *The Rattling Chains: Slave Unrest and Revolt in the Antebellum South* (New York, 1966); R. H. Taylor, "Slave Conspiracies in North Carolina," *NCHR*, V (1928), 20–34; Davidson Burns McKibben, "Negro Slave Insurrections in Mississippi, 1800–1865," *JNH*, XXXIV (1949), 73–90; William W. White, "The Texas Slave Insurrection of 1860," *SWHQ*, LII (1949), 259–285; Wendell G. Addington, "Slave Insurrections in Texas," *JNH*, XXXV (1950), 408–434; Edwin A. Miles, "The Mississippi Slave Insurrection Scare of 1835," *JNH*, XLII (1957), 48–60.

7. For these insurrections see Aptheker and Carroll, cited in note 6. Also, John M. Lofton, *Insurrection in South Carolina: The Turbulent World of Denmark Vesey* (Yellow Springs, Ohio, 1964); Richard C. Wade, "The Vesey Plot: A Reconsideration," *JSH*, XXX (1964), 143–161; Carl N. Degler, *Neither White nor Black: Slavery and Race Relations in Brazil and the United States* (New York, 1971), pp. 47–51; John W. Cromwell, "The Aftermath of Nat Turner's Insurrection," *JNH*, V (1920), 208–234; F. Roy Johnson, *The Nat Turner Slave Insurrection* (Murfreesboro, N.C., 1966); Herbert Aptheker, *Nat Turner's Slave Rebellion* (New York, 1966); Kenneth Wiggins Porter, "Florida Slaves and Free Negroes in the Seminole War, 1835–1842," *JNH*, XXVIII (1943), 390–421; Porter, "Negroes and the Seminole War, 1817–1818," *JNH*, XXXVI (1951), 249–280.

8. Aptheker, *American Negro Slave Revolts*.

isolated plantations, and in districts where blacks heavily outnumbered whites, the peril seemed a constant one. Every sign of restlessness in the slave quarters, every stranger seen along a lonely road, every withdrawn or cryptic look on a slave face, even the omission of some customary gesture of deference, might be the forewarning of nameless horrors lurking just beneath the placid surface of life.

This pervasive apprehension explains much, of course, about southern reaction to the antislavery movement. The southerners were not deeply concerned with what the abolitionists might persuade Congress or the northern public to do—indeed the whole elaborate territorial controversy had many of the aspects of a charade—but with what they might persuade the slaves to do. Southerners were acutely sensitized to direct abolitionist efforts at incitation, such as Henry H. Harnett's speech at a national Negro convention in 1843 in which he urged slaves to kill any master who refused to set them free.[9] It was seldom difficult to make an equation between abolitionist exhortation and slave violence. Thus southerners tried to link Nat Turner's revolt in August 1831 with the first appearance of the *Liberator* eight months previously, but in truth it appears likely that Turner was more influenced by an eclipse of the sun in February than by William Lloyd Garrison in January. Twenty-eight years later, however, John Brown made the equation explicit: a white abolitionist was caught trying to rouse the slaves to revolt. Brown's tying of the bond between abolition and slave revolt gave electrifying importance to what might otherwise have been dismissed as an act of suicidal folly.

This concern about antislavery propaganda as a potential cause of slave unrest also explains in part why white southerners seemed so oblivious to the great difference between the moderate attitude of an "ultimate extinctionist" like Lincoln and the flaming abolitionism of an "immediatist" like Garrison. When southerners thought of extinction it was in terms of Santo Domingo and not in terms of a gradualist reform to be completed, maybe, in the twentieth century. From their standpoint, the election to the presidency of a man who stated flatly that slavery was morally wrong might have a more

9. Benjamin Quarles, *Black Abolitionists* (New York, 1969), pp. 225–235; Howard H. Bell, "National Negro Conventions of the Middle 1840's: Moral Suasion versus Political Action," *JNH,* XLII (1957), 247–260; Bell, "Expressions of Negro Militancy in the North, 1840–1860," *JNH,* XLV (1960), 11–20.

inciting effect upon the slaves than denunciatory rhetoric from the editor of an abolitionist weekly in Boston.[10]

Since the determination to keep blacks in subordination took priority over other goals of southern society, the entire socio-economic system had to be conducted in a way that would maximize the effectiveness of racial control. This went far beyond the adoption of slave codes and the establishment of night patrols in times of alarm.[11] It meant also that the entire structure of society must be congruent with the objective, and no institutional arrangements should be countenanced which would weaken control. The blacks should live on plantations not only because plantations were efficient units for cotton production, but because in an era prior to electronic and bureaucratic surveillance, the plantation was a notably effective unit of supervision and control. Also, it provided maximum isolation from potentially subversive strangers. Slaves should be illiterate, unskilled, rural workers not only because the cotton economy needed unskilled rural workers for tasks in which literacy would not increase their usefulness, but also because unskilled rural workers were limited in their access to unsupervised contacts with strangers, and because the illiterate could neither read seditious literature nor exchange surreptitious written communication. In fact, the conditions of employment in the cotton culture seemed to fit the needs of a slave system as neatly as the conditions of slavery fitted the needs of employment in the cotton culture, and if cotton fastened slavery upon the South, it is also true that slavery fastened cotton upon the South.

Even beyond these broad relationships, the system of subordination reached out still further to require a certain kind of society, one in which certain questions were not publicly discussed. It must give

10. When John Slidell made a farewell address before leaving the Senate after Louisiana's secession, he declared that Lincoln's inauguration would have been regarded by the slaves as "the day of their emancipation." *Congressional Globe*, 36 Cong, 2 sess., pp. 720–721. Also, on Dec. 12, 1860, John Bell wrote a public letter in which he said the "simple announcement to the public that a great party at the North, opposed to slavery, has succeeded in electing its candidate to the Presidency, disguise it as we may, is well calculated to raise expectations among slaves, and might lead to servile insurrection in the Southern States." Quoted in Mary Emily Robertson Campbell, *The Attitude of Tennesseans toward the Union 1847–1861* (New York, 1961), pp. 147–148.

11. H. M. Henry, *The Police Control of the Slave in South Carolina* (Emory, Va., 1914); Ulrich Bonnell Phillips, *American Negro Slavery* (New York, 1918), pp. 489–502, on slave codes and the policing of slaves.

blacks no hope of cultivating dissension among the whites. It must commit the nonslaveholders to the unquestioning support of racial subordination, even though they might suffer certain disadvantages from a slave system in which they had no economic stake. This meant that books like *The Impending Crisis* must not circulate, and, indeed, universal education, extending literacy indiscriminately to all lower-class whites, need not be encouraged. In a mobile society it would be harder to keep slaves firmly fixed in their prescribed positions; therefore, the society must be relatively static, without the economic flexibility and dynamism of a money economy and a wage system. The more speculative a society became in its social thought, the more readily it might challenge the tenets of the established order. Therefore the South tended toward a religion which laid major emphasis on personal salvation and on a Bible-based orthodoxy; toward an educational system which stressed classical learning; and toward reforms of a pragmatic kind, such as better care for the blind, rather than reforms associated with ideology.[12] In short, the South became increasingly a closed society, distrustful of isms from outside and unsympathetic toward dissenters. Such were the pervasive consequences giving top priority to the maintenance of a system of racial subordination.[13]

By 1860, southern society had arrived at the full development of a plantation-oriented, slaveholding system with conservative values, hierarchical relationships, and authoritarian controls. No society is complete, of course, without an ethos appropriate to its social arrangements, and the South had developed one, beginning with a conviction of the superior virtues of rural life. At one level, this conviction embodied a Jeffersonian agrarianism which regarded landowning cultivators of the soil as the best kind of citizens, because their landownership and their production for use gave them self-sufficiency and independence, uncorrupted by commercial avarice—

12. Clement Eaton, "The Resistance of the South to Northern Radicalism," *NEQ*, VIII (1935), 215–231.

13. "The whole social institutions of the people in the slave-holding states rested as they then supposed upon the stability of the right, which was involved with the ownership of slaves. It was reputed to be the cornerstone of that society, which for ages had rested upon it, and which it was supposed would be overthrown at its removal. All of the transactions of life were based upon it; all of the arrangements for the progress of society were made with reference to it; . . . It was thus that the remarkable unanimity was produced in all of these states. No other cause would have produced it." A. G. Magrath, Nov. 20, 1865, quoted in Charles Edward Cauthen, *South Carolina Goes to War* (Chapel Hill, 1950), p. 72. See also Eaton, *Freedom of Thought*, pp. 280–332.

and also because their labor had dignity and diversity suitable to well-rounded men. But at another level, the commitment to rural values had led to a glorification of plantation life, in which even slavery was idealized by the argument that the dependence of the slave developed in the master a sense of responsibility for the welfare of the slaves and in the slaves a sense of loyalty and attachment to the master. This relationship, southerners argued, was far better than the impersonal, dehumanized irresponsibility of "wage slavery," which treated labor as a commodity.

From an idyllic image of slavery and plantation conditions, it was but a short step to the creation of a similar image of the planter as a man of distinctive qualities. Thus, the plantation virtues of magnanimity, hospitality, personal courage, and loyalty to men rather than to ideas held a social premium, and even the plantation vices of arrogance, quick temper, and self-indulgence were regarded with tolerance. From materials such as these, in an era of uninhibited romanticism and sentimentality, the southern upper class built a fully elaborated cult of chivalry, inspired by the novels of Sir Walter Scott and including tournaments, castellated architecture, a code of honor, and the enshrinement of women. Thus, with a mixture of self-deception and idealism, the South adopted an image of itself which some men used as a fiction to avoid confronting sordid reality, while others used it as a standard toward which to strive in order to develop, as far as they were able, the better aspects of human behavior that were latent even in a slaveholding society.[14]

14. Ulrich Bonnell Phillips, *Life and Labor in the Old South* (Boston, 1929); William E. Dodd, *The Cotton Kingdom* (New Haven, 1919); Avery O. Craven, *The Coming of the Civil War* (New York, 1942), pp. 17–38; J. G. Randall and David Donald, *The Civil War and Reconstruction* (2nd ed.; Boston, 1969), pp. 29–49; Eaton, *Growth of Southern Civilization*, pp. 295–324; Eugene Genovese, "Marxian Interpretations of the Slave South," in Barton J. Bernstein (ed.), *Toward a New Past: Dissenting Essays in American History* (New York, 1968), pp. 90–125; Genovese, "The Slave South: An Interpretation," in his *The Political Economy of Slavery* (New York, 1965), pp. 13–39; William E. Dodd, "The Social Philosophy of the Old South," *American Journal of Sociology*, XXIII (1918), 735–746; Wilbur J. Cash, *The Mind of the South* (New York, 1941), Book I; Rollin G. Osterweis, *Romanticism and Nationalism in the Old South* (New Haven, 1949); Louis Hartz, *The Liberal Tradition in America* (New York, 1955), part IV, "The Feudal Dream of the South"; John Hope Franklin, *The Militant South, 1800–1861* (Cambridge, Mass., 1956); William R. Taylor, *Cavalier and Yankee: The Old South and the American National Character* (New York, 1961); Clement Eaton, *The Mind of the Old South* (Baton Rouge, 1964); David Donald, "The Proslavery Argument Reconsidered," *JSH*, XXXVII (1971), 3–18; Jay B. Hubbell, "Cavalier and Indentured Servant in Virginia Fiction," *SAQ*, XXVI (1927), 23–39; Esther J. Crooks and Ruth W. Crooks, *The Ring Tournament in the United States* (Richmond, 1936); William O. Stevens, *Pistols at Ten Paces: The Story of the Code of Honor in America* (Boston, 1940); Guy A. Cardwell,

One other belief shared by the men of the South in 1860 was especially important because they felt just uncertain and insecure enough about it to be almost obsessively insistent and aggressive in asserting it. This was the doctrine of the inherent superiority of whites over Negroes. The idea was not distinctively southern, but it did have a distinctive significance in the South, for it served to rationalize slavery and also to unite slaveholders and nonslaveholders in defense of the institution as a system, primarily, of racial subordination, in which all members of the dominant race had the same stake.

This racial prejudice against Negroes cannot, of course, be dismissed as nothing but a rationalization to justify their subordination of the blacks, for in fact it was in part just such prejudice which had originally made blacks and Indians subject to enslavement, while servants of other races were not. Initially, the prejudice may have stemmed from the superiority which technologically advanced societies feel over less advanced societies; it may have reflected something of the attitude of Christians toward the "heathen"; it may have reflected the universal antagonism of in-groups and out-groups or the universal distrust of the unfamiliar. In these aspects, prejudice may even be regarded as a relatively innocent form of ethnocentrism, uncorrupted by consideration of self-interest. But once it became firmly tied to slavery, prejudice began to have certain functional uses which added immeasurably both to the strength of slavery and also to its brutalizing quality. Racial prejudice and slavery together created a vicious circle in which the assumed inferiority of the blacks was used as justification for their enslavement, and then their subordination as slaves was used to justify the belief that they were inferior. The stigma of race increased the degradation of slavery, and servile status, in turn, reinforced the stigma of race.[15]

Doctrines of race not only served to minimize the potentially

---

"The Duel in the Old South: Crux of a Concept," *SAQ*, LXVI (1967), 50–69; David Donald, "The Southerner as Fighting Man," in Charles G. Sellers (ed.), *The Southerner as American* (New York, 1966), pp. 72–88; Grace Warren Landrum, "Sir Walter Scott and His Literary Rivals in the South," *American Literature*, II (1930), 256–276; George Harrison Orians, *The Influence of Walter Scott upon America and American Literature before 1860* (Urbana, Ill., 1929); Orians, "Walter Scott, Mark Twain and the Civil War," *SAQ*, XL (1941), 342–359.

15. Two major studies, neither of which is wholly responsible for the interpretation above, but which dominate the extensive literature dealing with the origins and nature of racial prejudice and racial subordination in the Negro-white context, are Jordan, *White Over Black*, and David Brion Davis, *The Problem of Slavery in Western Culture* (Ithaca, N.Y., 1966).

serious economic divisions between slaveholders and nonslaveholders, but also furnished southerners with a way to avoid confronting an intolerable paradox: that they were committed to human equality in principle but to human servitude in practice. The paradox was a genuine one, not a case of hypocrisy, for though southerners were more prone to accept social hierarchy than men of other regions, still they responded very positively to the ideal of equality as exemplified by Jefferson of Virginia and Jackson of Tennessee. In their politics, they had moved steadily toward democratic practices for whites, and in fact it was argued, with a certain plausibility, that the system of slavery made for a greater degree of democracy within that part of the society which was free, just as it had made for democracy among the freemen of ancient, slaveholding Athens.[16] Still, this only made the paradox more glaringly evident, and no doubt it was partly because of the psychological stress arising from their awareness of the paradox that southern leaders of the late eighteenth and early nineteenth centuries had played with the idea of some day eliminating slavery. That was, in part, why the South had acceded to the exclusion of slavery from the Northwest Territory in 1787 and to the abolition of the African slave trade in 1808. It was why a limited number of southerners had emancipated their slaves, especially during the half-century after the Declaration of Independence, and why a greater number had indulged themselves in a rhetoric which deplored slavery without exactly condemning it. Some had even joined antislavery societies, and southerners had taken the lead in emancipating slaves and colonizing them in Liberia. Thus, for a generation, the great paradox had been masked by the vague and pious notion that at some remote future, in the fullness of time and God's infinite wisdom, slavery would pass away.[17]

16. Eaton, *Growth of Southern Civilization,* pp. 152, 158, 307–309; Fletcher M. Green, "Democracy in the Old South," *JSH,* XII (1946), 3–23; Frank L. and Harriet C. Owsley, "The Economic Basis of Society in the Late Ante-Bellum South," *JSH,* VI (1940), 24–45; Blanche H. Clarke, *The Tennessee Yeoman, 1840–1860* (Nashville, 1942); Herbert Weaver, *Mississippi Farmers, 1850–1860* (Nashville, 1945); Frank L. Owsley, *Plain Folk of the Old South* (Baton Rouge, 1949); Fabian Linden, "Economic Democracy in the Slave South: An Appraisal of Some Recent Views," *JNH,* XXXI (1946), 140–189; James C. Bonner, "Profile of a Late Ante-Bellum Community," *AHR,* XLIX (1944), 663–680.

17. On antislavery in the early South, see citations in chap. 2, note 26. Also, John Spencer Bassett, *Anti-Slavery Leaders of North Carolina* (Baltimore, 1898); Ruth Scarborough, *The Opposition to Slavery in Georgia Prior to 1860* (Nashville, 1933); H. M.

By the 1830s, however, this notion had begun to lose its plausibility, for even the most self-deceiving of wishful thinkers could not completely ignore the changes under way. In the lower South the great cotton boom was extending slavery westward across Georgia, Alabama, Mississippi, and Louisiana, and into Arkansas and Missouri. Texas had set up as an independent slaveholding republic. The traffic in slaves between these new states and the older centers of slavery was probably greater in magnitude than the traffic from Africa to the thirteen colonies had ever been.[18] Compared to the birth rate of new slaves, the rate of emancipation was as nothing. Meanwhile, the New England states, New York, Pennsylvania, and New Jersey had abolished slavery.[19] Concurrently, northern antislavery men had begun to abandon their tone of gentle, persuasive reproachfulness in discussing slavery and had fallen not only to denouncing slavery as a monstrous sin, but also to castigating slaveholders, as hideous sinners.[20] One should not accept the apologia that the South would itself have got rid of slavery if this indiscriminate onslaught had not compromised the position of the southern emancipationists,[21] but it does seem valid to say that, in the face of such bitter condemnation, white southerners lost their willingness to concede that slavery was an evil—even an inherited one, for which Yankee slave sellers and the southern slave buyers of the eighteenth century shared responsibility. Instead they responded by defending slavery as a positive good.[22] But this made all the more

---

Wagstaff (ed.), *North Carolina Manumission Society, 1816–1834* (Chapel Hill, 1934); Early Lee Fox, *The American Colonization Society, 1817–1840* (Baltimore, 1919); P. J. Staudenraus, *The African Colonization Movement, 1816–1865* (New York, 1961); Beverley B. Munford, *Virginia's Attitude toward Slavery and Secession* (New York, 1909); Asa Earl Martin, *The Anti-Slavery Movement in Kentucky Prior to 1850* (Louisville, 1918); Merton L. Dillon, *Benjamin Lundy and the Struggle for Negro Freedom* (Urbana, Ill., 1966); Matthew T. Mellon, *Early American Views on Negro Slavery* (Boston, 1934); Richard Beale Davis, *Intellectual Life of Jefferson's Virginia, 1790–1830* (Chapel Hill, 1964).

18. Frederic Bancroft, *Slave-Trading in the Old South* (Baltimore, 1931), pp. 269–364, 382–406.

19. Arthur Zilversmit, *The First Emancipation: The Abolition of Slavery in the North* (Chicago, 1967).

20. For changes in the tone of abolitionist literature, see note 66, below.

21. Hilary A. Herbert, *The Abolition Crusade and Its Consequences* (New York, 1912).

22. William Sumner Jenkins, *Pro-Slavery Thought in the Old South* (Chapel Hill, 1935); William B. Hesseltine, "Some New Aspects of the Proslavery Argument," *JNH*, XXI (1936), 1–14; Eric L. McKitrick (ed.), *Slavery Defended: The Views of the Old South* (Englewood Cliffs, N.J., 1963); Harvey Wish, *George Fitzhugh, Propagandist of the Old South* (Baton Rouge, 1943); Eugene D. Genovese, *The World the Slaveholders Made* (New York, 1969), part II: "The Logical Outcome of the Slaveholders' Philosophy,"

stark the contradiction between equality in theory and servitude in practice, and their only escape was to deny that the blacks were qualified for equality on the same basis as other men. Some theoreticians of race even denied that blacks were the descendants of Adam, which was a long step toward their exclusion not only from equality but also from the brotherhood of man.[23]

With the theory of race thus firmly linked to the theory of slavery, the belief in Negro inferiority was as functional and advantageous psychologically as slavery itself was economically. The belief could be used to justify a certain amount of ill treatment of the blacks and even hostility toward them, since, lacking full humanity, they did not deserve fully human treatment and might justifiably be despised for their inherent deficiencies. By maintaining slavery, the South had violated its own ideal of equality, but by adopting racist doctrine it had both perverted and rejected the ideal, as the only way, other than emancipation, to escape from their dilemma.

All these shared institutions, practices, attitudes, values, and beliefs gave to southern society a degree of homogeneity and to southerners a sense of kinship.[24] But a sense of kinship is one thing, and an impulse toward political unity is another. If one searches for explicit evidence of efforts to unify the South politically because of cultural homogeneity, common values, and other positive influences, rather than as a common negative response to the North, one finds relatively little of it.

Yet any separatist movement in the middle of the nineteenth century could scarcely fail to absorb some of the romantic nationalism that pervaded the Western world. At the Nashville convention in 1850, Langdon Cheves of South Carolina had appealed to all the

---

pp. 115–244; Wilfred Carsel, "The Slaveholders' Indictment of Northern Wage Slavery," *JSH*, VI (1940), 504–520; Robert Gardner, "A Tenth Hour Apology for Slavery," *JSH*, XXVI (1960), 352–367; Ralph E. Morrow, "The Proslavery Argument Revisited," *MVHR*, XLVIII (1961), 79-94; Alan Dowty, "Urban Slavery in Pro-Southern Fiction of the 1850's," *JSH*, XXXII (1966), 25–41; Lewis M. Purifoy, "The Southern Methodist Church and the Pro-Slavery Argument", *JSH*, XXXII (1966), 325–341; Jeannette Reid Tandy, "Pro-Slavery Propaganda in American Fiction of the Fifties," *SAQ*, XXI (1922), 41–50, 170–178.

23. William R. Stanton, *The Leopard's Spots: Scientific Attitudes toward Race in America, 1815–1859* (Chicago, 1960), pp. 110–112, 155–160.

24. "There is a community of interest and feeling between the fifteen Southern States, fully as great, perhaps greater, than existed between the original thirteen." Augusta, Georgia, *Daily Chronicle and Sentinel*, Nov. 13, 1860, quoted in Dumond, *Southern Editorials*, p. 232.

slaveholding states, "Unite, and you shall form one of the most splendid empires in which the sun ever shone, one of the most homogeneous populations, all of the same blood and lineage [note that to Cheves the black population was invisible], a soil the most fruitful and a climate the most lovely."[25] At about the same time, another South Carolinian had declared that as long as the South was in the Union, it occupied a false and dangerous position as "a nation within a nation."[26]

During the fifties, the spirit of southernism continued to grow. For example, in 1852, the governor of South Carolina spoke of "our place as a Southern Confederacy amongst the nations of the earth."[27] Near the end of the decade a Virginia Unionist complained that Alabamians denounced "anyone who professes the smallest love of the Union as a traitor to his country, namely the South."[28] When secession came, many southerners who favored it held back from separate state action because they wanted the South to act as a unit. Thus the principal opponent of immediate secession in Alabama wrote to a friend in Tennessee, "I resisted the secession of Alabama to the last moment, not because I doubted that it must come sooner or later, but because I preferred to wait until you in Tennessee were ready to go with us."[29] Even more, some southerners who chose to remain with the Union at the same time prepared to defend other southerners who might choose to go out of it. A Missouri newspaper declared that the border states, "while they are devoted to the Union, . . . will not stand idly by and see their sister States—bone of their bone and flesh of their flesh—trampled in the dust. They will not do it."[30]

Even when men in the southern states saw their political destiny as being outside the American union, they did not necessarily visualize a southern republic as the alternative. In 1832, John Pendleton

25. Quoted in Nathaniel W. Stephenson, "Southern Nationalism in South Carolina in 1851," *AHR*, XXXVI (1931), 314–335.

26. William H. Trescot, *The Position and Course of the South* (Charleston, 1850), pp. 6–18.

27. Message of Governor John H. Means in 1852, quoted in Cauthen, *South Carolina Goes to War*, p. 6.

28. Eugene Blackford to Mary L. Minor, Nov. 9, 1860 quoted in Crenshaw, *Slave States in the Election of 1860*, p. 251.

29. Jeremiah Clemens to Solon Borland, quoted in Durward Long, "Unanimity and Disloyalty in Secessionist Alabama," *CWH*, XI (1965), 259.

30. St. Louis *Missouri Republican*, Nov. 21, 1860, in Dumond, *Southern Editorials*, p. 259.

Kennedy declared, "Virginia has the sentiments and opinions of an independent nation," but he meant independence of the Gulf Coast states as well as the Yankees.[31] Twenty-eight years later, Kennedy denounced South Carolina's secession as "a great act of supreme folly and injustice passed by a set of men who have inflamed the passions of the people."[32]

State loyalty no doubt gave ground to regional loyalty between the 1830s and the 1860s, but localism by no means ceased to compete with southernism. It is significant that Robert E. Lee, who was opposed to secession, had no thought of resigning his commission in the United States army until Virginia seceded, but then he "went with his state." It is perhaps also significant that the vice-president of the Confederacy, who had repeatedly hampered its power by his localistic objections, when imprisoned at Fort Warren after the war wrote, "My native land, my country, the only one that is country to me, is Georgia."[33]

The "set of men" whom Kennedy denounced as inflaming the passions of the people might have included at least four well-known southern figures. Two of these, Edmund Ruffin of Virginia and William Lowndes Yancey, might well be labeled southern nationalists, for they both had the vision of a South united by shared distinctive qualities, and both seemed to care more for the South as a whole than for their own states. The other two, Robert Barnwell Rhett of South Carolina and James D. B. De Bow of Louisiana, were also major actors in the secession movement, but for them a united South was primarily an alliance against the North. If nationalism means something more than bitterness against another country, it would be difficult to show that Rhett and De Bow were southern nationalists.

Ruffin, as early as 1845, had declared, "We shall have to defend our rights by the strong hand against Northern abolitionists and perhaps against the tariffites," and he had formed an intense aversion to all Yankees. He later boasted that he was "the first and for

31. Jay B. Hubbell, "Literary Nationalism in the Old South," in David Kelley Jackson (ed.), *American Studies in Honor of William Kenneth Boyd* (Durham, N.C., 1940), p. 177.

32. Hubbell, *The South in American Literature* (Durham, N.C., 1954), p. 487; also, William P. Trent, *English Culture in Virginia* (Baltimore, 1889).

33. Myrta Lockett Avary (ed.), *Recollections of Alexander H. Stephens* (New York, 1910), p. 253.

some years the only man in Virginia who was both bold and disinterested enough to advocate the dissolution of the Union." After working steadily for more than a decade to publicize the cause of southern rights, Ruffin came forward early in 1858 with a proposal for a League of United Southerners, of which he besought Yancey to assume the leadership. The League was to consist of citizens who would pledge themselves to defend and secure the constitutional rights and interests of the southern states. Members might form local clubs or chapters, which could send delegates to a general council. "By discussion, publication, and public speeches," the League would have its impact upon the public mind of the South and would offset the excessive individualism with which many southerners approached public questions. In 1860, after Lincoln's election, Ruffin wrote, "If Virginia remains in the Union under the domination of this infamous, low, vulgar tyranny of Black Republicanism, and there is one other state in the Union that has bravely thrown off the yoke, I will seek my domicile in that state and abandon Virginia forever." True to his word, Ruffin went to South Carolina to encourage secession there in December, and to Georgia and to Florida during the weeks that followed, for the same purpose. In April, this sixty-seven-year-old champion of secession was given the distinction of firing one of the first shells in the bombardment of Fort Sumter. In the spring of 1865, utterly broken and unwilling to survive the Confederacy, he took his own life by shooting himself.[34]

Ruffin's friend and associate Yancey was another southerner who had proven extremely jealous in the assertion of southern rights, both in Congress in 1845–1847 and in his refusal to support the Democratic party in 1848 because it would not affirm the rights of slavery in the territories. But his public advocacy of southern separatism came much later and was somewhat inhibited until 1861 by the general stigma attached to the idea of disunion. By 1858, however, he seems to have become fully committed to the idea of a southern republic. In that year, he followed up Ruffin's proposal by organizing at Montgomery the first chapter of the League of United Southerners. In the same year, at a meeting in Montgomery

34. Avery Craven, *Edmund Ruffin, Southerner: A Study in Secession* (New York, 1932), quotations from pp. 107, 162, 198. The text of the constitution of the League of United Southerners is in John Witherspoon Du Bose, *The Life and Times of William Lowndes Yancey* (2 vols.; Birmingham, Ala., 1892), I, 377–378.

of one of the annual commercial conventions, he sought to stress the southern rather than the purely states' rights theme by addressing his auditors as "My Countrymen of the South," and by suggesting that their gathering was "a foreshadow of a far more important body" which must "ere long assemble upon Southern soil" if injustice and wrong should "continue to rule the hour and the councils of the dominant section of this country." It was at this time also that Yancey put the rhetorical question: "Are you ready, countrymen? Is your courage up to the highest point? Have you prepared to enter upon the great field of self-denial as your fathers did, and undergo, if necessary, another seven years of war in order that you and your posterity may enjoy the blessings of liberty?" But perhaps his most straightforward statement was in a letter to a fellow Alabamian, also in 1858: "No national party can save us; no sectional party can do it. But if we could do as our fathers did, organize Committees of Safety all over the cotton states, (and it is only in them that we can hope for any effective movement), we shall fire the Southern heart—instruct the Southern mind—give courage to each other, and at the proper moment, by one, organized, concerted action, we can precipitate the cotton states into a Revolution."[35]

James D. B. De Bow made his contribution to the southern cause primarily by serving as editor, from 1846, of *De Bow's Review,* the most vigorous and effective of antebellum southern periodicals. With an enlightened feeling for breadth of coverage, he made the *Review* a vehicle for distributing information about the South as a whole and, especially, the southern economy. In the late fifties, he became one of the foremost advocates of secession. No southern nationalist exceeded him in zeal, but under analysis, his southernism seems to be about one part concern for the unity and cultural integrity of the South and nine parts hostility toward abolition and the economic hegemony of the North. If there had been no abolitionists, it appears that De

35. Du Bose, *Life of Yancey,* I, 358–360, 376; Joseph Hodgson, *The Cradle of the Confederacy; or the Times of Troup, Quitman, and Yancey* (Mobile, 1876); Austin L. Venable, "William L. Yancey's Transition from Unionism to State Rights," *JSH,* X (1944), 331–342; Venable, "The Public Career of William Lowndes Yancey," *AR,* XVI (1963), 200–212; Eaton, *The Mind of the Old South,* pp. 202–221; Alto L. Garner and Nathan Stott, "William Lowndes Yancey: Statesman of Secession," *AR,* XV (1962), 190–202; Malcolm C. McMillan, "William L. Yancey and the Historians: One Hundred Years," *AR,* XX (1967), 163–186; *De Bow's Review,* XXIV (1858), 578–588.

Bow might have remained, as he began, an exultant spokesman of expansionist American nationalism.[36]

Robert Barnwell Rhett, affiliated with the Charleston *Mercury* since 1830, had been demanding separation from the North intermittently for a generation. His speech at Grahamville, South Carolina, on July 4, 1859, was a prelude to the final push by the fire-eaters, for Rhett declared that the South should either prevent the election of a Republican president in 1860 or secede. In his peroration he began with the incredible statement that he had spent twenty years trying to preserve the Union, and then, he said, "I turned at last to the salvation of my native land—the South—and in my latter years did all I could to dissolve her connection with the North, and to establish for her a Southern Confederacy." A toast was drunk that day to "The election of a Black Republican President—the signal for the dissolution of the Federal Union and the establishment of a Southern Confederacy."[37]

For all the electrifying rhetoric on this and other occasions there were no committees of correspondence, and the League of United Southerners apparently never expanded beyond three towns in Alabama. There was, however, one widespread southern organization, though it never did anything effective for the cause of southern nationalism. In 1859, at Louisville, Kentucky, a somewhat itinerant promoter and self-styled general, George F. Bickley, launched a fraternal enterprise which he called the Knights of the Golden Circle. Whether Bickley really had gorgeous dreams of a tropical empire or was merely selling such dreams to earn a living is not clear, but during 1860 he spent much of his time perambulating the South recruiting knights. The KGC, taking shrewd advantage of the spirit of filibusterism both in the South and in other parts of the country, and also of the rising tide of southernism, proposed that Mexico be acquired for the United States through negotiation with Benito

36. Ottis Clark Skipper, "J. D. B. De Bow, the Man," *JSH*, X (1944), 404–423; Skipper, *J. D. B. De Bow, Magazinist of the Old South* (Athens, Ga. 1958); Robert F. Durden, "J. D. B. De Bow: Convolutions of a Slavery Expansionist," *JSH*, XVII (1951), 441–461.

37. Laura A. White, *Robert Barnwell Rhett, Father of Secession* (New York, 1931), is standard and excellent. A more recent study is H. Hardy Perritt, "Robert Barnwell Rhett, South Carolina Secession Spokesman" (Ph.D. dissertation, University of Florida, 1954). Perritt, "Robert Barnwell Rhett's Speech, July 4, 1859," in J. Jeffery Auer (ed.), *Anti-Slavery and Disunion, 1858–1861: Studies in the Rhetoric of Compromise and Conflict* (New York, 1963), pp. 98–107, is especially pertinent to the paragraph above.

Juárez. This annexation would solve the difficulties of the South as a minority section by bringing twenty-five new slave states into the Union. But if the North should spurn this glorious opportunity, or if sectional antagonisms should lead to disruption of the Union, the South could make the annexation alone and create a great tropical empire extending in a golden circle from the tip of Florida, around the shores of the Gulf of Mexico, to the Yucatán Peninsula. The southern press lavished an astonishing amount of favorable attention upon this hare-brained scheme, and General Bickley, a man not given to understatement, claimed a membership of 65,000 Knights in September 1860, and 115,000 in November. Probably one figure was as reliable as the other. In any case, the Knights played no significant part in 1861 in either forming or upholding the southern Confederacy.[38]

The southern commercial conventions provided perhaps the best opportunities to coordinate the impulses of southern nationalism. At first, they had carefully disavowed any spirit of sectional antagonism, even toasting the North and proclaiming a purpose to emulate the enterprise of their northern brothers. But at Charleston in 1854, Albert Pike of Arkansas advocated a program of southern joint action in the form of a corporation, chartered and financed by the fifteen slave states collectively, to build a Pacific railroad by the southern route. Pike also introduced overtly, perhaps for the first time at any of these conventions, the theme of disunion. The South, he said, should seek equality with the North within the Union, but if the South "were forced into an inferior status, she would be better out of the Union than in it." The following year at New Orleans, one delegate proposed the reopening of the African slave trade, another complained that the monopoly of northern textbooks in southern schools made for an education that was "unsouthern," and the St. Louis *Democrat* denounced the conventions as disunionist. At Richmond in 1856, a toast was offered which for the first time defined the boundaries of a prospective southern republic: "on the North by the Mason-Dixon line, and on the South by the Isthmus of Te-

38. Ollinger Crenshaw, "The Knights of the Golden Circle: The Career of George Bickley," *AHR*, XLVII (1941), 23–50; C. A. Bridges, "The Knights of the Golden Circle: A Filibustering Fantasy," *SWHQ*, XLIV (1941), 287–302; Jimmie Hicks, "Some Letters Concerning the Knights of the Golden Circle in Texas, 1860–1861," *SWHQ*, LXV (1961), 80–86; Roy Sylvan Dunn, "The KGC in Texas, 1860–1861," *SWHQ*, LXX (1967), 543–573.

huantepec, including Cuba and all other lands on our Southern shore which threaten Africanization."

In the last four conventions, held at Savannah, Knoxville, Montgomery, and Vicksburg from 1856 to 1859, politicians and fire-eaters had largely replaced businessmen as the dominant delegates, and the meetings had become to a great extent rallies in support of disunion and a southern nation. At Savannah, the chairman spoke of "the beloved Southern section" and addressed his auditors as "free citizens of the South." The *New York Times* concluded that the primary object of the conventions was "to separate in the public mind of the South, Southern interests from national interests," while the Louisville *Journal* denounced most of the members of the convention as "brazen-faced disunionists . . . as thoroughly treasonable as the vilest conclave that ever polluted the soil of South Carolina."[39]

The convention at Montgomery in 1858 marked a high tide of militant southernism. With Ruffin, Yancey, and Rhett all in attendance, disunion had a field day, but the discussions also revealed the lack of a united South and the dilemma which secessionists would face: if they forced the issue, they might destroy the southern unity they were seeking to create; if they waited for such unity to become complete, they might never act. Yancey spoke eloquently of "a unity of climate, a unity of soil, a unity of production, and a unity of social relations." The business committee harmoniously endorsed the League of United Southerners. But when Yancey called for reopening the African slave trade, Roger Pryor of Virginia charged that his real purpose was dissolution of the Union. He, Pryor, would not dissolve it on this ground. Questioned as to the ground on which he would be willing to dissolve it, he replied, "Give me a case of oppression and tyranny sufficient to justify a dissolution of the Union, and give me a united South, and then I am willing to go out of the Union." A delegate retorted that if Pryor had to wait for an undivided South, he would never secede, but Pryor, not at all abashed, told him to remember that, in case of war, "the first onset would have to be met by Virginia, and one must not expect of her the same inordinate enthusiasm felt by others not as vulnerably situated as she." The impasse was partially broken by a more con-

39. Herbert Wender, *Southern Commercial Conventions, 1837–1859* (Baltimore, 1930), pp. 123–129, 155–156, 162–228.

servative Alabamian, Henry Hilliard, who suggested that the election of a Black Republican to the presidency would result in the subversion of the government and the dissolution of the Union—with the implication that the former would justify the latter. Pryor agreed that the election of a Republican president would probably be sufficient grounds for secession, adding that in such an event Virginia would be as ready to act as Alabama.[40]

Since nationalism is frequently as much a negative phenomenon as a positive one, it does not disprove the reality of southern nationalism to say that the southern movement arose primarily from antagonism to the North. Yet one is left with a feeling that the South did not want a separate destiny so much as it wanted recognition of the merits of southern society and security for the slave system, and that all the cultural ingredients of southern nationalism would have had very little weight if that recognition and that security had been forthcoming. Southern nationalism was born of resentment and not of a sense of separate cultural identity. But the cultural dissimilarities of North and South were significant enough to turn a campaign for the protection of southern interests into a movement with a strong color of nationalism. This does not mean that there was *never* a deeply felt southern nationalism. There was. But it resulted from the shared sacrifices, the shared efforts, and the shared defeat (which is often more unifying than victory) of the Civil War. The Civil War did far more to produce a southern nationalism which flourished in the cult of the Lost Cause than southern nationalism did to produce the war.

Even the manifestoes of the self-appointed custodians of southernism do not reflect the impulse to fulfill the unique potentialities of a unique society. Their complaint was not that the Union inhibited a robust but repressed culture struggling to be born, but rather that their cultural dependence upon the Yankees was humiliating. Why must southern children study textbooks written and published in the North, and incompatible with southern values? Why must southern readers subscribe to northern magazines instead of supporting southern journals which published southern authors? The

40. *Ibid.*, pp. 208–209, 214–217, 220–222. Robert R. Russel, *Economic Aspects of Southern Sectionalism, 1840–1861* (Urbana, Ill., 1924), pp. 123–150; John G. Van Deusen, *The Ante-Bellum Southern Commercial Conventions* (Durham, N.C., 1926); Weymouth T. Jordan, *Rebels in the Making: Planters' Conventions and Southern Propaganda* (Tuscaloosa, Ala., 1958).

frequency and the plaintiveness of this question is evidence of the rather self-conscious literary irredentism of a very small number of southern writers, but it also affords striking proof of the lack of cultural self-consciousness on the part of a large number of southern readers who ignored these pleas and continued to get their reading matter from the North. What the South's struggling authors wanted was not separation from the North but recognition by the North. Why must northern critics insist, they wanted to know, on lauding the doggerel of John Greenleaf Whittier while ignoring the genius of William Gilmore Simms? It was intolerable to have the *Atlantic Monthly* characterizing the South as a coarse and sordid oligarchy unhallowed by antiquity and unadorned by culture. But instead of separation, what they wanted was to escape the condescension of the metropolis toward the provinces, to attain some literary triumph which would force the North to acknowledge southern merit. Meanwhile, they retorted in kind, disparaging northern society as mercenary, materialistic, hypocritical, Godless, ill-mannered, and lacking in any class of gentlemen.[41] In 1858 a prominent Tennessee historian declared, "The high-toned New England spirit has degenerated into a clannish feeling of profound Yankeeism. . . . The masses of the North are venal, corrupt, covetous, mean, and selfish." But "the proud Cavalier spirit of the South," he added, not only remained but had become "intensified."[42] Early in 1860, Robert Toombs remarked in the Senate, "The feeling of a common interest and a common destiny, upon which foundations alone society can securely and permanently rest, is . . . rapidly passing away."[43] Later in the same year, conditions reminded Francis Lieber of what Thucydides had said of Greece at the time of the Peloponnesian War: "The Greeks did not understand each other any longer, though they spoke the same language."[44] As the secession movement got under way, the antitheses were drawn sharper and the

41. On literary and cultural nationalism in the South, see titles by Hubbell, above in notes 31, 32; by Osterweis and Eaton, cited in note 14; and also: Merle Curti, *The Growth of American Thought* (New York, 1943), pp. 427–453; John S. Ezell, "A Southern Education for Southrons," *JSH*, XVII (1951), 303–327; Howard R. Floan, *The South in Northern Eyes, 1831–1861* (Austin, 1958).

42. J. G. Ramsey to L. W. Spratt, quoted in James Welch Patton, *Unionism and Reconstruction in Tennessee* (Chapel Hill, 1934), p. 3.

43. *Congressional Globe*, 36 Cong. 1 sess., appendix, pp. 88–93.

44. Thomas Sergeant Perry (ed.), *The Life and Letters of Francis Lieber* (Boston, 1882), p. 314.

stereotypes became caricatures. The "Yankee-Union" was "vile, rotten, infidelic, puritanic, and negro-worshipping."[45] The people of the South were descended from Cavaliers, the people of the North from Roundheads; the people of the South from the conquering Normans of 1066; the people of the North from the subjugated race of Saxons.[46] With such dualisms as these, it was an easy step to the view that the day for brotherhood was "past, irrevocably past," or that the North and the South must separate, not because of Lincoln's election, but because of "the incompatibility growing out of two systems of labor, crystallizing about them two forms of civilization."[47]

In December 1860, when South Carolina seceded, she gave formal affirmation to all these ideas in an Address of the People of South Carolina. "The Constitution of the United States," it declared, "was an experiment. The experiment consisted, in uniting under one Government, peoples living in different climates, and having different pursuits and institutions." In short, the experiment failed. Instead of growing closer together, the sections grew farther apart. By 1860, "their institutions and industrial pursuits, have made them, totally different peoples. . . . All fraternity of feeling between the North and the South is lost, or has been converted into hate; and we, of the South, are at last, driven together, by the stern destiny which controls the existence of nations."[48]

During the secession winter, the South produced a ceaseless flow of statements such as these—all affirmed with such intensity that they suggest the rise of southern nationalism to a fully matured, triumphant, and unchallenged fulfillment.[49] If antipathy toward the Yankees and antipathy toward the American Union could be equated, this inference might be valid. But feelings of anger and fear which part of a society may feel toward another part are not the

45. Joseph Carlyle Sitterson, *The Secession Movement in North Carolina* (Chapel Hill, 1939), p. 238. See also William Howard Russell, *Pictures of Southern Life* (New York, 1861), pp. 5–8.

46. Osterweis, *Romanticism and Nationalism*, pp. 110, 148.

47. Clarence Phillips Denman, *The Secession Movement in Alabama* (Montgomery, Ala., 1933), p. 89; Cauthen, *South Carolina Goes to War*, p. 40. See also Henry T. Shanks, *The Secession Movement in Virginia, 1847–1861* (Richmond, 1934), p. 166.

48. Edward McPherson (ed.), *Political History of the United States during the Great Rebellion* (Washington, 1876), pp. 12–15.

49. "We separated because of incompatibility of temper; we are divorced, North from South, because we have hated each other so." Mary Boykin Chesnut, *A Diary from Dixie* (Boston, 1949), p. 20.

same as the cultural differences between two distinct civilizations. Nor did hostility toward other elements in the Union necessarily imply hostility toward the Union itself. There was still a vigorous Union nationalism remaining in the South, and in spite of all the emotional fury, there was probably more cultural homogeneity in American society on the eve of secession than there had been when the Union was formed, or than there would be a century later. Most northerners and most southerners were farmer folk who cultivated their own land and cherished a fierce devotion to the principles of personal independence and social equalitarianism. They shared a great pride in the Revolutionary heritage, the Constitution and "republican institutions," and an ignorance about Europe, which they regarded as decadent and infinitely inferior to the United States. They also shared a somewhat intolerant, orthodox Protestantism, a faith in rural virtues, and a commitment to the gospel of hard work, acquisition, and success. Southern aristocrats might disdain these latter attributes, but the cotton economy was itself prime evidence of southern possession of them. The development of steamboats and railroads and the telegraph had generated an internal trade which bound the sections increasingly closer economically and had generated a nationwide faith in American progress and in the greatness of America's destiny. The South participated in all of these experiences, and the crisis of 1860 resulted from a transfer of power, far more than from what some writers have called the divergence of two civilizations.[50]

The degree to which southern nationalism still fell short of a culmination was evident from the continued devotion to the Union of a large part of the population of the South. In the election of 1860, southern voters had had a choice between two stout defenders of the Union—Douglas and Bell—and one candidate who denied that he favored disunion. The Unionist candidates carried 49 percent of the vote in the seven states of the original Confederacy.[51] Even after Lincoln's election, Unionism survived in those states and maintained dominance in the upper South. A high proportion of

50. For a classic statement of the homogeneity of American culture in all parts of the country in the generation before the Civil War, see Allan Nevins, *Ordeal of the Union* (2 vols.; New York, 1947), I, 34–112. Also, J. G. Randall and David Donald, *The Civil War and Reconstruction* (2nd ed.; Lexington, Mass., 1969), pp. 1–28 and esp. p. 29; Randall, *Lincoln: The Liberal Statesman* (New York, 1947), pp. 41–43, 49–54; Carl Bode, *The Anatomy of American Popular Culture, 1840–1861* (Berkeley, 1959).

51. Thomas B. Alexander, "Persistent Whiggery in the Confederate South, 1860–1877," *JSH*, XXVII (1961), 307.

former Whigs, who had supported Bell in the election, boldly reaffirmed their Unionism. The Vicksburg *Whig* declared, "It is treason to secede." It also predicted the consequences of secession: "strife, discord, bloodshed, war, if not anarchy." Disunion would be a "blind and suicidal course."[52] The Unionists also castigated the secessionists for their irresponsibility. The governor of Louisiana said regretfully that the dissolution of the Union was spoken of, "if not with absolute levity, yet with positive indifference"; and Alexander H. Stephens complained that the secessionists really did not want redress for their grievances; they were "for breaking up" merely because they were "tired of the govnt."[53] In the upper South, the Unionists reminded one another of the importance of their material ties with the North. Senator Crittenden of Kentucky had pointed out in 1858 that "the very diversity of . . . resources" of the two sections led to interdependence and was "a cause of natural union between us." In 1860 a Tennessee newspaper declared: "We can't do without their [the North's] productions, and they can't do without our Rice, Sugar, and Cotton."[54]

Besides, the Union itself remained a priceless asset, an "empire of freemen," in the words of one southern president, with "the most stable and permanent Government on earth."[55] "As a nation," said a North Carolina newspaper, "We possess all the elements of greatness and power. Peace smiles upon us from all quarters of the globe; a material prosperity, unparalleled in the annals of the world, surrounds us; our territory embraces almost the entire continent; we enjoy wide-spread intelligence and universal plenty; we are happy; WE ARE FREE."[56]

Southern nationalism had arrived, but Union nationalism had by

52. Quoted in Percy Lee Rainwater, *Mississippi: Storm Center of Secession, 1856–1861* (Baton Rouge, 1938), p. 164.

53. Jefferson Davis Bragg, *Louisiana in the Confederacy* (Baton Rouge, 1941), p. 2; Ulrich Bonnell Phillips (ed.), *The Correspondence of Robert Toombs, Alexander H. Stephens, and Howell Cobb,* AHA *Annual Report,* 1911, II, 526.

54. *Congressional Globe,* 35 Cong., 1 sess., pp. 1153–1159; Campbell, *Attitude of Tennesseans,* p. 140.

55. Zachary Taylor's first annual message, Dec. 4, 1849, James D. Richardson (ed.), *A Compilation of the Messages and Papers of the Presidents* (11 vols.; New York, 1907), V, 9.

56. Dumond, *Southern Editorials,* p. 227. James L. Orr of South Carolina said, "When this government is destroyed, neither you, nor I, your children nor my children will ever live to see so good a government reconstructed." Quoted in Laura A. White, "The National Democrats in South Carolina, 1850 to 1860," *SAQ,* XXVIII (1929), 381.

no means departed. Sometimes, indeed, a man might declare his allegiance to both, in the same breath. Thus, Alexander H. Stephens as early as 1845 had said, "I have a patriotism that embraces, I trust, all parts of the Union, . . . yet I must confess that my feelings of attachment are most ardent towards that with which all my interests and associations are identified. . . . The South is my home—my fatherland."[57]

To one who thinks of nationalism as a unique and exclusive form of loyalty, the divisions of the South between Union nationalism and southern nationalism, and the movement of individuals from one camp to the other, will look like some sort of political schizophrenia. But if one thinks of nationalism instead as but one form of group loyalty, it becomes easier to see that the choice between Union nationalism and southern nationalism was basically a question of means—a question of whether the slaveholding society would be safer in the Union or in a southern Confederacy. The South had fared extremely well in the Union of 1787, with its bisectional balance, its lack of centralized power, and most of all, its indulgent attitude toward slavery. As these advantages dwindled, men began to speak of the Union of 1787 as the "Old Union," and the South cherished its memory with nostalgia and reverence, as "the Union of our Fathers." In 1861 the New Orleans *Picayune* opposed secession and called instead for "the reconstruction of the old Union."[58]

Not only did southerners think fondly of a nation within which the southern social system had been secure. They also said quite directly that the security of their system was the criterion by which they should choose between the existing nation and the incipient one. Many recognized that even if a southern confederacy were successfully formed, its existence would not prevent slaves from fleeing north to freedom, would not silence abolitionist attacks on slavery, and would probably mean an abandonment of the rights which slavery, under the Dred Scott decision, enjoyed in the territories. The South would have to resist antislavery in any case, and so perhaps it could fight more effectively inside the Union than out-

57. *Congressional Globe,* 28 Cong., 2 sess., appendix, p. 314.

58. Howell Cobb to Absalom H. Chappell, Feb. 7, 1851, in Phillips (ed.), *Toombs, Stephens, Cobb Correspondence,* p. 221; New Orleans *Picayune,* quoted in Willie Malvin Caskey, *Secession and Restoration of Louisiana* (University, La., 1938), p. 36. See also Campbell, *Attitude of Tennesseans,* p. 171; Richard Harrison Shryock, *Georgia and the Union in 1850* (Durham, N.C., 1926), pp. 293–294.

side. Abolitionists might be more dangerous as foreign neighbors than as fellow citizens. The Union, said Benjamin F. Perry of South Carolina, "should be saved as a bulwark against abolition."[59] Secession would endanger slavery more than Lincoln would. Alexander H. Stephens warned that nothing was more dangerous for the South than "unnecessary changes and revolutions in government." He considered "slavery much more secure in the Union than out of it" and thought that Lincoln would make "as good a President as Fillmore did."[60] Herschel V. Johnson, a secessionist in 1850 but converted to Unionism by 1860, offered a simple and pragmatic explanation for his change: "I had become satisfied that Slavery was safer in than out of the Union."[61] The *North Carolina State Journal* denied that the essential problem was a conflict of loyalties. "The question," it said, "is not union or disunion, but what shall she [North Carolina] do to protect herself."[62]

As long as the North and the South had remained fairly equal in economic and political power, and as long as slavery had been immune to serious attack, the two sections had coexisted in a reasonably harmonious way. They could differ and even quarrel fiercely over various political questions without placing the Union in great danger. But as time passed, the sections ceased to be evenly balanced, and slavery lost its immunity. These simultaneous developments had an overpowering effect in the South. They generated a feeling of being on the defensive, the psychology of a garrison under siege.[63]

At the beginning of the century, the population of the slave states had been equal to that of the North, and the South had had 40 percent of the total white population. But by 1860, northerners outnumbered southerners in a ratio of 6:4 in total population and 7:3 in white population. At the beginning of the century, Virginia and Kentucky might talk about the power of individual states to prevent enforcement of the Alien and Sedition Acts, but they really

59. Benjamin Franklin Perry, *Biographical Sketches of Eminent American Statesmen* (Philadelphia, 1887), pp. 171–180.

60. Stephens to J. Henly Smith, July 10, 1860, in Phillips (ed.), *Toombs, Stephens, Cobb Correspondence*, pp. 486–487.

61. Percy Scott Flippin, *Herschel V. Johnson of Georgia: State-Rights Unionist* (Richmond, 1931), p. 93.

62. Sitterson, *Secession Movement in North Carolina*, p. 213.

63. Rich in relevant data and quotations is Jesse T. Carpenter, *The South as a Conscious Minority, 1789–1861* (New York, 1930).

did not need to resort to such minority devices, for they still had enough political muscle to put a Virginian into the White House in 1801 and to keep the presidency in the hands of southerners for forty-two of the next fifty years. But by 1860, a man might win the presidency without even being on the ticket in most of the southern states. The increasing discrepancies in wealth, productive capacity, and technological advancement were equally apparent. William L. Yancey told a New York audience in 1860, "You have power in all the branches of the government to pass such laws as you like. If you are actuated by power, or prejudice, or by the desire of self-aggrandizement, it is within your power . . . to outnumber us and commit aggression upon us." The South was not only in a minority but, more ominously, in a permanent and dwindling minority.[64]

Its power was dwindling, moreover, at the very time when the South found itself exposed to increasingly sharp attack by antislavery spokesmen. During the first forty years of the republic, slavery had certainly been criticized, but it had virtually never been threatened. Antislavery men had been gradualists, who proposed no sudden action; emancipationists, who relied on reasoned appeals to the slaveholders to practice voluntary manumission; colonizationists, whose program looked to the removal of the blacks along with the removal of slavery. Slavery had been respectable, and eight of the first twelve men who reached the presidency were slaveholders. Until 1856, no major political party had ever, at the national level, made a public pronouncement against slavery, and in northern cities, mobs that included "gentlemen of property and standing" had hounded and harassed the abolitionists.[65] But in the 1830s, the abolitionists had captured the antislavery movement, demanding immediate, involuntary emancipation enforced by law, denouncing all slaveholders with unmeasured invective, and even sometimes proclaiming the equality of the blacks.[66] Antislavery parties had

64. Emerson David Fite, *The Presidential Campaign of 1860* (New York, 1911), pp. 301–329.

65. Leonard L. Richards, *Gentlemen of Property and Standing: Anti-Abolition Mobs in Jacksonian America* (New York, 1970).

66. For the mild tone of earlier abolitionist literature, see Locke and Adams, as cited in chap. 2, note 26. For the increasingly militant tone after 1831, see titles in chap. 2, note 28; also, Herbert Aptheker, "Militant Abolitionism," *JNH,* XXVI (1941), 438–484; Bell, "Expressions of Negro Militancy," pp. 11–12; Bell, "National Negro Conventions," pp. 247–260; Quarles, *Black Abolitionists;* John Demos, "The Antislavery Movement and the Problem of Violent Means," *NEQ,* XXXVII (1964),

appeared for the first time in the 1840s, and a major antislavery party in the mid-fifties. In 1856 the Republicans had branded slavery as a relic of barbarism, and in 1860 they had elected to the presidency a man who said that slavery must be put on a course toward ultimate extinction. In 1859 many northerners had mourned the hanging of a would-be leader of slave insurrection. Meanwhile, slavery had been disappearing from the Western world and remained significant only in Brazil, Cuba, and the southern United States.

If the government of the United States should pass into the control of opponents of slavery, as it seemed about to do in 1860, the South had realistic reason to fear the consequences, not so much because of legislation which the dominant party might adopt, but because the monolithic, closed system of social and intellectual arrangements upon which the South relied for the perpetuation of slavery might be disrupted. Once Lincoln was in office, he could appoint Republican judges, marshals, customs collectors, and postmasters in the South. This would strike a heavy blow at the mystique of planter control which had been vital to the maintenance of the southern system. With their political domination challenged, the planter class might lose some of their social ascendancy also. More explicitly, Lincoln might appoint abolitionists or even free Negroes to public office in the South. And even if he did not do this, the new Republican postmasters would refuse to censor the mails or to burn abolitionist papers.[67] The temptation of postmasterships might attract some nonslaveholding southerners and make them the nucleus of an antislavery force in the South. For a slave system vitally depen-

---

501–526; Martin Duberman (ed.), *The Antislavery Vanguard: New Essays on the Abolitionists* (Princeton, 1965), pp. 71–101, 270–298, 417–451; James B. Stewart, *Joshua R. Giddings and the Tactics of Radical Politics* (Cleveland, 1970); Stewart, "The Aims and Impact of Garrisonian Abolitionism, 1840–1860," *CWH*, XV (1969), 197–209; Lewis Curtis Perry, "Antislavery and Anarchy: A Study of the Ideas of Abolitionism before the Civil War" (Ph.D. dissertation, Cornell University, 1967). For the impact of this militancy on the South, Arthur Y. Lloyd, *The Slavery Controversy, 1831–1860* (Chapel Hill, 1931); Henry H. Simms, "A Critical Analysis of Abolitionist Literature," *JSH*, VI (1940), 368–382; Simms, *A Decade of Sectional Controversy, 1851–1861* (Chapel Hill, 1942), pp. 146–168; Simms, *Emotion at High Tide: Abolition as a Controversial Factor* (n.p., 1960).

67. For the importance of postal censorship to the South see Clement Eaton, "Censorship of the Southern Mails," *AHR*, XLVIII (1943), 266–280; William Sherman Savage, *The Controversy over the Distribution of Abolition Literature, 1830–1860* (Washington, 1938).

dent upon the solidarity of the whites, this loomed as a frightful menace. It was irrelevant to say that the Republicans did not constitute a threat because they still lacked the majorities which would enable them to enact legislation in Congress. They did not need to enact legislation.[68]

By 1860, southerners were acutely conscious of their minority status and their vulnerability to abolitionist agitation. After Harpers Ferry, a wave of fear swept through the South, subsiding somewhat in the spring and then rising again during the presidential campaign. There were reports and more reports of dark conspiracies for slave revolts, engineered by abolitionist incendiaries infiltrating the South in guises such as peddlers and itinerant piano tuners. Though seldom verified, the rumors were usually rich in details of plots uncovered; murder, rape, and arson prevented; and malefactors punished. For a time, the atmosphere was such that any fire of unknown origin or any white southerner's death of obscure causes might set off a report of arson or poisoning. And editors, no more immune than their readers, transformed into "news items" the fantasies of a society obsessed with fears of slave insurrection and with apocalyptic visions of terrible retaliation.[69]

When Lincoln's election came at last, the people of the slaveholding states were not united in any commitment to southern nationalism, nor to a southern republic, nor even to political separatism. But they *were* united by a sense of terrible danger. They were united, also, in a determination to defend slavery, to resist abolitionism, and to force the Yankees to recognize not only their rights but also their status as perfectly decent, respectable human beings. "I am a Southern man," a Missouri delegate had asserted in the Baltimore convention, "born and raised beneath the sunny sky of the South. Not a drop of blood in my veins ever flowed in veins north of Mason's and Dixon's line. My ancestors for 300 years sleep beneath the turf that shelters the bones of Washington, and I thank God that they rest in the graves of honest slaveholders."[70]

68. Roy F. Nichols, *The Disruption of American Democracy* (New York, 1948), pp. 352–353. For southern anxiety about the nonslaveholding whites, see Hesseltine, "Some New Aspects of the Proslavery Argument"; Channing, *Crisis of Fear*, pp. 254–256.

69. Crenshaw, *Slave States in the Election of 1860*, pp. 89–111; Nichols, *Disruption*, pp. 351, 367.

70. N. C. Claiborne, June 22, 1860, in Murat Halstead, *Caucuses of 1860* (Columbus, 1860), p. 239.

Motivated by this deeply defensive feeling, the people of the South also tended to accept an interpretation of the Constitution maximizing the autonomy of the separate states. According to this view, each state, when ratifying the Constitution, had retained its full sovereignty. The states had authorized the federal government, as their agent, to administer for them collectively certain of the functions which derive from sovereignty, but they had never transferred the sovereignty itself, and they could resume the exercise of all sovereign functions at any time by an act of secession, adopted in the same kind of state convention that had ratified the Constitution. However arid, and antiquarian it may now seem, the acceptance of this doctrine by a majority of the citizens of the Old South gave to it a historical importance independent of its validity as a constitutional theory. It is impossible to understand the rift between North and South without recognizing that one factor in this rift was a fundamental disagreement between the sections as to whether the American republic was a unitary nation in which the states had fused their sovereign identities or a pluralistic league of sovereign political units, federated for certain joint but limited purposes. Perhaps the United States is the only nation in history which for seven decades acted politically and culturally as a nation, and grew steadily stronger in its nationhood, before decisively answering the question of whether it was a nation at all. The framers of the Constitution had purposely left this question in a state of benign ambiguity. They did so for the best possible reason, namely that the states in 1787 were in hopeless disagreement about it, and some would have refused to ratify an explicitly national Constitution. Thus, the phrase "*E pluribus unum*" was a riddle as well as a motto. The utmost which the nationalists of 1787 could accomplish was to create the framework within which a nation might grow, and to hope it would grow there. But the legal question of the nature of the Union had been left in doubt and became a subject of controversy. The leading spokesmen on both sides were lawyers who confined themselves largely to drawing refined inferences from the exact wording of the Constitution and following every clue as to the intentions of the framers. In this kind of deductive reasoning, as it turned out, the defenders of state sovereignty had quite a strong case, made up essentially of five arguments:

First, at the time of the Articles of Confederation, proposed in 1777 and ratified in 1781, the states had explicitly included a statement that "each state retains its sovereignty, freedom and independence," and the treaty by which Britain recognized independence in 1783 named each of the thirteen states individually and acknowledged them to be "free, sovereign and independent states."[71]

Second, when the Constitution was framed in 1787, it was ratified by each state, acting separately and for itself only, so that the ratification of the requisite number of states (nine) would not have made any other state a member of the "more perfect union" under the Constitution unless that other state ratified.[72] It was true that the preamble said, "We the people of the United States . . . do ordain and establish this Constitution," and Daniel Webster, the Great Expounder of the Constitution and the great oracle of nationalism, had rung the changes on "We the people" as a proof that the citizens of all the states were merged into a consolidated Union.[73] But the term "people" was not used to indicate that one people instead of thirteen peoples were ratifying, but rather to distinguish between action by state governments and action by citizens exercising their ultimate power. Under the Articles, the central government had derived its power from the state governments and they in turn had derived their power from the people. Hence the central government could act only upon the state governments and not upon any citizens directly. But under the state constitutions *and* the Constitution of 1787, the people of each state (or the peoples of the thirteen states), by two separate acts, established for themselves two separate governments —a state government, operating locally for that state only, and a central government, operating collectively for the states together. Neither government had created the other; neither was subordinate to the other; they were coordinate governments, both sanctioned directly by the action of citizens, both operating directly on citizens without having to mediate through the machinery of the other gov-

71. Henry Steele Commager (ed.), *Documents of American History* (7th ed., 2 vols.; New York, 1963), I, 111, 117.

72. North Carolina did not ratify the Constitution until Nov. 21, 1789, nor Rhode Island until May 29, 1790, and neither was a member of the Union when Washington became president.

73. E.g., speech of Webster in Senate, Jan. 27, 1830, in *Register of Debates in Congress,* 21 Cong., 1 sess., cols. 74, 77. Webster argued that the powers conferred on the new government were perfectly well understood to be conferred, (1) not by any state, or (2) the people of any state, but (3) by the people of the United States. Since the only thing that made the Constitution applicable in any state was ratification by a convention of that state, it is very difficult to see any rational basis for Webster's rejection of (2).

ernment, and both subject to the ultimate authority, not of one or the other,[74] but of the constituencies which had established them. It was a truly dualistic system. This was the real implication of the term "We the people," and in the Convention the framers had originally planned a phrasing which would have avoided the confusion that later arose. They had agreed to list by name, one after another, "the people" of each of the thirteen states, severally, as the ordaining and establishing parties. But recognizing the awkwardness that would result if the Constitution should name as a member of the Union a state whose people later refused to ratify, they substituted the term "We the people of the United States," using it as a plural and not as a singular term.[75]

Third, the proceedings of the Convention showed clearly that its members had deliberately taken up the question of whether the federal government could coerce a state government, and had positively refused to confer any such power.[76]

74. It has been argued, of course, that this coordinate status is eliminated and federal ascendancy is established by Article VI of the Constitution, which declares that the Constitution and "the Laws of the United States which shall be made in pursuance thereof . . . shall be the Supreme Law of the Land." But under the theory of a dualistic system federal supremacy would not follow. Instead the argument would run that each state, having made two constitutions—one by itself, for local matters controlling only the state government, and the other jointly with other states, for general matters, controlling both state governments and the federal government—does not diminish its autonomy by providing that in cases of conflict between the two, the latter shall take priority. The "supreme law" clause gives a federal statute force only if it is "in pursuance of" the Constitution. Since both constitutions emanated from the same authority, the people of the states, acting severally in one case and jointly in another, the ultimate question was not which constitution should control in cases of conflict, but which government—federal or state—should act as arbiter for the people of the state in construing the constitution in question. Should each state act as its own arbiter, as Jefferson and Madison had argued in the Virginia and Kentucky resolutions of 1798; or should the federal judiciary act as arbiter, as in the cases of Fletcher *v.* Peck (1810) and Cohens *v.* Virginia (1821), when the Supreme Court asserted the power to declare the act of a state legislature void if contrary to the federal Constitution, and to reverse the decision of the highest court of a state?

75. Clinton Rossiter, *1787: The Grand Convention* (New York, 1966), p. 229; for the preamble as reported by the committee on detail, Aug. 6, 1787, and as reported back by the committee on style, see Charles C. Tansill (ed.), *Documents Illustrative of the Formation of the Union of the American States* (Washington, 1927), pp. 471, 989.

76. On May 29, 1787, Edmund Randolph introduced resolutions in the constitutional convention, including a provision that "the National Legislature ought to be empowered . . . to call forth the force of the Union against any member of the Union failing to fulfill its duty under the articles thereof." This was taken up on May 31. Madison was apparently the only speaker. He opposed it on the ground that "the use

Fourth, at the time of ratification, three states had specifically reserved their right to resume the powers which they were granting by their acts of ratification.[77]

Fifth, the continued integrity of the states was reflected in the structural features of the new government which provided that the states should be represented equally in the Senate, that the states alone could cast electoral votes for president, that the states alone could ratify amendments to the Constitution, and that, under the Tenth Amendment in the Bill of Rights, "The powers not delegated to the United States by the Constitution nor prohibited by it to the States are reserved to the states respectively, or to the people."[78]

From these arguments, the political theorists of the South had developed the doctrine of state sovereignty and, from it, of the right of secession. The Virginia and Kentucky Resolutions of 1798, written by Jefferson and Madison, had asserted state sovereignty and had declared each state to be "the judge . . . of the mode and measure of redress" in cases in which the federal government might violate the Constitution. In 1803, St. George Tucker of Virginia, in a treatise on the Constitution, had asserted that each state, "still sovereign, still independent . . . is capable . . . to resume the exercise of its functions to the most unlimited extent." Later, Spencer Roane and John Taylor of Virginia and Robert Y. Hayne of South Carolina, in his famous debates with Webster, all added cogent affirmations

---

of force against a state would look more like a declaration of war than an infliction of punishment, and would probably be considered by the party attacked as a dissolution of all previous compacts by which it might be bound." He moved to postpone, and the motion was adopted unanimously. Tansill, *Documents,* pp. 117, 131. Andrew C. McLaughlin, *A Constitutional History of the United States* (New York, 1935), p. 598, offers an argument that the action on this resolution did not mean what it appears to mean.

77. Virginia's ratification, June 27, 1788, specified that "the powers granted under the Constitution being derived from the people of the United States may be resumed by them whensoever the same shall be perverted to their injury or oppression." New York's ratification, July 26, specified, "We the Delegates declare and make known that the Powers of Government may be resumed by the People, whensoever it shall become necessary to their happiness"; Rhode Island, May 29, 1790, adopted the same provision as New York. Tansill, *Documents,* pp. 1027, 1034, 1052.

78. It became a truism that the government under the Constitution was neither wholly federal nor wholly national, but mixed. The best analysis of the nature of this mixture is by Madison in *The Federalist,* number 39. William Paterson declared that "as in some respects, the States all to be considered in their political capacity, and in others as districts of individual citizens the two ideas—instead of being opposed to each other, ought to be combined; that in *one* branch the people ought to be represented; in the other, the states." Tansill, *Documents,* p. 297.

of state sovereignty. In 1832, Calhoun's state papers on Nullification gave classic formulation to the same doctrine. Calhoun did not want to secede, and did not emphasize a doctrine of secession, but he made explicit his view that a state's ultimate recourse was to withdraw from the Union. In 1840, Abel P. Upshur of Virginia published a treatise which a modern critic has called "perhaps the strongest historical analysis for the support of state sovereignty . . . ever . . . written." Three years later, Henry St. George Tucker, a law professor like his father, compiled the existing arguments and added some of his own. Before his death in 1850, Calhoun again discussed the nature of the Union in his *Discourse on the Constitution.* By this time, the doctrine of secession had become, for a majority of politically minded southerners, a fundamental tenet of southern orthodoxy.[79]

There were, to be sure, a good many southerners who preferred to claim instead a right of revolution, as asserted in the Declaration of Independence. But the southern majority had committed itself to the right of secession during the crisis of 1846–1850, and James M. Mason, in 1860, was able to say, "Fortunately for the occasion and its consequences, this is not an open question in Virginia. Our honored state has ever maintained that our Federal system was a confederation of sovereign powers, not a consolidation of states into one people. . . . Whenever a state considered the compact broken, and in a manner to endanger her safety, such state stood remitted, as in sovereign right, to determine for herself . . . both the mode and measure of redress."[80]

Against the defenders of this doctrine, the defenders of nationalism did not come off as well as they might have, partly because they accepted the assumption that the nature of the Union should be determined by legal means, somewhat as if it were a case in the law

79. The historical literature on the development of the doctrine of secession is scanty. A brief but very able account is Carpenter, *The South as a Conscious Minority,* pp. 171–220. Also see Ulrich Bonnell Phillips, "The Literary Movement for Secession," in *Studies in Southern History and Politics Inscribed to William Archibald Dunning* (New York, 1914), pp. 33–60; William E. Dodd, "John Taylor: Prophet of Secession," in *John P. Branch Historical Papers,* 1908 (Ashland, Va., 1908), pp. 214–252; articles and correspondence of Spencer Roane, *ibid.,* 1905, pp. 51–142, and 1906, pp. 78–183.

80. Carpenter, *South as a Conscious Minority,* pp. 194–200, discusses the preference of some southerners for claiming a right of revolution rather than a right of secession. The quotation from Mason, taken from Richmond *Enquirer,* Nov. 23, 1860, is at p. 200.

of contracts. Yet in fact, the nature of the Union had been changing constantly, as the states increased in number until those which had created the Union were outnumbered by those which the Union had created. Between 1804 and 1865, the Constitution was not once amended, the longest such interval in American history. But while the text of the charter remained the same, the republic itself was transformed.

A thousand forms of economic and cultural interdependence had developed. Such changes do not occur without corresponding changes in the attitude of the people, and in a century of rampant nationalism throughout the Western world, there were probably no people who carried national patriotism and self-congratulation to greater lengths than the Americans, and this included the South. Regardless of arrangements made in 1787, nationalism changed the nature of the Union and began to answer the riddle of *pluribus* or *unum.* But nationalism grew at different rates and in different ways in North and South, and by 1860, the sections found themselves separated by a common nationalism. Each was devoted to its own image of the Union, and each section was indistinctly aware that its image was not shared by the other. The South had no idea how ruthlessly its northern Democratic allies were prepared to deal with anyone who tried to tamper with the Union. The North had no idea how fiercely southern Unionists who valued the Union for themselves would defend the right of other southerners to reject it for *them*selves and to break it up without being molested.

The dual focus of southern loyalties, even as late as 1860, has led one author to say very aptly that the South by then had become a kingdom, but that it did not become a nation until thrust into the crucible of the Civil War.[81] Within this kingdom there was sharp disagreement between the advocates of a southern Confederacy and those who favored remaining in the Union. Yet underneath the disagreement was consensus on two important points. Most southern Unionists shared with secessionists the conviction that no state should be forced to remain in the Union, and most of them also believed in secession as a theoretical right. Whether it was justified or opportune could still be debated. But for southerners generally, the *right* of a state to secede, if it chose to do so, had become an article of faith.

81. Henry Savage, Jr., *Seeds of Time: The Background of Southern Thinking* (New York, 1959), pp. 49–136.

CHAPTER 18

The Lower South Secedes

ELECTION day in 1860 fell on November 6, and by midnight the public knew that Lincoln had been elected. On November 8, the Charleston *Mercury* announced, "The tea has been thrown overboard; the revolution of 1860 has been initiated."[1]

But if the analogy was appropriate, what had happened at this point was that the tea had been brought into Charleston harbor. It remained to be seen whether anyone would throw it overboard, and if so, when and how. Lincoln's election forced southerners to come to grips with questions they had failed to resolve for almost a generation. It provoked an internal crisis in the South before secession presented a crisis for the nation.

Southerners might agree that they shared a common culture, faced a common enemy, and had dire need of a common defense. They might also agree in accepting as sound doctrine the right of secession or the right of revolution. But the southern consensus ended abruptly at the point of transition from generalities to specifications. There was continuing disagreement, of course, over whether the South could defend itself better inside or outside the Union. There would always be disagreement over whether the *time had come* for secession, however justifiable it might be in theory. But even beyond those hurdles there was the highly controversial question of how to execute a separation from the Union without placing intolerable stresses upon their own regional unity. If each of the

1. Quoted in Charles Edward Cauthen, *South Carolina Goes to War, 1860–1865* (Chapel Hill, 1950), p. 30.

southern states acted separately, they might take divergent courses and become estranged from one another, in which case, isolation could lead to impotence. But if they waited to act together, the inertia of some might become the paralysis of the others, and they might never take any initiative at all.

This dilemma of separate state action or cooperation, as the alternatives were called, had become familiar and painful by 1860. Technically, the problem was not a difficult one, for the southern states needed only to meet together in a convention, decide in concert what to do, and then individually execute the decision. The real problem, however, was not procedure but how to set it in motion and, in the case of South Carolina, whether to act alone or wait for the other southern states. From painful experience, South Carolina knew the risks in both alternatives. In 1832 she had acted alone in nullifying the tariff, had been left in solitude by the other southern states, and had been brought to heel by Andrew Jackson, though not without some gratifying concessions on the tariff. In February 1851, she had again made a move toward separate state action, but this time more warily, by electing a state convention committed to voting for secession *after* the meeting of a proposed southern convention. But the southern convention never met, and the disunion movement withered away. Yet, if separate action had proved futile, cooperative action had succeeded no better. In 1848–1849, Calhoun had failed to get united support from southerners in Congress for his "Southern Address." In 1850, the Nashville Convention had proved an obstacle to secession rather than an instrument for it. Further, South Carolina had experimented in vain with no fewer than three efforts to operate the machinery of cooperation in the ten months preceding Lincoln's election.

During the late fifties, South Carolina's fire-eating tendencies appeared to have burned themselves out, and a faction of "National Democrats," headed by James L. Orr, gained the ascendency.[2] But in the winter of 1859–1860, after Harpers Ferry and in the midst of the bitter contest over the speakership of the House, the old impulses flared up again. On December 22, 1859, the state legislature voted to send a special commissioner to Virginia with a proposal

2. Joel H. Silbey, "The Southern National Democrats, 1845–1861," *Mid-America*, XLVII (1965), 176–190. Laura A. White, "The National Democrats in South Carolina, 1852 to 1860," *SAQ*, XXVIII (1929), 370–389.

that the two states cooperate in measures of common defense, and also to invite other southern states to initiate a conference for the purpose of considering their common dangers and of planning common action.[3] Both of these proposals clearly looked to the possibility of cooperative steps leading to secession.

The governor of the state appointed Christopher C. Memminger as commissioner to Virginia. Not one of South Carolina's fire-eaters and therefore more acceptable as an envoy, Memminger left for Virginia on January 11, was received with full courtesy by Governor John Letcher, and addressed the Virginia legislature on January 19. His speech cautiously avoided any plain talk about secession and focused instead upon urging Virginia to participate in the conference of southern states which the Carolina legislature had proposed. He did, however, suggest that "if the worst must come, and we must take our destinies into our own hands, a Southern conference is the necessary step to such arrangements as are requisite to take our place among the nations of the earth." After this address he lingered in Richmond for nearly three weeks, hoping that the legislature would respond to his appeal, but he found that two prominent Virginians, Robert M. T. Hunter and Henry A. Wise, both hoped to receive the Democratic presidential nomination two months later. He concluded that the supporters of both men wanted to avoid jeopardizing the availability of their candidates by doing anything that might smack of disunion, and he departed without an answer.[4]

South Carolina, acutely conscious that other southern states distrusted her disunionist proclivities, had modestly refrained from naming either a time or a place for the conference. Therefore, in February, Mississippi proposed a meeting at Atlanta in June. But in March the legislatures, first of Virginia and then of Tennessee, declined to participate; expressions of support from Florida and Alabama came to naught; and the conference never materialized.[5]

These were two rebuffs to cooperation. A third one followed the

3. Steven A. Channing, *Crisis of Fear: Secession in South Carolina* (New York, 1970), pp. 102, 112.

4. Ollinger Crenshaw, "Christopher G. Memminger's Mission to Virginia, 1860," *JSH*, VIII (1942), 334–349; Henry D. Capers, *The Life and Times of C. G. Memminger* (Richmond, 1893), with text of Memminger's address at pp. 247–278; Channing, *Crisis of Fear*, pp. 17–18, 112–130.

5. Robert W. Dubay, "Mississippi and the Proposed Atlanta Convention of 1860," *Southern Quarterly*, V (1966–67), 347–362.

southern bolt from the Democratic convention at Charleston in April. After the main convention adjourned to Baltimore, to meet on June 18, the bolting delegations from eight states agreed to meet at Richmond on June 11. To the Democrats of South Carolina, the division at Charleston had been final. They expected to organize a party of southern Democrats at Richmond and had no thought of going to Baltimore to reenter the organization from which they had so dramatically withdrawn a few weeks earlier. They urged the other delegations also to remain aloof and were shocked when six of them—from Georgia, Alabama, Mississippi, Louisiana, Texas, and Arkansas—left the Richmond convention for Baltimore, where even the Yancey group clamored to be readmitted to the "regular" convention. Once more, the apparent unity of a southern group had quickly evaporated, leaving South Carolina almost isolated. Virginia had rebuffed her completely; only Mississippi had supported her call for a southern convention; and now only Florida had stood with her in refusing to "crawl" back into the national Democratic conclave.[6]

In October, South Carolina made one more attempt at cooperation when Governor William H. Gist wrote to the governors of the other states of the deep South: "It is the desire of South Carolina that some other state should take the lead, or at least move simultaneously with her. She will unquestionably call a convention as soon as it is ascertained that a majority of the electors will support Lincoln. If a single state takes the lead, she will follow her. If no other state secedes, South Carolina will secede (in my opinion) alone if she has any assurance that she will be soon followed by another, or other states; otherwise, it is doubtful."[7]

The only satisfactory answer came from the governor of Florida, who said that his state would not take the lead but would "assuredly . . . follow the lead of any single Cotton State" that might secede. Alabama and Mississippi were quite ready to resist Black Republican control, ready perhaps to follow one or two other southern states, ready also to resist federal coercion of any state, but both spoke favorably of a southern convention, which must have seemed ominous indeed to the South Carolinians. Georgia, Louisiana, and

6. Cauthen, *South Carolina Goes to War*, pp. 20–25.

7. Gist's letter in John G. Nicolay and John Hay, *Abraham Lincoln: A History* (10 vols.; New York, 1890), II, 306–307.

North Carolina gave virtually no encouragement, and did not believe that the election of Lincoln alone would be regarded as grounds for secession. If Lincoln, as president, should commit an overt act, that would be different.[8]

On the day of Lincoln's election, efforts to create a united South appeared to have resulted in complete failure. Ever since Harpers Ferry, southern orators and editors had been proclaiming that the election of a Black Republican president would be the signal for southern action.[9] Yet now the trumpet voices were muted, and people were talking about waiting for consultation, or for "overt acts." No mouse was ready to bell the cat, and the South would be exposed once again, quite justifiably, to the humiliating charge of bragging and blustering but doing nothing.[10]

Even within South Carolina itself, profound divisions existed. All of the state's congressmen had announced in advance their support of secession in response to Lincoln's election,[11] but one of the senators, James Chesnut, was saying nothing,[12] and the other, James H. Hammond, once a fire-brand of southern rights, was privately sabotaging the immediate secessionists. For more than a year, he had felt a growing skepticism about the willingness of the southern people to secede. In 1858 he had privately written that "999 in every 1000" of southern voters would go for the Union until it pinched them, and that "with cotton at .10¢ and negroes at $1000," the South would know no pinch.[13] Two days after Lincoln's election

8. *Ibid.*, II, 307–314. My reading of this correspondence does not agree with that of Cauthen, *South Carolina Goes to War*, p. 52, which says that the replies "were on the whole very reassuring."

9. These were, of course, renewals of threats made in 1856. See above, p. 262.

10. The sensitivity of South Carolinians to their reputation for threatening without acting is ably discussed in Laura A. White, *Robert Barnwell Rhett, Father of Secession* (New York, 1931), pp. 177–178, quoting William Porcher Miles to James Henry Hammond, Aug. 5, 1860, "I am sick and disgusted with all the bluster and threats and manifestoes and 'Resolutions' which the South has for so many years been projecting and hurling with such force at the devoted heads of 'our base oppressors' " (see also speech of Miles at Charleston, quoted in Cauthen, *South Carolina Goes to War*, p. 69, n. 36); White also quotes the British consul at Charleston, "It would appear certain that South Carolina must either secede at all hazards, on or before the inauguration of Mr. Lincoln, or be content to have herself exhibited to the ridicule of the world."

11. White, *Rhett*, p. 172, gives the exact dates at which each of the congressmen announced in favor of secession.

12. *Ibid.*, p. 175; Cauthen, *South Carolina Goes to War*, pp. 49–50, 53–54.

13. Hammond to William Porcher Miles, Nov. 23, 1858, quoted in Channing, *Crisis of Fear*, p. 144.

he sent a letter to the South Carolina legislature asserting that the position of the South in the Union was by no means desperate, advising against rash remedies, and warning that the other southern states would leave South Carolina in the lurch.[14]

The South Carolina legislature met on November 5, 1860, to cast the state's electoral votes. Immediately after learning the election result, it passed a bill providing for a state convention to be elected on January 8 and to meet on January 15. The significance of the measure lay in the dates. South Carolina would wait two months for some other state to act first, at the risk that none might do so, in which case the whole secession impulse would be dissipated. The immediate secessionists had reluctantly agreed to these dates because of the urgent need for harmony within the state and because of their fear of seeing their state left isolated once again.[15]

But on November 9 a decisive change occurred which may have altered the course of history. False information had reached Columbia that Robert Toombs of Georgia had resigned from the Senate. Correct information had arrived that the governor of Georgia had urged his legislature to call a state convention. Senator Chesnut, in a speech at Columbia, had abandoned his equivocation, announced his support of secession, and offered to drink all the blood that might be shed as a result of secession.[16] At this crucial point, a large delegation of Georgians arrived in Charleston from Savannah to celebrate the completion of a railroad between the two cities, and on the night of November 9 a grand secession rally took place, and delegates were elected to bring pressure on the South Carolina legislature for a convention "at the earliest possible moment." Large secession rallies had already taken place at Montgomery and Mobile in Alabama, and at Jackson, Mississippi.[17] The seces-

14. Quoted in White, *Rhett,* pp. 178–179.

15. *Ibid.*, p. 179.

16. Chesnut was not alone. In Feb. 1861, A. W. Venable of North Carolina promised to "wipe up every drop of blood shed in the war with this handkerchief of mine." Quoted in Joseph Carlyle Sitterson, *The Secession Movement in North Carolina* (Chapel Hill, 1939), p. 218. E. Merton Coulter, *The Confederate States of America, 1861–1865* (Baton Rouge, 1950), p. 15, quotes, as a common saying, "A lady's thimble will hold all the blood that will be shed."

17. Cauthen, *South Carolina Goes to War*, p. 58; John Witherspoon Du Bose, *The Life and Times of William Lowndes Yancey* (2 vols.; Birmingham, Ala., 1892), II, 539; Percy Lee Rainwater, *Mississippi: Storm Center of Secession, 1856–1861* (Baton Rouge, 1938), pp. 198–200 lists 25 county meetings between Nov. 8 and Nov. 24, at all of which resolutions were adopted advocating some form of resistance.

sionists of South Carolina decided that this was as good a chance as they would ever have, and on November 10 they hastily put through both houses a bill for the election on December 6 of a convention to meet on December 17. South Carolina would, once again, take the risk of unilateral action.[18]

There can be little doubt that the speed of South Carolina's action gave crucial encouragement to secessionists throughout the South and accelerated the tempo of the disunion movement in a decisive way. The first results appeared in Alabama, where the legislature had previously enacted a bill authorizing the governor to call an election of a state convention if a Republican should be elected to the presidency. After November 6, the governor suddenly became coy and suggested that he could not act under this authorization until Lincoln had been formally chosen by the electoral college or perhaps even until the electoral votes had been counted by Congress. But under pressure from Alabama citizens, he announced on November 14 that on December 6 (after the electoral college vote), he would call an election, December 24, for a convention to meet January 7.[19] Also on November 14, the governor of Mississippi called the legislature into extraordinary session on November 26. When it met, he recommended a state convention, and on November 29 the legislature voted for one, to be elected on December 20 and to meet on January 7. Meanwhile, the Georgia legislature had already adopted a bill on November 18, calling for a convention to be elected on January 2 and to meet on January 16. On November 22 the governor of Louisiana called a special session of the legislature, to meet on December 10. The Florida legislature on November 28 passed an act calling a convention. In Texas, the secessionist timetable was upset by Governor Sam Houston, who opposed disunion and refused to call the legislature into session. Otherwise the cotton states had moved with a rapidity that no one could have foreseen. Within twenty-three days after Lincoln's election, five of them had called state conventions and a sixth (Louisiana) had sum-

18. Charles Edward Cauthen, "South Carolina's Decision to Lead the Secession Movement," *NCHR*, XIX (1941), 360–372; Cauthen, *South Carolina Goes to War*, pp. 49–61.

19. Clarence Phillips Denman, *The Secession Movement in Alabama* (Montgomery, Ala., 1933), pp. 89–92. On the alarm in South Carolina because of the uncertainty about the date of meeting of the Alabama convention, see White, *Rhett*, p. 175, n. 39.

moned a special session of the legislature for the purpose of calling such a convention. None of them had made any demands for a prior meeting of the southern states in a general convention.

South Carolina took the lead at each stage, however. Her convention was elected on December 6 and assembled on December 17, before any other state had held an election. On December 20 the convention unanimously adopted an ordinance of secession.[20] On that same day, Mississippi elected her convention; two days later the Florida election took place; and two days after that, the election in Alabama. Again, there can be little doubt that the swift action of South Carolina had encouraged the secessionists elsewhere and increased their support against Unionists and advocates of delay.

Another important element of support came from Washington on December 14. Congress had met on December 3, and a day later the House appointed a Committee of Thirty-three (i.e., one from each state) to consider "the present perilous condition of the country" —which meant, in effect, to consider proposals for compromise. But the appointment of a similar committee in the Senate was delayed by acrimonious debate; thirty-eight Republicans cast the only votes in opposition to the House committee; the committee did not meet until December 11; and on December 13, Republican members divided eight to eight on a resolution that whether the discontent of the South was justified or not, "guarantees of their peculiar rights and interests as recognized by the Constitution . . . should be promptly and cheerfully granted."[21] That night, seven senators and twenty-three representatives from nine southern states issued a public address to their constituents stating, "The argument is ex-

20. On secession in South Carolina, Chauncey Samuel Boucher, *South Carolina and the South on the Eve of Secession, 1852 to 1860,* in "Washington University Studies," Humanistic Series, VI (1919), 85–144; White, *Rhett;* Lilliam Adele Kibler, *Benjamin F. Perry, South Carolina Unionist* (Durham, N.C., 1946); Cauthen, *South Carolina Goes to War;* Harold S. Schultz, *Nationalism and Sectionalism in South Carolina, 1852–1860* (Durham, N.C., 1950); Channing, *Crisis of Fear,* does not deal with the actual procedure of secession, but is excellent. The texts of the convention's "Address of the People of South Carolina" (by Rhett) and its "Declaration of the Causes which Justify the Secession of South Carolina" (by Memminger) are in Edward McPherson (ed.), *Political History of the United States . . . during the Great Rebellion* (Washington, 1876), pp. 12–16.

21. Establishment of Committee of Thirty-three, *Congressional Globe,* 36 Cong., 2 sess., p. 6; proceedings in *House Reports,* 36 Cong., 2 sess., No. 31 (Serial 1104), "Journal of the Committee of Thirty-three." For analysis of membership and votes, David M. Potter, *Lincoln and His Party in the Secession Crisis* (New Haven, 1942), pp. 89–98.

hausted. . . . We are satisfied the honor, safety, and independence of the Southern people are to be found only in a Southern Confederacy—a result to be obtained only by separate State secession."[22] Robert Toombs did not sign this communication, but ten days later, after the Senate committee had been created and had disappointed his expectations, he issued a manifesto of his own to the people of Georgia. The committees, which might have been agencies of compromise, he said, were "controlled by Black Republicans, your enemies, who only seek to amuse you with delusive hope. . . . I tell you upon the faith of a true man that all further looking to the North for security for your constitutional rights in the Union ought to be instantly abandoned. . . . Secession, by the fourth of March next, should be thundered from the ballot-box by the unanimous voice of Georgia on the second day of January next."[23]

Thus, even before southern voters outside of South Carolina had cast their ballots on the secession issue by voting for delegates to the conventions, the secession process had already gained substantial momentum. Theoretically, each southern state was acting independently, but in fact there was already a network of commissioners who maintained liaison between the states, and the southern members of Congress, meeting frequently in caucus, served as a kind of ready-made coordinating body to assure that the disparate action of the several states would converge as a "stroke for national independence."[24] In South Carolina, Robert Barnwell Rhett had proposed inviting other southern states to a conference to form a government and had suggested Montgomery as the place. On December 31 the South Carolina convention voted to elect commissioners to every other southern state which called a convention and authorized them to propose a meeting for the erection of a provisional government. On January 3 the commissioners accordingly proposed a meeting at

22. Text in McPherson, *History of the Rebellion*, p. 37.

23. Text, *ibid.*, pp. 37–38.

24. For a list of the commissioners of South Carolina, Georgia, Alabama, and Mississippi to other states, *ibid.*, p. 11; on the role of the commissioners, Dwight L. Dumond, *The Secession Movement, 1860–1861* (New York, 1931), pp. 134 n., 135–136, 195–196; on the coordinating activities of southern congressmen at Washington, Roy F. Nichols, *The Disruption of American Democracy* (New York, 1948), pp. 392, 399–404, 436, 441, 446–448. Ulrich Bonnell Phillips, "The Literary Movement for Secession," in *Studies in Southern History and Politics Inscribed to William Archibald Dunning* (New York, 1914), p. 59, said, "State sovereignty was used to give the insignia of legality to a stroke for national independence."

Montgomery on February 4. This proposal was the cornerstone of the Confederate States of America, and it was laid before the convention of any southern state, other than South Carolina, had even met, though the Florida convention did assemble on the same day.[25]

Between December 20 (the day South Carolina seceded) and January 8, the voters of six other southern states elected delegates to the conventions that would decide whether they should remain in the Union. Party labels were not used in these elections, and the citizens of the southern states, who were fairly united in their abstract commitment to the defense of southern rights, suddenly found themselves facing the concrete question of how their rights should be defended at the operative level. They tended to gravitate toward one of two positions, which acquired the labels of "immediate secession" and "cooperation." This antithesis was not at all the same as simple disunionism versus unionism, for there were very few voters in the South who wanted disunion for its own sake—almost all would have preferred to remain in the Union with satisfactory guarantees, which, however, they had no hope of receiving; and there were equally few Unionists whose loyalty to the Union took an unqualified priority over the defense of southern rights. The alignment therefore took the form of a division between those who believed prompt secession by the states separately was necessary to the defense of southern rights and those who believed that southern rights could best be defended by all of the slaveholding states acting in concert, through a southern conference, first to present collective demands which the North, faced with such a solid phalanx, might concede, or, failing that, to secede with a degree of southern unity which would assure their success.

In some respects, the two groups seemed fairly close together; in other respects, far apart. Both avowed their commitment to southern rights and to the use of secession if necessary as a device to secure these rights—in this sense, their disagreement seemed merely tactical. But at an operative level, the secessionists were ready to act, using the machinery of the state conventions to implement their decision; the "cooperationists" had not decided to secede, and were consistently opposed to secession by the means available and at hand. This meant that, in fact, the secessionists were all somewhat united on one clear program; but the cooperationists

25. Cauthen, *South Carolina Goes to War,* pp. 84–85; Armand J. Gerson, "The Inception of the Montgomery Convention," AHA, *Annual Report,* 1910, pp. 179–187.

represented a spectrum of positions ranging from genuine secessionism, firmly linked to the belief that action through a southern convention was the best policy, to strong Unionism masquerading as cooperative secessionism for tactical reasons. It is, of course, impossible to determine the distribution of voters along this spectrum. Some historians, emphasizing the theoretical willingness of both groups to secede—one by separate state action, the other by action coordinated through a southern convention—regard them basically as two different varieties of secessionists, and see the entire lower South as overwhelmingly secessionist. But viewed instrumentally, "separate state secessionists" and "cooperationists" were far apart. The latter set, as a prerequisite to secession, such a high degree of unanimity among the southern states that to their opponents, they seemed unwilling to secede at all. In short, the secessionists regarded the cooperationists not simply as more prudent secessionists, but as Unionists who found the cause of union too unpopular to support outright, and who resorted to "cooperation" as an obstructive device to prevent any action.[26]

The election canvasses in which these partially indeterminate groups opposed one another did little to clarify the distinctions, for they turned out to be unsystematic and rather poorly coordinated affairs in which there were no regular party divisions to give an overall pattern to the contests and no centralized secessionist or cooperationist organizations to put statewide tickets into the field. As a result, local districts approached the issue in various ways: sometimes with opposing slates; sometimes, in an effort at harmony, with mixed slates; sometimes with the candidates committed to a position, but in other cases with them running simply as influential local leaders who would make up their minds as events developed; and sometimes with secessionists or cooperationists, as the case might be, running unopposed.

All these circumstances have made it remarkably difficult to evaluate either the campaigns or their results. It seems clear that the

26. The problem of the meaning of cooperationism is highly significant, for whether secession was a popular movement depends very much upon whether cooperation was primarily a prudent form of secessionism or an intimidated form of unionism. See Dumond, *Secession Movement*, pp. 121–134; Denman, *Secession Movement in Alabama*, pp. 93–115; Rainwater, *Mississippi*, pp. 180–193; White, *Rhett*, pp. 173–174. Willie Malvin Caskey, *Secession and Restoration of Louisiana* (University, La, 1938), pp. 20–40, quotes the New Orleans *Daily Crescent*, Jan. 5, 1861: "Here in New Orleans nobody knows exactly what cooperation means. With some it means delay, with some conference with other states, with some it means submission."

campaigns took place in an atmosphere of steadily rising excitement. While they were in progress, vigilance organizations were being formed to guard against nefarious abolitionist schemes; military companies with picturesque names were organizing and arming; flags were being sewn; young men, especially, were reveling in warlike preparations; cooperationists were denouncing secessionists for their reckless destruction of priceless union; and secessionists were retorting with accusations that the cooperationists were old women and abject "submissionists." It appears that the cooperationists lost ground during most of these campaigns.[27] But the election returns, so far as they can be analyzed, show that in a number of states the results were remarkably close. This was less true of Mississippi and of Florida than of the states which followed, but even in these two, the cooperationists made a strong showing. In Mississippi, on December 20, about 41,000 votes were cast, of which some 12,000 were for candidates whose positions were not specified or are now unknown, but of the remaining 29,000, some 16,800 were for secessionists and 12,218 for cooperationists. In Florida, two days later, the cooperationists showed a strength amounting to between 36 percent and 43 percent of the vote. In Alabama, two days after that, the secessionists cast 35,600 ballots, but the cooperationists cast 28,100.[28]

In the January elections, the secessionists won—if they won at all—by even narrower margins. In Georgia, on January 2, their advantage was only 44,152 to 41,632, at the most generous estimate, and there is good evidence that the cooperationists may have held a very narrow majority.[29] In Louisiana, on January 7, the secessionists

27. For general conditions and developments accompanying secession, see Dumond, *Secession Movement,* and the works on secession in various states, cited in footnotes 20, 31, 32. For the secession conventions, the standard work is Ralph A. Wooster, *The Secession Conventions of the South* (Princeton, 1962).

28. Rainwater, *Mississippi,* pp. 196–200; Dorothy Dodd, "The Secession Movement in Florida, 1850–1861," *Florida Historical Quarterly,* XII (1933–34), 3–24, 45–66, esp. pp. 52–54.

29. On April 25, 1861, Governor Joseph E. Brown, in response to an inquiry, declared that "the delegates to the convention who voted for the ordinance of secession were elected by a clear majority . . . of 13,120 votes, or 50,243 for secession and 37,123 against secession." Historians have relied on these figures ever since. The reader will note, however, that Brown did not say that 50,243 votes were cast for candidates pledged to secession; he only says that they were cast for candidates who later voted for secession. But this figure is misleading because many delegates who had voted against immediate secession on preliminary votes decided on the final vote to acquiesce in the will of the majority, for the sake of harmony and solidarity.

prevailed by 20,214 to 18,451.[30] In Texas, the whole procedure was irregular, for Governor Sam Houston had refused to convene the legislature, and the election of delegates to a convention was formally called for by an informal group of secession leaders at Austin. Voting, beginning on January 8, apparently took place on various days in various places, and it is by no means certain that the cooperationists were always encouraged to put up a candidate. The resulting convention adopted an ordinance of secession by a vote of 166 to 8, but, perhaps doubting its own legitimacy, it then submitted the ordinance to the voters for ratification. In this election the secessionists won a more sweeping majority than they had gained anywhere except in South Carolina. The vote was 44,317 to 13,020.[31]

The problem is, first, to learn how voters voted in cases where a known secessionist was running against a known nonsecessionist; and second, in the numerous cases where the candidates' positions are not thus known, to learn the vote recorded for delegates who later voted for or against secession on preliminary test votes. Michael P. Johnson, "A New Look at the Popular Vote for Delegates to the Georgia Secession Convention," *Georgia Historical Quarterly*, LVI (1972), 259–275, has made a full analysis on the basis of these two criteria, arriving at the figures stated above. As a conclusion to this important analysis, Johnson states (p. 270): ". . . the most generous estimate of the popular sentiment for secession is just over 51% of those voting. An estimate that is probably more accurate places the majority for cooperation at just over 50% of the voters."

30. The full returns for the election of the convention in Louisiana were turned over to the convention when it met, but were not published. As early as Feb. 1861, public accusations were being made that the returns were being suppressed. In response to these accusations the New Orleans *Daily Delta*, March 27, 1861, published an inadequate record, which, in the absence of anything better, historians used from then until 1970. It gave a tally of 20,448 for the secessionists and 17,296 for cooperationists. Meanwhile, the records had been seized by federal authorities in 1865 and taken to Washington. They were in the War Department until 1934 and in the National Archives from then until 1961, at which time they were returned to Louisiana. Charles B. Dew has resurrected and analyzed these records, the interpretation of which is complicated by the overlap in votes between senatorial districts and representative districts. Dew concludes that "the actual returns show that in the representative races the vote was 20,557 secessionist versus 18,651 co-operationist, a difference of 1,906. When the totals are figured on the basis of the senatorial districts, probably the fairest tests, since there were fewer unopposed candidates in these races, . . . the results show the secessionists winning 20,214 votes, the co-operationists 18,451, for a radical majority of only 1,763." Dew, "The Long Lost Returns: The Candidates and Their Totals in Louisiana's Secession Election," *Louisiana History*, X (1969), 353–369; Dew, "Who Won the Secession Election in Louisiana?" *JSH*, XXXVI (1970), 18–32.

31. On secession in Texas, the fullest treatment is Edward R. Maher, Jr., "Secession in Texas" (Ph.D. dissertation, Fordham University, 1960). See also Maher, "Sam Houston and Secession," *SWHQ*, LV (1952), 448–458; Earl Wesley Fornell, *The Galveston Era: The Texas Crescent on the Eve of Secession* (Austin, 1961), esp. pp. 267–302; Wooster, *Secession Conventions*, pp. 121–135, a valuable revision of previous analyses;

Under the American system of majority representation rather than proportional representation, a narrow popular majority is often translated into a heavy majority in the body being elected, and this was true of the state conventions which met between January 3 and January 28. The immediate or "straight-out" secessionists controlled them all, and the procedures followed were somewhat similar. Commissioners from other states delivered addresses urging secession. In Alabama and Louisiana, messages from the state's representatives in Congress asserted that the Republicans refused to make any concession. In all of the states except Texas, the cooperationists sought to pass measures calling for some sort of southern conference to make final demands upon the Republicans or to arrange for concerted action by the southern states—or both; and in every state except Georgia, the cooperationists also made efforts to refer any act of secession to the voters for ratification. It was on this sort of question that the opponents of immediate secession developed their maximum strength, but even so, they were defeated in every state: in Mississippi, 74 to 25; in Florida, 39 to 30; in Alabama, 54 to 46; in Georgia, 164 to 133; and in Louisiana, 84 to 43. Texas alone voted to submit the ordinance to popular ratification, and this decision was made by the secessionists themselves (145 to 29) and was not forced upon them by the opposition. After these skirmishes, when the final votes on secession were taken, the cooperationist delegates showed their strong belief in the importance of presenting a united front, and the majorities in favor of secession were overwhelming. Mississippi, the second state to secede, did so on January 9 by a vote of 85 to 15; Florida on January 10, by 62 to 7; Alabama on January 11, by 61 to 39; Georgia on January 19, by 208 to 89; Louisiana on January 26, by 113 to 17; and Texas on February 1, by 166 to 8. Within a space of forty-two days, seven states, from South Carolina to Texas, had seceded.[32] They all accepted the invitation of

---

Ben H. Procter, *Not Without Honor: The Life of John H. Reagan* (Austin, 1962), pp. 118–129; Llerena Friend, *Sam Houston, The Great Designer* (Austin, 1954), pp. 321–354; Anna Irene Sandbo, "Beginnings of the Secession Movement in Texas," *SWHQ*, XVIII (1914), 51–73; Charles William Ramsdell, "The Frontier and Secession," in *Studies in Southern History and Politics Inscribed to William Archibald Dunning* (New York, 1914), pp. 61–79; Ramsdell, *Reconstruction in Texas* (New York, 1910), pp. 11–20; Charles A. Culberson, "General Sam Houston and Secession," *Scribner's Magazine*, XXXIX (1906), 586–587.

32. On secession in Mississippi, see Rainwater, *Mississippi;* Rainwater, "Economic Benefits of Secession: Opinions in Mississippi in the 1850's," *JSH*, I (1935), 459–474.

the South Carolina commissioners to meet at Montgomery on February 4 (the Texas delegates did not arrive until later). On February 7 these delegates, with full powers from their states, adopted for the Confederate States of America a provisional constitution based on the Constitution of the United States. On February 9 they elected Jefferson Davis as president, and on February 18 they inaugurated him. Forty days after Mississippi followed South Carolina out of the Union, the Southern Republic was in existence.[33]

On Florida: Dorothy Dodd, "The Secession Movement in Florida"; Dodd (ed.), "Edmund Ruffin's Account of the Florida Secession Convention, 1861: A Diary," *Florida Historical Quarterly*, XII (1933), 67–76; John E. Johns, *Florida During the Civil War* (Gainesville, 1963), pp. 1–22; John F. Reiger, "Secession of Florida from the Union: A Minority Decision?" *Florida Historical Quarterly*, XLVI (1968), 358–368; Herbert J. Doherty, Jr., *Richard Keith Call, Southern Unionist* (Gainesville, 1961), pp. 154–160; Arthur W. Thompson, "Political Nativism in Florida, 1848–1860: A Phase of Anti-Secessionism," *JSH*, XV (1949), 39–65.

On Alabama: Denman, *Secession Movement in Alabama;* Lewy Dorman, *Party Politics in Alabama from 1850 through 1860* (Wetumpka, Ala., 1935); David L. Darden, "The Alabama Secession Convention," *Alabama Historical Quarterly*, III (1941), 269–451; William Brantley, "Alabama Secedes," *AR*, VII (1954), 165–185; Hugh C. Bailey, "Disloyalty in Early Confederate Alabama," *JSH*, XXIII (1957), 522–528; Bailey, "Disaffection in the Alabama Hill Country, 1861," *CWH*, IV (1958), 183–193; William Stanley Hoole, *Alabama Tories* (Tuscaloosa, Ala., 1960), Durward Long, "Unanimity and Disloyalty in Secessionist Alabama," *CWH*, XI (1965), 257–273, is a good general discussion and argues effectively that much of the reluctance toward secession in northern Alabama arose not from any qualified Unionism, but rather from that region's close affinity with Tennessee. See also citations on William L. Yancey in chap. 17, note 35.

On Georgia: Ulrich Bonnell Phillips, "Georgia and State Rights" in AHA *Annual Report*, 1901, II, 193–210; Thomas Conn Bryan, "The Secession of Georgia," *Georgia Historical Quarterly*, XXXI (1947), 89–111; William M. Bates, "The Last Stand for the Union in Georgia," *Georgia Review*, VII (1953), 455–467; N. B. Beck, "The Secession Debate in Georgia, November, 1860–January, 1861," in J. Jeffrey Auer (ed.), *Antislavery and Disunion, 1858–1861* (New York, 1963), pp. 331–359—especially useful for indicating where to find texts of major speeches by participants.

On Louisiana: Caskey, *Secession and Restoration;* James Kimmias Greer, "Louisiana Politics, 1845–1861," *LHQ*, XII (1929), 381–425, 555–610; XIII (1930), 67–116, 257–303, 444–483, 617–654; Lane Carter Kendall, "The Interregnum in Louisiana in 1861," *ibid.*, XVI (1933), 175–208, 374–408, 639–669; XVII (1934), 339–348, 524–536; Roger Wallace Shugg, "A Suppressed Co-operationist Protest against Secession," *ibid.*, XIX (1936), 199–203: Shugg, *Origins of Class Struggle in Louisiana* (Baton Rouge, 1939), pp. 157–170; Jefferson Davis Bragg, *Louisiana in the Confederacy* (Baton Rouge, 1941), pp. 1–33. For all of these states, see the pertinent chapters in Wooster, *Secession Conventions.*

33. Gerson, "Inception of the Montgomery Convention"; Albert N. Fitts, "The Confederate Convention: The Provisional Constitution," and "The Confederate Convention: the Constitutional Debate," *AR*, II (1949), 83–101, 189–210; Charles R. Lee, Jr., *The Confederate Constitutions* (Chapel Hill, 1963); Ralph Richardson, "The Choice of Jefferson Davis as Confederate President," *Journal of Mississippi History*,

South Carolina's gamble, on November 10, of going full speed ahead with unilateral action had paid off. This time, the Palmetto State was not left humiliated and alone, and Henry Timrod, in his poem "Ethnogenesis," could write

> At last we are
> a nation among nations; and the world
> shall soon behold in many a distant port
> another flag unfurled.[34]

Yet, behind the façade of united action, it had been a far closer thing than appeared. If one or two states—especially Georgia and Alabama—had gone the other way, the magic spell would have been broken, and the situation of the Louisiana and Texas conventions would have been entirely different from what they were when these states seceded at the end of January. The vote in Georgia, Alabama, and Louisiana was so close that a limited change in the situation might have tipped the balance the other way. In this situation, a relatively small number of resolute secessionists were able to guide a confused and excited electorate into a program for dissolving the Union. Something of the confusion was reflected in the low voter participation in the elections of the conventions. Though this was one of the most important political decisions the voters ever had to make, the balloting was extremely light compared with the vote in the November presidential election. In Georgia the total vote was only 82 percent of that in the presidential election; in Louisiana, 75 percent; in Alabama, 70 percent; in Mississippi, 60 percent. In no state did the secessionists poll a vote large enough to have been a majority in the November election.

The crucial fact, as the secessionists clearly realized, was that all of the states were acting in an atmosphere of excitement approaching hysteria, first generated by John Brown's attempted slave insurrection and surging up again in the latter stages of the presidential campaign. This excitement still prevailed when the secessionists went into action. They stimulated and perpetuated it in frequent public meetings, with a ceaseless barrage of speeches, by organizing volunteer military units known as "minutemen" or the like, and by the denunciation and in some cases the physical intimidation of cooperationists. This mood was pervasive, and it even swept over

---

XVII (1955), 161–176; Wilfred Buck Yearns, *The Confederate Congress* (Athens, Ga., 1960), pp. 1–41; Coulter, *Confederate States*, pp. 1–56.

34. *Poems of Henry Timrod* (Boston, 1899), p. 150.

the churches, so that clergymen from the pulpit were almost as vocal as politicians from the stump in warning of the danger to the South, exhorting the people to declare their independence, and keeping emotions at a high pitch.[35] It is hard to believe that this mood of apprehension would have continued if the South had waited until Lincoln had come to office and been given a chance to show his Whiggish moderation.

But the secessionists knew that their iron was hot, and they struck. A secessionist in South Carolina wrote, "I do not believe the common people understand it; but who ever waited for the common people when a great movement was to be made. We must make the move and force them to follow." South Carolina's commissioner to Florida, defending the swift action of his state, said with remarkable frankness, "I . . . believe that if . . . South Carolina had stated some distant day for future action, to see if other states would join us, and had thus allowed the public feeling to subside, she herself would have lost the spirit of adventure and would have quailed from the shock of this great controversy."[36] Christopher Memminger, writing in November, said, "Our great point is to move the other Southern States before there is any recoil." Clearly, Howell Cobb, alarmed by South Carolina's rapid action, was correct when he said, "It looks as if they were afraid that the blood of the people will cool down."[37]

The secessionists realized that although their cause was a popular one, its ascendency was transient. Delay, from their standpoint, was almost worse than opposition. They seized the momentum of a popular emotional reaction to Lincoln's election and rode it through with astonishing speed.

In ninety days, they won ten legislative decisions to hold elections for state conventions, held seven such elections, gained majorities in each, assembled seven conventions, voted seven ordinances of secession, and also took the first steps toward formation of a southern confederacy.

In this achievement, the secessionists completely bewildered the

35. See titles cited in notes 31 and 32; also James W. Silver, *Confederate Morale and Church Propaganda* (Tuscaloosa, Ala., 1957), esp. pp. 7–41.

36. A. P. Aldrich to James Henry Hammond, Nov. 25, 1860; Leonidas Spratt, commissioner to Florida, speech in Charleston *Mercury*, Jan. 12, 1861, both quoted in White, *Rhett*, pp. 177, 180.

37. Christopher Memminger to John Rutherfoord, Nov. 27, 1860; Howell Cobb to "My Judge," Nov. 11, quoted in Channing, *Crisis of Fear*, pp. 283, 248.

cooperationists by insisting that *they* were the true cooperationists. As Robert Barnwell Rhett expressed it, he had supported unilateral action by South Carolina because he believed that once this action was taken, other states would follow.[38] As others did follow, Rhett's position became progressively more a functional form of cooperationism, and by the time the Louisiana and Texas conventions met, they faced the question of whether they would "cooperate" with five other states of the lower South. Thus, as one historian has stated it, "secession calmly paraded as cooperation." Or, as a secessionist member of the Georgia convention declared, he too favored cooperation, "but with the states which are determined to secede," while the antisecessionists "favor cooperation with the states which are disposed to remain in the Union."[39] In all this, there was no conspiracy to thwart the expressed will of any majority in any state.[40] In fact, the populace was clamoring for action. But the secessionists did take care to move before an opposition could organize; to minimize the prospects of a deadly war; to elicit a decision while emotions were high; and to create a situation which would ultimately force the people in all the slaveholding states—a majority of whom opposed secession—to make a hateful choice between leaving the Union and fighting against the South.[41]

As the secessionists well knew, there were serious political divisions within the seceding states, and these divisions represented a realignment which, if it developed further, would open a dangerous

38. White, *Rhett,* p. 176.

39. "With what states is Louisiana to co-operate? . . . Must she leave her five sisters [who had already seceded]?" New Orleans *Bee,* Dec. 24, 1860, in Dwight Lowell Dumond (ed.), *Southern Editorials on Secession* (New York, 1931), p. 367; White, *Rhett,* p. 176; William M. Bates, "The Last Stand for the Union in Georgia," *Georgia Review,* VII (1953), 459.

40. William J. Donnelly, "Conspiracy or Popular Movement: The Historiography of Southern Support for Secession," *NCHR,* XLII (1965), 70–84. Donnelly's title illustrates a fallacy which has done much to confuse the understanding of Gulf state secession. He assumes that if secession was not a conspiracy, it was therefore a popularly supported movement, but if not a popularly supported movement, then it must have been a deep-dyed conspiracy. There is much evidence that it was neither, but was rather a program put through in an open and straightforward manner by a decisive minority, at a time when the majority was confused and indecisive.

41. On the assertions that there would be no war see p. 489 and note 16 above. David Hoke, a representative in the South Carolina legislature from Greenville, wrote on Nov. 8, 1860, that Rhett's haste was designed to "make a speedy work of the whole matter, and that too, before two parties can be formed in South Carolina." Rosser H. Taylor (ed.), "Letters Dealing with the Secession Movement in South Carolina," *Furman University Faculty Studies Bulletin,* XVI (1934), 3–12.

breach between slaveholding and nonslaveholding whites. As late as the election of 1860, the southern electorate continued to vote in traditional patterns. The Jacksonian Democratic organization had been the party of the plain people, the nonslaveholders, the residents of pine barren counties, mountain counties, and backwoods counties, while the Whigs and their successors had been strongest among the planter class and in the rich, cotton-producing, slave-populated counties of the Black Belt.[42] Over time, the Democratic party had grown steadily less Jacksonian, but the hill folk, the "red necks," the "peckerwoods," had continued to vote Democratic. Thus, if the 537 counties of Virginia, North Carolina, Tennessee, Georgia, Alabama, Mississippi, and Louisiana in 1860 are divided into three groups according to whether a county ranked high, medium, or low in its proportion of slaves as compared with other counties in the same state, it appears that Breckinridge carried 64 percent of the counties with a low proportion of slaves, 56 percent of those with a medium proportion, and 52 percent of those with a large proportion. Since Breckinridge was closer to being a disunionist than Bell or Douglas, it appears that the nonslaveholders were more receptive to the idea of disunion than the large slaveholders—or at least the counties in which they lived were more receptive. But in 1861, the counties with the lowest ratios of slave

42. For the classic identification of the Whigs with the planter class, see Ulrich B. Phillips, "The Southern Whigs, 1834–1854," in *Essays in American History Dedicated to Frederick Jackson Turner* (New York, 1910), pp. 203–229; Arthur Charles Cole, *The Whig Party in the South* (Washington, 1913), with copious and impressive evidence. Charles G. Sellers, "Who Were the Southern Whigs," *AHR*, LIX (1954), 335–346, put forward the thesis that the Whigs were associated with towns, finance, and commerce, and not primarily with plantation agriculture. Thus, he, too, regarded them as a party of property, but a different kind of property. Grady McWhiney, "Were the Whigs a Class Party in Alabama?" *JSH*, XXXIII (1957), 510–522, analyzed large numbers of Whig and Democratic officeholders in Alabama, found no significant dissimilarities in wealth, education, occupation, or background, and hence concluded, "It cannot be proved by the men who sat in Congress and in the Alabama legislature that great social differences existed between the two parties." McWhiney is correct, no doubt, in showing that leaders of both parties came from the elite, but this does not mean that large slaveholders and nonslaveholders were equally prone to vote Whig. In fact, McWhiney shows that in six presidential elections in Alabama, counties with less than 30 percent slave population gave Whig majorities only 23 times out of 117, while counties with over 50 percent slave population gave Whig majorities 49 times out of 83. It seems valid to conclude that there was a political alignment in the South in the 1850s with propertied, slaveowning, or commercial and financial interests tending to vote Whig, and nonslaveholding, nonpropertied small farmers and others tending to vote Democratic.

population (in this same group of 537) gave only 37 percent of their vote for immediate secession, while the counties with the highest ratio of slaves were 72 percent for secession. Among the counties with a low slave ratio, there were 130 which had voted for Breckinridge, but only 65 of these later voted for secession. Among the counties with a high slave ratio, there were 87 which had voted for Bell or Douglas, but only 34 of these voted against unilateral state secession. Among the counties with low slave ratios, half of the Breckinridge counties changed sides and would not support secession, while among the counties with high slave ratios, 53 out of 87 of the Bell or Douglas counties changed sides, and would not support the Union. To a much greater degree than the slaveholders desired, secession had become a slaveholders' movement, toward which the people of the counties with few slaves showed a predominantly negative attitude.[43]

The coolness toward secession of the people in the counties with

43. These county results are shown in Seymour Martin Lipset, *Political Man* (New York, 1960), pp. 344–354. The reader should note that these data do not include Florida, Louisiana, or Texas (nor South Carolina, whose presidential vote was cast by the legislature), and they do include Virginia, North Carolina, and Tennessee (but not Arkansas or Missouri). Also see, David Y. Thomas, "Southern Non-Slaveholders in the Election of 1860," *Political Science Quarterly*, XXVI (1911), 222–237. Wooster, *Secession Conventions*, does not compare the presidential election results with the results in the secession conventions, but he shows that counties with a heavy slave population were much more prone to support secession in the conventions than counties with a low slave population. For the lower South, among counties with less than 25 percent slave population, 70 favored immediate secession, 39 were for conditional union or cooperation, and 9 were divided. Among counties with above 50 percent slave population, 113 favored immediate secession, 15 were for conditional union or cooperation, and 14 were divided. For the upper South, among counties with less than 25 percent slave population, 72 were secessionist, 109 antisecessionist, and 9 divided; among counties with more than 50 percent slave population, 35 were secessionist, 13 antisecessionist, and 4 divided.

As an exception to these general tendencies, it is important to note that Louisiana and Mississippi did not conform to the pattern. In Mississippi there were 31 counties with above 50 percent slave population of which 19 voted for secession and for the defeat of all amendments, while 12 voted otherwise at one time or another; there were 29 counties with under 50 percent slave population, of which 20 voted for secession and the defeat of all amendments, while only 9 voted otherwise at one time or another. (Compiled from Rainwater, *Mississippi*, pp. 198–210.) In Louisiana the Bell and Douglas tickets had carried 12 parishes in the Nov. election, and the cooperationists carried 11 of these same parishes (plus 8 others) in the Jan. election. (Compiled from Caskey, *Secession and Restoration*, map 1 and map 3.) Wooster makes a distinction between different classes of planters in Louisiana: "The parishes supporting immediate secession were the heavily slave-populated cotton-producing parishes in which per capita wealth was the highest in the state. These parishes felt the economic system of the South could best be preserved outside of the Union. Opposing immediate disruption . . . were half of the rich sugar parishes of Southern

low slave ratios may have offered a threat to the secessionists over the long term, but what troubled them in the first two months of 1861 was not local dissension. It was the coolness of the upper South.

In fact, with the secession ordinance in Texas, the momentum of secession was spent. Though seven slave states had left the Union, eight others had not. The slaveholding states were far indeed from forming a politically united South, and though South Carolina was not stranded and alone, the Gulf Coast Confederacy lacked the population, resources, and wealth of the slave states that were still in the Union. From the outset, secessionists had faced the dilemma that any given state might find itself impotent if it acted alone, or might be paralyzed if it waited for joint action with other states. When the Gulf states met at Montgomery on February 4 to form a confederacy, it was by no means certain that they had escaped the hazards of impotence, though they were seven and not one. No one was much impressed with a Gulf Coast Confederacy. No one was really convinced that it would be economically or politically viable.

On February 4, however, the secessionists still had hopes of attracting other states within a month, for between January 12 and January 29 the legislatures of five additional states had called elections for conventions: Arkansas on January 12 for an election on February 18; Virginia, January 14 for February 4; Missouri, January 18 for February 18; Tennessee, January 19 for February 9; and North Carolina, January 29 for February 28. But all of these measures placed limits on the proposed convention: Virginia gave voters the option of requiring a popular referendum on any action the convention might take; the Missouri legislature required such a referendum without waiting for the voters to demand it; Arkansas and Tennessee empowered the voters to decide whether the convention should be held, as well as to elect delegates; and North Carolina provided both for a Virginia-Missouri style of referendum and for an Arkansas-Tennessee style of decision on whether the convention should meet.

The election contests for these conventions were fought out under circumstances quite different from those of the contests in the lower South. Basically, the upper South was not as obsessively committed to slavery as the lower South. Of the seven states that had

---

Louisiana." Wooster also believes that parishes with a large French Creole population tended to oppose secession. Wooster, pp. 265, 46, 117–120.

seceded, only Texas had a Negro population of less than 40 percent; in the five states that were about to act, the Negro population averaged less than 30 percent. Further, the states of the upper South knew that they had stronger economic links with the North than with the lower South, and that breaking them might cause serious economic dislocation.[44] Further still, they had a long and strong tradition of Unionism. In 1860, Bell and Douglas together had received 234,000 votes in the five states about to act, compared with 206,000 for Breckinridge; but in the states that had already seceded, Breckinridge had received 220,000 votes, compared with 171,000 for Bell and Douglas. There were also vigorous Unionist leaders in Tennessee and Virginia, such as Andrew Johnson, Emerson Etheridge, William G. Brownlow, and John Minor Botts.[45] Many moderates in these states bitterly resented the precipitancy of South Carolina, which was "a pestiferous grumbler," "a nuisance anyway," a state given up to "frenzy [which] surpasses in folly and wickedness, anything which fancy in her wildest mood has yet been able to conceive." The Wilmington, North Carolina, *Herald* asked its readers: "Will you suffer yourself to be spit upon in this way? Are you *submissionists* to the dictation of South Carolina . . . are you to be called cowards because you do not follow the crazy lead of that crazy state?" The Charlottesville, Virginia, *Review* said it "hated South Carolina for precipitating secession."[46]

44. Mary Emily Robertson Campbell, *The Attitude of Tennesseans toward the Union, 1847–1861* (New York, 1961), pp. 153–154; Senator Garret Davis of Kentucky, Jan. 23, 1862, said, "Why Mr. President, Kentucky has almost peopled the northwestern states, especially Indiana and Illinois. . . . They are bone of our bone and flesh of our flesh. When you offer to the Union men of Kentucky the choice whether they will remain united forever with Indiana and Ohio and Illinois, or go with Georgia and South Carolina and Florida, they will answer 'A thousand fold will we be united rather with the Northwest than with those distant states.'" *Congressional Globe*, 37 Cong., 2 sess., pp. 452–453; see E. Merton Coulter, *The Civil War and Readjustment in Kentucky* (Chapel Hill, 1926), pp. 1–56.

45. On the border state Unionists, see especially Albert D. Kirwan, *John J. Crittenden: The Struggle for the Union* (Lexington, 1962); LeRoy P. Graf, "Andrew Johnson and the Coming of the War," *Tennessee Historical Quarterly*, XIX (1960), 208–221; E. Merton Coulter, *William G. Brownlow: Fighting Parson of the Southern Highlands* (Chapel Hill, 1937); Thomas B. Alexander, *Thomas A. R. Nelson of East Tennessee* (Nashville, 1956); Joseph Howard Parks, *John Bell of Tennessee* (Baton Rouge, 1950); John Minor Botts, *The Great Rebellion: Its Secret History, Rise, Progress, and Disastrous Failure* (New York, 1866), pp. 230–232.

46. Resentment by other southern states of the precipitancy of South Carolina has been somewhat neglected, but the evidence is copious. See Coulter, *Civil War in*

The secessionists faced all of these obstacles as they sought to carry their movement into the upper South, but their greatest handicap lay in revived hopes of concessions by the North. Senators Crittenden and Douglas had resumed a kind of peace offensive, and on January 19, after a conference with Republican Senators Seward and James Dixon of Connecticut, Crittenden sent a telegram to North Carolina expressing optimism about the prospects of an adjustment and urging delay. On January 25 he and Douglas sent a telegram of similar purport to Virginia.[47] Meanwhile, the Virginia legislature had been incubating a plan for a peace conference. On January 19 it invited all states, both slave and free, to send delegates to meet at Washington on February 4 and there explore "every reasonable means to avert" a dissolution of the Union.[48]

During December and early January, secessionists had kept up a vigorous activity and had developed appreciable strength in Virginia, but the promises of Crittenden and Douglas, as well as the hopes for the peace conference, and the obligation to give it a fair chance as Virginia's own creation—all tended to immobilize the secessionists. Virginians went to the polls on February 4, the same day on which the incipient Confederate Congress assembled at Montgomery and on which a limping peace conference, with only twenty-one of the thirty-four states in attendance, convened at Washington. When the votes were counted, the secessionists had incurred a stunning defeat. The question of whether any action by the convention must be submitted to the voters for ratification was regarded as a kind of test case, with the secessionists opposed. They were defeated, 100,536 to 45,161. Further, only about 32 immedi-

---

Kentucky, p. 45; James Welch Patton, *Unionism and Reconstruction in Tennessee, 1860–1869* (Chapel Hill, 1934), pp. 5, 8, 9 (quoting Knoxville *Whig*, Dec. 8, 1860, which said to South Carolina, "You may leave the vessel [the Union], you may go out in the rickety boats of your little state and hoist your miserable *cabbage-leaf* of a Palmetto flag; but depend upon it, men and brethren, you will be dashed to pieces on the rocks"); Campbell, *Attitude of Tennesseans*, p. 141; Henry T. Shanks, *The Secession Movement in Virginia, 1847–1861* (Richmond, 1934), pp. 134–135, 145, 164; Sitterson, *Secession Movement in North Carolina*, pp. 193, 242; Coulter, *Confederate States*, p. 1; Dumond, *Southern Editorials*, pp. 228, 389—editorials from Wilmington *Daily Herald*, Nov. 9, 1860 ("There are no two adjoining states in the Union whose people have so little community of feeling as North and South Carolina"); Charlottesville *Review*, Jan. 4, 1861 ("We entertain towards South Carolina the most bitter resentment").

47. Potter, *Lincoln and His Party*, pp. 304–306; Nichols, *Disruption*, p. 456; Kirwan, *Crittenden*, p. 406; George Fort Milton, *The Eve of Conflict: Stephen A. Douglas and the Needless War* (Boston, 1934), pp. 532–533.

48. See below, pp. 545–547.

ate secessionists won seats in a convention which would have 152 members when it met on February 18.[49]

The impact of this overwhelming vote becomes more evident as one considers the surrounding circumstances. For three months the secessionists had enjoyed an unbroken string of quick victories. Since Lincoln's election, not a week had passed that some governor did not call a special legislative session, or that some legislature did not call a convention, or that some state did not elect a convention, or that some convention did not assemble, or, having assembled, vote for secession. After such a sequence, it seemed a great turning point when Virginia, with all her prestige as the "mother of states," the cultural capital of the South, and the most populous and economically important of the southern states, dealt the secessionists such a smashing blow. The defeat was indeed a significant one, but even so, the secessionists, in their dejection, and the people of the North, in their exultation, both exaggerated it. A Charleston newspaper lamented that "Virginia would never secede now,"[50] and a correspondent of William H. Seward jubilantly assured him, "We have scarcely left a vestige of secession in the western part of Virginia, and very little indeed in any part of the state. . . . The Gulf Confederacy can count Virginia out of their little family arrangement—*she will never* join them."[51]

As time passed, it became apparent that Virginia's burst of Unionism was no stronger than her hope of compromise and her faith in the peace conference. As these waned, Virginia's Unionism waned. But, as of February 4, many men agreed with Seward that secession was a temporary fever which had passed its climax.[52] Indeed the

49. Shanks, *Secession Movement in Virginia,* pp. 120–157; Beverley B. Munford, *Virginia's Attitude toward Slavery and Secession* (New York, 1909), pp. 248–260; James C. McGregor, *The Disruption of Virginia* (New York, 1922), pp. 99–123; Richard Orr Curry, *A House Divided: A Study of Statehood Politics and the Copperhead Movement in West Virginia* (Pittsburgh, 1964), pp. 28–32; James Elliott Walmsley, "The Change of Secession Sentiment in Virginia in 1861," *AHR,* XXXI (1925), 82–101; Henry T. Shanks, "Conservative Constitutional Tendencies of the Virginia Secession Convention," in Fletcher M. Green (ed.), *Essays in Southern History Presented to J. G. de R. Hamilton* (Chapel Hill, 1949), pp. 28–48; F. N. Boney, *John Letcher of Virginia* (University, Ala., 1966), pp. 104–108; Barton H. Wise, *The Life of Henry A. Wise of Virginia, 1806–1876* (New York, 1899), pp. 268–281; Craig Simpson, "Henry A. Wise in Antebellum Politics, 1850–1861" (Ph.D. dissertation, Stanford University, 1973).

50. McGregor, *Disruption of Virginia,* p. 116.

51. W. D. Moss, Moundsville, Virginia, to Seward, Feb. 6, 1861, in Frederic Bancroft, *The Life of William H. Seward* (2 vols.; New York, 1900), II, 533–534.

52. Unsigned letter of Henry Adams, Feb. 5, in Boston *Daily Advertiser,* Feb. 8, 1861.

next four elections appeared to justify his appraisal. On February 9, Tennessee voted 69,387 to 57,798 against calling a convention. At the same time, votes were cast for men who would have been delegates if this proposed convention had met, and the Tennesseans rubbed salt into the secessionist wounds with 88,803 votes for Unionists to 24,749 for secessionists.[53] On February 18 the secessionists sustained a double defeat. In Arkansas, the voters favored a convention, 27,412 to 15,826, but they elected a majority of Unionists as delegates.[54] In Missouri, Unionists, either conditional or unconditional, polled about 110,000 votes to the secessionists' 30,000, and not a single clear-cut secessionist gained election as a delegate for the convention which would meet on February 28.[55] Finally, on February 28, North Carolina completed the discomfiture of the secessionists. By this time, the peace conference had adjourned after making some weak recommendations with dubious prospects of fulfillment, but a North Carolina delegate sent home a telegram saying, "All is right. The compromise [proposed by the peace conference] will be endorsed by the national Congress." This maneuver probably accounted for the defeat of the convention by the narrow vote of 47,323 to 46,672. But if the convention had met, secession would have been crushed, for only 42 secessionists were among the 120 delegates chosen to sit in this convention if it were approved by the voters.[56]

In three other slave states, the secessionists fared even worse. In

53. The best account of the struggle over secession in Tennessee is Campbell, *Attitude of Tennesseans.* See also: Thomas Perkins Abernethy, *From Frontier to Plantation in Tennessee* (Chapel Hill, 1932); Patton, *Unionism and Reconstruction in Tennessee;* Robert Love Partin, "The Secession Movement in Tennessee" (Ph.D. dissertation, George Peabody College, 1935).

54. Elsie M. Lewis, "From Nationalism to Disunion: A Study in the Secession Movement in Arkansas, 1850–1861" (Ph.D. dissertation, University of Chicago, 1946); Jack B. Scroggs, "Arkansas in the Secession Crisis," *Arkansas Historical Quarterly,* XII (1953), 179–224; David Y. Thomas, "Calling the Secession Convention in Arkansas," *Southwestern Political and Social Science Quarterly,* V (1924), 246–254.

55. Walter Harrington Ryle, *Missouri: Union or Secession* (Nashville, 1931); William H. Lyon, "Claiborne Fox Jackson and the Secession Crisis in Missouri," *MHR,* LVIII (1964), 422–441; Arthur Roy Kirkpatrick, "Missouri on the Eve of the Civil War," *ibid.,* LV (1961), 99–108; Kirkpatrick, "Missouri in the Early Months of the Civil War," *ibid.,* LV (1961), 235–266; Kirkpatrick, "Missouri's Secessionist Government, 1861–1865," *ibid.,* XLV (1951), 124–137; Jonas Viles, "Sections and Sectionalism in a Border State," *MVHR,* XXI (1934), 3–22; Thomas L. Snead, *The Fight for Missouri from the Election of Lincoln to the Death of Lyon* (New York, 1886).

56. Sitterson, *Secession Movement in North Carolina,* pp. 177–229, is basic. See also William K. Boyd, "North Carolina on the Eve of Secession," AHA *Annual Report,* 1910, pp. 165–178.

Kentucky, Governor Beriah Magoffin gravitated toward support of the Confederacy and called a special session of the legislature. When it assembled on January 17, he recommended a state convention, but the legislature, by a vote of 54 to 36 in the lower house, refused to call one and adjourned on February 11 without taking any decisive action.[57] In Maryland, where there was strong secessionist sentiment, Governor Thomas H. Hicks, like Sam Houston in Texas, resisted pressure to call a special session of the legislature. Also as in Texas, the secessionists took extralegal steps to call a convention, but these efforts did not succeed in forcing the hand of Hicks as they did that of Houston.[58] In Delaware, the state legislature voted "unqualified disapproval" of secession as a remedy for southern grievances. The vote was unanimous in the lower house and 8 to 5 in the upper.[59]

Thus, the late winter of 1860–1861 proved as depressing for the secessionists as the early winter had been exhilarating. During the early winter they had not met with a single reverse. After February 4 they gained not a single success. The month which witnessed the birth of a seven-state confederacy also saw the hopes of a united southern republic completely demolished. This turn of events gave hope to Unionists everywhere, and many of the Unionists of the upper South now began to believe that the initiative had passed into their hands and that they might shape the destiny of the Republic. As loyal members of the Union, the states of the upper South seemed in a strong position to insist upon the concessions which would be necessary to bring the impulsive Gulf states back into the Union. As sisters of the other slaveholding states, they could appeal to those states to come back into the Republic over the bridge which they were building. As one Virginian expressed it, "Without submission to the North or desertion of the South, Virginia has that moral position *within the Union* which will give her power to arbitrate between the sections." Or, as three prominent Tennesseans as-

57. Coulter, *Civil War in Kentucky,* pp. 1–34; Kirwan, *Crittenden,* pp. 430–431; Edward C. Smith, *The Borderland in the Civil War* (New York, 1927); William T. McKinney, "The Defeat of the Secessionists in Kentucky in 1861," *JNH,* I (1916), 377–391; Thomas Speed, *The Union Cause in Kentucky, 1860–1865* (New York, 1907).

58. Charles Branch Clark, "Politics in Maryland during the Civil War," *Maryland Historical Magazine,* XXXVI (1941), 239–262; Carl M. Frasure, "Union Sentiment in Maryland, 1859–1861," *ibid.,* XXIV (1929), 210–224; George L. P. Radcliffe, *Governor Thomas H. Hicks of Maryland and the Civil War* (Baltimore, 1901), pp. 19–42.

59. Harold Hancock, "Civil War Comes to Delaware," *CWH,* II (1956), 29–46.

serted, their state was charged with the "grand mission of peacemaker between the states of the South and the general government." The North, it appeared, had been sobered and shaken by the Gulf state secession; the fire-eaters, on the other hand, had been brought to heel by five severe defeats. Perhaps the upper South could avert the crisis, redeem the Union, and save the country from war.[60]

So it must have appeared also to many crestfallen secessionists, who now doubted the ability of the Gulf Coast Confederacy to stand alone, just as South Carolina cooperationists had for three decades doubted the ability of the Palmetto State to stand alone. But any leaders in the upper South who thought they held control of the situation were forgetful of one vital fact: they were committed to resisting the coercion of any seceding state. The Virginia legislature had passed a resolution to that effect early in January. The Tennessee general assembly, informed of offers by New York to provide armed forces "to be used in coercing certain sovereign states of the South," voted its conviction that in the event of such forces being sent "the people of Tennessee, uniting with their brethren of the South, will as one man, resist such invasion of the soil of the South at any hazard and to the last extremity." Formal legislative resolutions were not voted in North Carolina or Kentucky, but in those states also numerous public statements were made, and were not challenged, that the people would resist the coercion of any southern state. Although the people of the South were badly divided on the issue of secession, they remained united in their belief that "southern rights" must be maintained and that no southern state could acquiesce in the use of federal force against any other southern state. To this extent, at least, southern nationalism was a reality.[61]

But if the upper South was committed to protecting the lower

60. B. J. Barbour of Virginia, quoted in Shanks, *Secession Movement in Virginia*, p. 151; "Address of John Bell and Others to People of Tennessee," in Frank Moore (ed.), *The Rebellion Record* (12 vols.; New York, 1861–68), I, 71–72; James Guthrie of Kentucky said that God had "chosen Kentucky to be the great mediator for the restoration of peace and the preservation of our country," *ibid.*, p. 73.

61. Resolutions passed in Jan. in Virginia house of delegates, 117 to 5, and in senate with only one negative vote, Shanks, *Secession Movement in Virginia*, pp. 144–145. Tennessee resolutions also in Jan., Campbell, *Attitude of Tennesseans*, pp. 161–162; Sitterson, *Secession Movement in North Carolina*, pp. 196–197; Coulter, *Civil War in Kentucky*, pp. 28, 29, 44.

South against coercion, was the lower South really so isolated after all? And did the upper South really hold such a controlling position? Was not the upper South rather more in a position similar to that of a moderate and powerful nation which has made an unlimited alliance to protect a weak but belligerent neighbor, and which has thus placed its own peace at the discretion of its trigger-happy ally? Did not the lower South, after all, hold the initiative, and could it not draw the upper South into the vortex by unilateral action even more drastic than South Carolina had used to draw in the Gulf states?

A year earlier, near the end of Christopher Memminger's futile mission to Richmond, the frustrated commissioner had written to a friend in South Carolina, "I am brought to the opinion that we farther South will be compelled to act, and drag after us these divided states."[62] If Memminger knew how to drag them, he failed to reveal it. But Robert Barnwell Rhett, in October 1860, had expressed the same idea in more definite terms. The states of the Upper South, he said, could "only be managed by the course pursued at [the Democratic Convention meetings of 1860] at Charleston & Richmond & Baltimore. . . . They must be made to choose between the North and the South, and then they will redeem themselves, but not before."[63]

The fact that they were committed to resist coercion of any southern state meant that their choice could be predicted—and could be forced. David Hamilton of South Carolina was probably aware of this when he wrote, "I am amused at the coolness with which the Southern States offer to march to the assistance of So Car—they must be sleeping in fancied security—why in less than a year it is more than likely that the whole South will be in a blaze from one end to the other."[64] Some of the men of the upper South, however, did not sleep completely in fancied security. In the Virginia convention, one of the Unionist delegates bitterly assailed South Carolina and the Gulf states for having brought about Lincoln's election by splitting the Democratic party and for having deserted the other slave states by seceding without consulting them.[65] Yet he, and most

62. Quoted in Cauthen, *South Carolina Goes to War,* p. 13.

63. Rhett to Edmund Ruffin, Oct. 20, 1860, quoted in Channing, *Crisis of Fear,* pp. 263–264.

64. Quoted in Cauthen, *South Carolina Goes to War,* p. 134.

65. S. M. Moore, in Virginia convention, Feb. 25, 1861, speech described in Shanks, *Secession Movement in Virginia,* pp. 163–164.

others, forbore to make explicit the truth which he must have perceived: that the Gulf states (or the northern states) could precipitate a war from which the upper South could not escape.

Most South Carolinians also forbore to describe this situation in blunt terms, but Congressman William Boyce, an old cooperationist who had come out for immediate secession as early as August 1860, explained why he no longer feared isolation as a consequence of unilateral action. If South Carolina should secede by herself, he said, "then only two courses remain to our enemies. First, they must let us alone; secondly, they must attempt to coerce us. . . . suppose they attempt to coerce us; then the Southern states are compelled to make common cause with us, and we wake up some morning and find the flag of a Southern Confederacy floating over us."[66] The South at the end of February 1861 was still a divided South, but it remained to be seen what would happen when the logic sketched by Boyce began to operate.

66. Quoted in Cauthen, *South Carolina Goes to War*, p. 26.

CHAPTER 19

Winter Crisis

AS a modern reader is swept along by the accelerating rush of events which led to war in the spring of 1861, it becomes difficult to grasp the long duration of the final interlude before combat. It was almost half a year from Lincoln's election until the bombardment of Fort Sumter. This was far longer than the "Hundred Days" during which Franklin Roosevelt secured the enactment of most of the New Deal. It was longer than the whole Spanish-American War, from declaration to armistice. For the South it was a period of frantic activity; for Washington, a period of intermittent paralysis; for the North, a time of slow awakening to the fact that what was going on in the South was action and not rhetoric.

This long interval gave the Gulf states time for the elaborate process of secession, with its convening of legislatures, its legislative decisions to hold elections of state conventions, its brief election campaigns, its assembling of the conventions, its ordinances of secession, and even its gathering of the seceded states to form a provisional government, elect Jefferson Davis president, and inaugurate him at Montgomery before Lincoln could be inaugurated at Washington. Such an interval was built into the Constitution by its provision for a "lame-duck" session of Congress and for a span of approximately 120 days between the time when a president was elected and when he was sworn in.

This constitutional anachronism was the principal factor making for paralysis in Washington in the winter of 1860–1861. Buchanan held official power but little real power; Lincoln had no official power and showed little desire to use his access to real power. He remained in Springfield all winter. Meanwhile the informal devices which enabled public men to operate the political mechanism were less effective because the network of personal familiarity and shared experience in Washington had been partly dismantled. The president-elect was not only absent; he was also an outsider—an ex-congressman from Illinois of but one term's service, and a man who had scarcely set foot in Washington for a decade. According to the tribal customs of the Whig party from which he came, he might remain a nonentity, even as president. Before the election campaign, Lincoln had not been personally acquainted with his vice-presidential running mate or with most of the men who were to form his cabinet.

All these circumstances contributed to a default of power in Washington at a vital time in American history. But back of these factors were some others even more basic. In a sense, the winter of 1860–1861 marked the last stand of the old Federal Union, state-centered rather than nation-centered in its orientation. The two major parties were still loose coalitions of state parties, and some of the strongest politicians—men like Thurlow Weed, John A. Andrew, Simon Cameron, and Henry S. Lane—made state power the foundation of their political strength.[1] The national campaign tended to be not a single contest but many simultaneous statewide contests with varying tactics and issues. This was especially true in 1860, when there were virtually two elections, one between Lincoln and Douglas in the North, the other between Bell and Breckinridge in the South. Each section conducted its campaign very much as if the other section simply was not there. It was easier to do so at that time, for presidential candidates did not customarily make speeches, and there was nothing to focus all local activities upon a single man or a single issue. Lincoln remained at home and made no public addresses at all during the campaign; he had never stayed in Springfield so constantly at any time in his life as he did between his nomination in May 1860 and his inauguration ten months later. Breckinridge made but one address, in which he adroitly skirted the issue of disunion. Bell ventured no farther from his home in Nash-

1. Roy F. Nichols, *Blueprints for Leviathan: American Style* (New York, 1963), pp. 86–87; William B. Hesseltine, "Abraham Lincoln and the Politicians," *CWH*, VI (1960), 48.

ville than Bowling Green, Kentucky, making a few public appearances but no speeches. Only Douglas tore precedent to tatters and campaigned extensively from Maine to New Orleans. He told the voters, both North and South, what the election was about; he warned northerners that the election of a purely sectional candidate would result in disunion, and he told southerners that secession would bring a deadly punishment which he himself would help to administer. But Douglas was doomed to a Cassandra role. The South misconstrued his warnings as merely a strategic device to frighten them away from voting for Breckinridge, while northerners thought he was trying to frighten them away from voting for Lincoln.[2]

Thus a four-alarm crisis had crept upon the country while the voters were, many of them, enjoying the high-spirited antics of an old-style campaign in which the issues were not important enough to think about, and the frivolities might as well be enjoyed.

With vision of hindsight, one can see the thirty-year crisis entering its final stage with the breakup of the Charleston convention a year before Fort Sumter. But in fact the very familiarity of crisis—its chronic presence during three decades—had bred a contempt for it. So many rumblings had been heard without a sequel that men began to take the frequency of warnings as a reassurance that nothing would ever happen, rather than as an indication that something ultimately must happen. Further, the North tended to forget that one reason why southern threats had never been executed was that they had never really been tested with defiance. The Wilmot Proviso had never been adopted; Kansas and Nebraska had not been organized as free territories according to the requirements of the Missouri Compromise; Frémont had not been elected. To be sure, the South had failed to obtain the admission of Kansas under the Lecompton constitution, but then Kansas had not been admitted under the Topeka constitution either. Also, southern representatives had failed to prevent the election of a Republican Speaker in 1859,

2. Roy F. Nichols, *The Disruption of American Democracy* (New York, 1948), pp. 346–347; John Howard Parks, *John Bell of Tennessee* (Baton Rouge, 1950), p. 368; Benjamin P. Thomas, *Abraham Lincoln* (New York, 1952), p. 220; J. G. Randall, *Lincoln the President* (2 vols.; New York, 1945), I, 178; Allan Nevins, *The Emergence of Lincoln* (2 vols.; New York, 1950), II, 290–296; George Fort Milton, *The Eve of Conflict: Stephen A. Douglas and the Needless War* (Boston, 1934), pp. 490–500; Robert W. Johannsen, *Stephen A. Douglas* (New York, 1973), pp. 778–802.

but they had succeeded in depriving John Sherman of the speakership as a penalty for his endorsement of *The Impending Crisis*. *Dred Scott* v. *Sandford* was the law of the land. John Brown had been hanged. Southern threats might have a theatrical ring, and even southerners themselves had become sensitive about southern bluster, but in fact, the only occasions when southerners had defaulted on their threats was when their fellow southerners had left them isolated, as with South Carolina in the Nullification crisis and again in 1852. Yet it remained true that there was an overwhelming impulse in the North to discount the signals from the South and to suppose that South Carolina was merely having another temper tantrum.[3]

The old Union in 1860–1861 lacked the national press services, the network of electronic media, the large corps of public information specialists, and the array of news magazines which today would saturate public attention with an issue as urgent as secession. But in 1860, Congress was the only agency that held national affairs in any kind of national focus. It was out of session in November 1860, when the secession crisis began, and the country was ill-prepared to understand the situation, even for many weeks after Congress convened in December.

Even then, no leadership emerged, for the defeated administration was discredited and in disarray; the victorious Republicans, never before in power, were unprepared to act; and the two national leaders who grasped the gravity of the situation and the urgent need for action were Stephen A. Douglas, who had been very nearly broken by the result of the election, and James Buchanan, the lame-duck occupant of the White House.

History has stereotyped Buchanan as a weak president—"Buchanan the little" in the words of Theodore Roosevelt. The stereotype is not without some validity, but at least Buchanan understood one thing that few northerners did in November—namely, that the danger presented by secession was great and immediate. Three days after the election, he met his cabinet in a session that he labeled the most important one held during his administration. He called their attention especially to the fortifications at Charleston, built to defend the city against the naval attack of a foreign enemy but now

3. David M. Potter, *Lincoln and His Party in the Secession Crisis* (New Haven, 1942), pp. 47–49, 76–80.

threatened from the rear by its own secessionist citizens. Most important were Fort Moultrie, guarding the entrance to the harbor from the northeast, and Fort Sumter, on a small island at the center of the entrance. The tiny federal garrison of fewer than a hundred men commanded by Colonel John L. Gardner had been concentrated at Moultrie, which was vulnerable to assault by land. Sumter, after decades of construction now almost completed, was much more defensible but virtually unoccupied except for workmen. This situation presented Buchanan with the first of several dilemmas that were to plague him: If he left the garrison where it was, he might lose the entire position, but if he tried to reinforce it or move it to Sumter, he might precipitate war, which otherwise did not seem imminent. After some argument in the cabinet, it was decided to make no movement of troops but to replace Gardner with a younger and more alert officer of southern background, Major Robert Anderson of Kentucky.[4]

If they could avoid a premature collision in Charleston harbor, Buchanan thought he might be able to take constructive action on a broader scale. He considered especially either promptly issuing a proclamation asserting his intent to enforce the law or waiting until his annual message, then scarcely three weeks away, in which he would urge Congress to call a constitutional convention, with a view to working out a compromise.[5] Such a proposal had the additional advantage that it would gain time by putting the secessionists and the Republicans both in the position of appearing intransigent if they rejected it outright.

The cabinet was of little help to the president at this juncture. Howell Cobb of Georgia and Jacob Thompson of Mississippi were merely marking time until their states seceded, though they remained personally loyal to Buchanan. John B. Floyd of Virginia, already revealed as incompetent and tainted by his financial improprieties, was a weak man at the wrong time in the critically important War Department. Isaac Toucey of Connecticut was a yes-man who "had no ideas of his own." Lewis Cass, the ancient secretary of state,

4. Philip Shriver Klein, *President James Buchanan* (University Park, Pa., 1962), pp. 354–358; Bruce Catton, *The Coming Fury* (Garden City, N.Y., 1961), pp. 141–145; Nevins, *Emergence*, II, 340–343. On Fort Sumter generally, see Samuel Wylie Crawford, *The Genesis of the Civil War: The Story of Sumter, 1860–1861* (New York, 1887); W. A. Swanberg, *First Blood: The Story of Fort Sumter* (New York, 1957).

5. The example of Andrew Jackson in the Nullification crisis perhaps made the idea of issuing a proclamation doubly attractive to Buchanan.

had begun to get his back up against secession in a doddering sort of way, and he was joined by Joseph Holt, a Kentucky Unionist, and the able attorney general, Jeremiah S. Black, in efforts to stiffen the president's attitude. With the cabinet angrily divided, more or less along sectional lines, Buchanan at last put aside the idea of issuing a proclamation and instead gave his official response to the secession crisis in his annual message to Congress on December 3. In it, he recommended the calling of a federal constitutional convention, and he did so with a curious combination of realism and fantasy. The most realistic aspect of his proposal was its recognition of what had really caused southern disaffection—not a concern for territorial abstractions or constitutional refinements, but rather a pragmatic fear that continued propagandizing on the slavery issue would lead to slave insurrection. "The incessant and violent agitation of the slavery question throughout the North for the last quarter of a century," the president declared, "has at length produced its malign influence on the slaves and inspired them with vague notions of freedom. Hence a sense of security no longer exists around the family altar. This feeling of peace at home has given place to apprehensions of servile insurrections. . . . Should this apprehension of domestic danger, whether real or imaginary, extend and intensify itself until it shall pervade the masses of the Southern people, then disunion will become inevitable."[6]

Such insight enabled Buchanan to cut to the very core of the sectional problem, but when it came to proposing remedies, he had nothing new to offer except a more dramatic procedure. If secession were to be averted, he said, further steps must be taken to assure the return of fugitive slaves and to make slavery secure in the states where it already existed and in the federal territories. These guarantees could be better achieved by constitutional amendments than by ordinary legislation, and so a constitutional convention should be called.

Although Buchanan had shown a statesmanlike perception of southern motives, and although he may have cherished the practical hope that the calling of a constitutional convention would disrupt

6. James D. Richardson (ed.), *A Compilation of the Messages and Papers of the Presidents* (11 vols.; New York, 1907), V, 626–627; Klein, *Buchanan*, pp. 357–363; Nichols, *Disruption*, pp. 375–387. On Buchanan and the secession crisis generally, see his own defense, *Mr. Buchanan's Administration on the Eve of the Rebellion* (New York, 1866); Philip Gerald Auchampaugh, *James Buchanan and His Cabinet on the Eve of Secession* (Lancaster, Pa., 1926).

the schedule of the secessionists, thus giving the Unionists more time to organize, his plan was conspicuously unrealistic in certain other respects. To begin with, what he recommended was not a compromise, but acceptance of the fire-eaters' utmost demands regarding the territories; from the northern view, it was more a proposal to surrender than to negotiate. Further, he forfeited his position as a neutral arbiter by openly taking sides against the North. It was, he said, "the long-continued and intemperate interference of the Northern people with the question of slavery in the Southern states" that had arrayed the different sections against each other. Further still, he dealt futilely with the paramount legal question posed by the threat of disunion. His conclusion that secession could neither be legitimately carried out by a state nor be legitimately prevented by the federal government, though argued with considerable ability, lent a scholasticism to his argument which greatly weakened its force. "Seldom," declared a Cincinnati editor, "have we known so strong an argument come to so lame and impotent a conclusion."[7]

Like most documents of its kind, the presidential message got a mixed and partisan reception, but voices of unqualified approval were notably scarce. Republicans found it doubly outrageous in putting the blame for the crisis on their party and in failing to meet head-on the threat of disunion.[8] Northern Democrats of the Douglas variety, though heartily in agreement with the rebuke to antislavery agitators, were no less displeased by the apparent fecklessness with which Buchanan viewed the prospect of secession.[9] Yet neither did the message inspire any joy among secessionists; for it acknowl-

7. Cincinnati *Enquirer* (a Douglas Democratic paper), quoted in Nevins, *Emergence,* II, 353.

8. Martin B. Duberman, *Charles Francis Adams, 1807–1886* (Boston, 1961), p. 227, quotes Adams as declaring that the message was "in all respects like the author, timid and vacillating in the face of slaveholding rebellion, bold and insulting towards his countrymen whom he does not fear." In the Senate, John P. Hale of New Hampshire said of Buchanan: "He has acted like the ostrich, which hides her head, and thereby thinks to escape danger." *Congressional Globe,* 36 Cong., 2 sess., p. 9. George Templeton Strong wrote in his diary: "That Buchanan might be hanged under lynch law almost reconciles me to that code." Allan Nevins and Milton Halsey Thomas (eds.), *The Diary of George Templeton Strong* (4 vols.; New York, 1952), III, 74.

9. Howard Cecil Perkins (ed.), *Northern Editorials on Secession* (2 vols.; New York, 1942), I, 138–140. Douglas himself, in his "Norfolk doctrine," had long since categorically refused to countenance peaceable secession. Johannsen, *Douglas,* pp. 788–799, 813–814.

edged the justice of their complaints but then pronounced their remedy illegal and "neither more nor less than revolution." Furthermore, the president's labored disavowal of authority to coerce a state was somewhat compromised by his reaffirmation of a sworn duty to "take care that the laws be faithfully executed," insofar as he was able. Alert southerners quickly perceived the danger here—no need to "coerce" a state if one could "enforce the law" upon each of its citizens. The distinction was one with which Republicans could easily live, and in 1864 a Sherman could devastate Georgia without ever once coercing it.[10] Thus, in a message plainly the work of an inveterate doughface, Buchanan had nevertheless "taken the first major step toward the alienation of the South."[11]

The annual message and the various responses to it merely strengthened a general assumption that little of the enterprise and inspiration needed to save the Union could be expected from the White House. With only three months of his term remaining, Buchanan no longer commanded much public respect or exercised much control over his defeated and divided party. His influence on Capitol Hill, displayed so forcefully during the Lecompton controversy, had now virtually melted away. To hold together his own cabinet, let alone the nation, seemed almost beyond his strength. He had in fact lost much of the moral sway and informal leverage that constitutes a major part of presidential power. In the weeks ahead, he could do little more than exert the authority and perform the duties of chief executive as they were formally specified in the Constitution. To be sure, even this limited role was a crucial one in the circumstances. Presidential decisions of a purely administrative nature might precipitate civil war or promote an irretrievable acquiescence in secession. Buchanan's purpose, it transpired, was to avoid both of these extremes until the arrival of March 4 released him from his responsibilities. This policy of holding things steady reflected his constitutional views of the crisis and was no mere passing of the buck to Lincoln; for the fundamental problem posed by secession was, Buchanan believed, beyond the power of any president and could be dealt with only by Congress.[12]

10. See the Senate speeches of Alfred Iverson (Georgia) and Judah P. Benjamin (Louisiana), *Congressional Globe,* 36 Cong., 2 sess., pp. 11, 215.

11. Kenneth M. Stampp, *And the War Came* (Baton Rouge, 1950), p. 57.

12. The president, Buchanan said, had "no authority to decide what shall be the relations between the Federal Government and South Carolina." While offering

After all, it was Congress, with or without presidential help, that had made all the great sectional compromises, and it was in Congress, rather than the presidency, that the best political talents had for many years been concentrated. However, the major crises of the past had arisen within Congress itself over proposed legislation, and their resolution had been largely a matter of internal management. The crisis of 1860–1861 was ominously different, for it originated outside the legislative process in an irrevocable decision of the people. Congress, no less than the president, had suffered a loss of control. A fairly won election could be compromised only by passing legislation offsetting or limiting its expected consequences. In short, compromise this time meant persuading or compelling the victorious party to renounce a considerable portion of its own platform—in the hope that such a sacrifice would be enough to arrest the progress of secession. "The crisis," said a southern senator, "can only be met in one way effectually. . . . and that is, for the northern people to review and reverse their whole policy upon the subject of slavery." There was, he added, no evidence of any such disposition.[13]

Thus the second session of the Thirty-sixth Congress assembled on December 3 to face a crisis not of its own making and of unprecedented gravity. Many of its members were lame ducks repudiated in the recent election, and many others had come merely to mark time until their states officially withdrew from the Union. There could, moreover, be none of the habitual dawdling over compromise because the pace of secession was swift, and the life of this Congress would end in four months. It is therefore not surprising that a mixture of urgency and resignation should have hung in the air, casting strange shadows upon the confused scene and contributing to a general sense of unreality. Compromise activities of the following weeks would often seem to be primarily gestures for the historical record, responding more to conventional expectations than to real hope, and there was something perfunctory about even the most desperately worded speeches. Of the proposal to create a special compromise committee, Senator James M. Mason of Virginia took a not uncommon view. "I shall vote for the resolution,"

---

opinion and advice to Congress, he added: "It is therefore my duty to submit to Congress the whole question in all its bearings." Richardson (ed.), *Messages and Papers*, V, 635.

13. Albert G. Brown (Mississippi), *Congressional Globe*, 36 Cong., 2 sess., p. 33.

he said, "but without an idea that it is possible for anything that Congress can do to reach the dangers with which we are threatened."[14]

This does not mean that strong sentiment and effective leadership were lacking for the cause of sectional adjustment. Indeed, the forces of conciliation, spurred by Union-saving meetings across the country, had probably never been more numerous or eloquent. But the time had passed when a formula of compromise could be manipulated through Congress by playing proslavery and antislavery extremes off against each other. No legislative action, however favorable to the South, would make a dent in the secession movement unless it had received solid support from the Republican party. What southerners wanted now was not legislation as such, but something amounting to ironclad guarantees from their enemies. Compromisers like Crittenden and Douglas were reduced from managerial roles to playing mediators, while trying to perform a political miracle.

If it existed at all, the power to halt the progress of secession rested with the Republicans, but they were not ready to make the kind of dramatic and concerted effort that the crisis demanded. The great new increment of Republican power was, after all, still potential rather than operational, and authority within the party was much too diffused for swift, united action. From Springfield, where the president-elect received a steady stream of visitors and wrestled with the problem of forming a cabinet, there came no help whatever for Union-savers. Lincoln turned aside all pleas for a public statement reassuring the South. His political principles were already plainly on record, he insisted, and a further pronouncement now would be misinterpreted. "It would make me appear as if I repented for the crime of having been elected, and was anxious to apologize and beg forgiveness. To so represent me, would be the principal use made of any letter I might now thrust upon the public."[15] In Washington, the Republican caucus likewise decided to maintain a low profile during the interregnum. It tried, though with only partial success, to impose a policy of "reticence" upon its members. Seward, still nursing his resentment at having been refused the presidential

14. *Ibid.*, p. 35.
15. Roy P. Basler (ed.), *The Collected Works of Abraham Lincoln* (8 vols.; New Brunswick, N.J., 1953), IV, 151–152.

nomination, was for the time being content to watch and wait. No other leader arose to take his place and activate the party. Meanwhile, a huge mixed chorus of Republican editors, state officials, and local politicians was struggling to define the party's position and purpose. Out of such a milieu as this, a crystallized program of action was unlikely to issue.[16]

Thus there was confusion of purpose in one part of the country and decisive action in another. For the moment, Republicans and other northerners as well simply could not match the initiative of the deep South. In thousands of speeches and editorials during the previous decade, southerners had worked out the rationale of secession and rehearsed its procedure. Northerners, in contrast, had never yet faced up to the question of precisely what should be done if the threat of disunion were to become reality. But while this initial confusion discouraged Republican action, it also fostered a temporary plasticity in Republican attitudes. For example, during the early weeks of the crisis, a number of party newspapers echoed Horace Greeley's New York *Tribune* in the suggestion that it might be best to allow the cotton states to "go in peace." This "good riddance" solution to the problem of slavery and sectional conflict, once regarded as a Garrisonian heresy, appears to have been either empty rhetoric or strategic maneuver, inspired less by pacifism than by hostility to compromise. Because of the conditions he attached to it, Greeley's design for "peaceable secession" was never anything more than a theoretical alternative, and it soon evaporated in the heat of the crisis.[17]

More significant were indications from certain influential Republicans of a willingness to explore the possibility of conciliation. Some of the strongest pressure in this direction came from certain

16. Stampp, *And the War Came*, pp. 64–65. For one senator who did not hold his tongue, see H. L. Trefousse, *Benjamin Franklin Wade, Radical Republican from Ohio* (New York, 1963), pp. 133–136.

17. New York *Tribune*, Nov. 9, 1860; Potter, *Lincoln and His Party*, pp. 51–57; Potter, *The South and the Sectional Conflict* (Baton Rouge, 1968), pp. 219–242, with a reply to Thomas N. Bonner, "Horace Greeley and the Secession Movement, 1860–1861," *MVHR*, XXXVIII (1951), 425–444; Stampp, *And the War Came*, pp. 22–25; Jeter Allen Isely, *Horace Greeley and the Republican Party, 1853–1861* (Princeton, 1947), pp. 304, 312. Bonner argued that Greeley really favored peaceable separation if it were accomplished in the proper manner. Isely concluded that Greeley was sincere in offering his proposal, believing that if tried it "would test the strength of unionist sentiment in the south, and if necessary, provide a tranquil means for the exit of that region from the national government." Stampp, on the other hand, labeled the plan "a fraud from the start."

northern businessmen who saw hardship and even ruin for themselves in a prolonged sectional conflict.[18] Wall Street's apprehensions no doubt had some influence on Thurlow Weed, whose Albany *Evening Journal* in late November and again in mid-December proposed restoration of the Missouri Compromise as a basis of settlement. Although Weed had taken this step without consulting Seward, their long record of collaboration made it difficult for the New York senator to dissociate himself from the proposal, and he was thereafter marked down as perhaps amenable to compromise.[19] John Sherman, the Republicans' first choice for Speaker of the House in the previous session, had a different plan. Let all of the remaining western territories be divided "into States of convenient size, with a view to their prompt admission into the Union."[20] In this way, presumably, the deadly territorial issue, which could not be settled, would simply be erased. There were other scattered indications of a Republican disposition to make concessions, such as calling for repeal of personal liberty laws and formally guaranteeing the security of slavery in the southern states.[21]

But one must also take note of the great variations in the degree of seriousness with which Republicans at first viewed the crisis. The early weeks of maximum fluidity were also a time of much incredulity about secession in Republican ranks. It had long been party canon that southern talk of disunion was largely bluster and bluff, aimed at extracting concessions from weak-kneed northerners. This misapprehension survived for a while through a series of

18. Potter, *Lincoln and His Party,* pp. 116–127; Philip S. Foner, *Business and Slavery: The New York Merchants and the Irrepressible Conflict* (Chapel Hill, 1941), pp. 169–322; Thomas H. O'Connor, *Lords of the Loom: The Cotton Whigs and the Coming of the Civil War* (New York, 1968), pp. 144–146.

19. New York *Tribune,* Nov. 27, 1860; Glyndon G. Van Deusen, *Thurlow Weed, Wizard of the Lobby* (Boston, 1947), pp. 266–267; Van Deusen, *William Henry Seward* (New York, 1967), pp. 238–239; Seward to Weed, Dec. 3, 1860, in Thurlow Weed Barnes, *Memoir of Thurlow Weed* (Boston, 1884), p. 308; Potter, *Lincoln and His Party,* pp. 69–72, 81–87, 165–166. Henry J. Raymond's *New York Times* also became favorable to compromise, especially in February. See Carl F. Krummel, "Henry J. Raymond and the *New York Times* in the Secession Crisis, 1860–61," *NYH,* XXXII (1951), 377–398.

20. *Congressional Globe,* 36 Cong., 2 sess., pp. 77–78. Yet Sherman can scarcely be classified as a compromiser. See his fiercely antisecessionist letter of Dec. 22 to a group of Philadelphians, in his *Recollections of Forty Years in the House, Senate and Cabinet* (2 vols.; Chicago, 1895), I, 203.

21. Stampp, *And the War Came,* pp. 21–22. Most compromise proposals, however, came from border slave states and from northern Democrats. See, e.g., the resolutions presented to the House of Representatives on Dec. 12, *Congressional Globe,* 36 Cong., 2 sess., pp. 76–79.

adaptations: first, no state would actually go so far as to secede; then, only South Carolina would go so far as to secede; then, only a few other states would follow South Carolina; then, Unionist elements in the South would soon arise and undo much of secession with a counterattack; finally, the southern confederacy was not intended to be permanent but rather to strengthen the hand of the South in negotiations for reunion. By the time the range and intensity of the secession movement were fully apparent, Republican attitudes had been firmly set in a mold of party orthodoxy.[22]

Nothing did more to discourage any incipient thoughts of compromise among Republicans than word, informally circulated, of Lincoln's frosty opposition, especially in reference to the territorial issue. "Let there be no compromise on the question of *extending* slavery," he warned in early December. "If there be, all our labor is lost, and, ere long, must be done again." To a southerner he wrote: "On the territorial question, I am inflexible."[23]

Lincoln appears to have been one of those Republicans who underestimated the seriousness of the crisis and expected too much of southern Unionists.[24] Yet it would be hazardous to conclude that a better understanding of the southern temper would have made him and certain other members of his party more amenable to compromise. Lincoln himself had predicted in 1858 that the sectional conflict would not subside until a crisis was "reached and passed." When the crisis actually arrived, he showed no disposition to back off. "The tug has to come," he declared, "and better now, than any time hereafter."[25]

This was the conclusion to which the great majority of Republi-

22. Potter, *Lincoln and His Party*, pp. 77–80. Seward was but one of many who thought that secession was the work of "a relatively few hotheads," Van Deusen, *Seward*, p. 242. See also Stampp, *And the War Came*, pp. 13–14, with a quotation from William Cullen Bryant dated Nov. 29: "As to disunion, nobody but silly people expect it will happen." For discussion of secession as a temporary stratagem, see Catton, *Coming Fury*, pp. 139–140; Nichols, *Blueprints*, pp. 143–147, 160–161. According to Nichols, the real motive of many secession leaders "was the creation of the Confederacy as a bargaining agency more effective than a minority group negotiating within the Union."

23. Basler (ed.), *Works of Lincoln*, IV, 149–150, 152; Nevins, *Emergence*, II, 394–397.

24. Potter, *Lincoln and His Party*, pp. 18, 245–248. Henry J. Raymond was another, for example, who expected a rising of southern Unionists. See Krummel, "Raymond," pp. 389, 395.

25. Basler (ed.), *Works of Lincoln*, IV, 150. Mark M. Krug, *Lyman Trumbull, Conservative Radical* (New York, 1965), pp. 174–175.

cans came, sooner or later. Many, like Senator Lyman Trumbull of Illinois, stood adamantly against compromise from the beginning, and such attitudes did not necessarily reflect radicalism on the slavery issue.[26] In fact, antislavery moderates often proved to be antisecession militants. The progress of disunion, far from frightening Republicans into offering concessions, gave them an additional reason for standing firm—namely, that any yielding to the secessionists would be a surrender to extortion and a subversion of popular government. Here is the key to understanding why many Republicans seemed to become more intractable as the danger of disunion became more palpable. Secession in actual operation tended to change the whole nature of the sectional conflict. The main problem at hand was no longer the expansion of slavery but the survival of the United States, and the most pressing moral issue was not now slavery but majority rule.[27] In other words, secession gave the Republicans a second noble cause and one that would ultimately command broader support; for on the issue of slavery the South had always been more united than the North, whereas the question of disunion tended to split the South and unify the North. So not only Lincoln's opposition but the very logic of the developing conflict discouraged growth of compromise sentiment within the Republican party. At any rate, men like Thurlow Weed found themselves under fierce attack in the party press and fell silent, beat a retreat, or explained their flirtation with appeasement as subtle strategy.[28]

Historians have commonly viewed the crisis of 1860–1861 as one presenting three distinct and mutually exclusive options to the American people: peaceable separation, compromise, or war. Within such a framework, given the momentum of secession and the fundamental set of Republicanism, it is probably safe to say that compromise was impossible from the start. The maximum that

26. *Ibid.*, pp. 177–178.

27. Richard Henry Dana declared that the North could not "buy the right to carry on the government" by making concessions to slavery, quoted in Eric Foner, *Free Soil, Free Labor, Free Men: The Ideology of the Republican Party before the Civil War* (New York, 1970), p. 220. See also Trefousse, *Wade*, p. 135. Lincoln in his first inaugural declared: "If the minority will not acquiesce, the majority must, or the government must cease. Plainly, the central idea of secession is the essence of anarchy. A majority, held in restraint by constitutional checks and limitations, and always changing easily, with deliberate changes of popular opinions and sentiments, is the only true sovereign of a free people." Basler (ed.), *Works of Lincoln*, IV, 268.

28. Stampp, *And the War Came*, pp. 172–173.

Republicans might conceivably yield fell far short of the minimum that secessionists might conceivably accept as a basis of reconciliation. No action within the power of Congress would be forceful enough, or could even be completed soon enough, to stem the initial tide of secession—certainly not in South Carolina and probably not in the other states of the deep South. And as long as the secession movement continued, creating points of severe friction like the Charleston forts, war remained an imminent danger. There was, indeed, no way to choose "compromise" and in so doing make the other two options disappear. As an *exclusive* alternative to separation or war, "compromise" simply did not exist in the winter of 1860–1861.

If, however, one can break away from an illusion produced by scholarly logic infused with scholarly hindsight and can then view the crisis, as contemporaries did, in all its disorderly and changing variety of options and potential consequences, the unpromising efforts at compromise during that winter take on a different historical significance.[29] In the South, at mid-December, the secession movement was just about to enter its formal phase. How far it would proceed no one knew—perhaps through all of the slave states and even beyond; for there was talk of further fragmentation, of an independent Pacific coast republic and a free city of New York.[30] In the North, where only scattered voices were supporting a "go in peace" policy, fewer still were demanding the prompt dispatch of an army to suppress the rebellious South Carolinians.[31] Between these two extremes there seemed to be a great range of possible

29. For some discussion of the handicaps of hindsight in studying the secession crisis, see David M. Potter, "Why the Republicans Rejected Both Compromise and Secession," in George Harmon Knoles (ed.), *The Crisis of the Union, 1860–1861* (Baton Rouge, 1965), pp. 90–106.

30. Perkins (ed.), *Northern Editorials*, I, 389–390, 396–398; Samuel Augustus Pleasants, *Fernando Wood of New York* (New York, 1948), pp. 102–119; Joseph Ellison, *California and the Nation, 1850–1869* (Berkeley, 1927), pp. 178–188; William Henry Ellison, *A Self-Governing Dominion: California, 1849–1860* (Berkeley, 1950), pp. 309–314. The future copperhead leader, Clement Vallandigham, proposed to save the Union by dividing it into four parts, each with a degree of veto power over the acts of the federal government—a variation on Calhoun's concurrent majority. See *Congressional Globe*, 36 Cong., 2 sess., pp. 794–795; Frank L. Klement, *The Limits of Dissent: Clement L. Vallandigham and the Civil War* (Lexington, Ky., 1970), pp. 53–56.

31. "The idea that the free states intend to march armies into the seceding states to force their return to loyalty seems too monstrous for serious denial." Springfield, Massachusetts, *Republican*, in Perkins (ed.), *Northern Editorials*, I, 225. But see Stampp, *And the War Came*, pp. 25–28, for some advocacy of military coercion.

responses to secession, including nonresistance combined with official nonacquiescence; temporary acquiescence in the hope of subsequent reconciliation; nonviolent maintenance of some federal authority (such as collecting duties on ships stationed outside harbors); defense of federal property, by military action if necessary; economic sanctions (such as an embargo); limited coercion (such as imposing a blockade); and so forth.

These were contingent aspects of the secession crisis, which, it should be remembered, was a crisis over secession and not, in any direct way, over slavery. Yet all of the efforts at compromise in Congress dealt with the issue of slavery and only obliquely with the problem of secession. Support for compromise came primarily from those groups that were still uncommitted on the issue of secession, especially the border slave states and the northern Democrats. Insofar as they hoped by their efforts to make the secession crisis go away, the compromisers were bound to fail, but the crucial question was how they would behave in the face of failure. Which side would they blame more, the secessionists or the Republicans? What would they do, these compromisers, as they were pushed irresistibly from dealing with the question of slavery to committing themselves on the issue of secession? How would they answer, these peacemakers, if they were compelled to take sides in a war? To cancel secession and prevent civil war was no doubt beyond their powers, but they and their compromise movement could have a determining influence on how far secession would proceed and on the nature of the war that might have to be fought. Considered thus, compromise in 1860–1861 was something more than a great nonevent.

Certain secessionists and Republicans in Congress were well aware of the part that compromise negotiations might play in determining the ultimate allegiance of the compromisers themselves and of other more or less neutralist elements in the secession crisis. Without any expectation of successful compromise, they could see the advantage of making the other side appear to be the ones preventing it. Such was the case, for example, with Charles Francis Adams, who emerged as one of the Republican leaders in the House of Representatives. This son and grandson of presidents, a man of antislavery commitment and conservative temperament, said that he was prepared to yield "every doubtful point in favor of the Union" as long as there was no abandonment of Republican principles. He had no hope at all of coming to terms with the secession leaders, but he came to favor limited concessions as a means of dividing the border slave states from the deep South.[32]

Adams stepped into prominence almost immediately as a key figure on the special committee of the House created at the beginning of the new session to deal with the "perilous condition of the country." This "Committee of Thirty-three" (one member from each state) was not only ill-chosen and cumbersome but overshadowed in the public eye by a "Committee of Thirteen" which the Senate, after two weeks of oratory and quarreling, established for the same purpose.[33] Much more was expected of the Senate, which, after all, had long been the matrix of sectional compromise. The leader of the Senate committee (though not its official chairman) was John J. Crittenden, a Kentucky Whig in the Henry Clay tradition, ready with his own "omnibus" proposals. The committee membership, announced on the very day that South Carolina seceded, included political chieftains like Seward, Douglas, and Jefferson Davis. In contrast, the House committee is for the most part a roster of forgotten names. Its chairman, Thomas Corwin of Ohio, was a veteran Whig-Republican of some ability and distinction who nevertheless played a lesser role than Crittenden in committee work.[34] Furthermore, he and his colleagues made no effort to capture the attention of the country with a dramatic package of compromise proposals like the Crittenden plan. Yet, for exploring the potentialities of compromise and its limits in 1860–1861, the record of the House committee may be the more useful of the two.

For one thing, the Committee of Thirteen at the outset adopted a rule of procedure that invited failure. On the motion of Jefferson Davis, it was agreed that no action would be taken except by a dual majority of the five Republicans and of the other eight committee members. This injection of Calhoun's concurrent-majority principle into the legislative process made good sense in a way; for no measures of compromise, and certainly none requiring amendment

32. Duberman, *Charles Francis Adams*, pp. 224–226.

33. *Congressional Globe*, 36 Cong., 2 sess., pp. 6–7, 19, 22, 117, 158; Nevins, *Emergence*, II, 390, 405.

34. The chairman of the Senate committee was Crittenden's Kentucky colleague, Lazarus W. Powell, who had introduced the resolution creating it. On Crittenden, see Albert D. Kirwan, *John J. Crittenden: The Struggle for the Union* (Lexington, Ky., 1962). On Corwin, see Daryl Pendergraft, "Thomas Corwin and the Conservative Republican Reaction, 1858–1861," *Ohio State Archeological and Historical Quarterly*, LVII (1948), 1–23.

of the Constitution, had much chance of success without solid bipartisan and bisectional support.[35] Yet it should be remembered that such a rule in 1820 or in 1850 would have spelled defeat for compromise, and its adoption in 1860 virtually limited the Senate committee to dramatizing once again the incompatibility of Republicans and fire-eaters.

Crittenden followed Clay's example in laying a neat package of compromise proposals before his committee, although he must have remembered that the omnibus bill had not worked in 1850. The package consisted of six constitutional amendments and four supplementary resolutions. Only one of these ten items could be considered a concession to the antislavery element, a fact lending credence to Republican complaints that the whole thing was not a compromise but a surrender.[36] Still, much of the plan was probably negotiable, being either of marginal significance or of merely declaratory effect. One amendment, for example, would forbid the abolition of slavery on federal property located within slaveholding states; another proposed to compensate the owners of runaway slaves; and one of the resolutions called upon northern states to repeal their personal liberty laws.

The critical item in the package was an amendment restoring the Missouri Compromise line:

> In all the territory of the United States now held, or hereafter acquired, situated north of latitude 36°30′, slavery or involuntary servitude, except as a punishment for crime, is prohibited while such territory shall remain under territorial government. In all the territory south of said line of latitude, slavery of the African race is hereby recognized as existing, and shall not be interfered with by Congress, but shall be protected as property by all the departments of the territorial government during its continuance.

Several features of this proposal should be carefully noted. First, it repudiated the oldest and most important plank of the Republican platform. Second, it went beyond the Missouri Compromise by extending positive federal protection to slavery and would in fact introduce the very word "slavery" into the Constitution for the first

35. *Senate Reports*, 36 Cong., 2 sess., no. 288 (Serial 1090), p. 2; Potter, *Lincoln and His Party*, p. 171.

36. The only concession to the North was in one of the resolutions which somewhat modified the provisions of the Fugitive Slave Law. *Congressional Globe*, 36 Cong., 2 sess., p. 114.

time. Third, the phrase "or hereafter acquired" seemed to invite expansion southward for the incorporation of more slave territory. Fourth, the guarantee thus offered to slaveholders would itself be absolutely guaranteed against subsequent change.

This extra security was to be achieved by what must be regarded as the most unusual part of Crittenden's grand design. The last of his six amendments provided that the other five could never be affected by any future amendment, and it extended the same immunity to the three-fifths clause and the fugitive-slave clause of the Constitution. It also forbade any amendment authorizing Congress "to abolish or interfere" with slavery in any states where it was permitted. The idea of making some parts of the Constitution unamendable was perhaps illusive, whether as a matter of theory or of practice. But the proposal does aptly illustrate the widespread recognition that what the South wanted most was *reassurance,* and the widespread desire to make any sectional settlement a *final* one.

The deliberations of the Crittenden committee began on December 22 and came to an end just six days later. Some hope of success had been inspired by Weed's renewed advocacy of compromise, but it soon became clear that he did not speak for Seward. The latter joined the other four Republican members in voting against the most essential part of the Crittenden plan—that is, the proposal to extend the Missouri Compromise line. This in itself presumably spelled defeat because of the special rule requiring a dual majority. The two senators from the deep South, Jefferson Davis and Robert Toombs, nevertheless made it doubly certain by joining in the negative vote, although they had earlier indicated a willingness to support the measure if the Republicans would do likewise. Other proposals, including one from Douglas, were no more successful. The Republicans did more or less endorse two items in the Crittenden plan, the one against federal interference with slavery in the southern states and the one urging northern states to review their personal liberty laws. But these were minor concessions, and on New Year's Eve it was reported to the Senate that the committee had been unable to agree upon "any general plan of adjustment." This did not, of course, mean the end of all compromise efforts in the Senate. Yet one more door had obviously closed. "The day for the adjustment has passed," declared Judah P. Benjamin of Louisiana that same afternoon. "If you would give it now, you are too late. . . . within a few weeks we part to meet as Senators in one common council chamber of the nation no more forever. We desire, we beseech you, let this parting be in peace."[37]

Republicans were a minority of five on the Committee of Thirteen, and they stuck together, playing essentially a negative role in its deliberations. On the Committee of Thirty-three, however, Republicans constituted a virtual majority, which, as a group, displayed less unity and more interest in avoiding the appearance of mere obstructionism. For example, on a preliminary resolution giving qualified endorsement to the general principle of conciliation, Republican members divided evenly and thus contributed appreciably to an affirmative vote of 22 to 8.[38] They closed ranks, to be sure, on the critical issue of slavery in the territories, voting unanimously against a proposal to extend the Missouri Compromise line. But unlike the Senate Republicans, they at least offered an alternative concession—the immediate admission of New Mexico, presumably as a slave state. This was a new version of the familiar scheme for by-passing the territorial issue, advanced in this instance by Maryland's Henry Winter Davis with a specific purpose in mind. A Know-Nothing in the process of becoming a Republican, Davis explained to Charles Francis Adams that the New Mexico proposal was designed to please the border states and split them from the deep South, whose representatives would no doubt oppose it. Adams found this reasoning persuasive and agreed to sponsor the proposal, together with a resolution in favor of a constitutional amendment to protect slavery in the states. The latter won committee approval on December 28 by a vote of 21 to 3 (with Republicans voting 11 to 3). A day later, the committee accepted the New Mexico measure, 13 to 11, with Republicans 9 to 6 in favor and southerners 5 to 2 against.[39]

Here, then, were some curious developments. The House of Representatives had always been less amenable to compromise than the Senate, and Republicans in both houses had never wavered in their hostility to extension of the Missouri Compromise line. Yet now a

37. *Senate Reports,* 36 Cong., 2 sess., No. 288 (Serial 1090), pp. 5, 8–11; *Congressional Globe,* 36 Cong., 2 sess., pp. 211, 217.

38. *House Reports,* 36 Cong., 2 sess., no. 31 (Serial 1104), p. 8; Dubernan, *Charles Francis Adams,* p. 229. Adams voted against the resolution, however.

39. These were the votes on the preliminary but decisive resolutions. On the draft amendment, the vote was 20–5 in favor; on the New Mexico bill, 14–9 in favor. *House Reports,* 36 Cong., 2 sess., no. 31 (Serial 1104), pp. 19, 20–21, 35–37; Duberman, *Charles Francis Adams,* pp. 231–233, 236.

majority of Republicans on a House committee were endorsing the admission of a state to the Union, with the tacit understanding that it would be a slave state and with the knowledge that its boundary would extend well north of 36°30′. And most southerners, in turn, refused to accept this seemingly generous offer as a substitute for the Missouri formula. The anomaly is not inexplicable, however. Admitting New Mexico, unlike authorizing slavery in a federal territory, would have little symbolic value for the South, and it would offer no security for the institution in any territory subsequently acquired. Furthermore, there was considerable agreement on both sides that slavery would never flourish in New Mexico. Adams and his associates had been reassured on this point by a former federal judge in the territory.[40] Thus Republicans were actually yielding less, and slaveholders stood to gain less, than it appeared on the surface.

The Adams-Davis strategy was nevertheless beginning to work, as southerners on the committee found themselves maneuvered into the role of obstructing proffered compromise. Obviously some progress had been made toward separating the border states from the deep South. Later, Adams widened the breach when he won the support of four border-state members for a resolution declaring that it was the "high and imperative duty of every good citizen" to acquiesce in the election of a president.[41] The elements of a developing quid pro quo seemed plain enough—limited concessions to slavery in return for abstention from secession by some of the slaveholding states. Yet the strategy proceeded no further and soon began to disintegrate. Adams himself subsequently voted against his own resolutions, giving a lame excuse for doing so. Bitterly reproached in some Republican quarters for his flirtation with compromise, he apparently grew more cautious as a consequence. It had, in fact, become painfully clear that any strategy powerful enough to divide the South would probably also divide the Republican party. So meetings of the Committee of Thirty-three drifted downward into impotence, and it finished up most curiously in mid-January by submitting to the House a set of proposals that it did not expressly endorse.[42]

40. *Ibid.*, p. 235.

41. *Ibid.*, p. 246; *House Reports*, 36 Cong. 2 sess., no. 31 (Serial 1104), pp. 32–34.

42. Duberman, *Charles Francis Adams*, pp. 247–248; *House Reports*, 36 Cong., 2 sess. no. 31 (Serial 1104), pp. 39–40. The committee simply reported its actions. There were seven minority reports with a total of fourteen signatures.

From their unproductive committees, Corwin and Crittenden took the struggle for compromise to the floors of the House and Senate. During January and February the national legislature continued its efforts to meet the secession crisis by achieving some kind of breakthrough in the controversy over slavery. Meanwhile, the executive branch of the government had to respond directly to the problem of secession. There was, it should be added, not much collaboration between the president and Congress at this time. They proceeded along their separate courses, extending little help to each other.

As early as December 11, when the Louisiana legislature ordered a state convention, it was clear that at least six states of the deep South were likely to cut loose from the Union within the next two months. What remained to be seen was whether they could do so without precipitating a civil war and how many other states would follow their example. These two questions were closely connected, for anything resembling an effort at military coercion on the part of the North would almost certainly drive a number of still-uncommitted southern states (like Virginia) into the ranks of secession. The answer to both questions, as it turned out, depended heavily upon what the president decided to do about the small army garrison holding a fort at the entrance to Charleston Harbor.

With Buchanan, however, presidential decision making had usually been a collaborative activity. More often than not, he allowed himself to be governed by the collective will of his cabinet, in which southern views tended to predominate. But the cabinet was the first body in Washington to feel the shattering effect of the crisis. Within a period of little more than a month, four of its seven members resigned and two others moved to different positions. Only Secretary of the Navy Isaac Toucey continued insignificantly in his same place. Given the character of Buchanan's presidency and the kind of decisions he faced, this swift reorganization of the cabinet was probably a far more crucial development than anything occurring in Congress.

It began on December 8 with the resignation of Howell Cobb from the Treasury. A onetime Unionist now ready for secession, Cobb was one of the president's closer friends and the leading southerner in the cabinet. He took his departure earlier than necessary—a full six weeks before the secession of Georgia—and one cannot but wonder how different things might have been if he had chosen to stay on, playing a more Machiavellian game. It seems just possible that his added influence on Buchanan might have been enough to squeeze Major Anderson's garrison out of Charleston. But Cobb instead went off to strengthen his political position at home and later would serve as chairman of the convention at Montgomery, Alabama, that created the Confederate States of America. His successor, Philip F. Thomas of Maryland, was prosouthern but a lightweight who lasted only about a month.[43]

On the other hand, the highest-placed South Carolinian in the administration stayed on the job until his state seceded. William H. Trescot, the assistant secretary of state who was all the more influential because of the infirmities besetting his superior, Lewis Cass, acted openly as a secessionist and agent of South Carolina while continuing his official duties in November and December of 1860. Similarly, the Mississippian Jacob Thompson clung to his position as secretary of the interior while working as a "cooperationist" for simultaneous secession by all slaveholding states on March 4. In the interest of southern solidarity, Thompson even agreed to serve as an emissary from the governor of Mississippi to the state of North Carolina, and, most astonishing of all, with the plea that he would be trying to delay secession (until March 4, that is), he won presidential approval of his mission.[44]

The next person to leave the cabinet was Cass, the secretary of state. He roused himself momentarily and on December 15 became a Republican hero for a day by resigning in protest against Buchanan's failure to reinforce Anderson.[45] In his place the president named Attorney General Jeremiah S. Black, who, at this point if not earlier, emerged as the strong man of the administration. One could, indeed, argue that Black proved to be the pivotal figure of the entire secession crisis. This "Scotch-Irish son of thunder" from Pennsylvania was an expert and eloquent constitutional lawyer, as

43. Klein, *Buchanan*, pp. 370–371; Nichols, *Disruption*, pp. 402, 437–438.

44. Klein, *Buchanan*, pp. 368–369, 373; Auchampaugh, *Buchanan and His Cabinet*, pp. 118–119; Nichols, *Disruption*, p. 411; Swanberg, *First Blood*, pp. 52–54. Trescot, according to his own recollection, continued to work at the State Department until Dec. 17. See Gaillard Hunt (ed.), "Narrative and Letter of William Henry Trescot, Concerning the Negotiations between South Carolina and President Buchanan in December, 1860," *AHR*, XIII (1908), 528–556, and esp. p. 540.

45. Frank B. Woodford, *Lewis Cass, the Last Jeffersonian* (New Brunswick, N.J., 1950), pp. 325–329; Nichols, *Disruption*, pp. 408–410; Auchampaugh, *Buchanan and His Cabinet*, pp. 71–73.

well as a hard-headed but sometimes emotional politician.[46] No critic of slavery, he had been a loyal friend and lieutenant throughout Buchanan's administration, especially in the battles against Douglas and popular sovereignty. More recently, he had inspired that part of the annual message declaring that although secession was unconstitutional, the federal government had no power to coerce a seceded state back into the Union. Yet he also firmly believed that the government had a constitutional right and duty to protect its property and keep its laws in operation. Black, in short, did not rule out the use of force if it were clearly defensive, and in this respect he differed but little from a large number of Republicans.

As his own replacement, Black persuaded a reluctant Buchanan to appoint Edwin M. Stanton, a lawyer of great ability, erratic temperament, and antislavery convictions. The net result of these changes was a gain for Unionism—not in numbers but in emotional energy and strength of will. Together with Postmaster General Joseph Holt, a Kentucky Unionist, Black and Stanton soon began to get the upper hand in the cabinet.[47]

The high official under the most nervous strain at this time was the one occupying the seat of the greatest strategic importance. Nothing better illustrated Buchanan's weakness as a chief executive than the fact that John B. Floyd continued to serve as secretary of war. His careless and unprincipled administration of the department had already produced more than one public scandal, and now the worst of his derelictions was about to be fully revealed. He had endorsed the promissory paper of an army contractor to a total of several million dollars, persisting in the practice even after Buchanan instructed him to stop. Now, just before Christmas, it came to light that the contractor, William H. Russell, had persuaded a government clerk to accept $870,000 of these dubious bills in exchange for negotiable bonds in the Indian Trust Fund. This amounted to embezzlement, and the guilty clerk was a kinsman of Mrs. Floyd.[48]

46. Nichols, *Disruption*, p. 80. On Black generally, see William Norwood Brigance, *Jeremiah Sullivan Black* (Philadelphia, 1934).

47. Klein, *Buchanan*, pp. 358–360, 372; Nevins, *Emergence*, II, 342, 358–359; Benjamin P. Thomas and Harold M. Hyman, *Stanton: The Life and Times of Lincoln's Secretary of War* (New York, 1962), pp. 87, 90–91.

48. Nevins, *Emergence*, II, 373–374. The firm was Russell, Majors and Waddell, on which see Raymond W. and Mary Lund Settle, *War Drums and Wagon Wheels* (Lincoln, Neb., 1966).

Clearly, Floyd would have to go (though Buchanan lacked the nerve to tell him so directly), but for the moment he refused to resign, insisting that he must vindicate himself first. Now Floyd, a Virginian, had been regarded as a Unionist, one of the "conditional" variety at least. After Lincoln's election, however, he leaned more toward support of secession, and the tendency grew stronger as personal disgrace settled upon him. The cause of the South gave him a chance to leave with a belligerent flourish.[49]

As secretary of war, Floyd had of course helped shape administration policy regarding the Charleston forts. The pressures there mounted steadily as South Carolina moved briskly through the preliminaries of secession. Late in November, after having inspected his new command, Major Anderson began to call for reinforcements, arguing that the weakness of his garrison invited attack and thus constituted a threat to peace. Buchanan vacillated. At first he agreed with Cass and Black that Anderson's request should be met. Then Floyd persuaded him to wait until he could confer with Winfield Scott, the seventy-four-year-old commanding general of the army. But Scott, it transpired, was sick over in New York City, where he had set up his headquarters after quarreling with a previous secretary of war, Jefferson Davis. Taking advantage of the delay, southern members of the cabinet hastily fashioned a "gentlemen's agreement" and got it endorsed by the governor of South Carolina. If no reinforcements were sent to Anderson, there would be no efforts to dislodge him from Fort Moultrie. With much relief, Buchanan acceded, hoping to buy time for the promotion of a compromise movement. On December 8 a delegation of South Carolina congressmen sought to confirm the agreement. They asked the president for an explicit pledge not to reinforce Anderson. Buchanan responded with some characteristic double-talk in which he avoided making a literal promise while giving the impression that he had committed himself.[50]

Meanwhile, Anderson was asking for specific instructions in the face of expected attack on his highly vulnerable position. The answer decided upon in Washington was that he must do

49. Nichols, *Disruption*, pp. 423–427; Klein, *Buchanan*, pp. 377–378. Floyd subsequently became a Confederate general and died in 1863. See the essay by James Elliott Walmsley in the *Dictionary of American Biography*, VI, 482–483.

50. John Bassett Moore (ed.), *The Works of James Buchanan* (12 vols.; Philadelphia, 1908–11), XI, 56–57. Nichols, *Disruption*, pp. 382–383, 386–387, 408; Klein, *Buchanan*, pp. 368, 370–371; Catton, *Coming Fury*, pp. 145–146.

nothing provocative but defend himself vigorously if necessary. Floyd dispatched Major Don Carlos Buell of the adjutant general's office to deliver these orders in person and inspect Anderson's situation. Upon reaching Fort Moultrie, Buell soon came to agree with Anderson that it would be difficult to defend. Apparently going beyond his authority, he wrote out instructions that in effect authorized Anderson to move his troops to Fort Sumter for greater security whenever he had "tangible evidence of a design to proceed to a hostile act."[51] A copy of this memorandum, dated December 11, passed through Floyd's hands into the files of the War Department, having been endorsed by him but perhaps not very carefully read.

On December 12, General Scott arrived in Washington and added his voice to those calling for reinforcement of Anderson. It was at this point that Cass resigned, complaining of the administration's weakness. But pressure for evacuation likewise mounted, both in Charleston and in Washington, as the South Carolina convention assembled for its fateful purpose. The news flashed out over telegraph wires on December 20—an ordinance of secession had been approved without dissent. Soon three commissioners from the newly "independent commonwealth" were hastening northward to confer with Buchanan about the disposition of the Charleston forts. The situation had now changed. The gentlemen's agreement was probably no longer viable. Rumors in Charleston convinced Anderson that attacks on both Fort Sumter and Fort Moultrie were imminent. Obviously he needed new instructions, and they arrived on December 23, signed by Floyd, but written by Black, now secretary of state. Anderson must "exercise a sound military discretion," defending himself if attacked but making no "useless sacrifice" of lives. If the hostile force should be too strong, he must secure the best terms possible for yielding up his command. The key words here were "military discretion." Nothing in this dispatch contradicted the instructions of December 11 from Buell.[52]

51. Nichols, *Disruption*, p. 407; *Mr. Buchanan's Administration*, pp. 165–166; *The War of the Rebellion: A Compilation of the Official Records of the Union and Confederate Armies* (128 vols.; Washington, 1880–1901), series I, vol. I (hereafter cited as *Official Records*, I), 81–82, 89–90.

52. *Official Records*, I, 103; Nichols, *Disruption*, p. 423; Floyd's dispatch did nevertheless seem to make surrender of the fort easier, while leaving the decision to Anderson. It also contributed to the subsequent image of Floyd as a secessionist conspirator. See, for example, Roy Meredith, *Storm over Sumter* (New York, 1957), pp. 2, 19, 30–31, 199–200.

As a professional soldier Anderson had no stomach for surrendering a post so recently entrusted to him, and as a southerner he had no desire to precipitate hostilities between South Carolina and the United States. Yet the very weakness of Fort Moultrie invited an attack which he was expected to resist. His request for additional troops had been denied, but there remained the option of moving to a more defensible position. On an island at the harbor entrance stood Fort Sumter, its fortifications now almost completed, but still virtually unmanned—a power vacuum in the very center of the sectional storm. What followed was indisputably an exercise of "sound military discretion." During the night of December 26, Anderson spiked Moultrie's guns and with great skill transferred his entire command to Sumter.[53] The people of Charleston, awakening to the sounds of bitter discovery, were loud in their anger. Anderson rejected a curt demand that he return to Fort Moultrie, which was then seized, along with other federal property, by South Carolina troops. In Charleston, the Stars and Stripes now floated only over Fort Sumter.[54]

Anderson's maneuver may have been the most important one-man decision of the entire secession winter. While the political leaders of the nation orated and parleyed and disagreed, this middle-grade army officer in one swift move determined the place and the nature of the final confrontation between North and South. A roar of applause for his display of initiative echoed through the free states. Yet Anderson's bold and seemingly aggressive action was actually a conservative effort at disengagement. What drew him to Fort Sumter primarily was not so much a belief that it could repel attack as a hope that its obvious strength might deter attack altogether. A southerner with mixed feelings about the secession crisis, he wanted above all to avoid starting a war, and so he preferred to see the day of reckoning postponed. In this respect, the major did not differ greatly from his hand-wringing commander in chief. Anderson, in fact, had probably got Buchanan off the hook.

53. *Official Records,* I, 108–109; Abner Doubleday, *Reminiscences of Forts Sumter and Moultrie in 1860–61* (New York, 1876), pp. 58–67; Catton, *Coming Fury,* pp. 143–145, 149–150, 153–157; Swanberg, *First Blood,* pp. 95–101.

54. Crawford, *Genesis,* pp. 108–118; Swanberg, *First Blood,* pp. 104–108. "I can now breathe freely," Anderson wrote to his wife on Dec. 26. "The whole force of S. Carolina would not venture to attack us." Eba Anderson Lawton, *Major Robert Anderson and Fort Sumter* (New York, 1911), p. 8.

If he had stayed at the more vulnerable Fort Moultrie, the pressure upon him probably would have become unbearable well before March 4, and then Buchanan instead of Lincoln would have had to make the ultimate choice between withdrawing and fighting.

Furthermore, as long as Moultrie remained the focus, it would have been easier to choose withdrawal, perhaps after a faint gesture of resistance. Or, even if serious fighting did occur, it still would have been of a different kind from the formal, almost ritualistic assault upon Sumter launched three and one-half months later, not by a single rebellious state, but rather by a proud new confederacy. Moultrie in December had nothing like the symbolic meaning that Sumter acquired by April. Day by day, the emotional investment mounted, and the prestige at stake grew to be enormous, as Anderson's troops and the Charlestonians stared at each other across a little stretch of water. Fort Sumter in the end came to be the supreme symbol of the conflict between state and national sovereignty. Anderson succeeded in putting off the day of reckoning in Charleston Harbor, but at the cost of increasing its eventual impact. A maneuver that postponed war may at the same time have made war ultimately more unavoidable.

The news from Charleston caused dismay in the White House. Southern senators led by Jefferson Davis came buzzing like hornets around Buchanan, charging that solemn pledges had been broken and insisting that Anderson must be ordered back to Fort Moultrie. Thompson joined their chorus, and Floyd spouted indignation in one final display of bravado, but Black, Stanton, and Holt vehemently applauded the move to Fort Sumter. The unhappy president, though fervently wishing Anderson had stayed put, nevertheless concluded that there had been no disobedience of orders and refused to take any hasty action.[55]

At this point, December 28, the three South Carolina commissioners arrived to negotiate with Buchanan and promptly overplayed their hand with a peremptory demand for the immediate evacuation of all troops from Charleston.[56] If they had merely asked for Anderson's return to Moultrie, Buchanan might have acceded, but a complete withdrawal under threat was beyond considera-

55. Klein, *Buchanan*, pp. 378–380; Thomas and Hyman, *Stanton*, pp. 95–96; Auchampaugh, *Buchanan and His Cabinet*, p. 158; Jefferson Davis, *The Rise and Fall of the Confederate Government* (2 vols.; New York, 1881), I, 215–216.

56. Moore (ed.), *Works of Buchanan*, XI, 76–77; Nichols, *Disruption*, pp. 429–430.

tion.[57] Even so, he drafted a reply conceding so much that Black threatened to resign if it were delivered. Here was a critical moment, but the president, after some stubborn argument, gave in. His reworked and sterner reply included a statement of intention to defend Fort Sumter "against hostile attacks, from whatever quarter they may come."[58] The commissioners responded with an insulting letter and took their departure. Floyd had already resigned. Thomas and Thompson would soon do likewise. The last southern influence was being squeezed out of the cabinet. The Buchanan administration in early January presented a solid front of Unionism.[59]

A commitment to defend Fort Sumter was by implication a promise to reinforce its garrison. On December 31, the day after his firm reply to the South Carolina commissioners, Buchanan authorized preparation of a relief expedition using the sloop-of-war *Brooklyn.* General Scott preferred, however, to send a merchant vessel of lighter draft, arguing that it would be less conspicuous and more likely to make its way past harbor obstructions. Thus, on January 5, the chartered steamer *Star of the West* set out quietly from New York, loaded with supplies and about two hundred troops. Buchanan was painfully aware that the enterprise might precipitate war, and he soon had even greater reason for anguish. On that very same day, a message arrived from Anderson saying that he was well situated and in no immediate need of help. A countermanding order reached New York too late. The ship had already weighed anchor. For the president and his advisers, the next few days were a time of almost unbearable suspense.[60]

When the *Star of the West* approached Charleston on January 9, the South Carolinians were waiting for it. They had been warned by several southerners in Washington, including Jacob Thompson,

57. *Mr. Buchanan's Administration,* pp. 181–182. The significance of Governor Pickens's hasty seizure of the other Charleston forts should not be overlooked. This action, apparently taken to counteract public exasperation at his having allowed Anderson to occupy Sumter, made it far more difficult to negotiate simply for the latter's return to Moultrie, from which it would have been easier to evacuate him.

58. Moore (ed.), *Works of Buchanan,* XI, 79–91; Nichols, *Disruption,* pp. 429–432; *Mr. Buchanan's Administration,* p. 183; Thomas and Hyman, *Stanton,* pp. 97–102.

59. *Official Records,* I, 120–125; Moore (ed.), *Works of Buchanan,* XI, 100–104; *Mr. Buchanan's Administration,* pp. 183–184.

60. *Official Records,* I, 120, 131–132; Charles Winslow Elliott, *Winfield Scott, the Soldier and the Man* (New York, 1937), pp. 684–685, 688–689; *Mr. Buchanan's Administration,* pp. 189–191; Crawford, *Genesis,* pp. 174–177; Klein, *Buchanan,* pp. 388–389; Nichols, *Disruption,* p. 435.

even though he was still drawing his pay as secretary of the interior.[61] Shore batteries opened fire on the unarmed vessel, which retreated hastily without much damage and headed back to New York. During this little drama, the guns of Fort Sumter remained silent. Anderson, who had no orders covering such a contingency and, indeed, no official notification that reinforcements were on their way, elected not to return the South Carolinian fire. For this restraint he received formal commendation from Joseph Holt, now secretary of war in Floyd's place. The feeling at the White House was one of great relief that the risky and apparently unnecessary venture had ended without bloodshed.[62]

Northerners in large numbers and of all political varieties found the *Star of the West* episode humiliating. Many of them demanded immediate retaliation. The president might have closed out his term in heroic style by issuing a call for troops and organizing an assault in strength on Charleston. After all, the flag of the United States had been deliberately attacked. In fact, the first shots of the Civil War had actually been fired; and yet the war itself did not begin. Instead, Buchanan acquiesced passively in a new truce engineered by Major Anderson. The latter, confronted with another demand for surrender, persuaded the governor of South Carolina, Francis W. Pickens, to join him in referring the matter back to Washington. Pickens appointed Isaac W. Hayne as his emissary, and Anderson named an officer to accompany him. The situation in Charleston Harbor had thus been stabilized once more, pending the outcome of renewed negotiations in Washington.[63]

Hayne took with him a letter from Pickens demanding that Fort Sumter be yielded up, but he postponed delivering it to Buchanan at the urging of ten senators from the deep South. These were climactic hours in the cotton kingdom. Mississippi, Florida, and Alabama had just seceded on successive days (January 9–11); Georgia, Louisiana, and Texas would follow suit within three weeks. The more difficult task of achieving unity lay ahead, however. It was a time calling for bold statesmanship but military restraint. Secessionist leaders were as anxious as Buchanan to avoid bloodshed—at least until after the organization of their new confederacy. Hayne

61. Crawford, *Genesis*, pp. 179–183; Swanberg, *First Blood*, pp. 129–130.
62. *Official Records*, I, 9–10, 120; Doubleday, *Reminiscences*, pp. 102–104; Catton, *Coming Fury*, pp. 178–182; Crawford, *Genesis*, pp. 183–186.
63. *Official Records*, I, 134–140; Klein, *Buchanan*, p. 390.

did finally present the Pickens demand to the president at the end of January, receiving a firm letter of refusal from Holt on February 6. Then, like the three commissioners in December, Hayne fired off an offensive reply and headed for home. The truce arranged by Anderson had apparently ended. Yet Pickens, instead of preparing an assault on Fort Sumter, decided to hand the problem over to the new government just coming into existence at Montgomery, Alabama. Thus more weeks would elapse before any southern initiative threatened the status quo at Charleston.[64]

Of course any modus vivendi would end abruptly if the effort to reinforce Sumter were renewed. Black and Stanton nevertheless vigorously recommended the prompt dispatch of another expedition, but in this instance Buchanan's natural caution prevailed, partly because it was now fortified by the reassurances from Anderson. The commitment to hold Fort Sumter remained unchanged. In a special message to Congress on January 8, the president had emphatically reaffirmed his "right and duty to use military force defensively" against persons who resisted federal officers and attacked federal property.[65] He therefore agreed with Black and Stanton that a new relief expedition should be organized, but it was not to be sent until Anderson explicitly called for help. And such a call would never come; for Anderson was now utterly convinced that any attempt at reinforcement would mean war. On the secessionist side, meanwhile, Jefferson Davis and the new Confederate government were gradually taking over responsibility for the Sumter problem and warning the South Carolinians not to risk failure in a premature attack on the fort. Thus the way was clear for Buchanan to coast more or less peacefully through the remaining weeks of his term. All this time, however, the balance of power in Charleston Harbor had been shifting, as local troops day by day strengthened the ring of batteries confronting the Sumter garrison. Anderson, it seems plain, allowed personal feelings to color his periodic reports to Washington. He misled Buchanan about the strength of his position, but the president in turn was virtually asking to be deceived. Both men were motivated by the desire to avoid bloodshed and civil conflict. What they accomplished instead was the surrender of Fort

64. Moore (ed.), *Works of Buchanan,* XI, 126–141; Klein, *Buchanan,* pp. 393–395; Crawford, *Genesis,* pp. 226–234, 266–267.

65. Richardson (ed.), *Messages and Papers,* V, 656.

Sumter by inches over a period of several months.[66]

What made Buchanan doubly reluctant to force the issue in Charleston Harbor was his justifiable fear of giving added stimulus to secession and delivering a fatal blow to compromise at a time when, on both fronts, there appeared to be some grounds for renewed hope. On February 1, Texas brushed aside the opposition of its governor, Sam Houston, and became the seventh state to leave the Union. But for the moment, at least, no more departures were in sight. The secession movement had obviously lost much of its original momentum. Eight of the fifteen slave states would not be associated with the convention assembling in Montgomery on February 4 to create a new southern republic. The vote for secession had been surprisingly close in Georgia and certain other parts of the cotton kingdom. Throughout much of the upper South, neither Unionists nor disunionists had the upper hand as yet, and there was a strong disposition to reserve judgment, allowing time for one more attempt at sectional reconciliation. Prospects varied from state to state, but the most important decision, everyone knew, would be that of Virginia.

None of the seven seceding states had anything like the Old Dominion's historic ties with the Union. The "father of his country," the author of the Declaration of Independence, the "father of the Constitution," and the greatest chief justice—all had been Virginians. For more than half the years since 1789, someone born in Virginia had occupied the presidency. The luster of the past, though it could not in the end offset the pressures for secession, infused the crisis with a special poignancy for Virginians and gave them added reason to explore every alternative possibility.

So from Virginia there came a new compromise movement, with an ex-president as one of its principal sponsors. John Tyler, now seventy-one years old, was already at heart a secessionist and before long would be serving in the Confederate Congress. But with an ambivalence one might expect in a Virginia slaveholder who had held every elective federal office, he determined to make one last effort or one last gesture in defense of the Union. Tyler's plan,

66. Moore (ed.), *Works of Buchanan*, XI, 123–124; Doubleday, *Reminiscences*, pp. 129, 136; Stampp, *And the War Came*, pp. 100, 263–264; Swanberg, *First Blood*, pp. 203–204.

published January 17, called for a convention of border states, six free and six slave, because they were "most interested in keeping the peace." The Virginia legislature, having also received a recommendation from Governor John Letcher, promptly voted to sponsor such a conference but extended its invitation to all states North and South. The place designated was Washington; the date, February 4, was by no coincidence the same as the one chosen for the opening of the Montgomery convention.[67]

The deep South ignored Virginia's invitation, and the Pacific coast states were too far away to respond. Elsewhere, however, it stirred up hope and controversy and a flurry of mixed reactions. In general, the strongest support came from northern Democrats, from Whig-American elements in the border slave zone, and from commercial interests in certain eastern cities. Secessionists in the upper South tried unsuccessfully to have delegates sent to Montgomery instead of Washington. Wherever Republicans controlled the state government, there was sharp debate over whether to send a delegation. Even the radical wing of the party found itself divided on the question, but the argument for participation as a matter of strategy proved convincing. It would be wise, said Republican pragmatists, to control the conference instead of fighting it, and besides, some show of cooperation with the Union-savers might hold the border states—for a while at least. In the end, all the northern states were represented except Michigan, Wisconsin, and Minnesota.[68]

And so, at Willard's Hotel two blocks from the White House, there began "the last sad effort to avert war."[69] A total of 132 delegates from twenty-one states were eventually seated. Radical Republicans like Salmon P. Chase of Ohio and secession-minded southerners like James A. Seddon of Virginia had come to act merely as watchdogs. Perhaps a bare majority took their commission seriously and thought of themselves as in some respects a new version of the quasi-legal convention that had saved the Union by reconstructing it in 1787 at Philadelphia. They too put a distinguished Virginian in the chair, decided of necessity to vote by state, and closed their sessions to the press and public. A Vermont delegate named Lucius E. Chittenden, in conscious imitation of James

67. Robert Gray Gunderson, *Old Gentlemen's Convention: The Washington Peace Conference of 1861* (Madison, 1961), pp. 24–25.
68. *Ibid.*, pp. 33–42.
69. Nevins, *Emergence*, II, 411.

Madison, kept informally the fullest record of the proceedings.[70]

But John Tyler as chairman was no George Washington in influence or purpose, and whereas the framing of the Constitution had been dominated by men in their thirties and forties, too many compromise leaders in the assemblage at Willard's were past their political prime. In personnel and atmosphere it was indeed an "old gentlemen's convention," with nothing new to offer. After some three weeks of labor, the Peace Conference recommended a seven-part amendment to the Constitution which differed but little from the Crittenden Compromise. The crucial Section 1 extending the Missouri Compromise line had been defeated at first. Then later it passed by 9 votes to 8 as the Illinois delegation switched from negative to affirmative, thus giving momentarily the false impression that Lincoln had intervened in favor of compromise.[71] Presented to Congress on February 27, just three days before the end of the session, this proposed amendment could not possibly win acceptance. It was essentially a last ceremonial gesture, punctuated stylishly by General Scott with a one-hundred-gun salute to the adjourning delegates.

In Montgomery, Alabama, meanwhile, a convention of thirty-eight men representing six states was proceeding swiftly and efficiently to the task of creating a new American republic. There the sense of historic drama and the zest of revolution were apparent. In just one week of work, February 4–9, the delegates adopted a provisional constitution and elected a provisional president and vice-president. Jefferson Davis and Alexander H. Stephens were inaugurated by February 18, and, with the convention transformed into a provisional legislature, the Confederate States of America became a functioning government while Abraham Lincoln was still making his slow journey from Springfield to Washington.[72]

Vestiges of federal authority in the deep South were rapidly

70. L. E. Chittenden, *A Report of the Debates and Proceedings in the Secret Sessions of the Conference Convention, for Proposing Amendments to the Constitution of the United States, Held at Washington, D.C., in February, A.D. 1861* (New York, 1864). The roster of delegates is at pp. 465–466.

71. *Ibid.*, pp. 438, 441; Gunderson, *Old Gentlemen's Convention*, pp. 87–90, 107–109. See also Gunderson (ed.), "Letters from the Washington Peace Conference of 1861," *JSH*, XVII (1951), 382–392, for the revealing comments of an Ohio Republican delegate.

72. James D. Richardson (ed.), *A Compilation of the Messages and Papers of the Confederacy* (2 vols.; Nashville, 1905), I, 29–54; Davis, *Rise and Fall*, I, 229–240.

disappearing as state officials seized control of various forts, customhouses, post offices, and other government property. Besides Fort Sumter, the one important exception was Fort Pickens at Pensacola, Florida, where an informal truce had been arranged at the end of January. By its terms, the little federal garrison could receive supplies but not reinforcements, while Florida leaders promised not to launch an attack upon the defenders.[73] Most of the ties with the Union had nevertheless been cut, and as February gave way to March, the seceding states continued to hold the initiative, having displayed a clarity of purpose notably lacking farther north.

Uncertainty prevailed especially in the upper South, where not only states but many individuals were painfully divided in their sympathies. From Virginia and North Carolina to Missouri and Arkansas, secession had been rejected,[74] but these were interim and conditional decisions. Southern Unionism now subsisted largely on two expectations—that the North would offer substantial concessions on the slavery issue, and that the North would refrain from "coercive" measures against the lower South. If the first expectation should be disappointed, much of the upper South would probably drift toward secession. If the second expectation should be disappointed, secessionism, it appeared, would not only sweep instantly through much of the upper South but also win considerable support even in the free states.[75]

Thus, half of the slaveholding South hung in suspense, still tentatively part of the Union but waiting for something decisive to happen. Nothing better epitomized the state of affairs than the Virginia convention, which met on February 18 and showed no inclination either to act immediately or to adjourn immediately. Its wait-and-see majority, though considered a triumph of Unionism when

73. *Official Records*, I, 333–342; Grady McWhiney, "The Confederacy's First Shot," *CWH*, XIV (1968), 6–7; J. H. Gilman, "With Slemmer in Pensacola Harbor," in Robert Underwood Johnson and Clarence Clough Buel (eds.), *Battles and Leaders of the Civil War* (4 vols.; New York, 1887), I, 26–32.

74. See above, pp. 504–510.

75. On prosouthern sentiment in one part of the North, see William C. Wright, *The Secession Movement in the Middle Atlantic States* (Rutherford, N.J., 1973). At the Washington Peace Conference, delegates heard these words from Robert F. Stockton, retired naval commander and former senator: "Do you talk here about regiments for invasion, for coercion—you, gentlemen of the North? You know better; I know better. For every regiment raised there for coercion, there will be another raised for resistance to coercion." Chittenden, *Debates and Proceedings*, p. 116.

elected, seemed increasingly menacing as time went by. But with the problem of the forts for the moment stabilized and the progress of secession for the moment halted, there were sanguine souls who thought that Congress might yet nail together a framework of reconciliation.

Hope had continued to center on the figure of the seventy-four-year-old senator from Kentucky. Crittenden took his compromise plan to the Senate floor on January 3, supplementing it with the most innovative proposal of the entire secession winter. Acknowledging that there was little use of introducing the plan directly for consideration by the Senate, he offered instead a resolution calling for its submission to the electorate in a kind of advisory plebiscite. Thus Republicans were no longer asked to vote for the Crittenden compromise, but only to let the people pass judgment on it at the polls. Not surprisingly, this dramatic effort to invoke direct democracy received a warm endorsement from the great advocate of popular sovereignty. "Why not give the people a chance?" Douglas asked. Even Republican voters, he predicted, would "ratify the proposed amendments to the Constitution, in order to take this slavery agitation out of Congress, and restore peace to the country, and insure the perpetuity of the Union."[76]

Although we must depend upon impressions rather than scientific samplings of public opinion, it seems likely that the Crittenden compromise would have won approval in a popular vote. Even Horace Greeley later said so.[77] But the proposal for a national referendum was too much of a novelty and, in the eyes of some senators, unconstitutional. A bill designed to put it promptly into effect, introduced by William Bigler of Pennsylvania, made no headway at all.[78]

The substance of Crittenden's plan remained the principal issue before Congress, however, and by late January it had become the beneficiary of a remarkable outpouring of public sentiment. Simon Cameron, for example, acknowledged receiving "daily, by every

76. *Congressional Globe,* 36 Cong., 2 sess., p. 237; appendix, p. 42; Johannsen, *Douglas,* pp. 819–821.

77. Horace Greeley, *The American Conflict* (2 vols.; Hartford, 1864), I, 380. But Greeley did not think it would have any effect on the secessionists. See also Kirwan, *Crittenden,* p. 404; Dwight Lowell Dumond, *The Secession Movement, 1860–1861* (New York, 1931), p. 168 n.

78. *Congressional Globe,* 36 Cong., 2 sess., pp. 351–352.

mail, a large number of letters . . . all sustaining the proposition of the Senator from Kentucky." There was also a flood of petitions such as had not been seen since the early days of organized abolitionism. Seward presented one procompromise memorial from 38,000 citizens of New York City which, if fully extended, "would cross the Senate Chamber, in its extremest length, eighteen times."[79] It is little wonder that Crittenden and other peacemakers began to think the tide might be at last running in their favor.

Yet in Congress, the Republicans maintained an almost solid front against the Crittenden omnibus, and their relative numerical strength increased significantly with the withdrawal of delegations from the seceding states. Moderates like Seward and Cameron might talk a good deal about compromise, but on critical votes they consistently followed the lead of their radical colleagues. Despite all the Union-saving pressures upon them, most Republicans were more determined than ever to take over the government on March 4 without having ransomed their right to do so. "Inauguration first —adjustment afterwards," insisted Salmon P. Chase. Again and again, Republican tactics of delay prevented the Crittenden plan from coming to a vote in either House. As a consequence, advocates of compromise began shifting their hope to the Washington Peace Conference, which did not finish its uninspired work, however, until almost the eve of congressional adjournment.[80]

The seven-point proposal of the Peace Conference aroused scant enthusiasm in the end-of-session legislative flurry. It could not even be considered in the House without a two-thirds vote suspending the rules, and this its supporters failed to get when they were allowed to try on March 1. Two days earlier, the House had at last voted on the Crittenden compromise and rejected it, 113 to 80. A bill admitting New Mexico, nominally as a slave state, likewise met defeat. But Corwin did succeed in getting the necessary two-thirds vote for his constitutional amendment forbidding any amendment authorizing Congress to interfere with slavery in the states. Some forty-five Republicans supported this concession, knowing that it was acceptable to the president-elect.[81]

79. *Ibid.*, pp. 402, 657.

80. Chase to Lincoln, Jan. 28, 1861, in David C. Mearns (ed.), *The Lincoln Papers* (2 vols.; Garden City, N.Y., 1948), II, 424–425; Stampp, *And the War Came*, pp. 136–141; *Mr. Buchanan's Administration*, pp. 144–145.

81. *Congressional Globe*, 36 Cong., 2 sess., pp. 1258, 1285, 1327, 1333; Nichols, *Disruption*, pp. 476–477.

In the Senate, Crittenden welcomed the Peace Conference plan as a substitute for his own, but by the evening of Sunday, March 3, he had sadly concluded that nothing could be salvaged except concurrence in the Corwin constitutional amendment. Debate continued through the night, and toward dawn on inauguration day the amendment finally passed with the bare minimum of 24 votes to 12.[82] Then, with the House already adjourned, the Senate proceeded to a series of meaningless votes on the compromise proposals. The recommendations of the Peace Conference were rejected, 28–7, after which the Crittenden plan at last came to a vote and was defeated, 20–19.[83]

Legislative compromise had failed because most Republicans in Congress were unwilling to abandon the fundamental principle of their party, and doubly unwilling to do so under duress. In fact, they never gave a majority of their votes to *any* procompromise action in either house. Even the Corwin amendment received only 40 percent of their votes, although the principle involved had been endorsed by the Chicago platform. The platform, however, had not been written in the face of an overt secession movement. Although the crisis undoubtedly frightened a good many Republicans into the ranks of "Union-savers," the element of threat appears to have had the opposite effect on a larger number, hardening their resistance to compromise.

A substantial minority of Republicans did support certain secondary concessions, primarily as a matter of strategy, and many historians have accordingly exaggerated the possibility of a party schism, ignoring the extraordinary solidarity manifested on the key issue—extension of slavery. The crucial fact is that the Republicans in Congress never cast *a single vote* for the Crittenden plan.

It is true that the Republicans had cooperated fully in the creation of three new territories (Colorado, Nevada, and Dakota) without making any effort to forbid slavery in any of them. Douglas could not resist crowing a little that they had thus discarded the Wilmot Proviso at last and accepted instead his own much-abused territorial formula of "nonintervention." But, with some condescension, he also praised Republican "patriotism" in abandoning the principal party doctrine when "its effect would have been to disturb the peace of the country." Southerners, he continued, should accept this re-

82. *Congressional Globe,* 36 Cong., 2 sess., pp. 1374–1403.
83. *Ibid.*, p. 1405.

markable retreat, along with Republican willingness to guarantee slavery in the states where it already existed, as "evidence of a salutary change in public opinion at the North."[84] The argument, however, neither made much impression on southerners nor caused much disturbance among Republicans. For both sides knew that with a Republican president appointing the territorial officials, there was little likelihood of any slaves being taken into the new territories. Furthermore, the three organic acts, unlike the Crittenden plan, made no verbal concessions to slavery whatsoever. They were thus actually in harmony with the Chicago platform, which called for prohibitory federal legislation only if it should prove "necessary."[85]

The Republicans, in short, without compromising their central principle, could now take a more flexible attitude about implementing it because they would soon have control of the executive department. But the implications of that control had in turn replaced the territorial question as the focus of sectional conflict. Secession had begun, after all, not as a response to anything done or left undone by Congress, but rather as a response to the election of Lincoln. It was a new kind of national crisis, brought on by the people themselves instead of their legislators. Traditional congressional methods could deal only indirectly with this crisis; for the outstanding sectional issues were now less important than the northward shift of political power—and southern reaction to it.

One event especially signalized the ending of an era. Late in January, with little more than perfunctory southern opposition, Congress approved the admission of Kansas as a free state—Kansas, no longer bleeding and no longer a battle cry.[86] Thus, the most troublesome of territories had ceased to be a territory; three new territories had been created with scarcely a ripple of controversy; and there was not much interest, northern or southern, in what happened to New Mexico Territory. As a practical matter, the whole territorial issue seemed largely exhausted.

84. *Ibid.*, pp. 728–729, 766, 1003–1005, 1206–1208, 1334–1335, 1391.

85. Nevins, *Emergence,* II, 448; Foner, *Free Soil,* p. 133.

86. *Congressional Globe,* 36 Cong., 2 sess., pp. 487–489, 603–604. On Jan. 21, by a vote of 36 to 16, the Senate approved an amended version of an admission bill passed by the House in the preceding session. On Jan. 28 the House accepted the Senate amendment without a roll call, but on a preliminary test vote there were 119 in favor and 41 against.

Yet it was the territorial aspect of the Crittenden compromise that Republicans rejected most emphatically and that southerners demanded most insistently. At the same time, in their support of the Corwin amendment giving slavery within the slave states an everlasting constitutional guarantee, many Republicans acquiesced in what now seems an appallingly greater concession to the South; but southerners in Congress for the most part treated the concession as a "mere bagatelle."[87] There are numerous explanations for this double anomaly, including Republican fear of southern expansionism and southern hunger for a Republican disavowal of Republicanism. Perhaps also mere habit held both sides to the familiar line of controversy. But in addition, it appears that neither North nor South had anything more than a muddled understanding of what was at issue between them, or what they wanted from each other.

Whether the Republicans or the compromisers were wiser and more patriotic in their conduct remains a subject of scholarly dispute that has sometimes recaptured with admirable felicity the rancorous spirit of congressional debates in the secession winter. Much depends, of course, upon one's retrospective prediction of the results of a Crittenden success, and here the importance of timing must be emphasized. Given the amount of resistance that secessionists had to overcome in much of the deep South, it is not difficult to agree with the judgment of Douglas that "if the Crittenden proposition could have been passed early in the session, it would have saved all the States, except South Carolina."[88] For Congress to have acted so swiftly and so forcefully would have been spectacular enough in itself to give secessionists pause, aside from the substance of the concessions offered. But such expeditious behavior at the opening of a session would have been unnatural at any time and all the more unlikely in the extraordinary confusion of December 1860. Douglas of all people must have known that a Union-saving compromise achieved by New Year's Day was something too improbable to be regarded, in retrospect, as a viable historical alternative.

What seems to have been much more within the realm of possibility is a compromise achieved late in the session—after all doubts about the reality of disunion had been dispelled, after the outpour-

87. These were Douglas's words, *ibid.*, p. 1391.
88. *Ibid.*, p. 1391.

ing of northern petitions and editorials in support of Crittenden, after the influence of the border states and the Peace Conference had been felt. But such a compromise, one must recognize, would have been relatively limited in its immediate effects. The Crittenden formula, whatever it might have accomplished if adopted in December, was by March viewed primarily as a means of holding the upper South and thus halting the progress of secession. No one expected that it would be enough alone to turn back the clock and dissolve the new Confederacy.

The only hope of complete reunion without a war of subjugation lay in the vague movement for "voluntary reconstruction," to which congressional compromise was regarded as a logical preliminary. It is important to be precise about what was lost when the Crittenden plan failed of passage. In rejecting the plan, Republicans did not reject an alternative to war but merely the first step of an alternative approach to the crisis, an approach that would have lessened the risk of war at the cost of increasing the risk of permanent disunion.[89]

Arranging a peaceable reconstruction no doubt would have been extremely difficult, and yet one is impressed by the widespread and sustained interest in the idea, which extended from Lincoln's designated secretary of state through the various ranks of compromisers and border-state neutrals into the high councils of the Confederacy.[90] Whatever the chances of success, they were greatly reduced, though not entirely foreclosed, by the failure of the Crittenden compromise. That failure sent a chill through the upper South, set a grim mood for the inauguration, and made the confrontation in Charleston Harbor all the more complete. No one, of course, can say whether it would have been any different at Fort Sumter on April 12 if Congress had earlier taken that first step down the alternative path of "voluntary reconstruction."

89. It may be said that the course actually followed by the Republicans took the risk of war without entirely disposing of the danger of disunion, since the war that did indeed follow might have been lost; whereas the policy of "voluntary reconstruction" would have risked permanent disunion without entirely disposing of the danger of war, since conflict might have arisen later—over slavery again if reconstruction were successful, and over the territories or navigation of the Mississippi if it were not.

90. See Potter, *Lincoln and His Party*, pp. 219–248.

CHAPTER 20

Fort Sumter: End and Beginning

HOW many German-Americans voted for Lincoln in 1860 has been a subject of much scholarly argument, but in Illinois, at least, their numbers were considerable. They included, for example, the former lieutenant governor, Gustave Koerner, who had figured prominently in the nomination of Lincoln at Chicago and in the campaign that followed.[1] Less distinguished, but no less fervently Republican, was John G. Nicolay, in his late twenties and formerly a small-town journalist, but now clerk to the Illinois secretary of state. Nicolay's parents had emigrated to America when he was five years old. After settling first in Cincinnati, the family moved to Indiana, then Missouri, and finally Illinois. Here and there he had managed to pick up a few month's schooling but was largely self-taught, especially by intensive study of the Bible and Shakespeare. Lincoln may have recognized a kindred spirit and, needing a full-time private secretary after his nomination, offered the position to Nicolay. The quiet, methodical young German brought an increased orderliness into the affairs of his new employer, who had never held any kind of administrative office. They were strikingly representative of social mobility in American life—the next president and his secretary, with scarcely two years of formal education between them.[2]

1. On Koerner, see his illuminating *Memoirs*, ed. by Thomas J. McCormack (2 vols.; Cedar Rapids, Iowa, 1909).
2. On Nicolay, see Helen Nicolay, *Lincoln's Secretary: A Biography of John G. Nicolay* (New York, 1949).

It was Nicolay who directed the traffic of visitors to the temporary office put at Lincoln's disposal on the second floor of the state capitol. Their arrivals increased sharply after the election. "They descended on him in such numbers," says Benjamin P. Thomas, "that Springfield's hotels and boarding-houses were crammed and the overflow put up in sleeping-cars."[3] The volume of mail also became burdensome, and Nicolay was allowed to recruit one of his friends as an assistant. Thus, at the age of twenty-two, John Hay entered upon his long career of public service.[4]

Along with the importunate din of office seekers and their recommenders, there were letters and visits from many party leaders offering advice especially on two subjects—cabinet appointments and the secession crisis. Very early in his thoughts about it, apparently, Lincoln decided that a broadly representative cabinet was more essential than a doctrinally cohesive one. After months of working at the task, he would end by achieving some measure of balance between former Whigs and former Democrats, between radicals and conservatives, between easterners and westerners (although he failed to enlist the services of anyone regarded as a true southerner). As a further indication of his desire to incorporate all major factions of the Republican party in his administration, Lincoln in the end would fill four of the seven cabinet positions with the four men who had been his principal competitors for the presidential nomination. But these things were settled slowly and with much accompanying rumor and confusion. When he left Springfield for Washington on February 11, more than three months after his election, only two appointments had been formally announced: William H. Seward as secretary of state and Edward Bates as attorney general.[5]

Lincoln's first major decisions as president-elect were essentially negative ones. He refused to issue any public statement aimed at appeasing the South, and he made it clear in private correspon-

3. Benjamin P. Thomas, *Abraham Lincoln* (New York, 1952), p. 231. On office seekers generally, see Harry J. Carman and Reinhard H. Luthin, *Lincoln and the Patronage* (New York, 1943), pp. 3–109.

4. Tyler Dennett, *John Hay* (New York, 1933), pp. 33–35. Hay, unlike Nicolay, was not officially a secretary to Lincoln but rather a clerk in the Department of the Interior, detailed to special service at the White House.

5. J. G. Randall, *Lincoln the President* (2 vols.; New York, 1945), I, 256–272. On Lincoln as president-elect generally, see William E. Baringer, *A House Dividing: Lincoln as President-Elect* (Springfield, Ill., 1945).

dence that he opposed any compromise involving a retreat from the Republican platform. His reasons were summed up in a letter of November 16 to a St. Louis editor:

> I could say nothing which I have not already said, and which is [not] in print and accessible to the public. . . .
>
> I am not at liberty to shift my ground—that is out of the question. If I thought a *repetition* would do any good I would make it. But my judgment is it would do positive harm. The secessionists *per se,* believing they had alarmed me, would clamor all the louder.[6]

Hundreds of visitors and letter writers pressed their views upon him. Thurlow Weed and Duff Green (an unofficial emissary from President Buchanan) were perhaps the most notable advocates of some kind of appeasement; Horace Greeley, William Cullen Bryant, and Salmon P. Chase were among those who undertook to stiffen his resolution against what Greeley called "another nasty compromise."[7] He listened attentively, but there is good reason to believe that he had already independently made up his own mind.

Some Republican leaders warned him that any "caving in" to the South would probably mean disruption of the party. No doubt this consideration would have weighed heavily with Lincoln if he had been otherwise disposed to compromise, but since he was not, there is little basis for asserting that he deliberately chose to save his party instead of his country.[8] The critical elements in his decisionmaking at this point were his own reading of the crisis and his own conceptualization of the role that he must play. Strong emotion infused his thinking on the subject and may at times have dominated it. There are more displays of pride and anger, more indications of self-consciousness, than in any other phase of Lincoln's career. (Significantly, perhaps, it was during these months of interregnum that he

6. Roy P. Basler (ed.), *The Collected Works of Abraham Lincoln* (8 vols.; New Brunswick, N.J., 1953), IV, 139–140; also 149–155.

7. Randall, *Lincoln the President,* I, 248–249; John G. Nicolay and John Hay, *Abraham Lincoln: A History* (10 vols.; New York, 1890), III, 286–287; Jeter Allen Isely, *Horace Greeley and the Republican Party, 1853–1861* (Princeton, N.J., 1947), pp. 325–326; David C. Mearns (ed.), *The Lincoln Papers* (2 vols.; Garden City, N.Y., 1948), II, 349–350, 399, 424–425; Basler (ed.), *Works of Lincoln,* IV, 158, 162–163. Lincoln denied, however, that Weed, in their conference, pressed the subject of compromise. *Ibid.,* p. 163.

8. For an intelligent discussion of this question, see Kenneth M. Stampp, *And the War Came* (Baton Rouge, 1950), p. 186; also Stampp, "Lincoln and the Strategy of Defense in the Crisis of 1861," *JSH,* XI (1945), 300–301.

made the greatest change ever in his outward appearance by growing a beard.) He seemed almost neurotically sensitive about giving any impression of "weakness," "timidity," "sycophancy," or "cowardice"—all his words. To make public protestations of his conservatism, he believed, would encourage "bold bad men" to think that they were dealing with someone who could "be scared into anything."[9]

To Lincoln, in short, secession as a mass movement was incredible. He could understand it only as the conspiratorial action of a slaveholding minority, whose early advantage, he hoped, would eventually be offset by an ebullition of southern Unionism, and whose true purpose, he suspected, was not so much separation as blackmail. In his view, the question was not compromise but government by minority coercion:

> We have just carried an election on principles fairly stated to the people. Now we are told in advance, the government shall be broken up, unless we surrender to those we have beaten, before we take the offices. In this they are either attempting to play upon us, or they are in dead earnest. Either way, if we surrender, it is the end of us, and of the government. They will repeat the experiment upon us *ad libitum.* A year will not pass, till we shall have to take Cuba as a condition upon which they will stay in the Union.[10]

Here it must be noted again that as secession replaced slavery at the center of controversy, old distinctions between radical and conservative Republicans lost some of their meaning. Thus, Lincoln, who had been clearly less radical than Chase on the subject of slavery, was emphatically the more militant of the two on the subject of preserving the Union. In late December, when rumors reached Springfield that Buchanan had decided to surrender the Charleston forts, Lincoln is said to have exclaimed, "If that is true they ought to hang him!" In a letter to Lyman Trumbull, he proposed to counter any such move by announcing publicly his intention of retaking the forts after he had been sworn in to office. This willingness to promise the forcible recovery of lost federal property placed Lincoln among the more aggressive Republicans, and until shortly before his inauguration he was still planning to make the commitment.[11]

9. Basler (ed.), *Works of Lincoln,* IV, 132–133, 134–135, 138.
10. *Ibid.*, p. 172.
11. *Ibid.*, pp. 159, 162, 164; *Lincoln Day by Day: A Chronology* (3 vols.; Washington, 1960), II, 302.

In his militant attitude toward secession, Lincoln reflected the strong feelings of a region as well as a party. For people living in the upper Mississippi Valley, disunion presented the special threat of closing off their access to the sea. Railroad construction had reduced but by no means eliminated their dependence on river commerce, and in any case, the need for untrammeled passage was partly psychological. The very thought of returning to the days when foreign authority controlled the mouth of the mighty stream inspired a kind of claustrophobic alarm and much belligerent outcry. "There can be no doubt," a Milwaukee editor warned, "that any forcible obstruction of the Mississippi would at once lead to a war between the West and the South." The people of the Northwest, said the Chicago *Tribune,* would never negotiate with anyone for free navigation of the river. "It is *their right,* and they will assert it to the extremity of blotting Louisiana out of the map." These and similar threats, issuing from Democrats as well as Republicans, served to remind Americans that there were many places besides Fort Sumter where the friction of disunion could provide the spark for civil war.[12]

Another point of danger, it appeared, was the national capital itself. Lincoln's sense of being the target of a conspiracy was no doubt enhanced by the reports that he began to receive of plots to prevent the official counting of electoral ballots, to disrupt his inauguration, to assassinate him, or even to seize control of Washington by military force.[13] These were not fantasies from the lunatic fringe but the apprehensions of sober, responsible men. Charles Francis Adams, for instance, was utterly certain that disunionists would try to "take forcible possession of the Government" before March 4, and General Scott's military secretary informed an Illinois congressman that he had "overwhelming" evidence of a "widespread and powerful conspiracy to seize the capitol." Sandwiched between slaveholding Virginia and slaveholding Maryland, Washington was certainly vulnerable. Much depended upon the fate of Baltimore to the north, a city divided in its loyalties and full of talk about plots and counterplots.[14]

12. Howard Cecil Perkins (ed.), *Northern Editorials on Secession* (2 vols.; New York, 1942), II, 545, 558.

13. Mearns (ed.), *Lincoln Papers,* II, 354–357, 358–360, 363, 377, 398, 401, 407, 409, 424–425, 427–428.

14. *Ibid.*, pp. 434–435; David M. Potter, *Lincoln and His Party in the Secession Crisis* (New Haven, 1942), pp. 254–257.

In the circumstances, the journey from Springfield to Washington began to take on an air of suspense as well as a symbolic importance. Ignoring advice from some quarters that he should make the trip swiftly and unobtrusively, Lincoln chose a slow, roundabout itinerary with many stops en route—something not unlike a royal progress. His reasons were never explicitly stated, but he had received many invitations to visit specific communities, and, as a man risen suddenly from relative obscurity, he apparently felt an obligation to present himself before the people who had elected him. Nevertheless, it was in some respects a strange decision. The two weeks of travel would tire him needlessly when he should be conserving his strength and would expose him needlessly to crowds when he was under threat of assassination. Furthermore, still determined not to be premature in announcing the policies of his administration, he put himself in the position of having to make numerous speeches saying little more than that he had nothing to say.

And so, on February 11, 1861, Lincoln set out for Washington by way of Cincinnati, Pittsburgh, Cleveland, Buffalo, Albany, and New York City—a distance of nearly two thousand miles, requiring the use of more than twenty different railroads.[15] To the crowds assembled at each stop, he responded with brief remarks that often seemed pedestrian, awkward, and even downright trivial. His view of the secession movement as a conspiracy revealed itself in repeated assertions that the crisis was "an artificial one," such as could be "gotten up at any time by designing politicians." There was, he said, "nothing going wrong. . . . nothing that really hurts anybody." The crisis had "no foundation in facts. . . . Let it alone and it will go down of itself." Small wonder that such words struck many Americans as appallingly inadequate to the circumstances.[16]

Yet at times his studied reticence gave way, and in bits and pieces he did disclose the general set of his attitude and intentions. Thus, very early in the journey, he spoke for the first time in public about the possibility of "retaking" surrendered forts, as well as holding on to those still in federal hands. He spoke also of enforcing the laws, collecting import duties, and perhaps withholding mail service in areas where it was being interfered with. Furthermore, by arguing

15. On the trip generally, see Victor Searcher, *Lincoln's Journey to Greatness* (Philadelphia, 1960).

16. Basler (ed.), *Works of Lincoln,* IV, 204, 211, 215–216, 238.

that a state was essentially a "district of country with inhabitants," he struck a slashing blow at the southern doctrine of state sovereignty. "If a State, in one instance, and a county in another," he said, "should be equal in extent of territory, and equal in the number of people, wherein is that State any better than the county?"[17]

In Columbus, Ohio, on February 13, Lincoln received word by telegram that the counting of the electoral vote had formalized his election. Five days later, while his train was rolling eastward through the Mohawk Valley toward Albany, Jefferson Davis took the presidential oath and delivered his inaugural address in Montgomery, Alabama. The issue had been more plainly joined, and in his remarks during the remainder of the journey, Lincoln paid more explicit attention to the danger of war. He insisted upon his devotion both to peace and to the Union, but as an honest man he had to recognize that one or the other might have to be given priority. Thus, his commitment to preservation of the Union was very nearly unconditional, while his promises to preserve peace were ringed with qualifications. "There will be no blood shed unless it be forced upon the Government," he declared. "It shall be my endeavor to preserve the peace of this country so far as it can possibly be done, consistently with the maintenance of the institutions of the country." Perhaps most revealing were some words spoken before the New Jersey legislature and received with loud and sustained cheering: "The man does not live who is more devoted to peace than I am. None who would do more to preserve it. But it may be necessary to put the foot down firmly."[18]

Lincoln had in fact already written a first draft of his inaugural address before leaving Springfield. It is therefore not surprising that in his extemporary speeches along the way to Washington he should have given some clear indications of the policy he would enunciate on March 4. And the responses of the crowds tended generally to confirm his judgment and stiffen his determination.

Upon arriving in Philadelphia on February 21, Lincoln was warned of a plot to assassinate him when he passed through Baltimore, a city that in any case constituted hostile territory for a Republican. He nevertheless continued with his regular schedule of public appearances, including a side trip to Harrisburg. But then,

17. *Ibid.*, pp. 195–196.
18. *Ibid.*, pp. 233, 237, 240–241, 243–244.

the next day, a messenger directly from Seward and General Scott convinced him that the danger might be serious, and he agreed to change his plans. With just a single companion, he slipped quietly aboard a night train that carried him unnoticed through Baltimore to Washington in the hours before dawn on February 23. It was an anticlimactic and even ignominious ending to a journey that had been in some respects an extended celebration. Opposition newspapers seized gleefully on the episode and made the president-elect a target of ridicule in editorials and cartoons. His prestige, never extraordinarily high, sank probably to its lowest point since his election.[19]

There were many Americans, in any case, who believed that the fate of the country rested less with Abraham Lincoln than with William H. Seward, and Seward himself was emphatically one of their number. This first "Mr. Republican," shrewd, persuasive, and protean, seemed now to be, in the words of his biographer, "on a pinnacle of power."[20] As secretary of state serving under a man of much less experience and fame, he expected to become the virtual premier of the new administration, a role that he had practiced for a time at the elbow of Zachary Taylor. Having also carefully maintained his friendly personal ties with certain southern leaders, Seward more than half believed that he was the one indispensable man in the nation's hour of crisis. Some of his remarks, as a consequence, remind us of General George B. McClellan a year later. "I will try to save freedom and my country," he wrote his wife after accepting Lincoln's appointment. "It seems to me," he added in late January, "that if I am absent only three days, this Administration, the Congress, and the District would fall into consternation and despair. I am the only hopeful, calm, conciliatory person here." The young Henry Adams, who saw Seward frequently during these days, described him as the "virtual ruler of this country."[21]

19. Mearns, (ed.), *Lincoln Papers,* II, 442–443; Randall, *Lincoln the President,* I, 288–291; Nicolay and Hay, *Lincoln,* III, 302–316; Ward H. Lamon, *The Life of Abraham Lincoln* (Boston, 1872), pp. 511–527. It is uncertain whether the plot was real or the imaginative work of the detective Allen Pinkerton. Documents from the Pinkerton files are in Norma B. Cuthbert (ed.), *Lincoln and the Baltimore Plot* (San Marino, Calif., 1949); also, Edward Stanley Lanis, "Allen Pinkerton and the Baltimore 'Assassination' Plot Against Lincoln," *Maryland Historical Magazine,* XLV (1950), 1–13.

20. Glyndon G. Van Deusen, *William Henry Seward* (New York, 1967), p. 246.

21. *Ibid.*, pp. 241, 246; Worthington Chauncey Ford (ed.), *Letters of Henry Adams, 1858–1891* (Boston, 1930), p. 81. On Seward's character and outlook, see Frederic

Seward had a way of making inconsistency look like profundity, and his purposes were always somewhat obscured by his own deviousness, but he did apparently put together at least the vague outlines of a plan for saving the Union. Like Lincoln, he regarded secession as the work of a zealous minority and believed that southern Unionism would in time reassert itself. The crucial difference between the two men was in their respective estimates of the effect that conciliatory gestures would have upon the progress of secession. Lincoln feared that too much temporizing with disunion would tend to legitimate it in the deep South and to encourage its advocates in the border states. Seward, in contrast, persuaded himself that a policy of "forbearance, conciliation, magnanimity," and even of unspoken acquiescence in the withdrawal of seven states, would secure the loyalty of the upper South and thus halt the progress of secession. Then, in a matter of months, disunion would lose its glamor and momentum in the abortive republic of slaveholders. Unionists and disunionists there would "have their hands on each other's throats," and the process of "voluntary reconstruction" could begin. But just in case reconciliation should prove more difficult than expected, Seward was laying plans to stimulate it in a most dramatic way. Americans, he believed, would still close ranks against any threat from abroad. Stir up a crisis with Spain or France or England, even start a war with one or more of them, and the problem of disunion would melt away. If New York were attacked by a foreign enemy, he said publicly in December, "all the hills of South Carolina would pour forth their population for the rescue."[22]

"Conservative" scarcely seems the right label for a man nurturing such a scheme. Yet Seward, partly because of his own equivocal behavior and partly because of his continuing political intimacy with an avowed compromiser, Thurlow Weed, was now generally regarded as head of the conservative wing in the Republican party. Radicals like Charles Sumner had given him up as a lost soul and launched a drive to keep him out of the

---

Bancroft, *The Life of William H. Seward* (2 vols.; New York, 1900), II, 70–90; and Major L. Wilson, "The Repressible Conflict: Seward's Concept of Progress and the Free-Soil Movement," *JSH,* XXXVII (1971), 533–556.

22. Potter, *Lincoln and His Party,* pp. 240–245; Van Deusen, *Seward,* pp. 242, 246–248; Ford (ed.), *Letters of Henry Adams,* p. 87.

cabinet. Seward elements, in turn, were working just as hard to secure a cabinet compatible with the functioning of a premiership. That meant, above all, preventing the appointment of Chase, acknowledged leader of the radical wing, as secretary of the treasury. The fierce struggle reached its climax shortly before inauguration day when Seward threatened to withdraw. As it turned out, neither side won. Lincoln nominated a cabinet that included both Seward and Chase, together with Simon Cameron of Pennsylvania as secretary of war, Gideon Welles of Connecticut as secretary of the navy, Caleb B. Smith of Indiana as secretary of the interior, Edward Bates of Missouri as attorney general, and Montgomery Blair of Maryland as postmaster general.[23]

In his cabinetmaking, as well as in some of his speechmaking on the journey from Springfield, Lincoln gave indications of intending to be his own master. But he was flexible by nature; he had great respect for Seward; and he spent much time in the latter's company after arriving in Washington. It would have been surprising if the persuasive New Yorker had not had some influence on his thinking. In addition, a procession of distinguished visitors—including Crittenden, Douglas, and John Bell—pressed him to adopt a conciliatory policy. The crisis looked different from inside Washington. It was no longer remote and abstract but near and real. Complexities not visible from Illinois now became apparent, and the District of Columbia, a border-state enclave, was charged with the uncertainties and apprehensions of the border-state dilemma. Virginia—the critical importance of holding Virginia—seemed to dominate the very landscape.

By late February, despite the activity of the Peace Conference and the final flurry of effort in Congress, the issue had shifted from compromise to coercion. Virginia Unionists calling on Lincoln warned him that anything resembling use of force against the Confederacy would tip the delicate balance in their state irrevocably toward secession. Above all, they pleaded for the evacuation of Fort Sumter. In response, Lincoln apparently showed the first signs of weakening on this point, at least to the extent of proposing a bar-

23. Randall, *Lincoln the President,* I, 256–272; Thomas, *Lincoln,* pp. 232–235; Nicolay and Hay, *Lincoln,* III, 345–374; Van Deusen, *Seward,* pp. 249–254; Allan Nevins, *The Emergence of Lincoln* (2 vols.; New York, 1950), II, 438–446, 452–455.

gain: he would withdraw the garrison from Fort Sumter if Virginians would discontinue their state convention. Scholars differ about whether Lincoln was serious in making the offer—that is, whether he really thought there was any chance of its being accepted. Even as mere talk, however, it revealed a greater flexibility than he had previously displayed in considering the problem of the forts.[24]

The influence of Seward and other advocates of conciliation is also visible in the final version of the inaugural address, a somewhat more moderate document than the first draft which Lincoln had prepared in Springfield. For example, he eliminated passages declaring his intention to abide by the Republican platform. He reversed his previous view and endorsed the idea, favored by Seward and Buchanan, of calling a constitutional convention. He gave his blessing to the Corwin amendment forbidding federal interference with slavery in the states. At Seward's suggestion, he deleted the second half of the following sentence: "The government will not assail *you,* unless you *first* assail *it.*" He also accepted Seward's draft of a closing appeal to the bonds of Union, recasting it into one of the most eloquent and familiar paragraphs in political literature.[25]

Most important of all, Lincoln agreed to modify this highly provocative passage: "All the power at my disposal will be used to reclaim the public property and places which have fallen; to hold, occupy and possess these, and all other property and places belonging to the government, and to collect the duties on imports." Seward recommended striking out the entire sentence and replacing it with one couched in innocuous generalities. Lincoln was unwilling to go that far, but, at the suggestion of his friend Orville H. Browning, he did delete the pledge to "reclaim" federal property already in Confederate hands.[26] This was no small concession for a man who in December had notified General Scott to be ready "to either *hold,* or *retake,* the forts, as the case may require."[27] It meant a substantial reduction of the amount of coercive menace that southerners could read in the address. Still, the words that Lincoln

24. *Lincoln Day by Day,* III, 22–23; Potter, *Lincoln and His Party,* pp. 353–354; Stampp, *And the War Came,* pp. 274–275; Allan Nevins, *The War for the Union* (4 vols.; New York, 1959–71), I, 46–47; Richard N. Current, *Lincoln and the First Shot* (Philadelphia, 1963), pp. 34–35.

25. Basler (ed.), *Works of Lincoln,* IV, 249–271, containing the first and final texts with indications of revisions and their sources.

26. *Ibid.,* p. 254.

27. *Ibid.,* p. 159.

refused to change proved decisive in the end; for they officially committed his administration to the defense of Fort Sumter.

Perhaps as another gesture of good will, Lincoln went to the Senate on the evening of March 3 to hear Crittenden's farewell plea for conciliation. The next day at about noon, James Buchanan called for the president-elect at Willard's Hotel, and together in an open carriage they proceeded up Pennsylvania Avenue, lined as it was with cheering multitudes. A feeling of tension accompanied them, for rumors of assassination plots had continued to circulate. Besides some six hundred United States troops deployed by Scott, there were perhaps two thousand volunteers on hand in their varieties of uniform. The military display and the elaborate parade of celebrating Republicans made the whole affair, said the *National Intelligencer,* "in some respects the most brilliant and imposing pageant ever witnessed in this Capital."[28]

On a temporary platform at the east front of the Capitol, Lincoln delivered his inaugural address and took the presidential oath administered by Chief Justice Taney. He began with reassurances to the South, first disclaiming any purpose or lawful right to interfere with slavery in the states where it already existed. He endorsed, as a constitutional obligation, the principle of a fugitive slave law, though not without indicating some dissatisfaction with the notorious statute currently in force. There would be, he said, no invasion of the South, and there need be no bloodshed or violence. He intended to act "with a view and a hope of a peaceful solution of the national troubles, and the restoration of fraternal sympathies and affections."

But then, alongside his hope for peace he laid his adamant rejection of secession. "I hold," he sternly declared, "that in contemplation of universal law, and of the Constitution, the Union of these States is perpetual." This meant that no state, "upon its own mere motion," could lawfully break away from the Union; that secession ordinances were "legally void"; and that acts of violence against the authority of the United States constituted insurrection or revolu-

28. Albert D. Kirwan, *John J. Crittenden: The Struggle for the Union* (Lexington, Ky., 1962), p. 415; Washington *National Intelligencer,* March 5, 1861; Charles Winslow Elliott, *Winfield Scott, the Soldier and the Man* (New York, 1937), pp. 694–696. See also Charles P. Stone, "Washington on the Eve of the War," in Robert Underwood Johnson and Clarence Clough Buel (eds.), *Battles and Leaders of the Civil War* (4 vols.; New York, 1887), I, 7–25.

tion. The central idea of secession, he insisted, was "the essence of anarchy," for it rested on the ruinous principle that a minority might secede rather than acquiesce in the will of the majority—a process which, once established as a precedent, could be repeated ad infinitum. In any case, as president he had been given no constitutional power "to fix terms for the separation of the States." Instead, his duty was "to administer the present government, as it came to his hands, and to transmit it, unimpaired by him, to his successor."

But how, then, in the extraordinary circumstances of the hour, did he intend to perform that duty? He would "hold, occupy, and possess" federal property within the seceded states (meaning, for the most part, Forts Sumter and Pickens). Import duties would be collected (but on ships stationed offshore). The mails would be delivered everywhere in the country (that is, "unless repelled"). As for government appointments in the seceded states, here Lincoln offered still another concession designed to allay southern fears of being infested with Republican officeholders: "Where hostility to the United States, in any interior locality, shall be so great and so universal, as to prevent competent resident citizens from holding the federal offices, there will be no attempt to force obnoxious strangers among the people for that object." In short, while emphatically reasserting federal authority throughout the South, he would avoid, as much as possible, provocative efforts to enforce that authority.

His stance was thus firm but defensive and pacific. He showed no desire to force the issue, but rather urged Americans, one and all, to think "calmly and well" upon the whole subject, adding that nothing valuable could be lost by "taking time." The choice between peace and war rested, however, with his "dissatisfied fellow countrymen" of the South. "You can have no conflict," he said, "without being yourselves the aggressors. *You* have no oath registered in Heaven to destroy the government, while *I* shall have the most solemn one to 'preserve, protect and defend' it." Then came the final paragraph rough-drafted by Seward:

> I am loth to close. We are not enemies but friends. We must not be enemies. Though passion may have strained, it must not break our bonds of affection. The mystic chords of memory, stretching from every battlefield, and patriot grave, to every living heart and hearthstone, all over

this broad land, will yet swell the chorus of the Union, when again touched, as surely they will be, by the better angels of our nature.[29]

The reception of this First Inaugural, with its counterpoise of sternness and good will, reflected the sectional and partisan attitudes of a bitterly divided country. Historians ever since have likewise disagreed about its true meaning. One problem, according to many hostile critics, was Lincoln's literary style, which a New Jersey editor found to be "involved, coarse, colloquial, devoid of ease and grace, and bristling with obscurities and outrages against the simplest rules of syntax." Republican newspapers, on the other hand, insisted that Lincoln's "plain, terse, wire-woven sentences" were as "clear as a mountain brook," and indeed "strikingly adapted to the occasion."[30]

The address pleased Republicans primarily because of its "firmness." At the same time, its conciliatory features heartened many border-state Unionists and northern advocates of compromise (including Douglas, who called it a "peace offering"). But throughout much of the South and for a substantial minority in the North as well, Lincoln's words meant coercion, and coercion meant war. If his announced policies were carried out, said an Ohio editor, blood would "stain the soil and color the waters of the entire continent."[31]

Many people in all parts of the country who considered secession illegal or unjustified, or both, nevertheless believed that the existence of the Southern Confederacy must now be accepted as a fact of life and dealt with accordingly. "There stands secession—bold and palpable," declared a Rhode Island newspaper that had supported Douglas, "and if we refuse to recognize it today, we shall have to recognize it, with arms in our hands, tomorrow. It cannot long be dodged. There is an irrepressible conflict between the sim-

29. On the inaugural generally, see Randall, *Lincoln the President*, I, 293–302; Potter, *Lincoln and His Party*, pp. 319–329; Marie Hochmuth Nichols, "Lincoln's First Inaugural Address," in J. Jeffery Auer (ed.), *Antislavery and Disunion, 1858–1861* (New York, 1963), pp. 392–414.

30. Perkins (ed.), *Northern Editorials*, II, 618, 623, 625. The *Times* of London declared that the conclusion of the address was "scarcely conceived in a vein worthy of the occasion."

31. Perkins (ed.), *Northern Editorials*, II, 624, 629, 634, 639. For other newspaper comments, see Randall, *Lincoln the President*, I, 303–308; Nichols, "Lincoln's First Inaugural," pp. 409–411; Donald E. Reynolds, *Editors Make War: Southern Newspapers in the Secession Crisis* (Nashville, 1970), pp. 190–193.

ple fact which stares us in the face when we look Southward, and the execution of the laws as proposed by the President."[32]

Similar views were apparently gaining favor among many interested observers in Europe, as an earlier disapproval of secession gave way to the conviction that it had nevertheless become irreversible. "I do not see how the United States can be cobbled together again," wrote the British foreign secretary, Lord John Russell, in January. "The best thing *now,*" he added, "would be that the right to secede should be acknowledged. . . . But above all I hope no force will be used." The London *Times,* which had begun by welcoming Lincoln's election and denouncing southern intemperance, was soon shifting its emphasis from the indefensible evilness of slavery to the dreadful prospects of armed conflict. Having in January derided Buchanan's annual message as a pusillanimous acceptance of disunion, the *Times* in March frowned upon Lincoln's inaugural as "neither more nor less than a declaration of civil war."[33]

What puzzles one here is the fact that the policy announced by Lincoln did not differ all that much from the policy already settled upon by the outgoing administration. It was in each case the "strategy of defense," including retention of Fort Sumter and "enforcement of the laws," if possible. Buchanan, according to his biographer, carefully examined the inaugural address and found in it many parallels with his own messages to Congress.[34] Yet the diarist George Templeton Strong, no bitterly partisan radical, could hear "a clank of metal" in Lincoln's words, while dismissing Buchanan as "lowest . . . in the dirty catalog of treasonable mischief-makers."[35] Of course, the difference is partly one of context. Buchanan's policy during his final weeks in office has been judged against the background of his proslavery behavior in the preceding four years; Lincoln's conduct during his first weeks in office has been judged against the background of all that followed in the next four years; and as a consequence, the interval of continuity

32. Perkins (ed.), *Northern Editorials,* II, 647.

33. Ephraim Douglass Adams, *Great Britain and the American Civil War* (2 vols.; New York, 1925), I, 52–53; London *Times,* Jan. 9, March 18, 1861.

34. Philip Shriver Klein, *President James Buchanan* (University Park, Pa., 1962), pp. 405–406.

35. Allan Nevins and Milton Halsey Thomas (eds.), *The Diary of George Templeton Strong* (4 vols.; New York, 1952), III, 103, 106.

between the two administrations is often overlooked.

Lincoln's reticence as president-elect had been inspired to some extent by the realization that the rush of events could swiftly overtake any pronouncement and render it obsolete. Similarly, there was a strong note of contingency in the line of action announced in the inaugural. "The course here indicated will be followed," he said, "unless current events, and experience, shall show a modification, or change, to be proper."[36] Tentatively, then, but none the less clearly, Lincoln had set forth a policy that amounted to a limited version of peaceable reconstruction. He *would not* go so far as to surrender federal property voluntarily or to acknowledge the existence of the Confederacy, but he *would* try to avoid a confrontation at every point where one was likely to occur, thus giving time, as he said, for "calm thought and reflection." With such a display of restraint, he might encourage southern Unionists, who had already stemmed the tide of secession throughout the upper South, to rally in the lower South and bring their states back into the Union. It was a policy obviously depending upon extension of the status quo for a considerable period, but that, as Lincoln soon discovered, was going to be far more difficult than he had supposed.

The bad news arrived one day after his inauguration. From the outgoing secretary of war, Joseph Holt, he received a dispatch written by Major Anderson on February 28. Earlier, Anderson had discouraged reinforcement of Fort Sumter on the grounds that it was unnecessary; now he shifted to discouraging it on the grounds that it was impossible. No fewer than twenty thousand well-disciplined troops would probably be needed, he declared, to relieve the fort within the time limit fixed by the depletion of his provisions. That limit, according to a supplementary report, was in the neighborhood of four to six weeks.[37] Anderson, who preferred peace even at the price of disunion, apparently expected that the response to his gloomy diagnosis would be an order to evacuate. For Holt, the report was highly embarrassing. He pretended that it all came as a great surprise, but the Buchanan administration had in fact received more than enough detailed information to

36. Basler (ed.), *Works of Lincoln*, IV, 266; also 204, 210, 221, 226, 231, for explanations of his reticence.

37. Anderson's letter, once a subject of controversy because the original had not been found, is in Mearns (ed.), *Lincoln Papers*, II, 450–451. For estimates of supplies, *ibid.*, pp. 453–454, 477.

understand the true state of affairs in Charleston harbor. Over a period of several months, the continued peaceful occupation of Sumter had been bought by noninterference with South Carolinian construction of a "circle of fire" that made the ultimate downfall of the fort a little more certain each day. These developments were no secret. Everyone, Lincoln included, had good reason to recognize that time was running out for Anderson and his garrison. What shocked the president-elect was the professional, on-the-site military estimate of how very soon he must act and how much effort it would take to save Fort Sumter.[38]

Lincoln turned promptly to his commanding general for advice and got an utterly discouraging reply. Evacuation seemed "almost inevitable," Scott declared. As for reinforcing Anderson, that would require a fleet of war vessels and transports, together with twenty-five thousand additional troops and six or eight months in which to train them. Scott's judgment, which carried all the more weight because he had earlier favored sending a relief expedition, no doubt rested primarily on military considerations. But it also seems clear that by this time he had been fully converted to Seward's program of conciliation and peaceable reconstruction.[39]

Most members of the cabinet, still fitting themselves into their new offices, were likewise disposed to view Fort Sumter as a lost cause. A formal polling of their views on March 15 revealed that only one of the seven men, Montgomery Blair, stood unqualifiedly in favor of an attempt to provision the fort. Rumors of an impending order for evacuation were soon circulating, first in Washington and then throughout the country. Here and there, some Republicans expressed acquiescence in what appeared to be a military necessity, but the dominant mood within the party was one of rising anger and frustration. "The bird of our country," wrote George Templeton

38. Holt to Lincoln, March 5, 1861, *ibid.*, pp. 461–464. Anderson's estimate of troops needed was higher than the estimates of most of his officers. For example, his second in command, Captain Abner Doubleday, said 10,000. See *The War of the Rebellion: A Compilation of the Official Records of the Union and Confederate Armies* (128 vols.; Washington, 1880–1901), series I, vol. I (hereafter cited as *Official Records,* I), p. 202. For Doubleday's account, see his *Reminiscences of Forts Sumter and Moultrie in 1860–61* (New York, 1876).

39. Mearns (ed.), *Lincoln Papers,* II, 464–465, 477–478; Basler (ed.), *Works of Lincoln,* IV, 279; Elliott, *Scott,* pp. 697–701; Bancroft, *Seward,* II, 95–96, 124–125. Scott actually drafted an evacuation order and asked Cameron's permission to send it to Anderson. See Mearns (ed.), *Lincoln Papers,* II, 476.

Strong, "is a debilitated chicken, disguised in eagle feathers. . . . We are a weak, divided, disgraced people, unable to maintain our national existence."[40]

Under heavy pressure of grim fact and expert opinion, Lincoln for a time apparently gave serious thought to abandoning Fort Sumter, but he never could bring himself to the point of authorizing it. Instead, he sought additional information, sending first one man and then two more to survey the situation at Charleston.[41] He also ordered the immediate reinforcement of Fort Pickens as a partial offset, at least, to the loss of Sumter if evacuation should prove unavoidable.[42]

Meanwhile, Seward was taking hold as "premier," developing his own independent line of policy. The Davis government had sent three commissioners to negotiate for recognition of the Confederacy, and, as allegedly "foreign" emissaries, they naturally tried to

40. Basler (ed.), *Works of Lincoln,* IV, 284–85; Current, *Lincoln and the First Shot,* pp. 65–67; Bancroft, *Seward,* II, 97–101; Mearns (ed.), *Lincoln Papers,* II, 483–485; Nevins and Thomas (eds.), *Diary of Strong,* III, 109. For a cabinet memorandum on the pros and cons of supplying Sumter, see Basler (ed.), *Works of Lincoln,* IV, 288–290. Chase, it should be noted, actually voted for provisioning Sumter, but only because he thought that it would not start a war. See his later letters on the subject, printed in Samuel Wylie Crawford, *The Genesis of the Civil War: The Story of Sumter, 1860–1861* (New York, 1887), pp. 366–367. Louis T. Wigfall, Texas senator still in Washington, telegraphed Confederate authorities on March 11, "It is believed here in Black Republican circles that Anderson will be ordered to vacate Fort Sumter in five days." *Official Records,* I, 273.

41. The first of these visitors was Gustavus V. Fox, already planning and soon to head the Sumter expedition. Confederate authorities allowed him to visit Fort Sumter on March 21, thinking that he was making arrangements for its evacuation. Fox did not say much to Anderson about his relief plans but returned to Washington convinced that they were feasible. The other two visitors (March 25–27) were Illinois friends of Lincoln, Stephen S. Hurlbut and Ward H. Lamon (the man who had accompanied him on the secret night trip through Baltimore). Hurlbut took soundings in Charleston and reported that any effort to provision Sumter would surely meet resistance. Lamon gave the impression of speaking for the administration and assured Anderson as well as the Confederate leaders that evacuation was imminent. This provided one more basis for the charge of treachery leveled by the latter against the Lincoln government. *Official Records,* I, 208–209, 218, 222, 230, 280, 281, 282, 294; Current, *Lincoln and the First Shot,* pp. 71–74; Crawford, *Genesis,* pp. 369–373; Ari Hoogenboom, "Gustavus Fox and the Relief of Fort Sumter," *CWH,* IX (1963), 385–387; Nicolay and Hay, *Lincoln,* III, 389–392.

42. *Official Records,* I, 360; Nicolay and Hay, *Lincoln,* III, 393–394; "General M. C. Meigs on the Conduct of the Civil War," *AHR,* XXVI (1921), 300. Lincoln, it appears, had given verbal orders for the reinforcement of Fort Pickens shortly after his inauguration. They were not carried out, and he repeated them in writing. They were sent on March 12.

deal with the secretary of state. Seward, at Lincoln's insistence, refused to meet the three men face to face, but he did communicate with them through a series of intermediaries. One of the latter was Justice John A. Campbell of Alabama, who had not yet handed in his resignation from the Supreme Court. At a meeting with Campbell on March 15, Seward rashly indicated that Fort Sumter would be evacuated within a few days. Upon receipt of this welcome information, the commissioners agreed to suspend momentarily their demand for negotiations. A little delay, they thought, would in any case work to the advantage of their one-month-old republic. The days passed, however, with no signs of anything happening at Sumter. Pressed for an explanation, Seward on March 21 renewed his assurances to Campbell, though in somewhat less explicit terms.[43] Any uneasiness he may have felt at this point was probably still slight, but in another week it would be a different matter. What he had actually offered the commissioners was a prediction in which he had such great confidence that he allowed it to be accepted as an authoritative pledge. Events were about to make him a bad prophet at best and, in southern eyes, a master of duplicity.[44]

43. Henry G. Connor, *John Archibald Campbell* (Boston, 1920), pp. 122–127; Crawford, *Genesis,* pp. 325–332. In his interviews with Seward, Campbell was accompanied by his Supreme Court colleague Samuel Nelson.

44. The weight of scholarly opinion supports the conclusion that Seward, without Lincoln's knowledge, really did give the commissioners a pledge. See, e.g., Stampp, *And the War Came,* pp. 273–274; Bancroft, *Seward,* II, 113–117; Ludwell H. Johnson, "Fort Sumter and Confederate Diplomacy," *JSH,* XXVI (1960), 455–461; Burton J. Hendrick, *Lincoln's War Cabinet* (Boston, 1946), pp. 166–169. But see Nevins, *War for the Union,* I, 51 n., for a contrary argument. Nicolay and Hay, *Lincoln,* III, 404–413, treats Campbell harshly as a rebel agent, pointing out that he was in communication not only with the commissioners but also with Jefferson Davis. That Seward practiced deception seems clear enough; yet it should be noted: (1) Campbell's subsequent account of the affair, virtually the only primary source, was by its very nature self-serving. (2) Whatever one may think of his methods, Seward's purpose was to delay giving the commissioners the official rebuff that might mean war. (3) The commissioners were as anxious as Seward to prolong their stay in Washington and thought that *they* were cleverly deceiving Seward. (4) Technically, no pledge passed from Seward to the commissioners; for Campbell, in talking with the latter, refused to use Seward's name and said that the commissioners were "not authorized to infer" that he was "acting under any agency." (5) The commissioners had already convinced themselves before March 15 that Sumter would be evacuated, but Confederate leaders in Montgomery, according to the secretary of war, "at no time placed any reliance on assurances by the Government at Washington in respect to the evacuation of Fort Sumter." *Official Records,* I, 275, 285. There is, in short, no indication that Seward's double-talk had any effect that the Confederates themselves did not desire and connive at.

Late in March, Lincoln was still pondering the Sumter problem. He had felt the heat of Republican response to the rumors of impending withdrawal, and he had listened to some persuasive arguments for the feasibility of a relief expedition, especially from a former naval officer named Gustavus V. Fox. He had also lost some of his confidence in General Scott's pessimistic views on Sumter, perhaps because Scott was now on record as favoring the evacuation of Fort Pickens too, and for plainly political reasons.[45] But the critical influence may have been simply the pressure of the calendar. Presumably, Anderson could hold out only until mid-April. The time had come when the administration must decide whether to begin preparing a relief expedition. In the circumstances, a negative decision would be irrevocable; an affirmative decision would not. Lincoln elected to keep his options open for another week, and that meant taking action. On March 29 he again polled his cabinet about provisioning Fort Sumter. A striking change had occurred since the middle of the month. Seward now stood almost alone in opposition, while four members were firmly on the president's side. Later the same day, Lincoln issued orders for the preparation of an expedition ready to sail by April 6.[46]

Seward did not give up, however. He revised his strategy and launched a counterattack, with a resulting confusion that has never been entirely dispelled. Heretofore, the secretary of state had displayed no particular concern for the safety of Fort Pickens. Indeed, more than one member of the cabinet suspected him of having

45. Robert Means Thompson and Richard Wainwright (eds.), *Confidential Correspondence of Gustavus Vasa Fox* (2 vols.; New York, 1920), I, 7–9; *Official Records*, I, 200–201. There is a story, generally accepted though resting primarily on the subsequent reminiscence of Montgomery Blair, that Lincoln on the night of March 28 informed the cabinet of Scott's advice to evacuate both forts. This allegedly stiffened their backs for a stronger stand the next day. See, e.g., Nicolay and Hay, *Lincoln*, III, 394–395; Gideon Welles, *Lincoln and Seward* (New York, 1874), pp. 57–58, 64–65; Nevins, *War for the Union*, I, 55; Current, *Lincoln and the First Shot*, pp. 76–77; Bancroft, *Seward*, II, 123–124. Yet Lincoln presumably had Scott's memorandum in his hands by about March 16 or 17, since it was attached to Cameron's response to Lincoln's query of the 15th. See Hoogenboom, "Fox and Fort Sumter," p. 387 n., and Kenneth P. Williams, *Lincoln Finds a General* (5 vols.; New York, 1949–59), I, 387, for dissenting views. "Meigs on the Civil War," p. 300 (diary entry for March 31, 1861), lends support to the Blair version.

46. Basler (ed.), *Works of Lincoln*, IV, 301–302; Nicolay and Hay, *Lincoln*, III, 429–433; Howard K. Beale (ed.), *The Diary of Edward Bates* (Washington, 1933), p. 180. Caleb B. Smith somewhat less emphatically joined Seward on the negative side. Bates declared that the time had come for Sumter "to be either evacuated or relieved." All members of the cabinet agreed on reinforcing Pickens. It appears that Lincoln's determination influenced the cabinet vote, rather than vice versa.

inspired General Scott's suggestion that Pickens be evacuated along with Sumter. That the suspicion was accurate seems likely, and not only because of Seward's obvious influence over the physically decrepit general. The Seward plan for saving the Union required the avoidance of armed conflict everywhere, but especially at the two points of worst friction, Sumter and Pickens. Reinforcement of Pickens would be much easier strategically, but just a little less dangerous politically, than reinforcement of Sumter. Seward's friend John A. Gilmer of Virginia insisted that both forts must be given up if peace were to be preserved and further secessions prevented. So did border-state Unionists generally. So did Stephen A. Douglas and many other northern Democrats. Yet Seward could now plainly see that a policy of double withdrawal would get no support in the cabinet and very little from Republicans anywhere. He therefore decided to settle for half a loaf if possible. That is, he would try to shift attention from Fort Sumter to Fort Pickens, continuing to press for the peaceable evacuation of Sumter while taking personal charge of arrangements for a dramatic reinforcement of Pickens. In this manner, he might be able to keep his pledge to the southern commissioners and yet exonerate himself from charges of having become a Republican doughface.[47]

Thus, at the decisive cabinet meeting of March 29, Seward balanced his negative vote on the relief of Fort Sumter with an emphatic statement in favor of holding Fort Pickens. On the same day, he urged Lincoln to put Captain Montgomery C. Meigs, an army engineer, in command of an expedition to Pickens. Lincoln agreed two days later and issued the necessary orders.[48] From that date, two separate expeditions were under preparation. The one destined for Sumter, organized through regular cabinet channels, was placed in charge of Gustavus Fox, the former naval officer who had conceived and persuasively promoted his own plan for relief of the fort. Fox, a man of dynamic personality, had been the first of Lincoln's three recent visitors to Charleston, returning still confident that the plan would work. He also happened to be the brother-in-law of Montgomery Blair, the most militant member of the cabinet, with whom he kept in close touch. The Pickens expedition was organized more anomalously by Meigs and a naval officer under Seward's

47. Bancroft, *Seward,* II, 124–125, 545–548 (Gilmer letters to Seward); Robert W. Johannsen, *Stephen A. Douglas* (New York, 1973), pp. 847–858.

48. Van Deusen, *Seward,* pp. 279–280; "Meigs on the Civil War," pp. 299–300; Basler (ed.), *Works of Lincoln,* IV, 313–315.

supervision, without the knowledge of the secretary of war or the secretary of the navy. In a sense, the hawk and dove wings of the cabinet were each preparing their own expeditions.[49]

On April 1, Justice Campbell had another interview with Seward, who presented him with a written statement, ostensibly authorized by Lincoln, declaring that the government would not undertake to supply Fort Sumter without first notifying the governor of South Carolina. This, on its face, was a sharp deviation from the original "pledge," but Campbell rather gullibly allowed himself to be convinced that nothing had really changed.[50] Seward, in fact, was still willing to say that there would be no effort to relieve Sumter because he still believed that he could make the prediction come true. He hoped to persuade Lincoln that Fort Pickens, which was more defensible and less controversial than Fort Sumter, would serve just as well to symbolize the persistence of federal authority within the seceded states.

Yet the reinforcement of Pickens in conjunction with the evacuation of Sumter would probably not in itself make peace more likely than war. Something else was needed, and Seward now unveiled the spectacular strategy that he had been holding in reserve. On April 1 he sent Lincoln the notorious memorandum entitled "Some thoughts for the President's consideration." Opening with an assertion that the administration was still "without a policy either domestic or foreign," the paper offered three major proposals: First, Seward reiterated his conviction that Fort Sumter should be evacuated and Fort Pickens reinforced, arguing curiously that this would somehow change the whole sectional issue from one of slavery to

49. Howard K. Beale (ed.), *Diary of Gideon Welles* (3 vols.; New York, 1960), I, 14–26 (this part of the "diary" is actually memoir); "Meigs on the Civil War," pp. 300–301; Hoogenboom, "Fox and Fort Sumter," pp. 387–389; John Niven, *Gideon Welles: Lincoln's Secretary of the Navy* (New York, 1973), pp. 329–332.

50. Connor, *Campbell*, pp. 127–129. In this interview of April 1, according to Campbell, Seward told him that Lamŏn (who had promised evacuation of Sumter during his recent visit to Charleston) had not acted under any authority from Lincoln. Then Seward handed him a written statement to the effect "that the President may desire to supply Fort Sumter, but will not undertake to do so without first giving notice to Governor Pickens." After ostensibly conferring with Lincoln, Seward then revised the statement to read: "I am satisfied the Government will not undertake to supply Fort Sumter without giving notice to Governor Pickens." This was the only formal pledge that Campbell ever received, and it was kept five days later. Campbell's uncorroborated explanation was that these plain indications were offset by Seward's verbal assurances. *Ibid.*, pp. 132, 134.

one of union or disunion. Second, he proposed the initiation of an aggressive foreign policy aimed especially at Spain, which had recently taken steps to annex Santo Domingo, and at France, which was suspected of having similar designs in the Caribbean and Mexico. He would "demand explanations" from both nations and, if satisfactory replies were not forthcoming, "would convene Congress and declare war against them." Third, with a certain amount of transparent circumlocution, he suggested that the president delegate to him the responsibility for "energetic prosecution" of whatever policy should be adopted. This amounted to a formal bid for the informal role of premier.

Here were evacuation, foreign war, and partial abdication all wrapped up for Lincoln in one neat package. He responded with remarkable restraint but unmistakable firmness, exposing the weaknesses in Seward's argument and adding that he himself must do whatever must be done.[51] Yet the effrontery of the document may have strengthened his disposition to do something decisive. There was a rising clamor throughout the North for an end to hesitation and inaction. On April 3, for example, the Republican *New York Times* published a biting editorial entitled "Wanted—A Policy!" The new Confederacy, said the *Times,* was conducting itself with a "degree of vigor, intelligence, and success" that could not be detected anywhere in Washington. "The President," it warned, "must adopt some clear and distinct policy in regard to secession, or the Union will not only be severed, but the country will be disgraced."[52]

With the final Sumter deadline now almost upon him, Lincoln nevertheless continued to weigh his alternatives. By Seward's arrangement, he conferred on April 4 with a member of the Virginia convention. It is a matter of dispute whether he again raised the possibility of evacuating Fort Sumter in exchange for adjournment of the convention, but in any case the discussion proved fruitless.[53]

51. Basler (ed.), *Works of Lincoln,* IV, 316–318. Van Deusen, *Seward,* pp. 281–284, points out that some ideas in the memorandum "were later implemented." For indications that Seward was working in conjunction with Thurlow Weed and Henry J. Raymond, editor of the *New York Times,* see Patrick Sowle, "A Reappraisal of Seward's Memorandum of April 1, 1861, to Lincoln," *JSH,* XXXIII (1967), 234–239. On April 3 and 5 the three commissioners reported to Montgomery that the expedition being prepared might be destined for Santo Domingo. *Official Records,* I, 286.

52. Perkins (ed.), *Northern Editorials,* II, 660–664.

53. The Virginian involved, John B. Baldwin, afterward asserted that Lincoln made no offer of any kind. This was disputed, but not very credibly, by another Virginia

According to his own subsequent testimony, Lincoln also gave serious thought to Seward's plan for withdrawing from Fort Sumter while tightening the federal grip on Fort Pickens.[54] At just this time, however, the security of the latter seemed particularly doubtful. The fort was held by fewer than fifty soldiers under command of a lieutenant, while the Confederate general, Braxton Bragg, had more than a thousand troops in the Pensacola area, with many more on their way to join him. An artillery company arriving in early February to reinforce the fort had been kept on board its transport in accordance with a truce arranged between Florida secession leaders and the Buchanan administration. At Lincoln's direction, orders to land the reinforcements had gone forth from Washington on March 12, but after more than three weeks there was still no indication that the landing had been effected. Reports that did arrive told of multiplying Confederate batteries, a garrison suffering from fatigue, and depleted supplies on the naval vessels standing by. What, then, if the Lincoln government should voluntarily vacate Fort Sumter, only to learn that Fort Pickens had been taken by assault? The cause of Union might not survive such a double blow.[55]

On April 4, Lincoln drafted a letter to Anderson telling him that relief would soon be on its way. He also sent for Fox and said that "he had decided to let the expedition go." But the original plan was now modified in such a way as to indicate that he still cherished some hope of evading the dilemma presented by Sumter. That is, confronted with a choice between evacuation and war, he proposed to offer the other side instead a choice between war and continuation of the status quo. Fox, upon arriving at Charleston Harbor, must first try merely to provision Fort Sumter with unarmed boats. If they were fired upon, then he was to force his way into the fort

---

Unionist, John Minor Botts. See Potter, *Lincoln and His Party*, pp. 356–358; Current, *Lincoln and the First Shot*, pp. 94–96; Thompson and Wainwright (eds.), *Correspondence of Fox*, I, 39; Tyler Dennett (ed.), *Lincoln and the Civil War in the Diaries and Letters of John Hay* (New York, 1939), p. 30.

54. In his message of July 4, 1861, to the special session of Congress, Lincoln indicated that the Sumter expedition remained tentative until he learned (April 6) of the failure to reinforce Pickens. Basler (ed.), *Works of Lincoln*, IV, 424–425. This assertion has been strongly challenged, however. For example, see Stampp, "Lincoln and the Strategy of Defense," pp. 313–314; Richard N. Current, *The Lincoln Nobody Knows* (New York, 1958), pp. 121–124. For rebuttal, see the new preface in the 1962 paperback edition of Potter, *Lincoln and His Party*, pp. xxvi–xxvii.

55. *Official Records*, I, 352, 355–356, 363–365; Grady McWhiney, "The Confederacy's First Shot," *CWH*, XIV (1968), 8.

with both supplies and reinforcements. But if the provisioning should be allowed to proceed peacefully, then there would be no attempt to send in reinforcements, and the naval force would withdraw from the vicinity. Furthermore, South Carolina authorities would be notified well ahead of time that the expedition was intended to do nothing more than "feed the hungry."[56]

Still, Lincoln delayed sending the fateful notification until on April 6 there came further disturbing news from Florida. The naval commander at Pensacola had refused to cooperate in landing reinforcements at Fort Pickens on the grounds that he had received no direct orders to do so from his own superiors. Pickens was thus still precariously held by a handful of men, and its fate could not be determined for another week at least.[57] At this point, Lincoln waited no longer. On the same day, he dispatched a messenger to Charleston with the following notice for the governor of South Carolina: "I am directed by the President of the United States to notify you to expect an attempt will be made to supply Fort Sumter with provisions only; and that, if such attempt be not resisted, no effort to throw in men, arms, or ammunition will be made, without further notice, or in case of an attack upon the fort."[58]

These words do not sound like an ultimatum, but they were received as one by Confederate leaders and acted upon accordingly. Lincoln, it has been said, must have known that he was taking a step likely to precipitate war. All this really means, however, is that any line of action other than abject surrender was bound to be provocative, given the intemperate mood in Charleston and Montgomery. Lincoln, in short, did not choose to initiate hostilities, but he did refuse to accept the Confederate terms for peace. He made one last effort to thread his way between war and disunion. The mere provisioning of Fort Sumter, if it had been permitted, would have placed

56. Basler (ed.), *Works of Lincoln,* IV, 321–322; *Official Records,* I, 232–235, 248, 294. Anderson received the message on April 7 and recorded his emphatic disapproval of the expedition.

57. *Official Records of the Union and Confederate Navies in the War of the Rebellion* (30 vols.; Washington, 1894–1922), series I, vol. IV, 109–111, 115. New orders for reinforcement were sent the same day, and the additional troops were landed at Fort Pickens on April 12. The Meigs expedition arrived four days later. Pickens remained in Union hands throughout the war.

58. Basler (ed.), *Works of Lincoln,* IV, 323–324. The message was delivered to Governor Pickens and General Beauregard on the evening of April 8. *Official Records,* I, 291.

relations there on the same terms that had already been established at Fort Pickens. But Lincoln's strategy, besides offering an extension of the status quo, had a secondary purpose which, in the end, became the operative one. That is, while trying to fulfill his inaugural pledge that the forts would be held, he was also determined to keep his promise that the government would not use military coercion unless it were first attacked.

At this juncture, Lincoln began to pay the cost of having allowed his secretary of state such a free hand. The key vessel in the plans for the Sumter expedition was the warship *Powhatan,* but late at night on April 5, Secretary of the Navy Welles learned that it had been appropriated for the Pickens expedition. Seward, it transpired, had arranged the transfer four days earlier, securing Lincoln's signature on the order. The president, when confronted with the problem, admitted somewhat sheepishly that he had been confused but ordered Seward to have the *Powhatan* restored to the Sumter fleet. Seward took his time about sending off the telegram and then signed it with his own name. The message overtook the *Powhatan* after it had put to sea, but the ship's commander refused to turn back, declaring that he was under superior orders signed by the president and would proceed to Pensacola.[59]

Meanwhile, Fox was delayed in his final preparations and did not get under way for Charleston until the morning of April 9. Even then, incredibly, he left without having been told where the *Powhatan* was headed. Stormy weather delayed his arrival off Charleston until April 12, and then more time was spent waiting in vain for the *Powhatan.* Before he could organize any kind of effort to reach Fort Sumter, it had been pounded into submission. The fort was no doubt past saving in any case, but if the *Powhatan* had been at his disposal, Fox probably would have tried to force his way into the harbor. The odds, it now seems clear, were not in his favor. Thus Seward's high-handed interference perhaps helped produce a simple fiasco in place of a bloody, heroic failure.[60]

59. *Official Records, Navies,* IV, 108–109, 111–112; Niven, *Gideon Welles,* pp. 329–336. Montgomery Blair later charged that Seward had deliberately tried to sabotage the Sumter expedition (Welles, *Lincoln and Seward,* p. 66), and more than a few historians have agreed; but see rebuttal in Bancroft, *Seward,* II, 144; Van Deusen, *Seward,* p. 285.

60. Thompson and Wainwright (eds.), *Correspondence of Fox,* I, 31–36, 38–41; *Official Records, Navies,* IV, 248–251; Niven, *Gideon Welles,* p. 336.

The decision to inaugurate civil war at Charleston was made by Jefferson Davis and his cabinet in Montgomery on April 9, four years to the day before Lee surrendered at Appomattox Courthouse. Word of Lincoln's notice of intent had come to them the night before, after a week of confusion. By Seward and others they had been led to believe that Fort Sumter would be evacuated; yet all the while they were also receiving ominous reports of military and naval preparations. Now it appeared that they had been deliberately deceived by the Black Republican government in Washington. The unauthenticated notice from Lincoln might be another trick, masking a general attack on Charleston. Seward's jugglery and Lincoln's vacillation had utterly destroyed the administrations's credibility in Montgomery.[61]

Yet, even if the communications between Montgomery and Washington had been as direct, cordial, and mutually trustful as the relations between Major Anderson and General Beauregard at Charleston, it would have made no difference. Lincoln was prepared to accept war rather than acknowledge the dissolution of a Federal Union which in Davis's eyes had ceased to exist; Davis, in turn, was ready to make war for the territorial integrity of a Southern Confederacy which in Lincoln's eyes had never begun to exist. If a peaceable evacuation of Fort Sumter had somehow been arranged or compelled, Lincoln would only have redoubled his efforts to hold Fort Pickens, and the Davis government was as determined to have one as the other. In fact, Confederate leaders were much less diffident about commencing hostilities than their counterparts in Washington. To Davis and most of his cabinet, the forts were now essentially military problems. General Bragg already held instructions to attack Pickens whenever he thought it could be done successfully, and Sumter had been spared largely in the hope of obtaining it undamaged.[62] Narrow military considerations likewise dictated the critical

61. Johnson, "Sumter and Confederate Diplomacy," pp. 472–476, but compare with Nevins, *War for the Union,* I, 73.

62. Dunbar Rowland (ed.) *Jefferson Davis, Constitutionalist: His Letters, Papers, and Speeches* (10 vols.; Jackson, Miss., 1923), V, 61; McWhiney, "Confederacy's First Shot," pp. 10–12. Davis, in his *The Rise and Fall of the Confederate Government* (2 vols.; New York, 1881), I, 292, declared: "He who makes the assault is not necessarily he that strikes the first blow or fires the first gun." The charge that Lincoln deliberately chose war and then "maneuvered the Confederates into firing the first shot" has never won widespread support. For the debate on the subject, see Charles W. Ramsdell, "Lincoln and Fort Sumter," *JSH,* III (1937), 259–288; J. G. Randall, *Lincoln the Liberal Statesman* (New York, 1947), pp. 88–117; Stampp, "Lincoln and the Strategy of Defense," pp. 311–323; McWhiney, "Confederacy's First Shot," pp. 5–6,

cabinet decision of April 9, with unfortunate consequences for the Confederacy.

At Charleston, Beauregard had standing orders to prevent any relief expedition from entering the harbor, and they scarcely needed repeating. But the Davis government went further and decided that Fort Sumter must be taken before the Fox expedition arrived. That would eliminate the danger of having to fight both Anderson and his seaborne rescuers at the same time. In order to obtain this military advantage of dubious worth, the Confederacy voluntarily accepted the role of aggressor, preparing to open fire, if necessary, on the American flag, on a fortress charged with deep symbolic meaning, and on a soldier who had become a national hero. It would have been difficult to devise a strategy better calculated to arouse and unite the divided, irresolute North. The primary significance of the southern attack on Fort Sumter is not that it started the Civil War, but rather that it started the war in such a manner as to give the cause of Union an eruptive force which it might otherwise have been slow to acquire.

On April 10, Beauregard received his orders from Montgomery: demand the evacuation of Sumter, and if Anderson refused, proceed to reduce the fort. Anderson did refuse but added wistfully that in just a few more days he would have been starved out. This remark inspired some further negotiations which eventually proved fruitless, and at 4:30 A.M. on April 12, the first Confederate shell arched through the sky toward the fort.

The bombardment lasted for some thirty-three hours. Then, with fire raging through his barracks, with his ammunition nearly exhausted, and with no help in sight, Anderson acknowledged defeat. "I accepted terms of evacuation offered by General Beauregard," he reported, "and marched out of the fort Sunday afternoon, the 14th instant, with colors flying and drums beating, bringing away company and private property, and saluting my flag with fifty guns."[63]

14; Richard N. Current, "The Confederates and the First Shot," *CWH*, VII (1961), 357–369; Current, *Lincoln and the First Shot*, pp. 182–208; Williams, *Lincoln Finds a General*, I, 56–57, 390; Potter, *Lincoln and His Party*, pp. 371–374; Potter, "Why the Republicans Rejected Both Compromise and Secession," in George Harmon Knoles (ed.), *The Crisis of the Union, 1860–1861* (Baton Rouge, 1965), pp. 90–106, with comment by Kenneth M. Stampp, pp. 107–113.

63. *Official Records*, I, 12–25, 28–35, 297, 300–302, 305–306, 309; Crawford, *Genesis*, pp. 421–448; Bruce Catton, *The Coming Fury* (Garden City, N.Y., 1961), pp. 302–324; W. A. Swanberg, *First Blood: The Story of Fort Sumter* (New York, 1957), pp. 285–325; T. Harry Williams, *P. G. T. Beauregard, Napoleon in Gray* (Baton Rouge, 1954),

During the battle, a combined total of nearly five thousand artillery rounds had been fired, miraculously without causing a single fatality on either side. It was a deceptively bloodless beginning to one of the bloodiest wars in history.[64]

Exactly four years after the surrender—that is, on April 14, 1865—Robert Anderson returned to raise his old flag over Fort Sumter. By then, the sounds of battle had given way to the stillness at Appomattox and the issues that inflamed the antebellum years had been settled. Slavery was dead; secession was dead; and six hundred thousand men were dead. That was the basic balance sheet of the sectional conflict.

---

pp. 51–61. For the bizarre role of a Texas senator in the fall of Sumter, see Alvy L. King, *Louis T. Wigfall, Southern Fire-Eater* (Baton Rouge, 1970), pp. 118–122.

64. There was an anticlimactic death, however, which exemplified the mixture of tragedy and absurdity so often manifested in the Civil War. Anderson, who had managed to get off about a thousand shots or so in defense of Fort Sumter, insisted on firing another hundred in the ceremony of surrender. An explosion midway killed one soldier on the spot and wounded several others, one of whom soon died. Doubleday, *Reminiscences*, p. 171; Crawford, *Genesis*, pp. 446–447; Oliver Lyman Spaulding, Jr., "The Bombardment of Fort Sumter, 1861," AHA *Annual Report*, 1913, I, 198–199.

Bibliography

Abels, Jules, *Man on Fire: John Brown and the Cause of Liberty.* New York, 1971.

Abernethy, Thomas Perkins, *From Frontier to Plantation in Tennessee.* Chapel Hill, 1932.

Adams, Alice Dana, *The Neglected Period of Anti-Slavery in America, 1808–1831.* Boston, 1908.

Adams, James Truslow, *The Epic of America.* Boston, 1931.

Addington, Wendell G., "Slave Insurrections in Texas," *JNH,* XXXV (1950), 408–434.

Alden, John Richard, *The First South.* Baton Rouge, 1961.

Alexander, Thomas B., "Persistent Whiggery in the Confederate South, 1860–1877," *JSH,* XXVII (1961), 305–329.

———, *Sectional Stress and Party Strength.* Nashville, 1967.

———, *Thomas A. R. Nelson of East Tennessee.* Nashville, 1956.

Allis, Frederick S., Jr., "The Dred Scott Labyrinth," in H. Stuart Hughes (ed.), *Teachers of History: Essays in Honor of Laurence Bradford Packard.* Ithaca, 1954, pp. 341–368.

Ambler, Charles Henry (ed.), "Correspondence of Robert M. T. Hunter, 1826–1876," *AHA* Annual Report, 1916, II.

Ames, Henry V., "John C. Calhoun and the Secession Movement of 1850," American Antiquarian Society *Proceedings,* New Series, XXVIII (1918), 19–50.

——— (ed.), *State Documents on Federal Relations,* No. VI, *Slavery and the Union, 1845–1861.* Philadelphia, 1906.

Ames, Seth (ed.), *Works of Fisher Ames.* 2 vols.; Boston, 1854.

Anderson, Godfrey Tryggve, "The Slavery Issue as a Factor in Massachusetts Politics, from the Compromise of 1850 to the Outbreak of the Civil War." Ph.D. dissertation, University of Chicago, 1944.

Angle, Paul M. (ed.), *Created Equal? The Complete Lincoln-Douglas Debates of 1858*. Chicago, 1958.

———, *Herndon's Life of Lincoln*. New York, 1930.

Aptheker, Herbert, *American Negro Slave Revolts*. New York, 1943.

———, "Militant Abolitionism," *JNH*, XXVI (1941), 438–484.

———, *Nat Turner's Slave Rebellion*. New York, 1966.

Auchampaugh, Philip Gerald, "The Buchanan-Douglas Feud," ISHS *Journal*, XXV (1932), 5–48.

———, *James Buchanan and His Cabinet on the Eve of Secession*. Lancaster, Pa., 1926.

———, "James Buchanan, the Court and the Dred Scott Case," *Tennessee Historical Magazine*, IX (1926), 231–240.

Auer, J. Jeffery (ed.), *Antislavery and Disunion*, 1858–1861. New York, 1963.

Avary, Myrta Lockett (ed.), *Recollections of Alexander H. Stephens*. New York, 1910.

Bailey, Hugh C., "Disaffection in the Alabama Hill Country, 1861," *CWH*, IV (1958), 183–193.

———, "Disloyalty in Early Confederate Alabama," *JSH*, XXIII (1957), 522–528.

———, *Hinton Rowan Helper, Abolitionist-Racist*. University, Ala., 1965.

Bain, Richard C., *Convention Decisions and Voting Records*. Washington, 1960.

Baird, Robert, *Religion in America*. New York, 1844.

Baker, George E. (ed.), *The Works of William H. Seward*. 5 vols.; Boston, 1887–90.

Bancroft, Frederic, "The Colonization of American Negroes, 1801–1865," in Jacob E. Cooke, *Frederic Bancroft, Historian*. Norman, Okla., 1957, pp. 145–258.

———, *The Life of William H. Seward*. 2 vols.; New York, 1900.

———, *Slave-Trading in the Old South*. Baltimore, 1931.

Barbee, David R., and Bonham, Milledge L., Jr. (eds.), "The Montgomery Address of Stephen A. Douglas," *JSH*, V (1939), 527–552.

Baringer, William E., "Campaign Techniques in Illinois—1860," ISHS *Transactions* (1932), 202–281.

———, *A House Dividing: Lincoln as President-Elect*. Springfield, Ill., 1945.

———, *Lincoln's Rise to Power*. Boston, 1937.

Barnes, Thurlow Weed, *Memoir of Thurlow Weed*. Boston, 1884.

Barry, Louise, "The Emigrant Aid Company Parties of 1854" and "The New England Emigrant Aid Company Parties of 1855," *Kansas Historical Quarterly*, XII (1943), 115–155, 227–268.

Bartlett, Irving H., "Wendell Phillips and the Eloquence of Abuse," *American Quarterly*, XI (1959), 509–520.

———, *Wendell Phillips, Brahmin Radical*. Boston, 1961.

Bartlett, Ruhl Jacob, *John C. Frémont and the Republican Party.* Columbus, Ohio, 1930.

Basler, Roy P. (ed.), *The Collected Works of Abraham Lincoln.* 8 vols.; New Brunswick, N.J., 1953.

Bassett, John Spencer, *Anti-Slavery Leaders of North Carolina.* Baltimore, 1898.

Bates, William M., "The Last Stand for the Union in Georgia," *Georgia Review,* VII (1953), 455–467.

Baxter, Maurice, *Orville H. Browning: Lincoln's Friend and Critic.* Bloomington, Ind., 1957.

Beale, Howard K. (ed.), *The Diary of Edward Bates.* Washington, 1933.

_____, *Diary of Gideon Welles.* 3 vols.; New York, 1960.

Bean, William G., "An Aspect of Know Nothingism—the Immigrant and Slavery," *SAQ,* XXIII (1924), 319–334.

_____, "Puritan versus Celt, 1850–1860," *NEQ,* VII (1934), 70–89.

Beard, Charles A. and Mary R., *The Rise of American Civilization.* 2 vols.; New York, 1927.

Bell, Howard H., "Expressions of Negro Militancy in the North, 1840–1860," *JNH,* XLV (1960), 11–20.

_____, "National Negro Conventions of the Middle 1840's: Moral Suasion versus Political Action," *JNH,* XLII (1957), 247–260.

Bemis, Samuel Flagg, *John Quincy Adams and the Union.* New York, 1956.

Bennett, Lerone, Jr., "Was Abe Lincoln a White Supremacist?" *Ebony,* Feb. 1968, 35–42.

Benson, Lee, *The Concept of Jacksonian Democracy: New York as a Test Case.* Princeton, 1961.

Benton, Thomas Hart (ed.), *Abridgment of the Debates of Congress.* New York, 1860.

_____, *Historical and Legal Examination of the Dred Scott Case.* New York, 1857.

_____, *Thirty Years' View . . . 1820 to 1850.* 2 vols.; New York, 1854–56.

Berger, Max, "The Irish Emigrant and American Nativism as Seen by British Visitors, 1836–1860," *PMHB,* LXX (1946), 146–160.

Bernstein, Barton J., "Southern Politics and Attempts to Reopen the African Slave Trade," *JNH,* LI (1966), 16–35.

_____ (ed.), *Towards a New Past: Dissenting Essays in American History.* New York, 1968.

Berwanger, Eugene H., *The Frontier Against Slavery: Western Anti-Negro Prejudice and the Slavery Extension Controversy.* Urbana, Ill., 1967.

Bestor, Arthur, "State Sovereignty and Slavery: A Reinterpretation of Proslavery Constitutional Doctrine, 1846–1860," ISHS *Journal,* LIV (1961), 117–180.

Beveridge, Albert J., *Abraham Lincoln, 1809–1858.* 4 vols.; Boston, 1928.

Billington, Ray Allen, *The Protestant Crusade, 1800–1860: A Study of the Origins of American Nativism.* New York, 1938, and Chicago, 1964.

Binkley, Wilfred E., *American Political Parties: Their Natural History.* 2nd ed.; New York, 1945.

Binkley, William Campbell, *The Expansionist Movement in Texas, 1836–1850.* Berkeley, 1925.

Blaine, James G., *Twenty Years of Congress.* 2 vols.; Norwich, Conn., 1884–86.

Bode, Carl. *The Anatomy of American Popular Culture, 1840–1861.* Berkeley, 1959.

Boney, F. N., *John Letcher of Virginia.* University, Ala., 1966.

Bonham, Milledge L., Jr., "New York and the Election of 1860," *NYH,* XXXII (1934), 124–143.

Bonner, James C., "Profile of a Late Ante-Bellum Community," *AHR,* XLIX (1944), 663–680.

Bontemps, Arna, *Free at Last: The Life of Frederick Douglass.* New York, 1971.

Boucher, Chauncey Samuel, "*In Re* That Aggressive Slavocracy," *MVHR,* VIII (1921), 13–79.

———, "The Secession and Cooperation Movements in South Carolina, 1848 to 1852," Washington University *Studies,* V (1918), 71–85.

———, *South Carolina and the South on the Eve of Secession, 1852 to 1860,* in "Washington University Studies," Humanistic Series, VI (1919), 85–144.

Boyd, William K., "North Carolina on the Eve of Secession," AHA *Annual Report,* 1910, pp. 165–178.

Bradley, Erwin Stanley, *Simon Cameron, Lincoln's Secretary of War.* Philadelphia, 1966.

———, *The Triumph of Militant Republicanism: A Study of Pennsylvania and Presidential Politics 1860–1872.* Philadelphia, 1964.

Bragg, Jefferson Davis, *Louisiana in the Confederacy.* Baton Rouge, 1941.

Brand, Carl Fremont, "The History of the Know Nothing Party in Indiana," *IMH,* XVIII (1922), 47–81, 177–206, 266–306.

Brantley, William, "Alabama Secedes," *Alabama Review,* VII (1954), 165–185.

Brauer, Kinley J., *Cotton versus Conscience: Massachusetts Whig Politics and Southwestern Expansion, 1843–1848.* Berkeley, 1963.

Brewerton, Douglas, *The War in Kansas: A Rough Trip to the Border.* New York, 1856.

Breyfogle, William, *Make Free: The Story of the Underground Railroad.* Philadelphia, 1958.

Bridges, C. A., "The Knights of the Golden Circle: A Filibustering Fantasy," *SWHQ,* XLIV (1941), 287–302.

Brigance, William Norwood, *Jeremiah Sullivan Black.* Philadelphia, 1934.

Brooks, Preston S., "Statement by Preston S. Brooks," Massachusetts Historical Society *Publications,* LXI (1927–28), 221–223.

Brooks, R. P., "Howell Cobb and the Crisis of 1850," *MVHR,* IV (1917), 279–298.

——— (ed.), "Howell Cobb Papers," *Georgia Historical Quarterly,* V (1921), 29–52.

Broussard, James H., "Some Determinants of Know-Nothing Electoral Strength in the South, 1856," *Louisiana History,* VII (1966), 5–20.

Brown, Everett Somerville, *The Constitutional History of the Louisiana Purchase.* Berkeley, 1920.

Brown, Margaret L., "Asa Whitney and His Pacific Railroad Publicity Campaign," *MVHR,* XX (1933–34), 209–224.

Brown, Salmon, "My Father, John Brown," *Outlook,* CIII (Jan. 25, 1913), 212–217.

Brown, Thomas N., "The Origins and Character of Irish-American Nationalism," *Review of Politics,* XVIII (1956), 327–358.

Bryan, Conn, "The Secession of Georgia," *Georgia Historical Quarterly,* XXXI (1947), 89–111.

Bryan, John A., "The Blow Family and Their Slave Dred Scott," Missouri Historical Society *Bulletin,* IV (1948), 223–231.

Buchanan, James, *Mr. Buchanan's Administration on the Eve of the Rebellion.* New York, 1866.

Buckmaster, Henrietta, *Let My People Go: The Story of the Underground Railroad and the Growth of the Abolition Movement.* Boston, 1959.

Burgess, John W., *The Civil War and the Constitution.* 2 vols.; New York, 1901.

Burnett, Edmund Cody, *The Continental Congress.* New York, 1941.

Burnham, W. Dean, *Presidential Ballots, 1836–1892.* Baltimore, 1955.

Byrne, Frank L., *Prophet of Prohibition: Neal Dow and His Crusade.* Madison, Wis., 1961.

Cain, Marvin R., "Edward Bates and the Decision of 1860," *Mid-America,* XLIV (1962), 109–124.

———, *Lincoln's Attorney General: Edward Bates of Missouri.* Columbia, Mo., 1965.

Caldwell, Robert Granville, *The López Expeditions to Cuba, 1848–1851.* Princeton, 1915.

Callahan, James Morton, *American Foreign Policy in Mexican Relations.* New York, 1932.

———, "The Mexican Policy of Southern Leaders under Buchanan's Administration," AHA *Annual Report,* 1910, pp. 133–151.

Campbell, Mary Emily Robertson, *The Attitude of Tennesseans toward the Union 1847–1861.* New York, 1961.

Campbell, Stanley W., *The Slave Catchers: Enforcement of the Fugitive Slave Law, 1850–1860.* Chapel Hill, 1968.

Capers, Gerald M., *John C. Calhoun, Opportunist.* Gainesville, Fla., 1960.

_____, *Stephen A. Douglas, Defender of the Union.* Boston, 1959.

Capers, Henry D., *The Life and Times of C. G. Memminger.* Richmond, 1893.

Cardwell, Guy A., "The Duel in the Old South: Crux of a Concept," *SAQ*, LXVI (1967), 50–69.

Carman, Harry J. and Luthin, Reinhard H., "The Seward-Fillmore Feud and the Disruption of the Whig Party," *NYH*, XLI (1943), 335–357.

_____, "Some Aspects of the Know-Nothing Movement Reconsidered," *SAQ*, XXXIX (1940), 213–234.

Carnathan, W. J., "The Attempt to Reopen the African Slave Trade in Texas, 1857–1858," Southwestern Political and Social Science Association *Proceedings*, 1925, pp. 134–144.

_____, "The Proposal to Reopen the African Slave Trade in the South, 1854–1860," *SAQ*, XXV (1926), 410–429.

Carpenter, Jesse T., *The South as a Conscious Minority, 1789–1861.* New York, 1930.

Carroll, Joseph Cephas, *Slave Insurrections in the United States, 1800–1865.* Boston, 1938.

Carsel, Wilfred, "The Slaveholders' Indictment of Northern Wage Slavery," *JSH*, VI (1940), 504–520.

Carter, Clarence Edwin, (ed.), *The Territorial Papers of the United States.* 26 vols.; Washington, 1934–62.

Cary, Rita McK., *The First Campaigner: Stephen A. Douglas.* New York, 1964.

Cash, W. J., *The Mind of the South.* New York, 1941.

Caskey, Willie Malvin. *Secession and Restoration of Louisiana.* University, La., 1938.

Catterall, Helen T., "Some Antecedents of the Dred Scott Case," *AHR*, XXX (1924), 56–71.

Catton, Bruce, *The Coming Fury.* Garden City, N.Y., 1961.

Cauthen, Charles Edward, *South Carolina Goes to War.* Chapel Hill, 1950.

_____, "South Carolina's Decision to Lead the Secession Movement," *NCHR*, XIX (1941), 360–372.

Chambers, William Nisbet, *Old Bullion Benton, Senator from the New West: Thomas Hart Benton, 1782–1858.* Boston, 1956.

Channing, Edward, *A History of the United States.* 6 vols.; New York, 1905–25.

Channing, Steven A., *Crisis of Fear: Secession in South Carolina.* New York, 1970.

Chesnut, Mary Boykin, *A Diary from Dixie.* Boston, 1949.

Chittenden, L. E., *A Report of the Debates and Proceedings in the Secret Sessions of the Conference Convention, for Proposing Amendments to the Constitution of the United States, Held at Washington, D.C., in February, A.D. 1861.* New York, 1864.

Claiborne, J. F. H., *Life and Correspondence of John A. Quitman.* 2 vols.; New York, 1860.

Clark, Blanche H., *The Tennessee Yeoman, 1840–1860.* Nashville, 1942.

Clark, Charles Branch, "Politics in Maryland during the Civil War," *Maryland Historical Magazine,* XXXVI (1941), 239–262.

Clusky, M. W., *The Political Text-Book.* Philadelphia, 1860.

Cochran, William C., *The Western Reserve and the Fugitive Slave Law.* Cleveland, 1920.

Coffin, Levi, *Reminiscences of Levi Coffin, the Reputed President of the Underground Railroad.* Cincinnati, 1876.

Cole, Arthur C., *The Era of the Civil War, 1848–1870.* Chicago, 1922.

———, *The Irrepressible Conflict, 1850–1865.* New York, 1934.

———, "The South and the Right of Secession in the Early Fifties," *MVHR,* I (1914–15), 376–399.

———, *The Whig Party in the South.* Washington, 1913.

Coleman, Mrs. Chapman, *The Life of John J. Crittenden.* 2 vols.; Philadelphia, 1871.

Colton, Calvin, *The Last Seven Years of the Life of Henry Clay.* New York, 1856.

Commager, Henry Steele (ed.), *Documents of American History.* 7th ed., 2 vols.; New York, 1963.

———, *Theodore Parker.* Boston, 1936.

——— (ed.), *Theodore Parker: An Anthology.* Boston, 1960.

Commons, John R., "Horace Greeley and the Working Class Origins of the Republican Party," *Political Science Quarterly,* XXIV (1909), 468–488.

Connor, Henry G., *John Archibald Campbell.* Boston, 1920.

Corwin, Edward S., "The Doctrine of Due Process of Law before the Civil War," *Harvard Law Review,* XXIV (1911), 366–385, 460–479.

———, "The Dred Scott Decision in the Light of Contemporary Legal Doctrines," *AHR,* XVII (1911), 52–69.

Cotterill, Robert S., "Early Agitation for a Pacific Railroad, 1845–1850," *MVHR,* V (1919), 396–414.

———, "Memphis Railroad Convention, 1849," *Tennessee Historical Magazine,* IV (1918), 83–94.

———, "The National Railroad Convention in St. Louis, 1849," *MHR,* XII (1918), 203–215.

Coulter, E. Merton, *The Civil War and Readjustment in Kentucky.* Chapel Hill, 1926.

———, *The Confederate States of America, 1861–1865.* Baton Rouge, 1950.

Cox, Samuel S., *Three Decades of Federal Legislation.* Providence, R.I., 1888.

Craik, Elmer LeRoy, "Southern Interest in Territorial Kansas, 1854–1858," KSHS *Collections,* XV (1919–22), 334–450.

Crallé, Richard K. (ed.), *Works of John C. Calhoun.* 6 vols.; New York, 1854–57.

Crandall, Andrew Wallace, *The Early History of the Republican Party.* Boston, 1930.

Craven, Avery O., *The Coming of the Civil War.* New York, 1942.
______, *Edmund Ruffin, Southerner: A Study in Secession.* New York, 1932.
______, *The Growth of Southern Nationalism, 1848–1861.* Baton Rouge, 1953.
Crawford, Samuel Wylie, *The Genesis of the Civil War: The Story of Sumter, 1860–1861.* New York, 1887.
Crenshaw, Ollinger, "The Knights of the Golden Circle: The Career of George Bickley," *AHR,* XLVII (1941), 23–50.
______, *The Slave States in the Presidential Election of 1860.* Baltimore, 1945.
______, "The Speakership Contest of 1859–1860," *MVHR,* XXIX (1942), 323–338.
______, "Urban and Rural Voting in the Election of 1860," in Eric F. Goldman (ed.), *Historiography and Urbanization: Essays in American History in Honor of W. Stull Holt.* Baltimore, 1941, pp. 43–66.
Crippen, Lee F., *Simon Cameron, Ante Bellum Years.* Oxford, Ohio, 1942.
Crocker, Lionel, "The Campaign of Stephen A. Douglas in the South, 1860," in J. Jeffery Auer (ed.), *Antislavery and Disunion, 1858–1861: Studies in the Rhetoric of Compromise and Conflict.* New York, 1963, pp. 262–278.
Cromwell, John W., "The Aftermath of Nat Turner's Insurrection," *JNH,* V (1920), 208–234.
Crooks, Esther J., and Crooks, Ruth W., *The Ring Tournament in the United States.* Richmond, 1936.
Culberson, Charles A., "General Sam Houston and Secession," *Scribner's Magazine,* XXXIX (1906), 584–591.
Curran, Thomas J., "Seward and the Know-Nothings," New York Historical Society *Quarterly,* LI (1967), 141–159.
Current, Richard N., "The Confederates and the First Shot," *CWH,* VII (1961), 357–369.
______, "John C. Calhoun, Philosopher of Reaction," *Antioch Review,* III (1943), 223–234.
______, *Lincoln and the First Shot.* Philadelphia, 1963.
______, *The Lincoln Nobody Knows.* New York, 1958.
Curry, Richard Orr, *A House Divided: A Study of Statehood Politics and the Copperhead Movement in West Virginia.* Pittsburgh, 1964.
Curti, Merle, "George N. Sanders, American Patriot of the Fifties," *SAQ,* XXVII (1928), 79–87.
______, *The Growth of American Thought.* New York, 1943.
______, *The Roots of American Loyalty.* New York, 1946.
______, " 'Young America,' " *AHR,* XXXII (1926), 34–55.
Curtis, Benjamin R., Jr., *Memoir of Benjamin Robbins Curtis.* 2 vols.; Boston, 1879.
Curtis, George Ticknor, *Constitutional History of the United States.* 2 vols.; New York, 1889–96.
______, *Life of Daniel Webster.* 2 vols.; New York, 1870.

———, *Life of James Buchanan.* 2 vols.; New York, 1883.

Cuthbert, Norma B. (ed.), *Lincoln and the Baltimore Plot: 1861.* San Marino, Calif., 1949.

Daniels, George H., "Immigrant Vote in the 1860 Election: The Case of Iowa," *Mid-America,* XLIV (1962), 146–162.

Darden, David L., "Alabama Secession Convention," *Alabama Historical Quarterly,* III (1941), 269–451.

Davis, David Brion, *The Problem of Slavery in Western Culture.* Ithaca, 1966.

———, *The Slave Power Conspiracy and the Paranoid Style.* Baton Rouge, 1969.

———, "Some Themes of Counter-Subversion: An Analysis of Anti-Masonic, Anti-Catholic, and Anti-Mormon Literature," *MVHR,* XLVII (1960), 205–224.

Davis, Jefferson, *The Rise and Fall of the Confederate Government.* 2 vols.; New York, 1881.

Davis, Richard Beale, *Intellectual Life of Jefferson's Virginia, 1790–1830.* Chapel Hill, 1964.

Degler, Carl N., *Neither White Nor Black: Slavery and Race Relations in Brazil and the United States.* New York, 1971.

Demos, John, "The Antislavery Movement and the Problem of Violent Means," *NEQ,* XXXVII (1964), 501–526.

Denman, Clarence Phillips, *The Secession Movement in Alabama.* Montgomery, Ala., 1933.

Dennett, Tyler, *John Hay.* New York, 1933.

——— (ed.), *Lincoln and the Civil War in the Diaries and Letters of John Hay.* New York, 1939.

Deutsch, Karl W., *et al., Political Community and the North Atlantic Area.* Princeton, 1957.

Dew, Charles B., "The Long Lost Returns: The Candidates and Their Totals in Louisiana's Secession Election," *Louisiana History,* X (1969), 353–369.

———, "Who Won the Secession Election in Louisiana?" *JSH,* XXXVI (1970), 18–32.

Diary and Correspondence of Salmon P. Chase, AHA *Annual Report,* 1902, II.

Dickerson, O. M., "Stephen A. Douglas and the Split in the Democratic Party," MVHA *Proceedings,* VII (1913–14), 196–211.

Dillon, Merton L., *Benjamin Lundy and the Struggle for Negro Freedom.* Urbana, Ill., 1966.

Dixon, Mrs. Archibald, *True History of the Missouri Compromise and Its Repeal.* Cincinnati, 1898.

Dodd, Dorothy (ed.), "Edmund Ruffin's Account of the Florida Secession Convention, 1861: A Diary," *Florida Historical Quarterly,* XII (1933), 67–76.

_____, "The Secession Movement in Florida, 1850–1861," *Florida Historical Quarterly,* XII (1933), 3–24.

Dodd, William E., *The Cotton Kingdom.* New Haven, 1919.

_____, "The Fight for the Northwest, 1860," *AHR,* XVI (1911), 774–788.

_____, *Jefferson Davis.* Philadelphia, 1907.

_____, "John Taylor: Prophet of Secession," in *John P. Branch Historical Papers, 1908.* Ashland, Va., 1908.

_____, "The Social Philosophy of the Old South," *American Journal of Sociology,* XXIII (1918), 735–746.

Doherty, Herbert J., Jr., "Florida and the Crisis of 1850," *JSH,* XIX (1953), 32–47.

_____, *Richard Keith Call, Southern Unionist.* Gainesville, Fla., 1961.

Donald, David, *Charles Sumner and the Coming of the Civil War.* New York, 1960.

_____, *Lincoln Reconsidered.* New York, 1956.

_____, *Lincoln's Herndon.* New York, 1948.

_____, "The Southerner as Fighting Man," in Charles G. Sellers (ed.), *The Southerner as American.* New York, 1966, pp. 72–88.

Donaldson, Thomas, *The Public Domain.* Washington, 1884.

Donnelly, William J., "Conspiracy or Popular Movement: The Historiography of Southern Support for Secession," *NCHR,* XLII (1965), 70–84.

Donovan, Herbert D. A., *The Barnburners.* New York, 1925.

Dorman, Lewy, *Party Politics in Alabama from 1850 through 1860.* Wetumpka, Ala., 1935.

Dorpalen, Andreas, "The German Element and the Issues of the Civil War," *MVHR,* XXIX (1942), 55–76.

Doubleday, Abner, *Reminiscences of Forts Sumter and Moultrie in 1860–61.* New York, 1876.

Douglass, Frederick, *Life and Times of Frederick Douglass Written by Himself.* 1881; rev. 1892; reprint, 1962.

Dowty, Alan, "Urban Slavery in Pro-Southern Fiction of the 1850's," *JSH,* XXXII (1966), 25–41.

Drew, Benjamin, *The Refugee: or, The Narratives of Fugitive Slaves in Canada* (sometimes titled *A Northside View of Slavery*). Boston, 1856.

Dubay, Robert W., "Mississippi and the Proposed Atlanta Convention of 1860," *Southern Quarterly,* V (1967), 347–362.

Duberman, Martin (ed.), *The Antislavery Vanguard.* Princeton, 1965.

_____, *Charles Francis Adams, 1807–1886.* Boston, 1961.

DuBois, W. E. Burghardt, *The Suppression of the African Slave Trade to the United States of America, 1638–1890.* New York, 1896.

Du Bose, John Witherspoon, *The Life and Times of William Lowndes Yancey.* 2 vols.; Birmingham, Ala., 1892.

Dudley, Thomas H., "The Inside Facts of Lincoln's Nomination," *Century Magazine,* XL (1890), 477–479.
Duignan, Peter, and Clendenen, Clarence, *The United States and the African Slave Trade, 1619–1862.* Stanford, 1963.
Dumond, Dwight Lowell, *Antislavery: The Crusade for Freedom in America.* Ann Arbor, Mich., 1961.
———, *The Secession Movement, 1860–1861.* New York, 1931.
——— (ed.), *Southern Editorials on Secession.* New York, 1931.
Dunn, Roy Sylvan, "The KGC in Texas, 1860–1861," *SWHQ,* LXX (1967), 543–573.
Durden, Robert F., "J. D. B. De Bow: Convolutions of a Slavery Expansionist," *JSH,* XVII (1951), 441–461.
Dusinberre, William, *Civil War Issues in Philadelphia.* Philadelphia, 1965.
Dyer, Brainerd, "The Persistence of the Idea of Negro Colonization," *Pacific Historical Review,* XII (1943), 53–65.
———, *Zachary Taylor.* Baton Rouge, 1946.
Eaton, Clement, "Censorship of the Southern Mails," *AHR,* XLVII (1943), 266–280.
———, "A Dangerous Pamphlet in the Old South," *JSH,* II (1936), 323–334.
———, *Freedom of Thought in the Old South.* Durham, N.C., 1940.
———, *The Growth of Southern Civilization, 1790–1860.* New York, 1961.
———, "Henry A. Wise and the Virginia Fire-Eaters of 1856," *MVHR,* XXI (1934–35), 495–512.
———, *The Mind of the Old South.* Baton Rouge, 1964.
———, "The Resistance of the South to Northern Radicalism," *NEQ,* VIII (1935), 215–231.
Edelstein, Tilden G., *Strange Enthusiasm: A Life of Thomas Wentworth Higginson.* New Haven, 1968.
Ehrlich, Walter, "History of the Dred Scott Case through the Decision of 1857." Ph.D. dissertation, Washington University, 1950.
———, "Was the Dred Scott Case Valid?" *JAH,* LV (1968), 256–265.
Eiselen, Malcolm Rogers, *The Rise of Pennsylvania Protectionism.* Philadelphia, 1932.
Elliott, Charles Winslow, *Winfield Scott, the Soldier and the Man.* New York, 1937.
Elliott, R. G., "The Big Springs Convention," KSHS *Transactions,* VIII (1903–04), 362–377.
Ellison, Joseph, *California and the Nation, 1850–1869.* Berkeley, 1927.
Ellison, William Henry, *A Self-Governing Dominion: California, 1849–1860.* Berkeley, 1950.
Emerson, Ralph Waldo, *Miscellanies.* Boston, 1904.
Emery, Charles Wilson, "The Iowa Germans in the Election of 1860,"

Annals of Iowa, 3rd series, XXII (1940), 421–453.

England, J. Merton, "The Free Negro in Ante-Bellum Tennessee," *JSH,* IX (1943), 37–58.

Ernst, Robert, *Immigrant Life in New York City, 1825–1863.* New York, 1949.

Ettinger, Amos Aschbach, *The Mission to Spain of Pierre Soulé, 1853–1855.* New Haven, 1932.

Ezell, John S., "A Southern Education for Southrons," *JSH,* XVII (1951), 303–327.

Faulk, Odie B., *Too Far North, Too Far South.* Los Angeles, 1967.

Fehrenbacher, Don E., *Chicago Giant: A Biography of "Long John" Wentworth.* Madison, 1957.

______, *Prelude to Greatness: Lincoln in the 1850's.* Stanford, 1962.

______, "The Republican Decision at Chicago," in Norman A. Graebner (ed.), *Politics and the Crisis of 1860* (Urbana, Ill., 1961), pp. 32–60.

Ferree, Walter L., "The New York Democracy: Division and Reunion, 1847–1852." Ph.D. dissertation, University of Pennsylvania, 1953.

Filler, Louis, *The Crusade Against Slavery, 1830–1860.* New York, 1960.

Finnie, Gordon E., "The Antislavery Movement in the Upper South before 1840," *JSH,* XXXV (1969), 319–342.

Fite, Emerson David, *The Presidential Campaign of 1860.* New York, 1911.

Fitts, Albert N., "The Confederate Convention" and "The Confederate Convention: The Constitutional Debate," *Alabama Review,* II (1949), 83–101; 189–210.

Fladeland, Betty, *James Gillespie Birney: Slaveholder to Abolitionist.* Ithaca, N.Y., 1955.

Fleming, Walter L., "The Buford Expedition to Kansas," *AHR,* VI (1900), 38–48.

Flippin, Percy Scott, "The Crisis of 1850 and Its Effect on Political Parties in Georgia," *Georgia Historical Quarterly,* XXIV (1940), 293–322.

______, *Herschel V. Johnson of Georgia, State-Rights Unionist.* Richmond, 1931.

Floan, Howard R., *The South in Northern Eyes, 1831–1861.* Austin, 1958.

Foner, Eric, *Free Soil, Free Labor, Free Men: The Ideology of the Republican Party before the Civil War.* New York, 1970.

______, "The Wilmot Proviso Revisited," *JAH,* LVI (1969), 262–279.

Foner, Philip S., *Business and Slavery: The New York Merchants and the Irrepressible Conflict.* Chapel Hill, 1941.

Foote, Henry S., *Casket of Reminiscences.* Washington, 1874.

Ford, Worthington Chauncey (ed.), *Letters of Henry Adams, 1858–1891.* Boston, 1930.

Fornell, Earl Wesley, *The Galveston Era: The Texas Crescent on the Eve of Secession.* Austin, 1961.

Foster, Herbert Darling, "Webster's Seventh of March Speech and the Secession Movement, 1850," *AHR,* XXVII (1922), 245–270.

Fox, Dixon Ryan, "The Negro Vote in Old New York," *Political Science Quarterly, XXXII (1917), 252–275.*

Fox, Early Lee, The American Colonization Society, 1817–1840. Baltimore, 1919.

Frank, John P. *Justice Daniel Dissenting: A Biography of Peter V. Daniel, 1784–1860.* Cambridge, Mass., 1964.

Franklin, John Hope, *The Militant South, 1800–1861.* Cambridge, Mass., 1956.

Frasure, Carl M., "Union Sentiment in Maryland, 1859–1861," *Maryland Historical Magazine,* XXIV (1929), 210–224.

Fredrickson, George M. (ed.), *The Impending Crisis of the South: How to Meet It,* by Hinton R. Helper. Cambridge, Mass., 1968.

Freedley, Edwin Troxell, *The Issue and Its Consequences.* N.p., n.d.—a campaign pamphlet, 1856.

Freehling, William W., *Prelude to Civil War: The Nullification Controversy in South Carolina, 1816–1836.* New York, 1965.

Friend, Llerena, *Sam Houston, the Great Designer.* Austin, 1954.

Frothingham, Octavius Brooks, *Theodore Parker.* Boston, 1874.

Fuess, Claude M., *Daniel Webster.* 2 vols.; Boston, 1930.

———, "Daniel Webster and the Abolitionists," Massachusetts Historical Society *Proceedings,* LXIV (1930–32), 28–42.

———, *The Life of Caleb Cushing.* 2 vols.; New York, 1923.

Furnas, J. C., *Goodbye to Uncle Tom.* New York, 1956.

———, *The Road to Harper's Ferry.* New York, 1959.

Galpin, W. Freeman, "The Jerry Rescue," *NYH,* XLIII (1945), 19–34.

Ganaway, Loomis Morton, *New Mexico and the Sectional Controversy, 1846–1861.* Albuquerque, 1944.

Gara, Larry, *The Liberty Line: The Legend of the Underground Railroad.* Lexington, Ky., 1961.

Garber, Paul Neff, *The Gadsden Treaty.* Philadelphia, 1923.

Gardiner, Oliver C., *The Great Issue or the Three Presidential Candidates.* New York, 1848.

Gardner, Robert, "A Tenth-Hour Apology for Slavery," *JSH,* XXVI (1960), 352–367.

Garner, Alto L., and Stott, Nathan, "William Lowndes Yancey: Statesman of Secession," *AR,* XV (1962), 190–202.

Garraty, John Arthur, *Silas Wright.* New York, 1949.

Garrison, Wendell P. and Francis J. (eds.), *William Lloyd Garrison, 1805–1879: The Story of His Life Told by His Children.* 4 vols., New York, 1885–89.

Gatell, Frank Otto, "Conscience and Judgment: The Bolt of the Massachusetts Conscience Whigs," *The Historian,* XXI (1958), 18–45.

———, *John Gorham Palfrey and the New England Conscience.* Cambridge, Mass., 1963.

Gates, Paul Wallace, *Fifty Million Acres: Conflicts over Kansas Land Policy, 1854–1890.* Ithaca, N.Y., 1954.

_____, "The Homestead Law in Iowa," *Agricultural History,* XXXVIII (1964), 67–78.

Genovese, Eugene D., "The Legacy of Slavery and the Roots of Black Nationalism," in *Studies on the Left,* VI, 6 (1966), 3–65.

_____, "Rebelliousness and Docility in the Negro Slave: A Critique of the Elkins Thesis," *CWH,* XIII (1967), 293–314.

_____, "The Slave South: An Interpretation," in his *The Political Economy of Slavery.* New York, 1965.

_____, *The World the Slaveholders Made.* New York, 1969.

Gerson, Armand J., "The Inception of the Montgomery Convention," AHA *Annual Report,* 1910, pp. 179–187.

Gihon, John H., *Geary and Kansas.* Philadelphia, 1857.

Goetzmann, William H., *Army Exploration in the American West, 1803–1863.* New Haven, 1959.

Going, Charles Buxton, *David Wilmot, Free Soiler.* New York, 1924.

Graebner, Norman A., *Empire on the Pacific.* New York, 1955.

_____, "1848: Southern Politics at the Crossroads," *The Historian,* XXV (1962), 14–35.

Graf, LeRoy P., "Andrew Johnson and the Coming of the War," *Tennessee Historical Quarterly,* XIX (1960), 208–221.

Greeley, Horace, *The American Conflict.* 2 vols.; Hartford, 1864.

Green, Fletcher M., "Democracy in the Old South," *JSH,* XII (1946), 3–23.

_____, "Listen to the Eagle Scream: One Hundred Years of the Fourth of July in North Carolina (1776–1876)," *NCHR,* XXXI (1954), 295–320, 529–549.

Green, Helen I., "Politics in Georgia, 1830–1854." Ph.D. dissertation, University of Chicago, 1946.

Greene, Laurence, *The Filibuster: The Career of William Walker.* Indianapolis, 1937.

Greer, James Kimmins, "Louisiana Politics, 1845–1861," *LHQ,* XII (1929), 381–425, 555–610; XIII (1930), 67–116, 257–303, 444–483, 617–654.

Griffin, Clifford S., *Their Brothers' Keepers: Moral Stewardship in the United States, 1800–1865.* New Brunswick, N.J., 1960.

Gunderson, Robert Gray (ed.), "Letters from the Washington Peace Conference of 1861," *JSH,* XVII (1951), 382–392.

_____, *Old Gentlemen's Convention: The Washington Peace Conference of 1861.* Madison, 1961.

Gusfield, Joseph R., *Symbolic Crusade: Status Politics and the American Temperance Movement.* Urbana, Ill., 1963.

Gibson, Florence E., *The Attitudes of the New York Irish toward State and National Affairs, 1848–1892.* New York, 1951.

Hagan, Horace H., "Ableman vs. Booth, Effect of Fugitive Slave Law on Opinions as to Rights of Federal Government and of States in the North and South," *American Bar Association Journal*, XVII (1931), 19–24.

Haines, Charles Grove, and Sherwood, Foster H., *The Role of the Supreme Court in American Government and Politics, 1835–1864*. Berkeley, 1957.

Halasz, Nicholas, *The Rattling Chains: Slave Unrest and Revolt in the Antebellum South*. New York, 1966.

Halstead, Murat, *Caucuses of 1860*. Columbus, 1860.

Hamer, Philip May, *The Secession Movement in South Carolina, 1847–1852*. Allentown, Pa., 1918.

Hamilton, Charles Granville, *Lincoln and the Know-Nothing Movement*. Washington, 1954.

Hamilton, Holman, "The 'Cave of the Winds' and the Compromise of 1850," *JSH*, XXIII (1957), 331–353.

———, "Democratic Senate Leadership and the Compromise of 1850," *MVHR*, XLI (1954), 403–418.

———, *Prologue to Conflict: The Crisis and Compromise of 1850* (Lexington, Ky., 1964).

———, "Texas Bonds and Northern Profits," *MVHR*, XLIII (1957), 579–594.

———, *Zachary Taylor: Soldier in the White House*. Indianapolis, 1951.

Hamilton, Luther (ed.), *Memoirs, Speeches and Writings of Robert Rantoul, Jr.* Boston, 1854.

Hamlin, Charles Eugene, *The Life and Times of Hannibal Hamlin*. Cambridge, Mass., 1899.

Hammond, Jabez D., *The History of Political Parties in the State of New York*. 2 vols.; Buffalo, 1850.

Hancock, Harold, "Civil War Comes to Delaware," *CWH*, II (1956), 29–46.

Handlin, Oscar, *Boston's Immigrants, 1790–1880: A Study in Acculturation*. Cambridge, Mass., 1941; rev. ed., 1959.

Hansen, Marcus Lee, *The Atlantic Migration, 1607–1860*. Cambridge, Mass., 1940.

———, *The Immigrant in American History*. Cambridge, Mass., 1960.

Harlow, Ralph Volney, *Gerrit Smith, Philanthropist and Reformer*. New York, 1939.

———, "The Rise and Fall of the Kansas Aid Movement," *AHR*, XLI (1935), 1–25.

Harmon, George D., "Douglas and the Compromise of 1850," ISHS *Journal*, XXI (1929), 453–499.

———, "President James Buchanan's Betrayal of Governor Robert J. Walker of Kansas," *PMHB*, LIII (1929), 51–91.

Harrington, Fred Harvey, *Fighting Politician: Major General N. P. Banks*. Philadelphia, 1948.

———, "The First Northern Victory," *JSH,* V (1939), 186–205.

———, "Fremont and the North Americans," *AHR,* XLIV (1939), 842–848.

Hart, Albert Bushnell, *Slavery and Abolition, 1831–1841.* New York, 1906.

Hart, James D., *The Popular Book: A History of America's Literary Taste.* Berkeley, 1961.

Hartz, Louis, *The Liberal Tradition in America.* New York, 1955.

Hasse, Adelaide R., "The Southern Convention of 1850," New York Public Library *Bulletin,* XIV (1910), 239–240.

Hearon, Cleo, "Mississippi and the Compromise of 1850," Mississippi Historical Society *Publications,* XIV (1914), 7–229.

Heck, Frank H., "John C. Breckinridge in the Crisis of 1860–1861," *JSH,* XXI (1955), 316–346.

Heckman, Richard Allen, *Lincoln vs. Douglas: The Great Debates Campaign.* Washington, 1967.

Helper, Hinton R., *Compendium of the Impending Crisis of the South.* New York, 1860.

Henderson, Gavin B. (ed.), "Southern Designs on Cuba, 1854–1857, and Some European Opinions," *JSH,* V (1939), 371–385.

Hendrick, Burton J., *Lincoln's War Cabinet.* Boston, 1946.

Hendrix, James Paisley, Jr., "The Efforts to Reopen the African Slave Trade in Louisiana," *Louisiana History,* X (1969), 97–123.

Henry, H. M., *The Police Control of the Slave in South Carolina.* Emory, Va., 1914.

Hensel, William Uhler, *The Christiana Riot and the Treason Trials of 1851.* Lancaster, Pa., 1911.

Henson, Josiah, *The Life of Josiah Henson, Formerly a Slave, as Narrated by Himself.* Boston, 1849.

Herbert, Hilary A., *The Abolition Crusade and Its Consequences.* New York, 1912.

Herndon, Dallas Tabor, "The Nashville Convention of 1850," Alabama Historical Society *Transactions,* V (Montgomery, 1906), 216–237.

Hertz, Frederick, *Nationality in History and Politics.* London, 1944.

Hesseltine, William B., "Abraham Lincoln and the Politicians," *CWH,* VI (1960), 43–55.

———, "Some New Aspects of the Proslavery Argument," *JNH,* XXI (1936), 1–14.

———, (ed.), *Three Against Lincoln: Murat Halstead Reports the Caucuses of 1860.* Baton Rouge, 1960.

——— and Fisher, Rex G. (eds.), *Trimmers, Trucklers and Temporizers: Notes of Murat Halstead from the Political Conventions of 1856.* Madison, 1961.

Hibben, Paxton, *Henry Ward Beecher: An American Portrait.* New York, 1927.

Hicken, Victor, "John A. McClernand and the House Speakership Struggle of 1859," ISHS *Journal,* LIII (1960), 163–178.

Hicks, Jimmie, "Some Letters Concerning the Knights of the Golden Circle in Texas, 1860–1861," *SWHQ*, LXV (1961), 80–86.

Higginson, Thomas Wentworth, *Cheerful Yesterdays.* Boston, 1898.

Higham, John, "Another Look at Nativism," *Catholic Historical Review*, XLIV (1958), 147–158.

Hilliard, Henry W., *Politics and Pen Pictures: At Home and Abroad.* New York, 1892.

Hinton, Richard J., *John Brown and His Men.* Rev. ed.; New York, 1894.

Hodder, Frank H., "The Authorship of the Compromise of 1850," *MVHR*, XXII (1936), 525–536.

———, "Genesis of the Kansas-Nebraska Act," State Historical Society of Wisconsin *Proceedings*, 1912, pp. 69–86.

———, "The Railroad Background of the Kansas-Nebraska Act," *MVHR*, XII (1925), 3–22.

———, "Some Aspects of the English Bill for the Admission of Kansas," AHA *Annual Report*, 1906, I, 199–210.

———, "Some Phases of the Dred Scott Case," *MVHR*, XVI (1929), 3–22.

Hodgson, Joseph, *The Cradle of the Confederacy: Or the Times of Troup, Quitman, and Yancey.* Mobile, 1876.

Hofstadter, Richard, *The Paranoid Style in American Politics and Other Essays.* New York, 1965.

Hollcroft, Temple R. (ed.), "A Congressman's [Edwin B. Morgan] Letters on the Speaker Election in the Thirty-Fourth Congress," *MVHR*, XLIII (1956), 444–458.

Holloway, J. N., *History of Kansas.* Lafayette, Ind., 1868.

Holst, H. E. von, *The Constitutional and Political History of the United States.* 8 vols.; Chicago, 1877–92.

Holt, Michael Fitzgibbon, *Forging a Majority: The Formation of the Republican Party in Pittsburgh, 1848–1860.* New Haven, 1969.

Holt, W. Stull, *Treaties Defeated by the Senate.* Baltimore, 1933.

Hoogenboom, Ari, "Gustavus Fox and the Relief of Fort Sumter," *CWH*, IX (1963), 383–398.

Hoole, William Stanley, *Alabama Tories.* Tuscaloosa, Ala., 1960.

Hopkins, Vincent C., *Dred Scott's Case.* New York, 1951.

Horner, Harlan Hoyt, *Lincoln and Greeley.* Urbana, Ill., 1953.

Howard, Warren S., *American Slavers and the Federal Law, 1837–1862.* Berkeley, 1963.

Howe, Samuel Gridley, *The Refugees from Slavery in Canada West.* Boston, 1864.

Hubbell, Jay B., "Cavalier and Indentured Servant in Virginia Fiction, *SAQ*, XXVI (1927), 23–39.

———, "Literary Nationalism in the Old South," in David Kelley Jackson (ed.), *American Studies in Honor of William Kenneth Boyd.* Durham, N.C., 1940, pp. 175–220.

———, *The South in American Literature.* Durham, N.C., 1954.

Hubbell, John T., "Three Georgia Unionists and the Compromise of 1850," *Georgia Historical Quarterly,* LI (1967), 307–323.

Hunt, Gaillard (ed.), "Narrative and Letter of William Henry Trescot Concerning the Negotiations Between South Carolina and President Buchanan in December, 1860," *AHR,* XIII (1908), 528–556.

Hurt, Peyton, "The Rise and Fall of the 'Know Nothings' in California," *California Historical Society Quarterly,* IX (1930), 16–49, 99–128.

Isely, Jeter Allen, *Horace Greeley and the Republican Party, 1853–1861: A Study of the New York Tribune.* Princeton, 1947.

Isely, W. H., "The Sharps Rifle Episode in Kansas History," *AHR,* XII (1907), 546–566.

Jackson, Margaret Y., "An Investigation of Biographies and Autobiographies of American Slaves Published Between 1840 and 1860." Ph.D. dissertation, Cornell University, 1954.

Jaffa, Henry V., *Crisis of the House Divided: An Interpretation of the Issues in the Lincoln-Douglas Debates.* New York, 1959.

———, *Equality and Poverty: Theory and Practice in American Politics.* New York, 1965.

——— and Johannsen, Robert W. (eds.), *In the Name of the People: Speeches and Writings of Lincoln and Douglas in the Ohio Campaign of 1859.* Columbus, 1959.

James, H. Preston, "Political Pageantry in the Campaign of 1860 in Illinois," *Abraham Lincoln Quarterly,* IV (1947), 313–347.

Jameson, J. Franklin (ed.), "Correspondence of John C. Calhoun," AHA *Annual Report,* 1899, II.

Janes, Henry L., "The Black Warrior Affair," *AHR,* XII (1907), 280–298.

Jenkins, William Sumner, *Pro-Slavery Thought in the Old South.* Chapel Hill, 1935.

Jensen, Merrill (ed.), *Regionalism in America.* Madison, 1951.

Johannsen, Robert W., "Douglas at Charleston," in Norman A. Graebner (ed.), *Politics and the Crisis of 1860.* Urbana, Ill., 1961.

———, "The Douglas Democracy and the Crisis of Disunion," *CWH,* IX (1963), 229–247.

———, *Frontier Politics and the Sectional Conflict: The Pacific Northwest on the Eve of the Civil War.* Seattle, 1955.

———, "The Kansas-Nebraska Act and Territorial Government in the United States," in *Territorial Kansas* (Lawrence, Kan., 1954), 17–32.

———, "The Kansas-Nebraska Act and the Pacific Northwest Frontier," *Pacific Historical Review,* XXII (1953), 129–141.

———, "The Lecompton Constitutional Convention: An Analysis of Its Membership," *Kansas Historical Quarterly,* XXIII (1957), 225–243.

_____ (ed.), *The Letters of Stephen A. Douglas.* Urbana, Ill., 1961.
_____, *Stephen A. Douglas.* New York, 1973.
_____, "Stephen A. Douglas, 'Harpers Magazine,' and Popular Sovereignty," *MVHR,* XLV (1959), 606–631.
_____, "Stephen A. Douglas' New England Campaign, 1860," *NEQ* (1962), 162–186.
_____, "Stephen A. Douglas, Popular Sovereignty, and the Territories," *Historian,* XXII (1960), 378–395.
Johns, John E., *Florida During the Civil War.* Gainesville, 1963.
Johnson, Allen, "The Constitutionality of the Fugitive Slave Acts," *Yale Law Journal,* XXXI (1921), 161–182.
_____, "Genesis of Popular Sovereignty," *Iowa Journal of History and Politics,* III (1905), 3–19.
_____, "The Nationalizing Influence of Party," *Yale Review,* XV (1907), 283–292.
_____, *Stephen A. Douglas.* New York, 1908.
Johnson, F. Roy, *The Nat Turner Slave Insurrection.* Murfreesboro, N.C., 1966.
Johnson, Hildegard Binder, "The Election of 1860 and the Germans in Minnesota," *Minnesota History,* XXVIII (1947), 20–36.
Johnson, Ludwell H., "Fort Sumter and Confederate Diplomacy," *JSH,* XXVI (1960), 441–477.
Johnson, Robert Underwood, and Buel, Clarence Clough (eds.), *Battles and Leaders of the Civil War.* 4 vols.; New York, 1887.
Johnson, Samuel A., *The Battle Cry of Freedom: The New England Emigrant Aid Company in the Kansas Crusade.* Lawrence, Kan., 1954.
Johnston, Richard Malcolm, and Browne, William Hand, *Life of Alexander H. Stephens.* Philadelphia, 1883.
Jones, Maldwyn Allen, *American Immigration.* Chicago, 1960.
Jordan, Philip D., "The Stranger Looks at the Yankee," in Henry Steele Commager (ed.), *Immigration and American History: Essays in Honor of Theodore C. Blegen.* Minneapolis, 1961.
Jordan, Weymouth T., *Rebels in the Making: Planters' Conventions and Southern Propaganda.* Tuscaloosa, Ala., 1958.
Jordan, Winthrop D., *White Over Black: American Attitudes toward the Negro.* Chapel Hill, 1968.
Jorgenson, Chester E. (ed.), *Uncle Tom's Cabin as Book and Legend.* Detroit, 1952.
Julian, George W., "The First Republican National Convention," *AHR,* IV (1899), 313–322.
_____, *Life of Joshua R. Giddings.* Chicago, 1892.
Keleher, William A., *Turmoil in New Mexico, 1846–1868.* Santa Fe, 1952.
Kemble, John Haskell, *The Panama Route, 1848–1869.* Berkeley, 1943.
Kendall, John S., "Shadow Over the City," *LHQ,* XXII (1939), 142–165.

Kendall, Lane Carter, "The Interregnum in Louisiana in 1861," *LHQ*, XVI (1933), 175–208, 374–408, 639–669; XVII (1934), 339–348, 524–536.

Kibler, Lillian Adele, *Benjamin F. Perry, South Carolina Unionist.* Durham, N.C., 1946.

Kilson, Marion D. de B., "Towards Freedom: An Analysis of Slave Revolts in the United States," *Phylon*, XXV (1964), 175–187.

King, Alvy L., *Louis T. Wigfall, Southern Fire-Eater.* Baton Rouge, 1970.

King, Willard L., *Lincoln's Manager, David Davis.* Cambridge, Mass., 1960.

Kirkpatrick, Arthur Roy, "Missouri in the Early Months of the Civil War," *MHR*, LV (1961), 235–266.

———, "Missouri on the Eve of the Civil War," *MHR*, LV (1961), 99–108.

———, "Missouri's Secessionist Government, 1861–65," *MHR*, XLV (1951), 124–137.

Kirwan, Albert D., *John J. Crittenden: The Struggle for the Union.* Lexington, Ky., 1962.

Klein, Philip Shriver, *President James Buchanan.* University Park, Pa., 1962.

Klem, Mary J., "Missouri in the Kansas Struggle," MVHA *Proceedings*, IX (1917–18), 393–413.

Klement, Frank L., *The Limits of Dissent: Clement L. Vallandigham and the Civil War.* Lexington, Ky., 1970.

Kleppner, Paul J., "Lincoln and the Immigrant Vote: A Case of Religious Polarization," *Mid-America*, XLVIII (1966), 176–195.

Knapp, Charles Merriam, *New Jersey Politics During the Period of the Civil War and Reconstruction.* Geneva, N.Y., 1924.

Knoles, George Harmon (ed.), *The Crisis of the Union, 1860–1861.* Baton Rouge, 1965.

Kohn, Hans. *American Nationalism.* New York, 1957.

———, *The Idea of Nationalism.* New York, 1944.

Korngold, Ralph, *Two Friends of Man: The Story of William Lloyd Garrison and Wendell Phillips.* Boston, 1950.

Kraditor, Aileen S., *Means and Ends in American Abolitionism: Garrison and His Critics on Strategy and Tactics, 1834–1850.* New York, 1969.

Krout, John Allen, *The Origins of Prohibition.* New York, 1925.

Krug, Mark M., *Lyman Trumbull, Conservative Radical.* New York, 1965.

Krummel, Carl F., "Henry J. Raymond and the *New York Times* in the Secession Crisis, 1860–61," *NYH*, XLIX (1951), 377–398.

Lader, Lawrence, *The Bold Brahmins: New England's War Against Slavery 1831–1863.* New York, 1961.

Lamon, Ward H., *The Life of Abraham Lincoln.* Boston, 1872.

Landon, Fred, "The Negro Migration to Canada after the Passing of the Fugitive Slave Act," *JNH*, V (1920), 22–36.

Landrum, Grace Warren, "Sir Walter Scott and His Literary Rivals in the South," *American Literature*, II (1930), 256–276.

Landry, Harral E., "Slavery and the Slave Trade in Atlantic Diplomacy, 1850–1861," *JSH*, XXVII (1961), 184–207.

Lawrence, Alexander A., *James Moore Wayne, Southern Unionist.* Chapel Hill, 1943.

Lawson, John D. (ed.), *American State Trials.* 17 vols.; St. Louis, 1921.

Lawton, Eba Anderson, *Major Robert Anderson and Fort Sumter.* New York, 1911.

Learned, Henry B., "The Relation of Philip Phillips to the Repeal of the Missouri Compromise in 1854," *MVHR*, VIII (1922), 303–317.

———, "William Learned Marcy," in Samuel Flagg Bemis (ed.), *The American Secretaries of State and Their Diplomacy.* 10 vols.; New York, 1927–29.

Lee, Arthur M., "The Development of an Economic Policy in the Early Republican Party." Ph.D. dissertation, Syracuse University, 1953.

Lee, Charles R., Jr., *The Confederate Constitutions.* Chapel Hill, 1963.

Leonard, Ira M., "The Rise and Fall of the American Republican Party in New York City, 1843–1845," New York Historical Society *Quarterly*, L (1966), 151–192.

Lerner, Gerda, *The Grimké Sisters from South Carolina.* Boston, 1967.

Leslie, William R., "The Fugitive Slave Clause, 1787–1842." Ph.D. dissertation, University of Michigan, 1945.

Levy, Leonard W., "The 'Abolition Riot': Boston's First Slave Rescue," *NEQ*, XXV (1952), 85–92.

———, *The Law of the Commonwealth and Chief Justice Shaw.* Cambridge, Mass., 1957.

———, "Sims' Case: The Fugitive Slave Case in Boston in 1851," *JNH*, XXXV (1950), 39–74.

Lewis, Elsie M., "From Nationalism to Disunion: A Study in the Secession Movement in Arkansas, 1850–1861." Ph.D. dissertation, University of Chicago, 1946.

Lincoln Day by Day: A Chronology, 1809–1865. 3 vols.; Washington, 1960.

Linden, Fabian, "Economic Democracy in the Slave South: An Appraisal of Some Recent Views," *JNH*, XXXI (1946), 140–189.

Lipscomb, Andrew A. (ed.), *The Writings of Thomas Jefferson.* 20 vols.; Washington, 1903–04.

Lipset, Seymour Martin, *Political Man: The Social Bases of Politics.* New York, 1960.

Litwack, Leon F., *North of Slavery: The Negro in the Free States, 1790–1860.* Chicago, 1961.

Lloyd, Arthur Young, *The Slavery Controversy, 1831–1860.* Chapel Hill, 1939.

Locke, Mary Stoughton, *Anti-Slavery in America, 1619–1808.* Boston, 1901.

Lofton, John M., *Insurrection in South Carolina: The Turbulent World of Denmark Vesey.* Yellow Springs, Ohio, 1964.

Long, Durward, "Unanimity and Disloyalty in Secessionist Alabama," *CWH,* XI (1965), 257–273.

Loomis, Nelson H., "Asa Whitney, Father of Pacific Railroads," MVHA *Proceedings,* VI (1912), 166–175.

Lowell, James Russell, *The Biglow Papers.* First series; Cambridge, Mass., 1848.

———, *Political Essays.* New York, 1904.

Ludlum, Robert P., "The Anti-Slavery 'Gag Rule': History and Argument," *JNH,* XXVI (1941), 203–243.

Luthin, Reinhard H., "Abraham Lincoln and the Massachusetts Whigs in 1848," *NEQ,* XIV (1941), 619–632.

———, "Abraham Lincoln and the Tariff," *AHR,* XLIX (1944), 609–629.

———, "Abraham Lincoln Becomes a Republican," *Political Science Quarterly,* LIX (1944), 420–438.

———, "The Democratic Split During Buchanan's Administration," *Pennsylvania History,* XI (1944), 13–35.

———, *The First Lincoln Campaign.* Cambridge, Mass., 1944.

———, "Organizing the Republican Party in the 'Border Slave' Regions: Edward Bates's Presidential Candidacy in 1860," *MHR,* XXXVIII (1944), 138–161.

———, "Pennsylvania and Lincoln's Rise to the Presidency, *PMHB,* LXVII (1943), 61–82.

———, "Salmon P. Chase's Political Career before the Civil War," *MVHR,* XXIX (1943), 517–540.

Lynch, John P., "The Higher Law Argument in American History, 1850–1860." M.A. thesis, Columbia University, 1947.

Lynch, William O., "Antislavery Tendencies of the Democratic Party in the Northwest, 1848–1850," *MVHR,* XI (1924), 319–331.

———, "Indiana in the Douglas-Buchanan Contest of 1856," *IMH,* XXX (1934), 119–132.

Lyon, William H., "Claiborne Fox Jackson and the Secession Crisis in Missouri," *MHR,* LVIII (1964), 422–441.

MacDonagh, Oliver, "Irish Emigration to the United States of America and the British Colonies During the Famine," in R. Dudley Edwards and T. Desmond Williams (eds.), *The Great Famine: Studies in Irish History, 1845–1852.* New York, 1957.

Magdol, Edward, *Owen Lovejoy, Abolitionist in Congress.* New Brunswick, N.J., 1967.

Maher, Edward R., Jr., "Sam Houston and Secession," *SWHQ,* LV (1952), 448–458.

———, "Secession in Texas." Ph.D. dissertation, Fordham University, 1960.

Malin, James C., *John Brown and the Legend of Fifty-Six.* Philadelphia, 1942.

———, "Judge LeCompte and the 'Sack of Lawrence,' May 21, 1856," *Kansas Historical Quarterly,* XX (1953), 465–494.

———, "The Motives of Stephen A. Douglas in the Organization of Nebraska Territory: A Letter Dated December 17, 1853," *Kansas Historical Quarterly,* XIX (1951), 351–352.

———, *The Nebraska Question, 1852–1854.* Lawrence, Kan., 1953.

———, "The Proslavery Background of the Kansas Struggle," *MVHR,* X (1923), 285–305.

———, "The Topeka Statehood Movement Reconsidered: Origins," in *Territorial Kansas: Studies Commemorating the Centennial.* Lawrence, Kan., 1954, pp. 33–69.

Malone, Dumas, *The Public Life of Thomas Cooper, 1783–1839.* New Haven, 1926.

Manning, William R. (ed.), *Diplomatic Correspondence of the United States: Inter-American Affairs, 1831–1860.* 12 vols.; Washington, 1932–39.

Martin, Asa Earl, *The Anti-Slavery Movement in Kentucky Prior to 1850.* Louisville, 1918.

Mason, Virginia, *The Public Life and Diplomatic Correspondence of James M. Mason.* New York, 1906.

Mason, Vroman, "The Fugitive Slave Law in Wisconsin, with Reference to the Nullification Sentiment," State Historical Society of Wisconsin *Proceedings,* 1895, pp. 117–144.

May, Samuel J., *The Fugitive Slave Law and Its Victims.* New York, 1861.

———, *Some Recollections of Our Anti-Slavery Conflict.* Boston, 1869.

Mayer, George H., *The Republican Party, 1854–1964.* New York, 1964.

McAvoy, Thomas T., "The Formation of the Catholic Minority in the United States, 1820–1860," *Review of Politics,* X (1948), 13–34.

McClure, Alexander H., *Abraham Lincoln and Men of War Times.* Philadelphia, 1892.

McClure, James R., "Taking the Census and Other Incidents in 1855," KSHS *Transactions,* VIII (1903–04), 227–250.

McColley, Robert, *Slavery and Jeffersonian Virginia.* Urbana, Ill., 1964.

McConville, Mary St. Patrick, *Political Nativism in the State of Maryland, 1830–1860.* Washington, 1928.

McCormac, Eugene Irving, *James K. Polk.* Berkeley, 1922.

———, "Justice Campbell and the Dred Scott Decision," *MVHR,* XIX (1933), 565–571.

McCormick, Thomas J. (ed.), *Memoirs of Gustave Koerner.* 2 vols.; Cedar Rapids, Iowa, 1909.

McCoy, Charles A., *Polk and the Presidency.* Austin, Tex., 1960.

McGann, Agnes Geraldine, *Nativism in Kentucky in 1860.* Washington, 1944.

McGregor, James C., *The Disruption of Virginia.* New York, 1922.

McKay, Ernest A., "Henry Wilson and the Coalition of 1851," *NEQ*, XXXVI (1963), 338–357.

———, "Henry Wilson: Unprincipled Know-Nothing," *Mid-America*, XLVI (1964), 29–37.

McKibben, Davidson Burns, "Negro Slave Insurrections in Mississippi, 1800–1865," *JNH*, XXXIV (1949), 73–90.

McKinney, William T., "The Defeat of the Secessionists in Kentucky in 1861," *JNH*, I (1916), 377–391.

McKitrick, Eric L. (ed.), *Slavery Defended: The Views of the Old South*. Englewood Cliffs, N.J., 1963.

McLaughlin, Andrew C., *A Constitutional History of the United States*. New York, 1935.

McMillan, Malcolm C., "William L. Yancey and the Historians: One Hundred Years," *AR*, XX (1967), 163–186.

McPherson, Edward (ed.), *Political History of the United States During the Great Rebellion*. Washington, 1876.

McPherson, James M., "The Fight Against the Gag Rule: Joshua Leavitt and Antislavery Insurgency in the Whig Party, 1839–1842," *JNH*, XLVIII (1963), 177–195.

McWhiney, Grady, "The Confederacy's First Shot," *CWH*, XIV (1968), 5–14.

———, "Were the Whigs a Class Party in Alabama?" *JSH*, XXIII (1957), 510–522.

Mearns, David C. (ed.), *The Lincoln Papers*. 2 vols.; Garden City, N.Y., 1948.

Meerse, David E., "Buchanan, Corruption, and the Election of 1860," *CWH*, XII (1966), 116–131.

Meigs, M. C., "General M. C. Meigs on the Civil War," *AHR*, XXVI (1921), 285–303.

Mellen, George F., "Henry W. Hilliard and William L. Yancey," *Sewanee Review*, XVII (1909), 32–50.

Mellon, Matthew T., *Early American Views on Negro Slavery*. Boston, 1934.

Meltzer, Milton, *Tongue of Flame: The Life of Lydia Maria Child*. New York, 1965.

Mencken, H. L., *The American Language*. 4th ed.; New York, 1936.

Mendelson, Wallace, "Dred Scott's Case—Reconsidered," *Minnesota Law Review*, XXXVIII (1953), 16–28.

Meredith, Roy, *Storm Over Sumter*. New York, 1957.

Merk, Frederick, *Manifest Destiny and Mission in American History*. New York, 1963.

Merkel, Benjamin C., "The Slavery Issue and the Political Decline of Thomas Hart Benton, 1846–1856," *MHR*, XXXVIII (1944), 388–407.

Merrill, O. N., *A True History of the Kansas Wars*. Cincinnati, 1856.

Merrill, Walter M., *Against Wind and Tide: A Biography of William Lloyd Garrison.* Cambridge, Mass., 1963.

Meyer, Howard N., *Colonel of the Black Regiment: The Life of Thomas Wentworth Higginson.* New York, 1967.

Miles, Edwin A., " 'Fifty-Four Forty or Fight'—An American Political Legend," *MVHR,* XLIV (1957), 291–309.

———, "The Mississippi Slave Insurrection Scare of 1835," *JNH,* XLII (1957), 48–60.

Miller, Hunter, *Treaties and Other International Acts of the United States of America.* 8 vols.; Washington, 1931–48.

Milton, George Fort, *The Eve of Conflict: Stephen A. Douglas and the Needless War.* Boston, 1934.

Mitchell, Stewart, *Horatio Seymour of New York.* Cambridge, Mass., 1938.

Monaghan, Jay, "Did Abraham Lincoln Receive the Illinois German Vote?" ISHS *Journal,* XXXV (1942), 133–139.

———, " 'The Lincoln-Douglas Debates': The Follett, Foster Edition of a Great Political Document," *Lincoln Herald,* XLV (1948), 2–11.

Montgomery, Horace, *Cracker Parties.* Baton Rouge, 1950.

Moore, Frank (ed.), *The Rebellion Record.* 12 vols.; New York, 1861–68.

Moore, Glover, *The Missouri Controversy, 1819–1821.* Lexington, Ky., 1953.

Moore, John Bassett (ed.), *The Works of James Buchanan.* 12 vols.; Philadelphia, 1908–11.

Morehouse, Frances M. I., *The Life of Jesse W. Fell.* Urbana, Ill., 1916.

Morrison, Chaplain W., *Democratic Politics and Sectionalism: The Wilmot Proviso Controversy.* Chapel Hill, 1967.

Morrow, Ralph E., "The Proslavery Argument Revisited," *MVHR,* XLVIII (1961), 79–94.

Mott, Frank Luther, *Golden Multitudes: The Story of Best Sellers in the United States.* New York, 1947.

Muldowny, John, "The Administration of Jefferson Davis as Secretary of War." Ph.D. dissertation, Yale University, 1959.

Munford, Beverley B., *Virginia's Attitude toward Slavery and Secession.* New York, 1909.

Neighbours, Kenneth F., "The Taylor-Neighbors Struggle over the Upper Rio Grande Region of Texas in 1850," *SWHQ,* LXI (1958), 431–463.

Nevins, Allan (ed.), *The Diary of Philip Hone, 1828–1851.* 2 vols.; New York, 1927.

———, *The Emergence of Lincoln.* 2 vols.; New York, 1950.

———, *Frémont, Pathmarker of the West.* New York, 1939.

———, *Ordeal of the Union.* 2 vols.; New York, 1947.

———, *The War for the Union.* 4 vols.; New York, 1959–71.

———, and Thomas, Milton Halsey (eds.), *The Diary of George Templeton Strong.* 4 vols.; New York, 1952.

Nichols, Alice, *Bleeding Kansas.* New York, 1954.

Nichols, Charles H., *Many Thousand Gone: The Ex-Slaves' Account of Their Bondage and Freedom.* Leiden, 1963.

Nichols, Roy F., *Blueprints for Leviathan: American Style.* New York, 1963.

———, *The Democratic Machine, 1850–1854.* New York, 1923.

———, *The Disruption of American Democracy.* New York, 1948.

———, *Franklin Pierce, Young Hickory of the Granite Hills.* Rev. ed.; Philadelphia, 1958.

———, "The Kansas-Nebraska Act: A Century of Historiography," *MVHR,* XLIII (1956), 187–212.

———, "Some Problems of the First Republican Presidential Campaign," *AHR,* XXVIII (1923), 492–496.

Nicolay, Helen, *Lincoln's Secretary: A Biography of John G. Nicolay.* Westport, Conn., 1949.

Nicolay, John G., and Hay, John, *Abraham Lincoln: A History.* 10 vols.; New York, 1909.

Niven, John, *Gideon Welles: Lincoln's Secretary of the Navy.* New York, 1973.

Nye, Russel B., *Fettered Freedom: Civil Liberties and the Slavery Controversy, 1830–1860.* East Lansing, Mich., 1949.

———, *William Lloyd Garrison and the Humanitarian Reformers.* Boston, 1955.

Oates, Stephen B., "John Brown and His Judges: A Critique of the Historical Literature," *CWH,* XVII (1971), 5–24.

———, *To Purge the Land with Blood: A Biography of John Brown.* New York, 1970.

O'Connor, Thomas H., *Lords of the Loom: The Cotton Whigs and the Coming of the Civil War.* New York, 1968.

Orians, George Harrison, *The Influence of Walter Scott upon America and American Literature before 1860.* Urbana, Ill., 1929.

———, "Walter Scott, Mark Twain and the Civil War," *SAQ,* XL (1941), 342–359.

Osterweis, Rollin G., *Romanticism and Nationalism in the Old South.* New Haven, 1949.

Overdyke, W. Darrell, *The Know-Nothing Party in the South.* Baton Rouge, 1950.

Owsley, Frank Lawrence, "The Fundamental Cause of the Civil War: Egocentric Sectionalism," *JSH,* VII (1941), 3–18.

———, *Plain Folk of the Old South.* Baton Rouge, 1949.

——— and Harriet C., "The Economic Basis of Society in the Late Ante-Bellum South," *JSH,* VI (1940), 24–45.

Parks, Joseph Howard, "John Bell and the Compromise of 1850," *JSH,* IX (1943), 328–356.

———, *John Bell of Tennessee.* Baton Rouge, 1950.

Parrish, William E., *David Rice Atchison of Missouri, Border Politician.* Columbia, Mo., 1961.

Partin, Robert Love, "The Secession Movement in Tennessee." Ph.D. dissertation, George Peabody College, 1935.

Patton, James Welch, *Unionism and Reconstruction in Tennessee.* Chapel Hill, 1934.

Paxton, W. M., *Annals of Platte County, Missouri.* Kansas City, Mo., 1897.

Pendergraft, Daryl, "Thomas Corwin and the Conservative Republican Reaction, 1858–1861," *Ohio State Archeological and Historical Quarterly,* LVII (1948), 1–23.

Perkins, Howard C., "The Defense of Slavery in the Northern Press on the Eve of the Civil War," *JSH,* IX (1943), 501–531.

_____ (ed.), *Northern Editorials on Secession.* 2 vols.; New York, 1942.

Perritt, H. Hardy, "Robert Barnwell Rhett, South Carolina Secession Spokesman." Ph.D. dissertation, University of Florida, 1954.

Perry, Benjamin F., *Biographical Sketches of Eminent American Statesmen.* Philadelphia, 1887.

_____, *Reminiscences of Public Men.* Greenville, S.C., 1889.

Perry, Lewis Curtis, "Antislavery and Anarchy: A Study of the Ideas of Abolitionism before the Civil War." Ph.D. dissertation, Cornell University, 1967.

Perry, Thomas Sergeant (ed.), *The Life and Letters of Francis Lieber.* Boston, 1882.

Persinger, Clark E., "The 'Bargain of 1844' as the Origin of the Wilmot Proviso," AHA *Annual Report,* 1911, I, 189–195.

Phillips, Ulrich Bonnell, *American Negro Slavery.* New York, 1918.

_____ (ed.), *The Correspondence of Robert Toombs, Alexander H. Stephens, and Howell Cobb.* AHA *Annual Report,* 1911, Vol. II.

_____, *The Course of the South to Secession.* New York, 1939.

_____, "Georgia and State Rights," in AHA *Annual Report,* 1901, II, 193–210.

_____, *Life and Labor in the Old South.* Boston, 1929.

_____, *The Life of Robert Toombs.* New York, 1913.

_____, "The Literary Movement for Secession," in *Studies in Southern History and Politics Inscribed to William Archibald Dunning.* New York, 1914, pp. 33–60.

_____, "The Southern Whigs, 1834–1854," in *Essays in American History Dedicated to Frederick Jackson Turner.* New York, 1910, pp. 203–229.

Phillips, Wendell, *Speeches, Lectures, and Letters.* Boston, 1864.

Phillips, William, *The Conquest of Kansas by Missouri and Her Allies.* Boston, 1856.

Piatt, Donn, *Memories of the Men Who Saved the Union.* New York, 1887.

Pitkin, Thomas M., "Western Republicans and the Tariff in 1860," *MVHR*, XXVII (1940), 401–420.

Pleasants, Samuel Augustus, *Fernando Wood of New York.* New York, 1948.

Poage, George Rawlings, *Henry Clay and the Whig Party.* Chapel Hill, 1936.

Porter, Kenneth Wiggins, "Florida Slaves and Free Negroes in the Seminole War, 1835–1842," *JNH*, XXVIII (1943), 390–421.

———, "Negroes and the Seminole War, 1817–1818," *JNH*, XXXVI (1951), 249–280.

Porter, Kirk H., and Johnson, Donald Bruce (eds.), *National Party Platforms, 1840–1960.* Urbana, Ill., 1961.

Potter, David M., "The Know-Nothing Party in the Presidential Election of 1856," M.A. thesis, Yale University, 1933.

———, *Lincoln and His Party in the Secession Crisis.* New Haven, 1942.

———, *The South and the Sectional Conflict.* Baton Rouge, 1968.

——— and Manning, Thomas G. (eds.), *Nationalism and Sectionalism in America, 1775–1877.* New York, 1949.

Potter, George W., *To the Golden Door: The Story of the Irish in Ireland and America.* Boston, 1960.

Pratt, Julius W., "John L. O'Sullivan and Manifest Destiny," *NYH*, XXXI (1933), 213–234.

Procter, Ben H., *Not Without Honor: The Life of John H. Reagan.* Austin, 1962.

Purifoy, Lewis M., "The Southern Methodist Church and the Pro-Slavery Argument," *JSH*, XXXII (1966), 325–341.

Quaife, Milo Milton (ed.), *The Diary of James K. Polk.* 4 vols.; Chicago, 1910.

———, *The Doctrine of Non-Intervention with Slavery in the Territories.* Chicago, 1910.

Quarles, Benjamin, *Black Abolitionists.* New York, 1969.

———, *Frederick Douglass.* Washington, 1948.

———, *Lincoln and the Negro.* New York, 1962.

Rabun, James Z., "Alexander H. Stephens, 1812–1861." Ph.D. dissertation, University of Chicago, 1948.

Radcliffe, George L. P., *Governor Thomas H. Hicks of Maryland and the Civil War.* Baltimore, 1901.

Rainwater, Percy Lee, "Economic Benefits of Secession: Opinions in Mississippi in the 1850's," *JSH*, I (1935), 459–474.

———, *Mississippi: Storm Center of Secession, 1856–1861.* Baton Rouge, 1938.

Ramsdell, Charles W., "The Frontier and Secession," in *Studies in Southern History and Politics Inscribed To William Archibald Dunning.* New York, 1914, 61–79.

———, "Lincoln and Fort Sumter," *JSH*, III (1937), 259–288.

———, "The Natural Limits of Slavery Expansion," *MVHR*, XVI (1929), 151–171.

______, *Reconstruction in Texas.* New York, 1910.

Ranck, James Byrne, *Albert Gallatin Brown, Radical Southern Nationalist.* New York, 1937.

Randall, James G., "The Civil War Restudied," *JSH,* VI (1940), 441–449.

______, *Lincoln, the Liberal Statesman.* New York, 1947.

______, *Lincoln the President.* 4 vols.; New York, 1945–55.

______, and Donald, David, *The Civil War and Reconstruction.* 2nd ed.; Boston, 1969.

Rauch, Basil, *American Interest in Cuba, 1848–1855.* New York, 1948.

Rawley, James A., *Race and Politics: "Bleeding Kansas" and the Coming of the Civil War.* Philadelphia, 1969.

Ray, Sister Mary Augustina, *American Opinion of Roman Catholicism in the Eighteenth Century.* New York, 1936.

Ray, P. Orman, "The Genesis of the Kansas-Nebraska Act," AHA *Annual Report,* 1914, I, 259–280.

______, *The Repeal of the Missouri Compromise.* Cleveland, 1909.

Rayback, Joseph G., "The American Workingman and the Antislavery Crusade," *Journal of Economic History,* III (1943), 152–163.

______, *Free Soil: The Election of 1848.* Lexington, Ky., 1970.

______, "The Presidential Ambitions of John C. Calhoun, 1844–1848," *JSH,* XIV (1948), 331–356.

Rayback, Robert J., *Millard Fillmore.* Buffalo, 1959.

Redpath, James, *Echoes of Harper's Ferry.* Boston, 1860.

Reiger, John F., "Secession of Florida from the Union: A Minority Decision?" *Florida Historical Quarterly,* XLVI (1968), 358–368.

Reynolds, Donald E., *Editors Make War: Southern Newspapers in the Secession Crisis.* Nashville, 1970.

Rhodes, James Ford, *History of the United States, 1850–1877.* 7 vols.; New York, 1892–1906.

Rice, Philip Morrison, "The Know-Nothing Party in Virginia, 1854–1856," *Virginia Magazine of History and Biography,* LV (1947), 61–75, 159–167.

Richards, Leonard L., *Gentlemen of Property and Standing: Anti-Abolition Mobs in Jacksonian America.* New York, 1970.

Richardson, James D. (ed.), *A Compilation of the Messages and Papers of the Confederacy.* 2 vols.; Nashville, 1905.

______, *A Compilation of the Messages and Papers of the Presidents.* 11 vols.; New York, 1907.

Richardson, Ralph, "The Choice of Jefferson Davis as Confederate President," *Journal of Mississippi History,* XVII (1955), 161–176.

Riddle, A. G., *The Life of Benjamin F. Wade.* Cleveland, 1888.

Riddle, Donald W., *Congressman Abraham Lincoln.* Urbana, Ill., 1957.

Riddleberger, Patrick W., *George Washington Julian, Radical Republican.* Indianapolis, 1966.

Rippy, J. Fred, "Diplomacy of the United States and Mexico Regarding the Isthmus of Tehuantepec, 1848–1860," *MVHR,* VI (1919–20), 503–531.

———, *The United States and Mexico.* New York, 1926.

Robert, Joseph Clarke, *The Road from Monticello: A Study of the Virginia Slavery Debate of 1832.* Durham, N.C., 1941.

Robinson, Charles, *The Kansas Conflict.* Lawrence, Kan., 1898.

———, "Topeka and Her Constitution," KSHS *Transactions,* VI (1897–1900), 291–305.

Robinson, Donald L., *Slavery in the Structure of American Politics, 1765–1820.* New York, 1971.

Robinson, Elwyn B., "The 'North American': Advocate of Protection," *PMHB,* LXIV (1940), 345–355.

Robinson, Sara T. D., *Kansas, Its Interior and Exterior Life.* Boston, 1856.

Roll, Charles, "Indiana's Part in the Nomination of Abraham Lincoln for President in 1860," *IMH,* XXV (1929), 1–13.

Rosenberg, Morton M., *Iowa on the Eve of the Civil War: A Decade of Frontier Politics.* Norman, Okla., 1972.

Rossiter, Clinton. *1787: The Grand Convention.* New York, 1966.

Rowland, Dunbar (ed.), *Jefferson Davis, Constitutionalist: His Letters, Papers and Speeches.* 10 vols.; Jackson, Miss., 1923.

Ruchames, Louis (ed.), *A John Brown Reader.* London, 1959.

——— (ed.), *John Brown: The Making of a Revolutionary.* New York, 1969.

Rusk, Ralph L., *The Life of Ralph Waldo Emerson.* New York, 1949.

Russel, Robert R., *Economic Aspects of Southern Sectionalism, 1840–1861.* Urbana, Ill., 1924.

———, *Improvement of Communication with the Pacific Coast as an Issue in American Politics, 1783–1864.* Cedar Rapids, Iowa, 1948.

———, "What Was the Compromise of 1850?" *JSH,* XXII (1956), 292–309.

Russell, William Howard, *Pictures of Southern Life.* New York, 1861.

Ryle, Walter Harrington, *Missouri, Union or Secession.* Nashville, 1931.

Salter, William, *The Life of James W. Grimes.* New York, 1876.

Sanborn, F. B. (ed.), *The Life and Letters of John Brown.* Boston, 1891.

———, *Recollections of Seventy Years.* 2 vols.; Boston, 1909.

Sandbo, Anna Irene, "Beginnings of the Secession Movement in Texas," *SWHQ,* XVIII (1914), 51–73.

Savage, Henry, *Seeds of Time: The Background of Southern Thinking.* New York, 1959.

Savage, William Sherman, *The Controversy Over the Distribution of Abolition Literature, 1830–1860.* Washington, 1938.

Scarborough, Ruth, *The Opposition to Slavery in Georgia Prior to 1860.* Nashville, 1933.

Schafer, Joseph, *Four Wisconsin Counties.* Madison, 1927.

———, "Know-Nothingism in Wisconsin, *Wisconsin Magazine of History*, VIII (1924), 3–21.

———, "Stormy Days in Court—The Booth Case," *Wisconsin Magazine of History*, XX (1936), 89–110.

———, "Who Elected Lincoln," *AHR*, XLVII (1941), 51–63.

Schlesinger, Arthur M., Jr., "The Causes of the Civil War: A Note on Historical Sentimentalism," *Partisan Review*, XVI (1949), 969–981.

———, *et al.* (eds.), *History of American Presidential Elections, 1789–1968*. 4 vols.; New York, 1971.

Schmeckebier, Laurence Frederick, *The History of the Know Nothing Party in Maryland*. Baltimore, 1899.

Schuckers, J. W., *The Life and Public Services of Salmon Portland Chase*. New York, 1874.

Schultz, Harold S., *Nationalism and Sectionalism in South Carolina, 1852–1860*. Durham, N.C., 1950.

Schwartz, Harold, "Fugitive Slave Days in Boston, *NEQ*, XXVII (1954), 191–212.

———, *Samuel Gridley Howe, Social Reformer, 1801–1876*. Cambridge, Mass., 1956.

Scisco, Louis Dow, *Political Nativism in New York State*. New York, 1901.

Scripps, John Locke, *Life of Abraham Lincoln*. Chicago, 1860.

Scroggs, Jack B., "Arkansas in the Secession Crisis," *Arkansas Historical Quarterly*, XII (1953), 179–224.

Scroggs, William O., *Filibusters and Financiers: The Story of William Walker and His Associates*. New York, 1916.

Scrugham, Mary, *The Peaceable Americans of 1860–1861*. New York, 1921.

Searcher, Victor, *Lincoln's Journey to Greatness*. Philadelphia, 1960.

Sears, Louis Martin, "New York and the Fusion Movement of 1860," ISHS *Journal*, XVI (1923), 58–62.

———, "Slidell and Buchanan," *AHR*, XXVII (1922), 712–724.

Sellers, Charles, *James K. Polk, Continentalist, 1843–1846* (Princeton, 1966).

———, "Who Were the Southern Whigs," *AHR*, LIX (1954), 335–346.

Sellers, James L., "Republicanism and State Rights in Wisconsin," *MVHR*, XVII (1930), 213–229.

Senning, John P., "The Know-Nothing Movement in Illinois, 1854–1856," ISHS *Journal*, VII (1914), 7–33.

Settle, Raymond W. and Mary Lund, *War Drums and Wagon Wheels*. Lincoln, Neb., 1966.

Seward, Frederick W., *Seward at Washington as Senator and Secretary of State*. 2 vols.; New York, 1891.

Sewell, Richard H., *John P. Hale and the Politics of Abolition*. Cambridge, Mass., 1965.

Shanks, Henry T., "Conservative Constitutional Tendencies of the Virginia

Secession Convention," in Fletcher M. Green (ed.), *Essays in Southern History Presented to J. G. de R. Hamilton.* Chapel Hill, 1949.

———, *The Secession Movement in Virginia, 1847–1861.* Richmond, 1934.

Shapiro, Samuel, "The Rendition of Anthony Burns," *JNH,* XLIV (1959), 34–51.

———, *Richard Henry Dana, Jr.* East Lansing, Mich., 1961.

Shenton, James P., *Robert John Walker: A Politician from Jackson to Lincoln.* New York, 1961.

Sherman, John, *Recollections of Forty Years in the House, Senate and Cabinet.* 2 vols.; Chicago, 1895.

Shryock, Richard Harrison, *Georgia and the Union in 1850.* Durham, N.C., 1926.

Shugg, Roger W., *Origins of Class Struggle in Louisiana.* Baton Rouge, 1939.

———, "A Suppressed Co-operationist Protest Against Secession," *LHQ,* XIX (1936), 199–203.

Seibert, Wilbur H., *The Underground Railroad from Slavery to Freedom.* New York, 1898.

Silbey, Joel H., "The Southern National Democrats, 1845–1861," *Mid-America,* XLVII (1965), 176–190.

Silver, James W., *Confederate Morale and Church Propaganda.* Tuscaloosa, Ala., 1957.

Simms, Henry H., "A Critical Analysis of Abolitionist Literature," *JSH,* VI (1940), 368–382.

———, *A Decade of Sectional Controversy, 1851–1861.* Chapel Hill, 1942.

———, *Emotion at High Tide: Abolition as a Controversial Factor.* N.p., 1960.

Simon, Donald E., "Brooklyn in the Election of 1860," New York Historical Society *Quarterly,* LI (1967), 249–262.

Simpson, Albert F., "The Political Significance of Slave Representation, 1787–1821," *JSH,* VII (1941), 315–342.

Sioussat, St. George L., "Memphis as a Gateway to the West," *Tennessee Historical Magazine,* III (1917), 1–27, 77–114.

———, "Tennessee, the Compromise of 1850, and the Nashville Convention," *MVHR,* II (1915), 316–326.

Sitterson, Joseph Carlyle, *The Secession Movement in North Carolina.* Chapel Hill, 1939.

Skipper, Ottis Clark, *J. D. B. De Bow, Magazinist of the Old South.* Athens, Ga., 1958.

———, "J. D. B. De Bow, the Man," *JSH,* X (1944), 404–423.

Smith, Charles W., Jr., *Roger B. Taney, Jacksonian Jurist.* Chapel Hill, 1936.

Smith, Donnal V., "The Influence of the Foreign-Born of the Northwest in the Election of 1860," *MVHR,* XIX (1932), 192–204.

Smith, Edward C., *The Borderland in the Civil War.* New York, 1927.

Smith, Theodore Clarke, *The Liberty and Free Soil Parties in the Northwest.* New York, 1897.

Smith, Willard H., *Schuyler Colfax.* Indianapolis, 1952.

Smith, William Ernest. *The Francis Preston Blair Family in Politics.* 2 vols.; New York, 1933.

Smith, William Henry. *A Political History of Slavery.* New York, 1903.

Snead, Thomas L., *The Fight for Missouri from the Election of Lincoln to the Death of Lyon.* New York, 1886.

Snyder, Louis L., *The Meaning of Nationalism.* New Brunswick, N.J., 1954.

Sorin, Gerald, *Abolitionism: A New Perspective.* New York, 1972.

Soulé, Leon Cyprian, *The Know-Nothing Party in New Orleans.* Baton Rouge, 1962.

Soulsby, Hugh G., *The Right of Search and the Slave Trade in Anglo-American Relations, 1814–1862.* Baltimore, 1933.

Sowle, Patrick H., "A Reappraisal of Seward's Memorandum of April 1, 1861, to Lincoln," *JSH,* XXXIII (1967), 234–239.

Sparks, Edwin Erle (ed.), *The Lincoln-Douglas Debates of 1858.* Springfield, Ill., 1908.

Spaulding, Oliver Lyman, Jr., "The Bombardment of Fort Sumter, 1861," AHA *Annual Report,* 1913, I, 179–203.

Speed, Thomas, *The Union Cause in Kentucky, 1860–1865.* New York, 1907.

Spencer, Ivor Debenham, *The Victor and the Spoils: A Life of William L. Marcy.* Providence, R.I., 1959.

Spring, Leverett Wilson, "The Career of a Kansas Politican," *AHR,* IV (1898), 80–104.

———, *Kansas: The Prelude to the War for the Union.* Boston, 1888.

Stampp, Kenneth M., *And the War Came.* Baton Rouge, 1950.

———, "The Fate of the Southern Antislavery Movement," *JNH,* XXVIII (1943), 10–22.

———, "Lincoln and the Strategy of Defense in the Crisis of 1861," *JSH,* XI (1945), 297–323.

———, *The Peculiar Institution: Slavery in the Ante-Bellum South.* New York, 1956.

Stanton, William R., *The Leopard's Spots: Scientific Attitudes toward Race in America, 1815–1859.* Chicago, 1960.

Stanwood, Edward, *American Tariff Controversies in the Nineteenth Century.* 2 vols.; Boston, 1903.

Starling, Marion W., "The Slave Narrative: Its Place in American Literary History." Ph.D. dissertation, New York University, 1946.

Staudenraus, P. J., *The African Colonization Movement, 1816–1865.* New York, 1961.

Stearns, Frank Preston, *The Life and Public Services of George Luther Stearns.* Philadelphia, 1907.

Steiner, Bernard C., "James Alfred Pearce," *Maryland Historical Magazine,* XVI (1921), 319–339; XVII (1922), 33–47, 177–190, 269–283, 348–363; XVIII (1923), 38–52, 134–150, 257–273, 341–357; XIX (1924), 13–29, 162–179.

———, "Some Letters from the Correspondence of James Alfred Pearce," *Maryland Historical Magazine,* XVI (1921), 150–179.

Stenberg, Richard R., "The Motivation of the Wilmot Proviso," *MVHR,* XVIII (1932), 535–541.

———, "Some Political Aspects of the Dred Scott Case," *MVHR,* XIX (1933), 571–577.

———, "An Unnoticed Factor in the Buchanan-Douglas Feud," ISHS *Journal, XXV (1933), 271–284.*

Stephens, Alexander H., *A Constitutional View of the Late War Between the States.* 2 vols.; Philadelphia, 1868–70.

Stephenson, N. W., "Southern Nationalism in South Carolina in 1851," *AHR,* XXXVI (1931), 314–355.

Stephenson, Wendell Holmes, "The Political Career of General James H. Lane," Kansas State Historical Society *Publications,* III (1930), 41–95.

Stern, Philip van Doren, *Uncle Tom's Cabin, an Annotated Edition.* New York, 1964.

Stevens, William O., *Pistols at Ten Paces: The Story of the Code of Honor in America.* Boston, 1940.

Stewart, James Brewer, *Joshua R. Giddings and the Tactics of Radical Politics.* Cleveland, 1970.

Still, William, *The Underground Railroad.* Philadelphia, 1886.

Stowe, Charles Edward, *Life of Harriet Beecher Stowe.* Boston, 1889.

Swanberg, W. A., *First Blood: The Story of Fort Sumter.* New York, 1957.

Swierenga, Robert P., "The Ethnic Voter and the First Lincoln Election," *CWH,* XI (1965), 27–43.

Swisher, Carl Brent, *Roger B. Taney.* New York, 1935.

Takaki, Ronald T., "The Movement to Reopen the African Slave Trade in South Carolina," *South Carolina Historical Magazine,* LXVI (1965), 38–54.

———, *A Pro-Slavery Crusade: The Agitation to Reopen the African Slave Trade.* New York, 1971.

Tandy, Jeannette Reid, "Pro-Slavery Propaganda in American Fiction of the Fifties," *SAQ,* XXI (1922), 41–50, 170–178.

Tansill, Charles C. (ed.), *Documents Illustrative of the Formation of the Union of the American States.* Washington, 1927.

Taylor, R. H., "Slave Conspiracies in North Carolina," *NCHR,* V (1928), 20–34.

Taylor, William R., *Cavalier and Yankee: The Old South and American National Character.* New York, 1961.

Thomas, Benjamin P., *Abraham Lincoln.* New York, 1952.

———, *Theodore Weld, Crusader for Freedom.* New Brunswick, N.J., 1950.

——— and Hyman, Harold M., *Stanton: The Life and Times of Lincoln's Secretary of War.* New York, 1962.

Thomas, David Y., "Calling the Secession Convention in Arkansas," *Southwestern Political and Social Science Quarterly,* V (1924), 246–254.

———, "Southern Non-Slaveholders in the Election of 1860," *Political Science Quarterly,* XXVI (1911), 222–237.

Thomas, John L., *The Liberator: William Lloyd Garrison.* Boston, 1963.

Thompson, Arthur W., "Political Nativism in Florida, 1848–1860: A Phase of Anti-Secessionism," *JSH,* XV (1949), 39–65.

Thompson, Robert Means, and Wainwright, Richard (eds.), *Confidential Correspondence of Gustavus Vasa Fox.* 2 vols.; New York, 1920.

Thompson, William Y., *Robert Toombs of Georgia.* Baton Rouge, 1966.

Thoreau, Henry David, *A Yankee in Canada.* Boston, 1866.

Thwaites, Reuben Gold (ed.), *Early Western Travels.* 32 vols.; Cleveland, 1904–07.

Torrey, Bradford and Allen, Francis H. (eds.), *The Journal of Henry D. Thoreau.* 14 vols.; Boston, 1906.

Towner, Lawrence W., "The Sewall-Saffin Dialogue on Slavery," *William and Mary Quarterly,* 3rd series, XXI (1964), 40–52.

Trefousse, Hans L., *Benjamin Franklin Wade, Radical Republican from Ohio.* New York, 1963.

———, *The Radical Republicans: Lincoln's Vanguard for Radical Justice.* New York, 1969.

Trent, William P., *English Culture in Virginia.* Baltimore, 1889.

Trescot, William H., *The Position and Course of the South.* Charleston, 1850.

Trimble, William, "Diverging Tendencies in the New York Democracy in the Period of the Loco Focos," *AHR,* XXIV (1919), 396–421.

Turner, Frederick Jackson, *The Significance of Sections in American History.* New York, 1932.

———, *The United States, 1830–1850.* New York, 1935.

Tyler, Alice Felt, *Freedom's Ferment: Phases of American Social History to 1860.* Minneapolis, 1944.

Tyler, Samuel, *Memoir of Roger Brooke Taney.* Baltimore, 1872.

Urban, C. Stanley, "The Abortive Quitman Filibustering Expedition, 1853–1855," *Journal of Mississippi History,* XVIII (1956), 175–196.

———, "The Africanization of Cuba Scare, 1853–1855," *Hispanic American Historical Review,* XXXVII (1957), 29–45.

———, "The Ideology of Southern Imperialism: New Orleans and the Caribbean, 1845–1860," *LHQ,* XXXIX (1956), 48–73.

———, "New Orleans and the Cuban Question during the López Expeditions of 1849–1851," *LHQ,* XXII (1939), 1157–1165.

Van Alstyne, Richard W. (ed.), "Anglo-American Relations, 1853–1857," *AHR*, XLII (1937), 491–500.

______, "British Diplomacy and the Clayton-Bulwer Treaty, 1850–1860," *Journal of Modern History*, XI (1939), 149–183.

______, "The British Right of Search and the African Slave Trade," *Journal of Modern History*, II (1930), 37–47.

Van Deusen, Glyndon G., *Horace Greeley, Nineteenth Century Crusader*. Philadelphia, 1953.

______, *Life of Henry Clay*. Boston, 1937.

______, *Thurlow Weed, Wizard of the Lobby*. Boston, 1947.

______, *William Henry Seward*. New York, 1967.

Van Deusen, John G., *The Ante-Bellum Southern Commercial Conventions*. Durham, N.C., 1926.

Van Tassel, David D., "Gentlemen of Property and Standing: Compromise Sentiment in Boston in 1850," *NEQ*, XXIII (1950), 307–319.

Venable, Austin L., "The Conflict Between the Douglas and Yancey Forces in the Charleston Convention," *JSH*, VIII (1942), 226–241.

______, "The Public Career of William Lowndes Yancey, *AR*, XVI (1963), 200–212.

______, "William L. Yancey's Transition from Unionism to State Rights," *JSH*, X (1944), 331–342.

Vilá, Herminio Portell, *Historia de Cuba en sus Relaciones con los Estados Unidos y España*. 4 vols.; Havana, 1938–41.

______, *Narciso López y su Epoca*. Havana, 1930.

Viles, Jonas, "Sections and Sectionalism in a Border State," *MVHR*, XXI (1934), 3–22.

Villard, Oswald Garrison, *John Brown, 1800–1859*. Boston, 1910.

Voss, Arthur, "Backgrounds of Lowell's Satire in 'The Biglow Papers,'" *NEQ*, XXIII (1950), 47–64.

Wade, Richard C., "The Vesey Plot: A Reconsideration," *JSH*, XXX (1964), 143–161.

Wagstaff, H. M. (ed.), *North Carolina Manumission Society, 1816–1834*. Chapel Hill, 1934.

Walker, William, *The War in Nicaragua*. Mobile, 1860.

Wallace, Edward S., *Destiny and Glory*. New York, 1957.

Walmsley, James Elliott, "The Change of Secession Sentiment in Virginia in 1861," *AHR*, XXXI (1925), 82–101.

Warden, Robert B., *An Account of the Private Life and Public Services of Salmon Portland Chase*. Cincinnati, 1874.

Warren, Charles, *The Supreme Court in United States History*. Rev. ed., 2 vols.; Boston, 1926.

Weaver, Herbert, *Mississippi Farmers, 1850–1860*. Nashville, 1945.

Webster, Noah, *Dissertations on the English Language*. Boston, 1789.

Weeks, Stephen B., "Anti-Slavery Sentiment in the South," Southern History Association *Publications,* II (1898), 87–130.

Weik, Jesse W., *The Real Lincoln.* Boston, 1922.

Weinberg, Albert K., *Manifest Destiny.* Baltimore, 1935.

Weisberger, Bernard A., *Reporters for the Union.* Boston, 1953.

Weisenburger, Francis P., *The Life of John McLean: A Politician on the United States Supreme Court.* Columbus, Ohio, 1937.

Weiss, John, *Life and Correspondence of Theodore Parker.* 2 vols., New York, 1864.

Wells, Damon, *Stephen Douglas, the Last Years, 1857–1861.* Austin, Tex., 1971.

Wells, Tom Henderson, *The Slave Ship Wanderer.* Athens, Ga., 1967.

Wender, Herbert, *Southern Commercial Conventions, 1837–1859.* Baltimore, 1930.

Wertenbaker, Thomas J., *Patrician and Plebeian in Virginia.* Charlottesville, Va., 1910.

White, Laura A., "The National Democrats in South Carolina, 1852 to 1860," *SAQ,* XXVIII (1929), 370–389.

———, *Robert Barnwell Rhett, Father of Secession.* New York, 1931.

White, Melvin Johnson, *The Secession Movement in the United States, 1847–1852.* New Orleans, 1916.

White, William W., "The Texas Slave Insurrection of 1860," *SWHQ,* LII (1949), 259–285.

Whitfield, Theodore M., *Slavery Agitation in Virginia, 1829–1832.* Baltimore, 1930.

Whitney, Asa, *A Project for a Railroad to the Pacific.* New York, 1849.

Whitney, Henry C., *Lincoln the Citizen.* New York, 1907.

Wilder, D. W., *The Annals of Kansas.* Topeka, 1886.

Williams, Edwin L., "Florida in the Union, 1845–1861." Ph.D. dissertation, University of North Carolina, 1951.

Williams, Elgin, *The Animating Pursuits of Speculation: Land Traffic in the Annexation of Texas.* New York, 1949.

Williams, Kenneth P., *Lincoln Finds a General.* 5 vols.; New York, 1949–59.

Williams, Mary Wilhelmine, *Anglo-American Isthmian Diplomacy, 1815–1915.* Washington, 1916.

Williams, T. Harry, *P. G. T. Beauregard, Napoleon in Gray.* Baton Rouge, 1955.

Wilson, Edmund, *Patriotic Gore: Studies in the Literature of the American Civil War.* New York, 1962.

Wilson, Forrest, *Crusader in Crinoline: The Life of Harriet Beecher Stowe.* Philadelphia, 1941.

Wilson, Henry, *History of the Rise and Fall of the Slave Power in America.* 3 vols.; Boston, 1872–77.

Wilson, Hill Peebles, *John Brown, Soldier of Fortune: A Critique.* Lawrence, Kan., 1913.

Wilson, Howard L., "President Buchanan's Proposed Intervention in Mexico," *AHR,* V (1900), 687–701.

Wilson, Major L., "Of Time and the Union: Webster and His Critics in the Crisis of 1850," *CWH,* XIV (1968), 293–306.

_____, "The Repressible Conflict: Seward's Concept of Progress and the Free-Soil Movement," *JSH,* XXXVII (1971), 533–556.

Wiltse, Charles M., *John C. Calhoun, Nationalist, 1782–1828.* Indianapolis, 1944.

_____, *John C. Calhoun, Sectionalist.* Indianapolis, 1951.

Winks, Robin W., *Canada and the United States: The Civil War Years.* Baltimore, 1960.

Wise, Barton H., *The Life of Henry A. Wise of Virginia, 1806–1876.* New York, 1899.

Wish, Harvey, "American Slave Insurrections before 1861," *JNH,* XXII (1937), 299–320.

_____, *George Fitzhugh, Propagandist of the Old South.* Baton Rouge, 1943.

_____, "The Revival of the African Slave Trade in the United States, 1856–1860," *MVHR,* XXVII (1941), 569–588.

_____, "The Slave Insurrection Panic of 1856," *JSH,* V (1939), 206–222.

Wittke, Carl, "The German Forty-Eighters in America," *AHR,* LIII (1948), 711–725.

_____, *The Irish in America.* Baton Rouge, 1956.

Woodford, Frank B., *Lewis Cass.* New Brunswick, N.J., 1950.

Woodham-Smith, Cecil, *The Great Hunger: Ireland, 1845–1849.* London, 1962.

Woodward, C. Vann, "The Antislavery Myth," *American Scholar,* XXXI (1962), 312–328.

_____, *The Burden of Southern History.* Baton Rouge, 1960.

Wooster, Ralph A., "An Analysis of the Texas Know-Nothings," *SWHQ,* LXX (1967), 414–423.

_____, *The Secession Conventions of the South.* Princeton, 1962.

Wright, Quincy, "Stephen A. Douglas and the Campaign of 1860," *Vermont History,* XXVIII (1960), 250–255.

Wright, William C., *The Secession Movement in the Middle Atlantic States.* Rutherford, N.J., 1973.

Wyatt-Brown, Bertram, *Lewis Tappan and the Evangelical War Against Slavery.* Cleveland, 1969.

Yearns, Wilfred Buck, *The Confederate Congress.* Athens, Ga., 1960.

Zilversmit, Arthur, *The First Emancipation: The Abolition of Slavery in the North.* Chicago, 1967.

Acknowledgments

One must begin with a general expression of gratitude to everyone who contributed in any way to the writing and publication of this book; the following list of specific acknowledgments is no doubt regrettably incomplete. Fred Pearlman, James Cone, and Richard Rekoon served in that order as research assistants, concentrating primarily on the verification and standardization of citations. Frank O. Gatell's critique of the first twelve chapters was remarkably thorough and of inestimable value. The final draft of Chapter 8, dealing with foreign affairs, incorporated a number of expert suggestions offered by Thomas A. Bailey. George Forgie and Robert M. Senkewicz undertook to verify certain election statistics. Advice and assistance were also lent at various stages by Carl N. Degler, George H. Knoles, and Richard W. Lyman. Much of the final typing was done by the office staff of the Department of History at Stanford, including Patricia Bernier, Betty Eldon, Nancy Ray, Barbara Richmond, and Loraine Sinclair. Until her death in 1969, Dilys M. (Mrs. David M.) Potter participated in the work of revising some of the earlier chapters. Catherine M. Potter's cooperation made completion of the book possible, and she, together with Virginia Fehrenbacher, lent assistance in the task of proofreading. The Institute of American History at Stanford provided a considerable amount of financial support.

D. E. F.

Index